Functional Foods of the East

NUTRACEUTICAL SCIENCE AND TECHNOLOGY

Series Editor
FEREIDOON SHAHIDI
Ph.D., FACS, FAOCS, FCIC, FCIFST, FIAFoST, FIFT, FRSC
University Research Professor
Department of Biochemistry
Memorial University of Newfoundland
St. John's, Newfoundland, Canada

1. Phytosterols as Functional Food Components and Nutraceuticals, *edited by Paresh C. Dutta*
2. Bioprocesses and Biotechnology for Functional Foods and Nutraceuticals, *edited by Jean-Richard Neeser and Bruce J. German*
3. Asian Functional Foods, *John Shi, Chi-Tang Ho, and Fereidoon Shahidi*
4. Nutraceutical Proteins and Peptides in Health and Disease, *edited by Yoshinori Mine and Fereidoon Shahidi*
5. Nutraceutical and Specialty Lipids and Their Co-Products, *edited by Fereidoon Shahidi*
6. Anti-Angiogenic Functional and Medicinal Foods, *edited by Jack N. Losso, Fereidoon Shahidi, and Debasis Bagchi*
7. Marine Nutraceuticals and Functional Foods, *edited by Colin Barrow and Fereidoon Shahidi*
8. Tea and Tea Products: Chemistry and Health-Promoting Properties, *edited by Chi-Tang Ho, Jen-Kun Lin, and Fereidoon Shahidi*
9. Tree Nuts: Composition, Phytochemicals, and Health Effects, *edited by Cesarettin Alasalvar and Fereidoon Shahidi*
10. Functional Foods of the East, *edited by John Shi, Chi-Tang Ho, and Fereidoon Shahidi*

Functional Foods of the East

Edited by
John Shi
Chi-Tang Ho
Fereidoon Shahidi

CRC Press is an imprint of the
Taylor & Francis Group, an informa business

CRC Press
Taylor & Francis Group
6000 Broken Sound Parkway NW, Suite 300
Boca Raton, FL 33487-2742

First issued in paperback 2019

ISBN-13: 978-1-4200-7192-4 (hbk)
ISBN-13: 978-0-367-38349-7 (pbk)

Library of Congress Cataloging-in-Publication Data

Functional foods of the East / editors, John Shi, Chi-Tang Ho, Fereidoon Shahidi.
p. cm. -- (Nutraceutical science and technology ; 10)
Includes bibliographical references and index.
ISBN 978-1-4200-7192-4 (hardcover : alk. paper)
1. Functional foods--East Asia. 2. Functional foods--India. 3. Functional foods--Arab countries. 4. Functional foods--Iran. I. Shi, John. II. Ho, Chi-Tang, 1944- III. Shahidi, Fereidoon, 1951- IV. Title. V. Series.

QP144.F85F865 2011
613.2095--dc22 2010029379

Visit the Taylor & Francis Web site at
http://www.taylorandfrancis.com

and the CRC Press Web site at
http://www.crcpress.com

Contents

Preface

The food we eat provides us with the essential nutrients that contribute to our physiological and biological well-being. Over the past several decades, dramatic changes have been observed in the types of food that are available and consumed. These changes are a direct reflection of the application of scientific discoveries and technological innovations by the food industry. With the increase in diet-related chronic diseases in modern society, diet is now considered to be strongly associated with three major diseases, namely, cardiovascular (heart and artery) diseases, cancer, and obesity.

In recent years, a great deal of attention has been paid to anticarcinogenicity, antimutagenicity, and antioxidative and antiaging properties of certain foods, and nutritional studies have revealed their potential health significance. These studies have also provided insight into the relationship between diet and optimal health, particularly with respect to age-related degenerative disease risk reduction such as cancer, heart disease, osteoporosis, diabetes, and stroke.

Given the extended life expectancy of the aging population in developed countries and the commensurate increase in healthcare costs associated with treating chronic diseases, it is very likely that in the future there will be more emphasis placed on preventative rather than prophylactic treatment of diseases. After reading reports and listening to discussions on the potential health benefits of functional foods, consumers around the world have heightened their interest in food selection and preparation as a means of protecting their health against diseases. These changes in the attitude of consumers, coupled with the continuous advances in food technology, have provided food companies with significant incentives to produce cutting-edge health-promoting foods and diets that address the needs of the increasingly health conscious consumers who are interested in self-administered health care. The Eastern tradition of using food as medicine is becoming more relevant and better accepted by people in the West.

Eastern countries are proud of their heritage and the ingenuity of their early scientific and cultural accomplishments. One of their most remarkable contributions to civilization undoubtedly has been the abundance of information and the detailed description of the uses of natural substances in treating illnesses. They have a long history of using plants and animal tissues to improve human physiology and health, maintain and improve health status, prevent diseases, help treat diseases, and facilitate rehabilitation.

"Health and Healing" foods have a long history in Eastern cultures, particularly in China, Japan, Korea, India, and Arab countries and Persia, where people believe that food and medicine come from the same source, are based on the same fundamental theories, and are equally important in maintaining and improving health status and preventing and curing diseases as well as facilitating rehabilitation. Many unique traditional functional foods were developed by combining food with herbal medicines. In China, Japan, Korea, Taiwan, and Southeast Asia, traditional herbal products are widely used as medicines in dietary supplements, including both daily foods (cereals, vegetables, and fruits) and functional foods, for replenishment and medical purposes. The concept is connected with immunopotentiation, the improvement of systemic circulation, disease prevention, and aging control. Either a single herb or multiple herbs may be used in formulating herbal foods, teas, wines, congees, and pills (or powders).

In the broad history of the East, many plants, such as medicinal herbs, have been used for thousands of years to maintain health and to treat diseases. Now, these same ancient remedies are experiencing renewed importance and should be reassessed in our modern age for possible use in the development of high-quality natural health products and dietary supplements for the twenty-first century.

The twenty-first century will be an era of new scientific horizons based on the twin forces of globalization and information-intensive industries. We are entering a period of unparalleled opportunities and intense competition. While the traditional Eastern functional foods steadily gain in popularity in the Western world, people around the globe are accepting the concept of food and medicine coming from the same source. Today more and more people believe that the traditional functional foods of the East can reduce disease risk, maintain health, and make their dreams of living a longer and healthier life come true. The long history of traditional functional foods, where herbal products are used as traditional medicines, and healthcare is based on natural products, has given a new worldwide meaning to functional foods. This in turn may generate opportunities for greater utilization of traditional functional foods of the East in the Western world.

The steady increase in consumer use and demand for functional foods and herbal medicines has prompted international health organizations and governmental agencies to publish guidelines for their proper use. Accordingly, the scientific community must apply modern technology to assure the efficacy and safety of these traditional remedies and to promote them as first-class dietary supplements and new medicines, through a concerted awareness campaign based on existing high-level scientific research and development in the world.

After the publication of our earlier book *Asian Functional Foods* in 2005, people from North America and Europe and elsewhere around the globe have become interested in the role that Asian functional foods might play in their own state of wellness and health. This current book provides complementary material to achieve this desired goal.

The contents of each chapter in this book focus on the traditional functional foods and ingredients as used in the Eastern countries. The functional properties of these foods are discussed in terms of their chemistry, biochemistry, pharmacology, and epidemiological and engineering principles. Some of the information in this book have not been previously disclosed. The information presented here will allow the Western world to better familiarize itself with the concepts and beliefs associated with traditional functional foods that have sustained the people of the East. Increased knowledge in this area will promote the merger of traditional Eastern concepts and beliefs on functional foods with the advanced science and technologies of the West.

The contributing authors are internationally renowned experts in this area and we are grateful to them for their thoughtful contributions. We trust that this book provides food scientists and technologists, nutritionists, biochemists, engineers, and entrepreneurs worldwide with recent advances in the field and an update to the existing knowledge. This book would also serve as a useful research tool for scientists of diverse backgrounds including biologists, biochemists, chemists, dieticians, food scientists, and nutritionists, medical doctors and pharmacologists from universities, research institutes, and food industries. It will further stimulate research and development in this emerging field, and provide consumers with information about products that could reduce disease risk and assist them in maintaining a healthier life style. We believe the scientific community will benefit from the overall summary of each area presented.

John Shi
Chi-Tang Ho
Fereidoon Shahidi

Editors

Dr. John Shi is a senior research scientist at the Federal Department of Agriculture and Agri-Food Canada, and an adjunct professor at the University of Guelph. He is coeditor of three books *Functional Foods II—Biochemical and Processing Aspects* (2002), *Asian Functional Foods* (2005), and *Functional Food Ingredients and Nutraceuticals: Processing Technology* (2006) by CRC Press, USA. He graduated from Zhejiang University, China, and received a masters degree in 1985, and a PhD in 1994 from the Polytechnic University of Valencia, Spain. As a postdoctoral fellow, he conducted research at North Dakota State University, Fargo; and as a visiting professor he conducted international collaborative research at the Norwegian Institute of Fishery and Aquaculture, Norway, and at Lleida University, Spain. Dr. Shi has been invited as a keynote speaker at a number of international conferences in the United States, Canada, Japan, China, Korea, Italy, Thailand, Spain, Columbia, and Brazil. He has published more than 100 research papers in international scientific journals, and 25 book chapters. His current research interests focus on value-added food processing, including development of new processing technologies and new products to enhance the functionality of health-promoting products from agricultural material, especially on "green" separation technology for developments of functional food ingredients, and application of nano(micro)technology to stabilize the bioactivity of health-promoting components in functional foods.

Dr. Chi-Tang Ho received his BS degree in chemistry from the National Taiwan University in Taiwan in 1968. He then went on to receive both his MA in 1971 and his PhD in 1974 in organic chemistry from Washington University in St. Louis, Missouri. After completing two years as a postdoctorate fellow at Rutgers University, he joined the faculty at Rutgers University as an assistant professor in the Department of Food Science. He was promoted to associate professor in 1983. In 1987 he was promoted to Professor I, and in 1993 he was promoted to Professor II. He has published over 670 papers and scientific articles, coedited 34 professional books and is an associate editor for the *Journal of Agricultural and Food Chemistry* and an editorial board member for a variety of publications, including *Molecular Nutrition & Food Research*. He has also won numerous awards including the ACS Award for the Advancement of Application of Agricultural and Food Chemistry, and the Stephen S. Chang Award in Lipid and Flavor Science from the Institute of Food Technology, and has served in the Division of Agricultural and Food Chemistry of the American Chemical Society in various positions including as division chair. His current research interests focus on flavor chemistry and the antioxidant and anti-cancer properties of natural products.

Dr. Fereidoon Shahidi is a university research professor in the Department of Biochemistry at Memorial University of Newfoundland (MUN), Newfoundland,

Canada. He is also cross-appointed to the Department of Biology, Ocean Sciences Centre, and the aquaculture program at MUN.

Dr. Shahidi is the author of over 600 research papers and book chapters, has authored or edited 48 books, and has given over 400 presentations at scientific conferences. Dr. Shahidi's current research interests include different areas of nutraceuticals and functional foods as well as marine foods and natural antioxidants. Dr. Shahidi serves as the editor-in-chief of the *Journal of Food Lipids* and *Journal of Functional Foods*, an editor of *Food Chemistry* as well as an editorial board member of the *Journal of Food Science, Journal of Agricultural and Food Chemistry, Nutraceuticals and Food*, and the *International Journal of Food Properties*. He is also on the editorial advisory board of *Inform*.

Dr. Shahidi has received numerous awards, including the 1996 William J. Eva Award from the Canadian Institute of Food Science and Technology in recognition of his outstanding contributions to food science in Canada through research and service. He also received the Earl P. McFee Award from the Atlantic Fisheries Technological Society in 1998, the ADM Award from the American Oil Chemists' Society in 2002, and the Stephen Chang Award from the Institute of Food Technologists in 2005. In 2006, Dr. Shahidi was inducted as a fellow of the International Academy of Food Science and Technology and was one of the most highly cited (seventh position) and most published (first position) individuals in the area of food, nutrition, and agricultural science for 1996–2006 as listed by ISI; the highly cited standing has now been revised to the fourth position. Dr. Shahidi was the recipient of the Advancement of Agricultural and Food Chemistry Award from the Agricultural and Food Chemistry Division of the American Chemical Society in 2007 and its Distinguished Service Award in 2008. He has served as an executive member of several societies and their divisions and organized many conferences and symposia. Dr. Shahidi served as a member of the Expert Advisory Panel of Health Canada on Standards of Evidence for Health Claims for Foods, the Standards Council of Canada on Fats and Oils, the Advisory Group of Agriculture and Agri-Food Canada on Plant Products, and the Nutraceutical Network for Canada. He was also a member of the Washington-based Council of Agricultural Science and Technology on Nutraceuticals. Dr. Shahidi is currently a member of the Expert Advisory Committee of the Natural Health Products Directorate of Health Canada.

Contributors

Syed Dilnawaz Ahmad
Faculty of Agriculture
University of Azad Jammu and Kashmir
Muzaffarabad A.K., Pakistan

In-Sook Ahn
Department of Food Science and Nutrition
Pusan National University
Pusan, Korea

Amir Al-Weshahy
Department of Nutritional Sciences
University of Toronto
Toronto, Ontario, Canada

Carani Venkatraman Anuradha
Department of Biochemistry and Biotechnology
Annamalai University
Tamil Nadu, India

Hideo Etoh
Faculty of Agriculture
Shizuoka University
Suruga-Ku, Shizuoka, Japan

Peter P. Fu
National Center for Toxicological Research
U.S. Food and Drug Administration
Jefferson, Arkansas

Gabriel Fuentes
School of Chinese Medicine
China Medical University
Taichung, Taiwan, Republic of China

Jung-Hee Jang
College of Oriental Medicine
Daegu Haany Uniersity
Daegu, South Korea

Bo Jiang
Department of Food Science and Nutrition
Jiangnan University
Wuxi, People's Republic of China

Yueming Jiang
South China Institute of Botany
Chinese Academy of Sciences
Guangzhou, People's Republic of China

Yearul Kabir
Department of Family Sciences
Kuwait University
Safat, Kuwait

Yukio Kakuda
Department of Food Science
University of Guelph
Guelph, Ontario, Canada

Frank S.C. Lee
Key Laboratory of Analytical Technology Development and the Standardization of Chinese Medicines
and
First Institute of Oceanography
State Oceanic Administration
QingDao, People's Republic of China

Lai Peng Leong
Department of Chemistry
National University of Singapore
Singapore

Donghong Liu
Department of Food Science and Nutrition
Zhejiang University
Hangzhou, People's Republic of China

Ying Ma
School of Food Science and Engineering
Harbin Institute of Technology
Harbin, People's Republic of China

Erin Shea Mackinnon
Calcium Research Laboratory
University of Toronto
Toronto, Ontario, Canada

Takashi Maoka
Division of Food Function and Chemistry
Research Institute for Production Development
Sakyo-ku, Kyoto, Japan

Wanmeng Mu
Department of Food Science and Nutrition
Jiangnan University
Wuxi, People's Republic of China

Mitsuo Namiki
Department of Food Science and Technology
Nagoya University
Nagoya, Japan

Wokadala Obiro
Department of Food Science and Nutrition
Jiangnan University
Wuxi, People's Republic of China

Kun-Young Park
Department of Food Science and Nutrition
Pusan National University
Pusan, Korea

A. Venket Rao
Department of Nutritional Sciences
University of Toronto
Toronto, Ontario, Canada

Leticia G. Rao
Calcium Research Laboratory
University of Toronto
Toronto, Ontario, Canada

Syed Mubasher Sabir
Faculty of Agriculture
University of Azad Jammu and Kashmir
Muzaffarabad A.K., Pakistan

Amber Sharma
Department of Chemistry
National University of Singapore
Singapore

Lee-Yan Sheen
Institute of Food Science and Technology
National Taiwan University
Taipei, Taiwan

John Shi
Guelph Food Research Center
Agriculture and Agri-Food Canada
Guelph, Ontario, Canada

Guanghou Shui
Department of Biochemistry
National University of Singapore
Singapore

Jiwan S. Sidhu
Department of Family Sciences
Kuwait University
Safat, Kuwait

Krishnapura Srinivasan
Department of Biochemistry and Nutrition
Central Food Technological Research Institute
Mysore, Karnataka, India

Young-Joon Surh
National Research Laboratory of Molecular Carcinogenesis and Chemoprevention
Seoul National University
Seoul, South Korea

Rong Wang
Department of Chemistry
National University of Singapore
Singapore

Xiaoru Wang
Key Laboratory of Analytical Technology Development and the Standardization of Chinese Medicines
and
First Institute of Oceanography
State Oceanic Administration
QingDao, People's Republic of China

Sophia Jun Xue
Guelph Food Research Center
Agriculture and Agri-Food Canada
Guelph, Ontario, Canada

Xingqian Ye
Department of Food Science and Nutrition
Zhejiang University
Hangzhou, People's Republic of China

Weibiao Zhou
Department of Chemistry
National University of Singapore
Singapore

1 *Yin Yang*, Five Phases Theory, and the Application of Traditional Chinese Functional Foods

Lee-Yan Sheen and Gabriel Fuentes

CONTENTS

1.1 THE DEVELOPMENT OF FUNCTIONAL FOODS AND MEDICATED DIETS IN CHINA

China has a long tradition of using medicinal foods to prevent and treat a whole range of maladies ranging from mild colds and flues to more severe afflictions such as liver or dermatological diseases. Since ancient times, the Chinese people have understood the interrelationship between human beings and their environment, as well as the repercussions of living out of harmony with nature. They believe that if humans obey the laws of nature, they will benefit by maintaining an optimal state of health, prevent disease, and most importantly extend their life span. This understanding of interrelationship is not limited to seasonal changes and their surrounding environment, but can also be extended to exercise, sexual practice, and dietetics (Harper, 1998).

Ancient Chinese scholars were astute observers of nature, the environment, and its influence on human beings. Taking careful notice of the constant fluctuations between states of health and disease, they accumulated a vast wealth of empirical knowledge. They also noted and compiled both the benefits and ill effects of consuming certain foods during a specific season and a certain state of health and ill health, which led to the eventual gathering and systematization of food categories that possessed a therapeutic effect on disease states.

One possible way in which the ancient Chinese might have developed their ideas of corresponding functions of foods in order to treat disease can be demonstrated through the Western disease known as goiter. The disease manifestation could be clearly seen through visual inspection. Physicians of ancient China utilized the existing paradigm of Chinese medical theory to postulate that the manifestation of this particular mass, what we would call goiter in Western medicine and its associated symptoms, must have arisen from a disturbance within the body's normal function, manifesting as an insufficiency or overabundance within the internal organ systems.

In Chinese medicine, the Western disease known as goiter is traditionally understood as an accumulation of phlegm combined with heat that evolves from a functional breakdown within the body's organ system; there are also associations with external pathogenic factors that can also interfere with the internal physiological functions of the organ system.

As for the treatment of this condition, they observed that by giving the afflicted individual kelp (*kunbu* in Chinese) the condition would be alleviated (Zeng and Zhang, 1984). They then arrived at the conclusion that the salty quality or taste of kelp had a therapeutic effect. When looking at this example, it is not hard to see how ancient Chinese scholars might have arrived at such conclusions. From these experiences and observations, Chinese scholars developed an intricate therapeutic system that has given them the ability to prevent and treat an array of diseases.

In the following discussion, we cover some of the basic concepts in Chinese medicine, and how they apply to functional foods and medicated diets.

1.2 THE FUNDAMENTAL THEORIES OF TRADITIONAL CHINESE MEDICINE: *QI, YIN,* AND *YANG,* AND THE FIVE PHASES, *ZANG FU* ORGAN (VISCERA) SYSTEM

When exploring any field in traditional Chinese medicine (TCM), one cannot proceed without first discussing some basic principles and concepts inherent within the practice of Chinese medicine. The concept of *Qi*, the theories of *Yin* and *Yang*, and the Five Phases are of course some of the most fundamental theories in the Chinese practices of healing. These theories were developed and utilized by ancient Chinese scholars in order to understand their environment and the universe at large. They became familiar with the constant changes of the seasons, and made associations to them utilizing the basic principles of *Yin Yang* and the Five Phases (Unschuld, 1986). Through astute observation of nature and constant seasonal cycles, these ancient scholars realized the duality and transformative quality of everything surrounding them in the natural world. The transformation from day to night, hot to cold, mist to water, as well as winter to summer, reflected the dualistic nature of *Yin Yang*, and the intrinsic changes of the Five Phases within the macrocosm. The theories of *Yin Yang* and the Five Phases were not however restricted only to medicine, but were also applied in every other field of science, including Cosmology, Calendrics, and Agriculture (Unschuld, 2003).

1.2.1 *Qi*

The concept of *Qi* is not only relegated to Chinese medicine but it also constitutes an intrinsic part of Chinese philosophy and culture. *Qi* cannot be easily defined because of its inherent multiplicity of meaning. It is best understood in relation to its functions and activities. *Qi* in Chinese medicine is associated with five different functions.

1. *Qi* is in charge of transformation and transportation, for example, the spleen *Qi* is responsible for the transformation of food into *Qi* and blood.
2. *Qi* is in charge of warming the body, allowing the body's organs and tissues to maintain and perform their functions.
3. *Qi* is in charge of defending the outer part of the body from pathogenic evils, therefore, if a person's *Qi* is depleted, they may be more prone to frequent bouts of illness.
4. *Qi* is in charge of activity, every physiological activity within the human body is attributed to *Qi*. *Qi* moves the blood through the vessels, giving rise to the saying, "*Qi* is the commander of blood." If *Qi* does not move there may be symptoms such as abdominal pain or fullness.
5. *Qi* is in charge of containment; *Qi* keeps the organs in their proper place, keeps blood within the vessels, and keeps body fluids inside the body. If *Qi* is depleted, it can lead to bleeding disorders, excessive sweating, and urination.

TABLE 1.1
Opposing Aspects of *Yin* and *Yang*

Yin	*Yang*
Dark	Light
Down	Up
Night	Day
Autumn, winter	Spring, summer
Earth	Heaven
Cold	Heat
Heaviness	Lightness
Moon	Sun
Stillness	Motion
Downward motion	Upward motion
Female	Male
Water	Fire

1.2.2 *YIN* AND *YANG*

The ancient Chinese scholars used the doctrine of *Yin* and *Yang* in order to describe and categorize opposing phenomena existing in the natural world, such as Table 1.1. They noticed through observation that everything in nature could be divided into two opposing categories, and then made correlations between them as night and day, darkness and light, as well as stillness and motion.

Yin represents the feminine, stillness, coldness, and darkness.

Yang represents the masculine, activity, warmth, and light.

1.3 THERAPEUTIC EFFECTS OF FOODS POSSESSING *YIN* AND *YANG* PROPERTIES

According to Chinese medical lore, all foods possess properties of *Yin* and *Yang* (Table 1.2). These properties of *Yin* and *Yang* are often represented by the thermal

TABLE 1.2
Foods Possessing *Yin* and *Yang* Properties

Foods Possessing *Yin* Properties	Foods Possessing *Yang* Properties
Bean sprout	Garlic
Cabbage	Beef
Watermelon	Chicken
Crab	Egg
Cucumber	Ginger
Tofu	Pepper
Alfalfa sprout	Shitake mushroom
Artichoke	Green onion
Asparagus	Raspberry
Kelp	Cassia fruit
Water chestnut	Carp
Seaweed	Ham
Banana	Mutton
Oyster	Walnut
Radish	Coriander

properties inherent in foods; therefore, it is always essential to balance these properties in order to maintain a harmonious balance between *Yin* and *Yang*. For example, depending on the individual's body constitution, the excess consumption of *Yang* or hot-type foods will manifest itself in a variety of symptoms, ranging from headache, abdominal discomfort, irritability, and dizziness. In turn, the over consumption of *Yin* or cold-type foods will also manifest itself in symptoms ranging from poor digestion and diarrhea to generalized weakness. With this knowledge, each individual can also treat disease and prevent occurrence of certain chronic diseases.

1.4 COLD/COOL FOOD ATTRIBUTES

Cold/cool food attributes belong to *Yin*, and possess the ability to clear heat, drain fire, and resolve toxins. As an example, an individual suffering from a sore throat due to a heat pathogen can find relief by simply drinking a cup of tea made out of honeysuckle.

As a preventive to heat stroke, many people in Asian countries eat watermelon. Watermelon, according to Oriental medical food lore, is considered to possess cold and cooling attributes.

1.5 HOT/WARM FOOD ATTRIBUTES

Hot/warm food attributes belong to *Yang*, and possess the ability to strengthen, dispel cold, and assist *Yang*. As an example of this theoretical concept applied to real life, let us take an individual who suffers from rheumatoid arthritis, which worsens during cold weather or winter months. This person could considerably reduce his/her symptoms by simply eating a bowl of chicken soup with the addition of ginger and garlic as condiments, as these foods are *Yang* in nature and warming.

1.6 THE FIVE PHASES IN TRADITIONAL CHINESE MEDICINE

The Five Phases is yet another system of correspondences developed by ancient scholars in China in order to better understand naturally occurring phenomena. The Five Phases theory postulates that wood, fire, earth, metal, and water are the basic building blocks in the material world. These elements are in a constant state of movement and transformation. The five phases interact with each other through engendering and restraining cycles of transformation. In TCM, the five phases are used to interpret pathophysiological functions within the human body and their relation to the natural world. The Five Phases theory is the theoretical model that has been used to explain many Chinese medical notions of the human body and its functions (Table 1.3).

1.7 THE CONCEPT OF ENGENDERING AND RESTRAINING WITHIN THE FIVE PHASES

The concept of engendering and restraining illustrates the interplay within nature; to have one without the other would mean disharmony and chaos.

According to Wiseman and Ellis (1996), in their *Fundamentals of Chinese Medicine*, "Engendering denotes the principle whereby each of the phases nurture, produces, and benefits another specific phase. Restraining refers to the principle by which each of the phases constrains another phase."

1.8 THE ENGENDERING CYCLE

According to the Five Phases theory, the engendering cycle is as follows (Figure 1.1): "Wood engenders Fire," as wood burns to make fire; "Fire engenders Earth," as fire consumes and reduces to ashes; "Earth engenders Metal," as Earth contains metal; "Metal engenders Water," as metal melts to form fluid; and "Water engenders Wood," as water nourishes wood.

TABLE 1.3
The Concept and Examples of Five Phases

Five Phases	Wood	Fire	Earth	Metal	Water
Taste	Sour	Bitter	Sweet-Umami	Pungent	Salty
Zang (solid organ)	Liver	Heart	Spleen	Lung	Kidney
Fu (hollow organ)	Gall bladder	Small intestine	Stomach	Large intestine	Bladder
Color	Green-blue	Red	Yellow	White	Black
Seasons	Spring	Summer	Long Summer	Autumn	Winter
Directions	East	South	Center	West	North
Sense organ	Eyes	Tongue	Mouth	Nose	Ears
Tissue	Sinew	Vessels	Flesh	Body hair	Bone
Mind	Anger	Joy	Thought	Sorrow	Fear

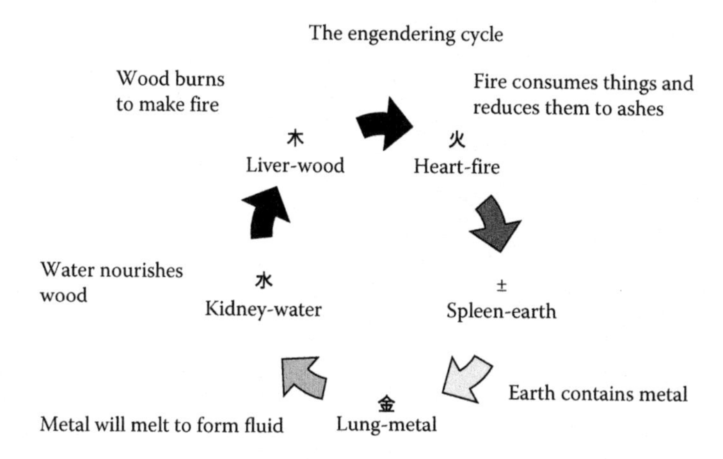

FIGURE 1.1 The engendering physiological function within the Five Phases.

When applying this theory in a clinical setting, one is better able to understand its reasoning and application. For example, let us take an average male complaining of poor digestion and bloating after meals. Upon questioning, he relates that he is constantly worrying about personal issues and always feels tired after a long day at the office. Utilizing the Five Phases theory framework, one can deduce that the earth phase is not being properly nourished by its mother phase. Through the recognition of associations between system functions, organs, and phases, the physician can decide upon a treatment plan that focuses on supplementing and strengthening the earth phase by employing foods and/or herbs that are known to have the desired effect on the earth phase.

1.9 THE RESTRAINING CYCLE

The restraining cycle is as follows: "Wood restrains Earth," as wood penetrates the earth; "Earth restrains Water," as earth dams water; "Water restrains Fire," as water douses fire; "Fire restrains Metal," as fire melts metal; and "Metal restrains Wood," as metal cuts wood (Figure 1.2). Each phase of the Five Phases correlates to a constant functional transformation and interaction that occurs in both the macrocosm and the microcosm.

1.10 THE *ZANG FU* ORGAN (VISCERA) SYSTEM

Within Chinese medical theory, the *Zang Fu* organ (viscera) systems are used to explain the physiological functions, pathological changes, and mutual relationships of every *Zang* and *Fu* organ. Within traditional Chinese medical theory, the *Zang* and *Fu* organs, though similar in anatomical structure to their western counterparts, carry to some extent dissimilar physiological functions within the body (Table 1.4).

The *Zang* and *Fu* organs consist of the five *Zang* and six *Fu* organs. The five *Zang* organs are the liver, heart, spleen, lung, and kidney. The six *Fu* organs are

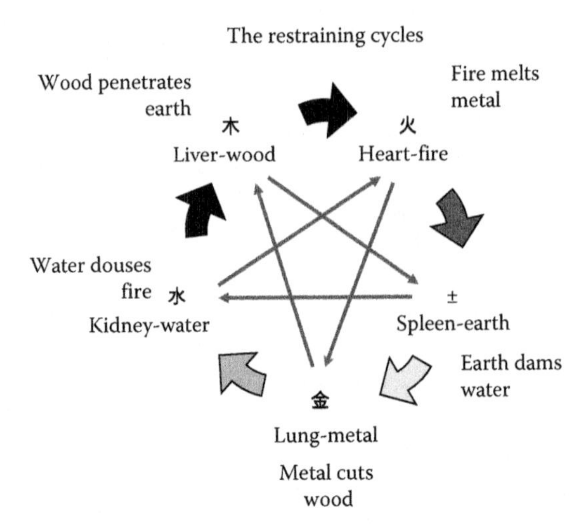

FIGURE 1.2 The restraining physiological function within the Five Phases.

the gall bladder, small intestine, stomach, large intestine, urinary bladder, and triple burner (the *Sanjiao* in Chinese). *Zang* and *Fu* organs are classified by their functional qualities. The five *Zang* organs are said to manufacture and store essence, *Qi*, blood, and body fluids. The six *Fu* organs are said to receive and digest food and absorb the essence of food, as well as transmit and excrete waste (Wiseman and Ellis, 1996).

Each *Zang* organ possesses an interior and exterior relationship to a *Fu* organ, amounting to five *Yin* and *Yang* interrelated organ systems. Each of these paired organ systems corresponds with one of the Five Phases.

Therefore, in order to maintain good health, it is necessary to maintain a harmonious relationship between the *Zang* and *Fu* organ systems. When there is disharmony between any of the *Zang Fu* organ systems, it will manifest as physical signs and symptoms that are associated to one or more of the *Zang Fu* organ systems affected. These signs and symptoms create patterns that can guide the practitioner to a diagnosis and the appropriate development of a treatment strategy.

TABLE 1.4
The Concept of *Zang Fu*

***Zang* Organs**	***Fu* Organs**
Yin	*Yang*
Relatively solid internal structures	Hollow structures
Responsible for the transformation and storage of vital substances	Responsible for the transportation and transformation of substances

1.11 THE TRADITIONAL DEFINITION OF THE FIVE *ZANG* ORGANS IN TRADITIONAL CHINESE MEDICINE

1.11.1 The Concept of the Liver in Chinese Medicine

- The liver belongs to "wood."
- The liver governs free coursing (such as blood circulation and metabolism).
- The liver stores the blood.
- The liver governs the tendon.
- The liver opens in the eyes.
- The health of the liver is reflected by the nails.

1.11.2 The Concept of the Heart in Chinese Medicine

- The heart belongs to "fire."
- The heart governs the spirit.
- The heart governs the blood and vessels.
- The heart opens at the tongue.
- The health of the heart is reflected by the facial complexion.

1.11.3 The Concept of the Spleen in Chinese Medicine

- The spleen belongs to "earth."
- The spleen governs movement and transformation.
- The spleen governs the engenderment of blood.
- The spleen governs the four limbs and muscle.
- The spleen opens at the mouth.
- The health of the spleen is reflected on the lips.

1.11.4 The Concept of the Lung in Chinese Medicine

- The lungs belong to metal.
- The lungs govern the *Qi*.
- The lungs govern down bearing.
- The lungs govern the skin and body hair.
- The lungs open at the nose.

1.11.5 The Concept of the Kidney in Chinese Medicine

- The kidneys belong to water.
- The kidneys store essence.
- The kidneys govern water.
- The kidneys govern the bones.
- The kidneys govern fertility (Ming Men fire, 命門之火).
- The kidneys open at the ears and two *Yins* (urethra and anus).
- The health of the kidneys is reflected by the head hair.

Five Phases theory is also used to describe the correlations of physiological functions between Zang Fu organ systems and body substances. The engendering relationships are as follows: the liver (represented by the wood phase) stores the blood in order to support the heart; heat from the heart (represented by the fire phase) warms the spleen; the spleen (represented by the earth phase) transforms and sends the finest essences to replenish the lungs; the dispersing and down bearing functions of the lungs (represented by the metal phase) assist the downward flow of kidney water; the essence of the kidneys (represented by the water phase) nourishes the liver in order to maintain its normal function of coursing and discharging.

The restraining relationships among the same organs are as follows: the unobstructed coursing of the liver (represented by the wood phase) is able to control hyperactivity of the spleen (represented by the earth phase); the transportation and transformation of the spleen is able to subdue overflowing of kidney water; the moistening function of the kidneys (represented by the water phase) can prevent hyperactivity of heart fire flaring upwards; *Yang* heat of the heart (represented by the fire phase) can control hyperactivity of the lungs' downbearing and dispersing functions; and the downbearing and descending function of the lungs (represented by the metal phase) can restrain hyperactivity of liver *Yang*.

By understanding the qualities and functions associated with a particular phase, ancient Chinese physicians could manipulate the state of the body by engendering or restraining a particular phase, which may have been in a state of repletion or depletion. For example, if one of the *Zang* or *Fu* organs was in a state of insufficiency or repletion, the physician would treat the phase in disharmony by supplementing or draining the phase that generates or controls it.

1.12 THE PREVENTIVE AND THERAPEUTIC FUNCTION OF THE FIVE TASTES OF FOODS

The five tastes of foods are sour (酸), bitter (苦), sweet and umami (甘), pungent (辛), and salty (鹹). Every taste corresponds to a particular phase within the Five Phases. Foods that do not possess any of these tastes are defined as bland (平). According to Bensky's *Materia Medica*, "taste" is defined as the subjective flavor given of when taken into the mouth (Bensky et al., 2004; Dang et al., 1999).

Different tastes of foods possess different therapeutic functions. For instance, the sour taste has the function to astringe and secure. It can treat vacuity sweating, diarrhea, and seminal emission. Some foods that are sour in taste include plum, hawthorn, litchi, and tomato. For example, the taste and temperature of plum are sour and warm, respectively. The physiological functions of plums include engendering fluids and stopping bleeding in humans.

The bitter taste has the function of drying dampness. It can treat heat signs, vexation, and agitation, as well as hyperventilation. Some foods that are bitter in taste include almond, bitter melon, and tea. For example, the taste and food attribute of bitter melon are bitter and cold, respectively. The physiological functions of bitter melons include relieving summer heat and vexation in humans.

The sweet and umami taste has the functions of relaxing and harmonizing, and of supplementing the stomach and spleen. Some foods that are sweet or umami in taste include eggplant, honey, mushroom, corn, soybean, peanut, and red date. For example, the taste and the temperature of eggplant is sweet and cool, respectively. The physiological functions of eggplant include stopping bleeding, reducing edema, and clearing heat, as well as eliminating dampness in humans.

The pungent taste has the function to effuse, scatter, and move *Qi* and blood. It can treat exterior signs, *Qi* and blood obstruction, and stagnation. Some foods that are pungent in taste include scallion, ginger, mint, red pepper, and black pepper. For example, the taste and food attributes of scallions are pungent and warm, respectively. The physiological functions of scallions include warming the stomach and moving the *Qi*, resolving toxins, and dispersing stasis in humans.

The salty taste functions to disperse gatherings and soften accumulations, and moisten the intestines. Some foods that are salty in taste are sea cucumber, kelp, sea weed, and scallop. For example, the taste and food attributes of sea cucumber are salty and warm, respectively. The physiological functions of sea cucumber include treating constipation and eliminating masses in the intestines in humans.

In conclusion, the doctrines of *Yin Yang* and the Five Phases have been important theories used in traditional Chinese dietary therapy in the past for over two millennia. These theories have been gathered and compiled from the daily life experiences of the Chinese people. If we, in our modern times, can utilize these theories in order to balance our lives and our nutrition, we would benefit by prolonged lifespan and improved quality of our lives.

REFERENCES

Bensky, D., S. Clavey, and E. Stoger. 2004. *Chinese Herbal Medicine Materia Medica*, 3rd ed. Eastland Press, USA.

Dang, Y., Y. Peng, and W. Li. 1999. *Chinese Functional Foods*. New World Press, Beijing, China.

Harper, D. 1998. *Early Chinese Medical Literature*. Kègan Paul International, UK.

Unschuld. P. U. 1986. *Nan-Ching*. University of California Press, USA.

Unschuld. P. U. 2003. *Huang Di Nei Jing Su Wen*. University of California Press, USA.

Wiseman, N., and A. Ellis. 1996. *Fundamentals of Chinese Medicine*. Paradigm Publications, Brookline, MA, USA.

Zeng, C., and J. Zhang. 1984. Chinese seaweeds in herbal medicine. *Hydrobiologia* 116/117 (1): 152–154.

2 Traditional Chinese Functional Foods

Bo Jiang, Wanmeng Mu, Wokadala Obiro, and John Shi

CONTENTS

2.1 INTRODUCTION

The research areas of functional foods and nutraceuticals are rapidly expanding throughout the world. Scientists are actively working on the health benefits of foods by identifying their functional constituents, elucidating their biochemical structures, and determining the mechanisms behind their physiological roles. These research findings contribute to a new nutritional paradigm, in which food constituents go beyond their role as dietary essentials for sustaining life and growth, to one of preventing, managing, or delaying the premature onset of chronic diseases later in life.

As early as 100 BC, *Huang-Di-Nei-Jing*, the bible of traditional Chinese medicine, stated that it is better to use food than drugs. This statement has established the value of Chinese alimentotherapy. Combining medicinal science with dietary practices, diverse ways of preparing dishes can be applied to alimentotherapy. In Chinese medical literature, there were several philosophical statements similar to those that characterize the modern functional foods paradigm stating that "medicine and food are of the same origin," and that "food therapy," "food supplement," and "food treatment" came to be an important part in maintaining health.

On June 1, 1996, China enacted the Measures of Functional Food Administration Law. Since then, modern Chinese functional foods have been legally approved, displaying a sky-blue-colored logo that is issued by the Ministry of Public Health. However, some traditional foods are also used as medicines, and do not need to be approved by the authorities. Thus, distinguishing between foods and traditional medicines is not an easy task. The difficulty apparently reflects the Chinese tradition that medicine and food share a common origin. Chinese health authorities have listed 77 traditionally consumed foodstuffs having medicinal effects, and which can be classified as traditional Chinese functional foods. These food items come from various fungal, plant (root, leaf, flower, fruit, and seed), and animal origins. This chapter discusses the origins, preparation methods, and functional aspects of some traditional Chinese functional foods.

2.2 TRADITIONAL CHINESE FUNCTIONAL SOY FOODS: *SUFU*

Soybeans are well recognized as an excellent source of high-quality protein and lipids. In Eastern Asia, they are grown as one of the most important protein resources. Traditional soybean foods have been consumed in Asia for many centuries, and remain popular. They are classified into two categories: nonfermented or fermented. Nonfermented types include soymilk, *tofu*, soy sprouts, and *tofu* pieces, whereas fermented ones include soy sauce, *sufu*, *douchi*, *miso*, *tempeh*, *onchom*, and *natto*.

The main benefits of soybean fermentation are improvements in sensory quality and nutritional value, rather than in preservation. The development of flavor and aroma through fermentation is a major characteristic of fermented soybean foods. Texture dramatically changes during the fungal fermentation of soybeans, leading to a cake-like product with a meat-like taste. Raw soybeans contain significant levels of antinutritional factors (ANFs) such as phytates, the majority of which are removed or destroyed during soaking, cooking, and fermentation of the soybeans (Nout and Rombouts, 1990). Soybean fermentation has been shown to improve the bioavailability of dietary zinc and iron (Hirabayashi et al., 1998; Kasaoka et al., 1997), and results in increased levels of vitamins (Denter et al., 1998; Sarkar et al., 1998). *Sufu*, *furu*, or fermented bean curd is an excellent vegetable protein food made from *tofu* by fungal solid-state fermentation. It is rich in nutritional values, with good protein levels, fat, and other nutrients. This food is a creamy cheese-type product with a mild flavor, fine texture, and good taste. Because of its characteristic salty flavor, *sufu* is consumed widely by Chinese as an appetizer. It may be eaten directly as a side dish, or cooked with vegetables or meats.

Sufu is a cheese-type product that is used in the same way as cheese. The resulting "pehtze" is salted, followed by ripening in a dressing mixture containing various ingredients (Ji and Li, 2005; Li et al., 2003). The merits of *sufu* are less known, let alone appreciated, outside the orient. Considering the increasing interest in nonmeat protein foods, *sufu* may become a popular commodity worldwide, especially if its

TABLE 2.1
Sensory Characteristics of *Sufu*

Item	Requirements			
	Red *Sufu*	White *Sufu*	Gray *Sufu*	Paste Sauce *Sufu*
Color	The exterior color of the *sufu* varies from red to purple, and the interior color varies from light yellow to orange	Even light yellow color inside and outside	Even gray color inside and outside	Almost the same color (reddish-brown or dark brown) interior and exterior
Taste and aroma	Fresh taste with good salinity, and the peculiar aroma of red *sufu* without the unpleasant smell	Fresh taste with good salinity and the peculiar aroma of white *sufu* without the unpleasant smell	Fresh taste with good salinity and the peculiar aroma of gray *sufu* without the unpleasant smell	Fresh taste with good salinity and the peculiar aroma of paste sauce *sufu* without the unpleasant smell
Structure and shape	Rectangle shape and smooth texture			
Purity	No visible foreign impurities			

organoleptic properties can be adjusted in accordance with regional food preferences. According to the standard SB/T 10170-2007 "Fermented bean curd," the sensory characteristics are described in Table 2.1.

2.2.1 History of *Sufu*

Sufu originated in China, as evidenced by ancient writings and archeological studies. When it was actually first made, and by whom, is still a mystery. *Sufu* is made by *tofu* fermentation. In the history of *tofu*, it has been widely reported since the Song Dynasty that the manufacture of *tofu* began during the era of the Han Dynasty (Yang, 2004). However, it is not known when *sufu* production began. Because of incomplete written records, little attempt has been made to search for its origin. The first historical record of *sufu* processing was in the late stage of the Northern Wei Dynasty (386–534 AD), where it was mentioned that *sufu* was made from dry *tofu* via salting and fermentation (Liu, 1988). *Sufu* became popular in the Ming Dynasty (1368–1644 AD), and many books describe its various processing technologies (Wang, 2006). Since then, the amount of *sufu* produced has steadily increased, and improved processing techniques have been developed. Depending on local customs, various *sufu* flavors were created, such as the Wangzhihe odoriferous *sufu* and paste sauce *sufu* in Beijing, Shaoxing *sufu* in Zhejiang Province, Guilin *sufu* in Guangxi Province, Shilin *sufu* in Yunnan Province, and Kedong *sufu* in Heilongjiang Province.

2.2.2 Varieties of *Sufu*

There are many kinds of *sufu*, which are produced by various processes in different localities in China (Han et al., 2001; Ji and Li, 2005). *Sufu* classification depends on the standards used, as shown below:

1. Based on the color and flavor, *sufu* is classified into four types that depend on the ingredients of the dressing mixtures used in the ripening stage. This classification type is recognized in the standard SB/T 10170-2007 "Fermented bean curd."

Red *sufu* (*Hong-fang*): The dressing mixture of red *sufu* mainly consists of salt, angkak (red kojic rice), alcohol sugar, flour-*koji*, paste sauce *koji*, and aroma condiments. The outer color of the *sufu* varies from red to purple, and the interior color varies from light yellow to orange. Because red *sufu* has a color considered attractive and a strong flavor, it is the most popular product consumed throughout China.

White *sufu* (*Bai-fang*): White *sufu* has ingredients similar to red *sufu* in the dressing mixture, but without the angkak. It has an even light yellow color both inside and outside. White *sufu* is a popular product in the south of China because it is less salty than red *sufu*. Based on flavor, white *sufu* can be further classified as mold-flavor *sufu* (*Mei-Xiang sufu*), lees *sufu* (*Zao-Fang*), alcohol *sufu* (*Zui-Fang*), oil *sufu* (*You-fang*), hot *sufu* (*La-fang*), and other types. Mold-flavor *sufu* is the most common. Lees *sufu* is characterized by the addition of rice wine and lees during ripening. The characteristics of alcohol *sufu* are achieved by adding pure rice wine or white wine.

Gray *sufu* (*Qing-Fang*, also called odoriferous *sufu*): The dressing mixture of gray *sufu* contains the soy whey by-product after *tofu* manufacture, salt, and some spices. Gray *sufu* is ripened with a special dressing mixture that contains abundant bacteria and mold enzymes and results in a product with a strong, offensive odor.

Paste sauce *sufu* (*Jiang-fang*): For this one, paste *koji*, such as soy sauce *koji* and flour *koji*, is added at the ripening stage. It has an almost uniform color (reddish-brown or dark brown) both inside and outside. As a distinction from red *sufu*, it does not contain angkak in the dressing mixture. Compared with white *sufu*, it has a stronger *miso* flavor but a weaker alcoholic flavor.

2. *Sufu* may also be classified into four types according to the processing technique used (Ji and Li, 2005; Li et al., 2003).

Natural fermented *sufu*: Four steps are normally involved in making this type of *sufu:* (i) preparation of the *tofu*, (ii) preparation of the pehtze (*pizi*) with natural fermentation, (iii) salting, and (iv) ripening.

Mold-fermented *sufu*: Four steps are normally involved in making this kind of *sufu*: (i) preparation of the *tofu*, (ii) preparation of the pehtze with pure culture mold fermentation, (iii) salting, and (iv) ripening.

Bacteria-fermented *sufu*: Five steps are normally involved in making this type of *sufu*: (i) preparation of the *tofu*, (ii) steaming and presalting, (iii) preparation of the pehtze with a pure culture bacterial fermentation, (iv) salting, and (v) ripening. This *sufu* is made in various places, such as Kedong (Heilongjiang) and Wuhan (Hubei).

Salting *sufu*: Only three steps are normally involved in making this kind of *sufu*: (i) preparing *tofu*, (ii) salting, and (iii) ripening. The flavor and aroma of salting *sufu* develop from the enzymatic reactions and the dressing mixture. The processing technique is simpler but with a lower degree of protein hydrolysis and a longer fermentation cycle. This type of *sufu* is produced only in a few areas of China, such as Taiyuan in Shanxi Province and Shaoxing in Zhejiang Province.

Sufu can also be classified according to size (big *sufu* and small *sufu*) and shape (square *sufu* and chess (round) *sufu*) (Ji and Li, 2005).

2.2.3 Manufacturing Process

Preparations of *sufu* vary with the different types of *sufu* and regions, but all involve four basic steps: (a) preparation of the *tofu*, (b) preparation of the pehtze, (c) salting, and (d) ripening. The schematic diagram for the production of *sufu* is outlined in Figure 2.1 (Ji and Li, 2005; Li et al., 2003).

Tofu preparation is an essential step in *sufu* production. The quality of the *tofu* has a significant effect on the sensory properties and other qualities of the *sufu*.

The traditional method for *sufu* production is a centuries-old household tradition involving natural fermentation, which mainly uses *Mucor* spp. from the air to ferment the *tofu* pehtze and produce a complicated mixture of enzymes and metabolites, thus creating a *sufu* with the desirable taste and special flavor. The process is still used today in Jiangsu and Zhejiang Provinces (Han et al., 2001).

In pehtze preparation, inoculation with microorganisms differentiates pure starter fermentation from natural fermentation. Using a pure mold strain, *sufu* can be prepared in any season and production efficiency is significantly increased. The fungal genera mainly involved are *Actinomucor*, *Mucor*, and *Rhizopus*, such as *Actinomucor elegans* and *A. taiwanensis*. These two strains seem to be the best molds for use in the commercial production of pehtze in Beijing and Taiwan, respectively (Han et al., 2001).

In the second fermentation, the recipe for the dressing varies, depending on the *sufu* kind and its manufacturer. For ripening, alternate layers of pehtze and dressing mixture are packed into jars. The jars are then sealed. The sealed jars can be aged either by natural fermentation or by manually controlling the fermentation temperature. Natural fermentation is adopted in places with high temperatures all year around, such as Southern China.

2.2.4 Chemical Composition, and Nutritional and Physiological Functions

2.2.4.1 Chemical Composition

The chemical and nutritional composition of *sufu* is shown in Table 2.2 (Han et al., 2001; Yang et al., 2006). In spite of their differences in color and flavor, most *sufus* have similar basic components.

The free amino acid (FAA) content of *sufu* with 11% (w/w) salt content after 80 days of ripening is shown in Table 2.3 (Han et al., 2001). Similar to all other soybean foods, methionine is a limiting amino acid in *sufu*. The absolute levels of FAA were also found to be higher in white *sufu* than in the red type. Glutamic acid is the most abundant acid, and other amino acids having high levels include aspartic acid

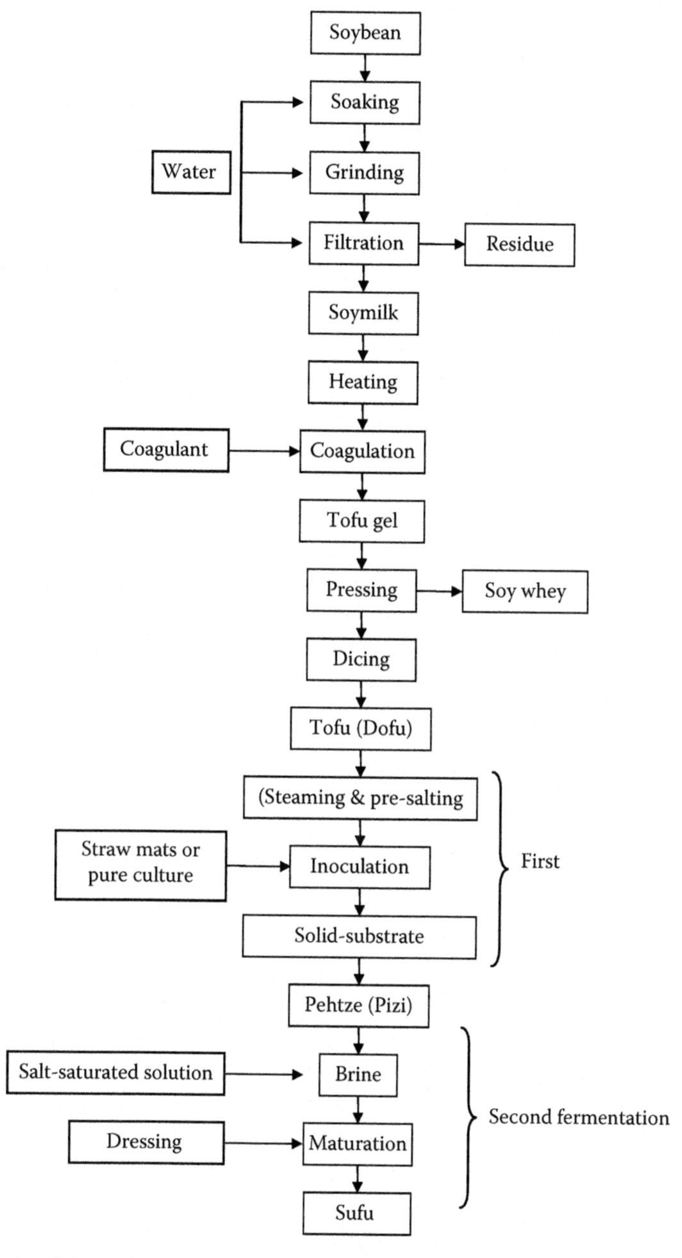

FIGURE 2.1 Schematic diagram for production of *sufu*.

and leucine. Glutamic acid, in combination with salt, contributes to the flavor of foods (also referred to as the umami taste). Since gray *sufu* evolves in a totally different way, compared with red and white *sufu*, it has a distinctively different amino acid profile. The content of hydrophilic amino acids (aspartic acid, glutamic acid, lysine, histidine, arginine, threonine, and serine) is quite low. This low level suggests

TABLE 2.2
Basic Components of Commercial *Sufu*

Component	Content[a]
Moisture (g)	50–70
Crude protein (g)	12–22
Crude lipid (g)	6–12
Crude fiber (g)	0.2–1.5
Carbohydrate (g)	6–12
Ash (g)	4–9
Calcium (mg)	100–230
Phosphorus (mg)	150–300
Iron (mg)	7–16
Thiamin (VB_1) (mg)	0.04–0.09
Riboflavin (VB_2) (mg)	0.13–0.36
Niacin (mg)	0.5–1.2
VB_{12} (mg)	1.7–22
Food energy (kJ)	460–750

Source: Adapted from Han B, Rombouts FM, Nout MJR: *International Journal of Food Microbiology* 65:1–10, 2001; Yang J: *Agricultural Archaeology* 1:217–226, 2004.

[a] Per 100 g sufu fresh weigh.

that they are somehow consumed or transformed at a greater rate than they were formed by proteolytic activity. This may result in the offensive odor characteristic of gray *sufu*.

2.2.4.2 Nutritional and Physiological Functions

1. *Isoflavones:* Soybean isoflavones fall into the class of phytoestrogens, which contribute to lower rates of osteoporosis (Ishida et al., 1998; Ishimi et al., 1999) and have a preventive effect on the development of cancers of the breast, intestines, liver, bladder, skin, and stomach (Adlercreutz et al., 1992; Wei et al., 1993). Although the *sufu* manufacturing process results in a loss of total isoflavones (Yin et al., 2004), the level of some isoflavones (i.e., genistein and daidzein) increases by about 20 times after 50 days of *sufu* ripening (Zhang et al., 2002). These changes in the composition of isoflavones during *sufu* manufacturing might be beneficial to the enhancement of physiological functions.

2. *Soybean oligopeptide:* Proteolysis is the principal and most complex biochemical event that occurs during the maturation of *sufu*. Yang et al. (2006) reported that microorganisms isolated in the ripening of *sufu* showed high endopeptidase activities, such as the Leu-aminopeptidase, Arg-aminopeptidase, dipeptidase, and carboxypeptidase activities of *Brevibacterium*, which contribute to its flavor, the removal

TABLE 2.3
Free Amino Acid Content (mg/g dry matter) of *Sufu* with 11% (w/w) Salt Content

FAA	Red *Sufu*	White *Sufu*	Gray *Sufu*
Asp	5.2	7.9	0.7
Thr	2.4	4.7	0.1
Ser	2.6	0.2	0.2
Glu	14.4	18.5	0.1
Gly	2.2	3.7	0.0
Ala	4.9	7.9	8.9
Val	3.5	6.5	6.8
Met	1.0	1.7	0.0
Ile	3.7	7.3	7.1
Leu	6.1	10.3	9.6
Tyr	4.1	0.9	3.7
Phe	4.8	7.4	6.5
Lys	4.0	7.5	0.2
His	1.0	1.8	0.1
Arg	0.2	0.0	0.0
Pro	2.9	5.2	0.0
Subtotal	63.0	91.5	44.0

of the bitterness of polypeptide, and the production of amino acids and bioactive oligopeptides.

The angiotensin I-converting enzyme (ACE) is a dipeptidyl carboxypeptidase associated with the regulation of blood pressure. It converts angiotensin I into the potent presser peptide, angiotensin II, and also degrades depressor peptide bradykinin. ACE inhibitors from various foods have recently been studied in terms of their ability to prevent or alleviate hypertension. ACE inhibitory activity was observed in *sufu* extract (Ni et al., 1997; Wang et al., 2003).

Zhang et al. (1998) investigated the hypocholesterolemic effect of the insoluble fraction of *sufu* as a dietary supplement. *Sufu* powder showed a much greater effect than soybean separated isolate.

3. *Antioxidative activities of sufu:* Varieties of fermented soybean foods, such as *natto* and *tempeh*, have been reported to exhibit a much stronger antioxidative activity than unfermented ones. *Sufu* not only has a peculiar, palatable taste and aroma, but also presents strong antioxidative activity (Ren et al., 2006; Wang et al., 2004). The antioxidative activity increases during the fermentation process. Wang et al. (2004) investigated the antioxidative activity of 14 commercial brands of fermented *tofu*. Their antioxidative activity units of total extracts ranged from 2.91 to 16.87 μg α-tocopherol equivalents/mg.

4. *Other bioactive substances:* As a fermented soybean product, *sufu* contains saponins, soy oligosaccharides, soybean phospholipids, phytic acid, and other

bioactive compounds present in soybeans. During the manufacturing process, many nutritional ingredients are also added. Angkak (red kojic rice), an important ingredient in the dressing mixture of red *sufu* as a natural colorant, is traditionally obtained by fungal solid-state fermentation of cooked rice, mainly with *Monascus* spp. Angkak is also considered to be a health-promoting food ingredient (Wang et al., 1997).

2.3 TRADITIONAL CHINESE FUNCTIONAL FISH SAUCE

Marine foods have traditionally been popular because of their varieties of flavor, color, and texture. More recently, seafoods have been recognized for their roles in health promotion, arising primarily from constituent long-chain omega-3 fatty acids. Nutraceuticals from marine sources and the potential applications are varied as listed in Table 2.4.

Processing of the catch results in a considerable amount of byproducts, accounting for 10–80% of the total landed weight. The components of interest include lipids, proteins, flavorants, minerals, carotenoids, enzymes, and chitin. Various materials from such sources may be isolated and used in different applications, including functional foods and nutraceuticals. The importance of omega-3 fatty acids in reducing the incidence of heart disease, certain types of cancer, diabetes, autoimmune disorders, and arthritis has been well recognized.

In addition, the residual protein in seafoods and their byproducts may be separated mechanically or through a hydrolysis process. The bioactive peptides so obtained may be used in a variety of food and nonfood applications. The bioactives from marine resources and their applications are generally diverse.

Fish sauce (*Xiajiang*) in China is a nutritious condiment made from a traditionally fermented shrimp and salt mixture. The abundance of amino acids, oligopeptides, short-chain fatty acids, and aldehydes, together with minerals, impart a characteristic cheesy and meaty aroma, apart from its sharp and salty taste. Having vitamin B_{12} as an indigenous constituent makes fish sauce a unique product in the class of condiments.

TABLE 2.4
Functional and Bioactive Components from Marine Resources and Their Application Areas

		Requirement for Brewing Water	
Analytical Item	**Unit**	**Ideal Requirement**	**Limits**
pH		6.8–7.2	6.5–7.8
Total rigidity	degree	2–7	<12
Nitric nitrogen	mg/L	<0.2	0.5
Total plate count	mL^{-1}	None	<100
Coliforms	L^{-1}	None	<3
Free chlorine	mg/L	<0.1	<0.3

Fish sauce can be added directly to dishes and is used for its saltiness and flavor, or it can be made into a dip together with other spicy ingredients. To date, genuine fish sauce manufacturing remains a traditional process. Heavily salted fresh and entire shrimp and fish are tightly packed into clay containers. Fish and shrimp protein hydrolysis and natural fermentation are allowed to proceed naturally at ambient conditions. A clear brown liquid is drawn from a spigot at the base of the fermenting broth and further aged before use. The total storage time is at least 6 months. In recent years, investigations on fish sauce with low salinity have been undertaken.

2.3.1 History

According to archeological evidence, in the early years of *Kang Xi* of the *Qing* Dynasty, in the south of *Jinzhou* of *Bo Hai* Bay, there was a small village called *Xiao Yan Guo*. In the village there was a family named *Li*, who made a living by catching shrimp. Every year around the time of *Xiao Man*, they would catch fish and shrimp near *Er Jie* channel. Because it was a place of convergence of sea water and fresh water, and was also a floodplain, the gray shrimp there were rich in flesh with thin shells and a strong taste but without a fishy smell. The family took the fish and shrimp to sell in *Jinzhou*. The remaining shrimps were kept in jars, with salt to avoid perishing. After some years, the shrimps were increasingly prepared in large quantities and were fermented with a faint scent. Thus, they evolved into a sticky sauce. The family added a bit of fish sauce with the rice and some of it was given to the neighbors. Once tasted, the fish sauce was praised by the people. Later, after careful management and stirring, the tastes became even fresher and better. The paste was then named "Salted Fish Sauce," and sold in the market, becoming locally popular.

Xiao Yan Guo was located in the northwest corner of *Er Dao Ling Zi*. It was the horse land in the *Qing* Dynasty, named *Xi Jin Jia*. The place could keep horses by the thousands. After the ninth year of the *Kang Xi* Dynasty, the land owner reported to *Beijing* about the status of the horses, and presented to the emperor a jar of fish sauce from the *Li* family. It happened that the *Kang Xi* Emperor did not have a good appetite. After tasting the fish sauce, he felt better and praised the paste. After that, the salted fish sauce that was presented to the emperor was named "*Yuan Feng* Royal Shrimp," and was sealed by soil in the jar, with each jar weighing 5–6 *jin* (500 g). The jars were sealed with pink paper, marking them as "*Yuan Feng* Royal Shrimp."

2.3.2 Nutritional Components of Salted Fish Sauce

Shrimp is rich in nutrients, and, notably, the amount of protein it contains is much higher than that in fish, egg, or milk (Table 2.5) (Shi and Wen, 2006). In addition, it has a tender texture that makes it easy to digest (Zhou et al., 2002). After the pickling treatment, significant changes occur to the nutrients. Particularly, there is a remarkable increase in the level of soluble calcium (Liu and Lin, 2003). Fish sauce is a good source for many micronutrients. Because it is made from animal protein, it contains vitamin B_{12}. Plant cells do not contain vitamin B_{12}-dependent enzymes, and therefore they do not produce cobalamine, the scientific name for vitamin B_{12}.

TABLE 2.5
Nutritional Ingredients in 100 g of Shrimp (Esculent 86 g)

	Content	Ingredient	Content	Ingredient	Content
Quantity of energy/kcal	87	Thiamin/mg	0.04	Ca/mg	325
Protein/g	16.4	Vitamin B_2/mg	0.03	Mg/mg	60
Fat/g	2.4	Niacin/mg	0	Fe/mg	4
Carbohydrate/g	0	Vitamin C/mg	0	Mn/mg	0.27
Dietary fiber/g	0	Vitamin E/mg	5.33	Zn/mg	2.24
Vitamin A/μg	48	Cholesterol/mg	240	Cu/mg	0.64
Carotenoid/μg	3.9	K/mg	329	P/mg	186
Retinol equivalent/μg	78.1	Na/mg	133.8	Se/mg	29.65

Source: From Shi M, Wen H: *Fisheries Science and Technology Information* 33:143–144, 2006. With permission.

2.3.3 Processing Technology

Traditional fish sauce fermentation is an ancient technology. Presumably fish sauce fermentation originated as a household recipe. With increasing demand, it gradually developed into cottage industries. Although the thousands of small-scale fish sauce producers spread all over China adhere to the traditional method of processing, medium- and large-scale plants, for instance in Southeast China, benefit from modern mechanization in many steps of the production line. These include salting, filtration, pasteurization, bottling, and packaging. Despite the modern facilities used in the processing, the actual fermentation is still generally carried out in large covered concrete vats that are built into the ground. Since traditional fish sauce fermentation is a time-consuming process, attempts to shorten the fermentation time have been a popular topic of research. Various innovations have involved largely the use of exogenous sources of enzymes, with varying degrees of success.

It is well known that salted fish sauce is a kind of fermented food; the raw material is always small *Acetes chinensis*, white, gray, and euphausiid shrimp. The shrimps have to be washed to remove impurities. After drying, 30% salt is added and the shrimps are put in a wooden jar or cask. After 5–7 days, the shrimp body turns red. This is the initial sign of fermentation and the salty liquid can then be compressed. The body is then mashed into a sauce. It is left in the jar for around 10 days, during which time the fish sauce becomes swollen. After several days of stirring (at least twice a day), fermentation takes place and an odor is released. After 1 month, the fermented product becomes the finished product.

Some regions such as *Haikou* of *Hainan* Province use sealed fermentation. The resulting shrimp sauce tastes different from that obtained by the traditional method.

The sealed fermentation process involves mixing the fresh and washed shrimps with 15–17% of salt (no more than 22%). After mashing the shrimp bodies into a sauce with an electric mashing machine, rice wine (3%) is then added as well as 2–3% of sugar. After stirring, the jar is sealed and stored at a temperature of 35–37°C. After several days, the finished product is obtained (Qi et al., 2007).

In the *Guangdong, Fujian*, and *Taiwan* regions, there is another processing technique. Shrimps are washed and dried and then put into a jar. Thirty to thirty-five percent of salt is added and stirred. The jar is then kept in an open area for the fermentation process. After maturing, the liquid becomes shrimp oil. The remaining substance is shrimp sauce.

2.3.3.1 Traditional Fish Sauce of the Dong Nationality

2.3.3.1.1 Preparation

The main procedures are summarized in Figure 2.2 (Shi and Wen, 2006).

2.3.3.1.2 Main Points in the Preparation of the Fish Sauce

Mature, live, and intact shrimps are selected. During the period from the White Dew to the autumnal equinox, shrimps are fed with rice blossoms and grain, and develop a good body size. The favorable weather is suitable for paste production.

The shrimps are placed in clean water and then starved for 2–3 days to get rid of any fishy smell. The shrimps are washed repeatedly with clean water to remove any mud and sand as well as any dead shrimps, and then drained using a fabric piece.

The full and nonmolded round sticky rice kernels (preferred *Xiang He* from *Jiang* County of *Guizhou* Province) are selected. After washing, the rice is soaked in hot water (60–80°C) and then at an ambient temperature for 8–10 h. The soaked rice is then cooked in a wooden steam box. The steamed rice is removed, cleaned, and

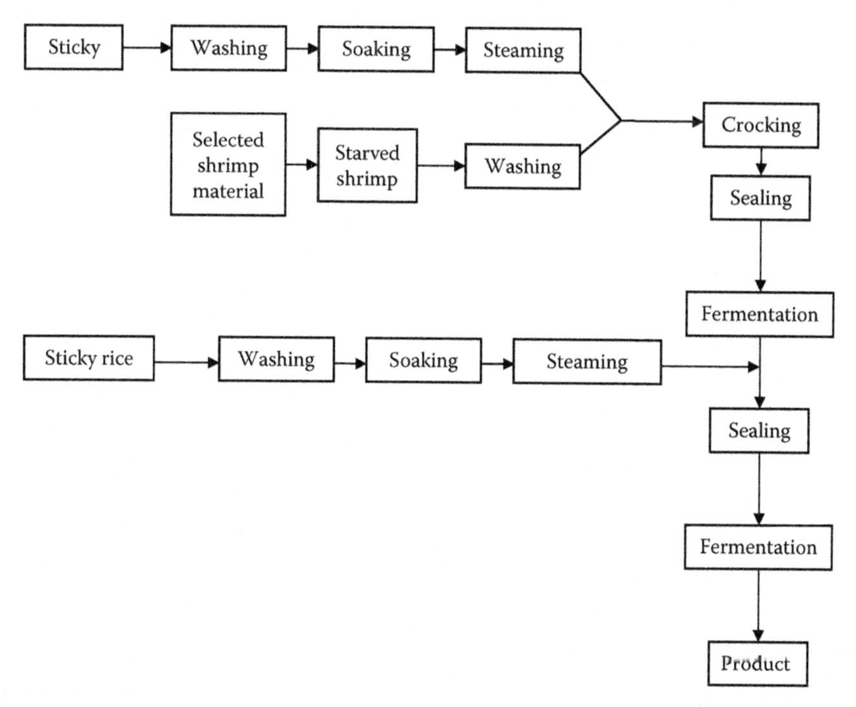

FIGURE 2.2 Flow chart for production of salted fish sauce using the *Dong* nationality method.

layered on the shrimps. The heat from the sticky rice cooks the shrimp. After the temperature of the rice and shrimp goes down to 30–40°C, *baijiu* (liquor) and salt are added and mixed according to the recipe. The containers for making shrimp sauce are usually porcelain jars. The jars are heated in hot water and cleaned. The shrimp mixture is then put into the jar and covered with a film. The lid is fitted onto the jar and clean water is added to the jar to maintain the seal.

2.3.3.1.3 Fermentation Cycle

The fermentation cycles last approximately 15–20 days in summer, 20–25 days in autumn, and 30–35 days in winter.

2.3.3.1.4 Compounding

The "original-taste" salted fish sauce is fermented using the traditional compounding method of the *Dong* nationality. The "improved pastes" with sour-sweet flavor, sour-pungent taste, and pepper-pungent taste are achieved by adding different compounds and with slight improvements in manufacturing (Table 2.6).

2.3.3.1.5 Disadvantages of the Processing Technology

Although it is very convenient for common families to brew fish sauce using the traditional technology, it has many disadvantages, such as a long fermentation cycle and a high salt content (25–30%). It is well known that traditionally fermented fish

TABLE 2.6
Recipes of Salted Fish Sauces of the *Dong* Nationality

Ingredient	Original	Sour-Sweet	Sour-Pungent	Pepper-Pungent	Remarks
Sticky rice/kg	12.50	12.50	12.50	12.50	
Sticky rice powder/kg	2.50	2.50	2.50	2.50	Parched and powdered
Salt/kg	5.00	5.00	5.00	5.00	Baked
Chili powder/kg	5.00	2.00	5.00	5.00	
Distillate spirit/kg	2.50	2.50	2.00	1.50	35–45°
Peanut/kg	2.50	0.02	0.02	0.02	Parched and powdered
Zanthoxyli powder/kg	0.75	0.50	0.75	2.50	Parched and powdered
Zingiberis recens/kg	2.50	2.50	3.00	3.00	
Semen oryzae cum monasci/kg	2.50	2.50	2.50	2.50	
Sesame/kg	2.50	2.50	2.50	2.50	Parched and powdered
Garlic/kg	1.50	0.75	1.50	1.50	
Tangerine pericarp/kg	–	–	0.15	0.25	

Source: From Shi M, Wen H: *Fisheries Science and Technology Information* 33:143–144, 2006. With permission.

sauce contains a high concentration of salt in the final product and this consequently limits its utilization. A high salt content always inhibits the decomposition of protein, leading to a low FAA content and less tasty flavor. Generally, salt is recognized as a basic flavor and plays a role in maintaining the osmolarity of extracellular fluid compartments of the human body. However, the consumption of a large amount of salt might be unhealthy. Case studies, which have been limited to industrialized societies, have often failed to find a conclusive relationship between blood pressure and sodium excretion (Pickering, 1980). Most individuals can eat large amounts of sodium without remarkably affecting blood pressure because they excrete sodium adequately (Kaplan, 2000; Luft, 1998; McCarron, 2000). In addition, the quality of the product is comparatively difficult to control.

2.3.3.2 Fish Sauce with Lower Salinity

2.3.3.2.1 Preparation

The fish sauce obtained by the traditional homemade processing technology can not meet modern expectations. To eliminate the disadvantages of natural fermentation technology, some improvements have been made to the production process.

Fish sauces with low salinity can be manufactured by proteolytic action. It is more suitable for mass consumption. The protease hydrolysis method, carried out under high temperatures, not only improves the flavor and quality but also shortens the production period. The hydrolysis products have a longer shelf life, greater nutritional value, are of good quality, and comply with the national sanitation standard.

2.3.3.2.2 Main Points in the Preparation

The critical parameters are shrimp quality and hydrolysis conditions. The selection of mature, live, and intact shrimps is recommended.

The optimum reaction conditions of alkaline protease and neutral protease are 0.5% enzyme, a hydrolysis temperature of 55°C, a salt content (NaCl) of 18%, and a pH of 7.0. The optimum reaction times for alkaline protease and neutral protease are 2 and 1 h, respectively (Liu and Lin, 2003).

2.3.3.2.3 Quality Control Tests

The free amino nitrogen (FAN) content is determined by the methanol aldehyde titration technique (Zhao and Gong, 1986). Total nitrogen is assayed by Kjeldahl determination (Li, 1994). Conway's plate law is adopted to measure the volatile base nitrogen (VBN) (Zhao and Gong, 1986), and the amount of salt-free solids is obtained by drying the sample at 105°C.

2.4 TRADITIONAL CHINESE FUNCTIONAL WALNUT KERNELS

2.4.1 Product Characteristics and Functionality

Walnut (*Juglans regia*) flesh is nutritionally valuable (USDA, 2007). This high-fat nut is relatively high in protein, potassium, phosphorus, and folate, with a good content of fiber and vitamin E, but low in sugar. One of the principal characteristics of the nuts is their fat composition, which is low in saturated fatty acids (lower than other tree nuts) and monounsaturated fatty acids, but rich in polyunsaturated omega-3 and omega-6 fatty acids (linoleic and linolenic acids). Being of vegetable origin, it does not contain any cholesterol, but other esters such as β-sitosterol, avenasterol, or campesterol that could lower the absorption of cholesterol in the intestine.

Walnuts also contain a great variety of antioxidant compounds such as α- and γ-tocopherols, lycopene, β-carotene, and others, which act as a sequential protection system against oxidation on the organism, preventing the formation of free radicals. The antioxidant potential of walnuts prevents the oxidation of low-density lipoprotein (LDL), thus preventing them from sticking to the artery walls (Chen et al., 2005; Samaranayaka et al., 2008).

In traditional Chinese medicine, walnut seeds are primarily considered a kidney tonic. They are also considered beneficial to the brain, back, and skin, and to relieve constipation if it is caused by dehydration (Chen et al., 2005).

2.4.2 History of Functional Food Development and Application

China's traditional consumption of walnuts is mainly dominated by unprocessed and simply processed nuts, such as specialty foods like sweet walnut flesh, walnut crisp, and honey walnut. However, with the development of processing techniques, the market now provides walnut oil, full walnut powder, and compounded walnut milk, which are all high value-added products. This shows that there is a great market potential for the walnut processing industry. A few companies in China have now applied for processing patents for walnut milk, walnut juice, walnut syrup, instant walnut powder, and walnut liquor. Walnut drinks are currently prevalent in the Chinese domestic market. Walnut milk yogurt with *Bacillus bifidus* is a new product, which benefits the brain, and will improve longevity. With 15% walnut milk, there is no harmful effect to the *B. bifidus*, and it can improve yogurt flavor and nutritional value. Walnut ice cream is also a newly developed flavor food. By adding flavor potentiators, the flavor can be made richer. Walnuts and red dates have been used as the raw materials to make a health drink claimed to improve immunity, reduce fatigue, enhance blood circulation, and strengthen the kidneys (Liu, 2004).

2.4.3 Distribution of Raw Material

There are three major walnut species: *Juglans regia* Linn, *J. nigra* Linn, and *J. cinerea* Linn. Other species of walnut trees include the Arizona (or Nogal) walnut, *J. major* Heller (New Mexico and Arizona), the California walnut, *J. californica* Walnut (southern California), the Chinese walnut, *J. cathayensis* Dode (central and western China), the Hinds walnut, *J. hindisii* Sarg (California), the Texas walnut, *J. rupestris* Engelm (southwestern United States to Mexico), the Manchurian walnut, *J. mandshurica* Maxim (eastern Commonwealth of Independent States and northern China), and the Japanese walnut, *J. sieboldiana* Maxim (Japan) (Prasad, 1993).

In China, walnuts are distributed across the south and north, but the sigillata walnut is mainly scattered in the southwest area (mainly Yunnan, Guizhou, West Sichuan, and Southern Tibet). It is grown in most areas, except Inner Mongolia, Zhejiang, and Fujian Provinces, on a relatively small scale (Xi and Zhang, 1995).

2.4.4 Processing Technologies for Homemade and Industrial Production Scales

2.4.4.1 Storage

Newly picked walnuts contain high levels of water and are perishable. After peeling and after the nuts are fully dried, they are usually put into a cloth bag or harl bag, or stored indoors on wood, on benches and bricks, about 40–50 cm above the ground. The storage room must be kept cool, dry, airy, and dark, and sealed to prevent rats getting in. The above methods are only suitable for short-term storage. Under normal conditions, walnuts can be preserved until summer comes. The large quantity long-term storage requires that the walnuts are put in harl bags or boxes and stored below 0–5°C. The shelf life can then reach 2 years (Li and Chen, 2000).

When long-term walnut storage is required but refrigeration is unavailable, a plastic film cover will be useful. In autumn, after the nuts are fully dried, they can be stored under plastic. The film must be sealed before the temperature starts to rise again in February, and low temperatures need to be maintained. Carbon dioxide (CO_2) should be pumped under the film cover. When the CO_2 level is above 50%, it can prevent oil oxidation and reduce pest invasion. Under the film cover, nitrogen will be absorbed and thus will prevent the walnuts from aging to a certain extent. Storage in nitrogen for weeks will improve the luster and flavor compared with storage indoors. After 25 weeks, good quality can still be maintained while the indoor stored walnuts tend to become rancid and the walnut peel broken (Li and Chen, 2000).

2.4.4.2 Stability Factors Affecting Storage Quality

The shell as an integral part of the walnut plays an important part in the growing, maturing, washing, delivering, and storage processes. In 1987, China issued the national standards for Walnut Fertility and Nut Quality. The standard stipulates that if 10% of the samples have shell seal openings, the walnuts can not be graded as good. Broken walnuts means the walnut has its shell seal open. During storage, this will affect the pest infection rates.

The water content in walnuts should be maintained at 6–8% (Li and Chen, 2000; Sze-tao et al., 2000; Xi and Zhang, 1995), similar to that in the United States. If it is lower than 8%, the water activity is usually less than 0.64. Under this condition, the growth or replication of microorganisms is inhibited. However, drying can not completely prevent walnut rancidity. If the water content is too low, the possibility of rancidity increases. A low-temperature environment is the best for long-term storage. Walnuts can remain fresh for 1 year if stored at 10°C under a relative humidity of 60%. The physical, chemical, and sensory quality indicators will remain within the stipulated scope. During storage at low temperatures for 1 year, the stability of the oils in the walnut will decrease during the first 4 months. This is attributed to a decrease in antioxidant factors. The stability of the oils does not change much after several months. The fatty acids in the oil remain the same. Walnuts stored at relatively high humidity are more prone to oil oxidation.

Baked walnuts with oxygen-barrier packaging or packed in nitrogen will have a prolonged shelf life. The radioactive dosage needed to control pests is quite low, usually 1 kGy, will not induce significant changes in nut composition, and will not have a negative influence on the sensory qualities (Al-Bachir, 2000). Meanwhile, it can effectively eliminate all the pests.

2.4.5 Production and Marketing

China's walnut production has increased for the last several years because of increased planting and bearing acreage, improved walnut cultivars, and better tree management. In 2003, China became the world leader in walnut production, followed by the United States. In the first three decades of the twentieth century, China exported 3000 tons of walnut kernel to Europe each year. By the 1970s and 1980s, China's walnut exporting volume accounted for half of the world's walnut trading volume. Since 1986, the United States has dominated the world walnut trade and taken the place of China in world walnut trading. At present, the total shelled walnut exportation is about 10,000 tons every trading year. Major destinations are Europe, Japan, Canada, New Zealand, and the Middle East.

This low export quantity is because walnut growing is not carried out intensively, and cultivar standardization was disregarded until recent times. Other factors are the insufficient number of growers associations, and the lack of a necessary marketing organization implemented by the government. However, the enhancement in the number of standard walnut orchards in the future will result in more profitable and valuable products for export (Zhao and Gong, 2006).

2.5 Traditional Chinese Functional *XiangGu*

XiangGu (*Lentinus edodes*, the Japanese named shiitake) is a nonpathogenic fungus that is grown on a variety of currently underutilized logs. It has a chocolate-brown top, fibrous stems, meaty texture, and a pleasant woody taste. In China, *XiangGu* has been cultivated on notched logs stacked in evergreen forests since ancient times (Wu, 2002; Wu and Chen, 2002). These edible fungi are generally not eaten raw because few people like the sharp, garlic-like flavor. Their most popular use is as flavoring for gravies, stews, stuffings, dips, and spreads.

Cooked *XiangGu* has a strong, wild mushroom flavor. This flavorful mushroom is a mainstay in the Chinese diet, and is highly valued as a centerpiece of Chinese cuisine.

2.5.1 History of *XiangGu*

The cultivation of *XiangGu* has been practiced for a thousand years, originating in China during the *Sung* Dynasty (960–1127). Both history and legend credit Wu San Kwung as the originator of *XiangGu* cultivation. Almost every mushroom-growing village in China has a temple in his honor (Chang and Miles, 1989). It is believed that Chinese growers introduced *XiangGu* cultivation techniques to Japanese farmers, who named the mushroom and were later responsible for its spread eastward from China.

2.5.2 Nutritional Value of *XiangGu*

XiangGu contains 18 amino acids, including 7 of the 8 essential amino acids. One unique amino acid, called eritadenine, is believed to be responsible for *XiangGu*'s ability to reduce cholesterol (Bai, 2007). It also contains more than 30 different enzymes. This fungus is the second most popular edible mushroom worldwide. Its extracts have health benefits.

2.5.3 Physiological Function and Clinical Research

There has been a considerable amount of research conducted on *XiangGu* in the last 20 years, to elucidate its role as a functional food. The polysaccharides in *XiangGu* have beneficial properties and are considered central to *XiangGu*'s role as an immune system adjuvant.

2.5.3.1 Antitumor Effect and Antimutagenic Effect

Lentinan, a β-glucan, was the first antitumor compound isolated from *XiangGu*. Although its mechanism of action is not completely clear, it inhibits tumorigenesis mainly by activating the immune system and inducing gene expression of immunomodulatory cytokines and their receptors (Liu et al., 2002). It was applied on diverse tumors in therapeutic methods, such as ancillary drug therapy, a combination of radiotherapy, chemotherapy, and recovery therapy. In clinical trials, lentinan inhibits or breaks down the tumor cells by regulating the function of the immune system and the activity of diverse immunocytes, and activating the immune system. Chihara et al. (1969) first reported on the antitumor properties of mushrooms, stating that lentinan "was found to almost completely regress the solid type of tumors such as Sarcoma 180 and several kinds of tumors including methylchloranthrene-induced fibrosarcoma in synergic host–tumor systems."

Duan et al. (1997) assayed the effects of *XiangGu* on the prevention and repair of genetic damage induced by radiation by using micronucleus tests on bone marrow polychromatoerythrocytes in mice. The results showed that the extract of *XiangGu* could hasten the repair of genetic damage induced by radiation. The effect of prevention and repair was better than the control when it was administered before and after radiotherapy (Duan et al., 1997). Several important bioactive compounds such as ergosterol have already been isolated from *XiangGu*. Studies have shown that some of the active substances in *XiangGu* exert a protective effect against mutagenesis (Chihara et al., 1970).

2.5.3.2 Antiviral Effect

Another important effect of *XiangGu* is its antiviral properties, including the ability to help prevent influenza. *XiangGu* has been believed to cure the common cold for hundreds of years in Asia (Dong et al., 2005). More recently, this opinion has been supported by some scientific evidence. Zhang et al. (2006a) used the method of methyl thiazolyl tetrazolium (MTT) to determine the antiviral activity of lentinan. Results showed that lentinan could reduce the level of affection of Madin–Darby canine kidney (MDC) cells by the influenza virus; the ratio of protection could be up to 95.6%.

Anti-HIV activities were also reported for mycelial culture medium of *XiangGu*. Sulfated lentinan from *XiangGu* completely prevented HIV-induced cytopathic effects. The protein-bound polysaccharides were also found to have an antiviral effect on HIV and cytomegalovirus *in vitro*. Besides immunostimulation, another effect of the polysaccharide–protein complex is a contribution to antiviral activity. Studies have demonstrated *XiangGu*'s ability to induce interferon production in the human body. Lentinan could enhance host resistance against infection of the agents of AIDS, and reduce the toxic effects of the drugs for treating AIDS patients (Chihara, 1993).

2.5.3.3 Anti-Inflammatory and Antioxidant Effect

A vast number of studies have demonstrated that free radicals such as superoxide, hydroxyl radicals, and high-energy oxidants such as peroxynitrite act as mediators of inflammation, shock, and ischemia/reperfusion injury. There is growing evidence that the production of reactive oxygen species (ROS) at the site of inflammation can contribute to tissue damage. Antioxidant compounds reduce the action of ROS in tissue damage. Natural products with antioxidant activity are used to aid the endogenous protective system, increasing interest in the antioxidative role of *XiangGu* (Li, 2005). Recently, some studies suggested that *XiangGu* plays an important role as an anti-inflammatory agent. Ou-Yang et al. (2006) made a model of acute inflammation in mice tumefaction to study the anti-inflammatory effects of lentinan. Given different doses, each inflammation action was clearly inhibited by lentinan.

As an anti-inflammatory agent, *XiangGu* also improves stomach and duodenal ulcers, neuralgia, gout, constipation, and hemorrhoids. Studies also alleged that it exhibited significant antioxidant activity. Lin et al. (2006a) obtained crude polysaccharide extracts from *XiangGu* and investigated the antioxidant activity in four different assays *in vitro*. The polysaccharide extracts strongly scavenged reactive oxygen species and eliminated the activities of NO_2^-.

2.5.3.4 Antimicrobial Activity

The antimicrobial activities of *XiangGu* extracts have also been investigated because edible fungi are considered as a source of natural antibiotics. Extracts of *XiangGu* were first examined for the inhibition of microbial growth by an agar dilution method. Lenthionine and its analogues extracted with chloroform inhibited the growth of both Gram-positive and Gram-negative bacteria. Lenthionine was found to be a more effective inhibitor than its analogues. Fungi were more sensitive to lenthionine than were bacteria. The disulfide derivative extracted with ethyl acetate had antibacterial and antifungal activities. Masatomo et al. (1999) extracted three kinds of antibacterial substances from dried *XiangGu*. These substances demonstrated effective antibacterial activities against *Streptococcus* spp., *Actinomycete* spp., *Lactobacillus* spp., *Prevotella* spp., and *Porphyrophyromonas* spp., while other general bacteria such as *Staphylococcus* spp., *Escherichia* spp., *Bacillus* spp., and *Candida* spp. were relatively resistant to these substances.

Hospital-derived postoperative microbial infections and antibiotic resistance were increased. Such studies must warrant serious consideration, and further work should be done to quantify their antimicrobial effects.

2.5.3.5 Blood Pressure-Lowering, Cholesterol-Lowering, Antifibrotic, Antidiabetic, Antifatigue, and Liver-Protective Activities

Cardiovascular (CV) disease causes the most mortality worldwide and high blood cholesterol levels play an important risk role in the development of CV problems. *XiangGu* is almost ideal for diets designed to prevent CV diseases, as first suggested by traditional Chinese medicine (Wu, 2002).

The active hypocholesterolemic substance in *XiangGu*, eritadenine, was isolated and identified as an adenosine derivative. It supposedly lowers all lipid components of serum lipoproteins in both animals and humans. A recent study has perhaps disproved the early claim that all lipid components of serum lipoproteins were reduced by *XiangGu* (Fulushima et al., 2001). Rats fed with a diet that contained *XiangGu* fiber showed decreased levels of serum LDL, intermediate-density lipoprotein (IDL), and very low-density lipoprotein (VLDL) compared to the control.

It has also been confirmed that *XiangGu* contains effective substances for antifatigue activity (Wu, 2002). Three doses of protein-bound polysaccharides from *XiangGu* were applied to mice to study their antifatigue effects. Administration of protein-bound polysaccharide showed remarkable effects on the reserves of muscle and liver glycogen, serum lactic dehydrogenase activity, and lowering of blood urea nitrogen before and after exercise (Yang et al., 2001). Lentinan produced favorable results when treating chronic persistent hepatitis and viral hepatitis B patients. A polysaccharide fraction from *XiangGu* could give liver-protective action to animals and it also improved liver function and enhanced the level of production of antibodies to hepatitis B. *XiangGu* polysaccharide was shown to be effective at decreasing the elevation of serum glutamic-pyruvic transaminase (SGPT) level in tetrachloride-, thioacetamide-, and prednisolone-intoxicated mice or rats (Lin and Huang, 1987).

More recently, *XiangGu* extracts have been used to treat patients with immune system disorders. It is clear that *XiangGu* is truly a beneficial plant that offers a natural avenue to better health.

2.5.4 Cultivation and Processing

XiangGu had been found in the wild only in China, Japan, and Korea until recently. This mushroom grows primarily in temperate climates, singly or in clusters, in declining or dead hardwoods. In nature, *XiangGu* is a saprophytic, white-rot fungus that degrades wood substrates containing recalcitrant and lignin components. It is due to this capacity that wood logs and sawdust are now used as substrates to cultivate this variety.

Cultivation of *XiangGu* includes selecting spores or strains, maintaining mycelial cultures, developing spawn, preparing growing medium, inoculating spawn, colonizing substrates, harvesting, and crop management. *XiangGu* is grown on synthetic as well as natural logs. Environmentally controlled houses allow for the adjustment of temperature, humidity, light, and the moisture content of the logs to produce the highest possible yields. Temperature is the most important factor in *XiangGu* cultivation. The optimum temperature for mycelium growth is about 25°C. At below 5°C or above 35°C, mycelial growth stops. Using modern methodologies, *XiangGu* produces 3–4 flushes of mushrooms per year on natural logs. Synthetic logs, composed of sawdust and supplemented with millet and wheat bran, may produce three to four times as many mushrooms as natural logs in one-tenth of the time. With synthetic substrates, each flush of *XiangGu* requires only 16–20 days. Sawdust cultivation may also produce 3–4 times as many mushrooms as natural logs in only one-tenth of the time. Recent trends suggest that in the future, most *XiangGu* will be cultivated on synthetic logs. The major advantages of producing *XiangGu* on synthetic logs rather than natural ones are a consistent market supply year-round, increased yields, and decreased time required to complete a crop cycle. These advantages far outweigh the major disadvantage of a relatively high initial investment cost to start a synthetic log manufacturing and production facility (Wu et al., 2002; Wu et al., 2002).

XiangGu is harvested when the caps are 50–80% open (He, 1994). At this stage the gills are exposed, but the cap edges are still rolled under. They are then cleaned, trimmed, and made ready for market. Fresh *XiangGu* is produced widely in China and is available in most supermarkets and groceries. As fresh *XiangGu* is a fast-respiring and highly perishable product, it should be stored under proper conditions after harvesting (Xue, 2007).

Traditionally, fresh *XiangGu* is dried under an appropriate temperature and then vacuum packed. Drying produces a quality product with a long shelf life. Sun drying is possible, but the quality is lower and there may be problems in humid climates. Most drying is done in air at 37–50°C for about 12 h. This method produces a weight reduction of approximately 7 g to 1 g. The dried *XiangGu* can be stored for up to a year (Tao et al., 2003).

Recently, the demand for freshly processed vegetables has grown and led to an increase in the quality and variety of produce available to customers. Low-temperature storage and the use of modified-atmosphere packaging (MAP) have been extensively reported as extending the shelf life of mushrooms. Under low temperatures, the shelf life of fresh *XiangGu* could be prolonged for up to 18 days. By combining the MAP method with low-temperature storage, a maximum shelf life of 25 days could be achieved (Xu et al., 1999).

XiangGu may be used in many different ways. It can be used as a nutritional supplement, or added to any number of herbal formulas. It is also a highly flavorful addition to dried soups or tonic drinks. Numerous *XiangGu*-containing products such as sauces, dried meats, pickled prawns (shrimp), milk candy, instant powder, instant noodles, and high-calcium sausages have been made in China (Li et al., 1998).

The popularity of *XiangGu* has grown even more because of the valuable therapeutic properties discovered in this fungus. Today, in China, the cultivation of *XiangGu* is an important agricultural industry. An estimated 18 million farmers are engaged in *XiangGu* production, and a steady increase in the annual output has occurred in recent years. In 1997, China produced 1,397,000 tons of *XiangGu*, which is about 89% of the total worldwide production. Much of China's crop yield is exported either fresh or dried. In 2002, the total production of *XiangGu* was estimated to be two million tons in fresh weight, of which 45,000 tons were exported to other countries and regions (Li, 2005).

As more consumers become aware of the special culinary characteristics offered by *XiangGu* and other specialty edible fungi, demand is likely to increase. The development of improved technologies to cultivate *XiangGu* more efficiently will allow the retail price to continue its decline.

2.6 TRADITIONAL CHINESE FUNCTIONAL *KOUMISS*

Koumiss, also called *Kumiss, Kumys*, or *Coomys*, is a traditional fermented dairy product originating in Central Asian areas. In Mongolia and China, *Koumiss* is also called *Airag* or *Chige*, which means fermented mare's milk (Zhang et al., 2006b). It is mildly alcoholic, sour-tasting, and usually made from the raw milk of mares or camels by a joint natural fermentation of microorganisms, that is, lactic acid bacteria and yeast. It is a popular beverage mainly produced and drunk in Eastern Europe, Central Asian Areas (Kazakhstan, Kyrgyzstan, etc.), Southeastern Russia, Mongolia, and China's Inner Mongolia and Xinjiang Provinces (Zhang et al., 2006c). According to archeological evidence, in prehistoric times, nomadic tribes living in the North China Plain often carried mare's milk in a bladder. The body temperature caused the mare's milk temperature to rise and the constant motion of the horse caused the milk to ferment evenly. This was the original *Koumiss* (Ha et al., 2003). Later, more practical and improved production techniques led to the invention of *Tongzhi* Fermentation Technology, which uses a specially made stick (*buluri* in Mongolian) to stir the milk upward and downward, resulting in an even fermentation. This technology is still used today.

2.6.1 HISTORY OF *KOUMISS*

As early as 2500 years ago, the nomads in Southeast Russia and the Scythia tribes in Central Asia drank *Koumiss* made from mare's milk. The famous Greek historian Herodotus also recorded its use in the fifth century BC. The Scythias drank and processed mare's milk. The French missionary William Lubuluqi, who came to China in the thirteenth century, gave a detailed account of the *Koumiss* production techniques and processes of the Mongolians in his work *Lubuluqi Journey to the East. The Customs of Northern Captives* also recorded *Koumiss* making procedures from fermented mare milk. A quote states "on the 1st day, the mare's milk taste is too sweet, but after 2–3 days, the taste is too sour. Only when it's made for alcohol, can the taste be the same as the liquor. First heat the mare's milk, then with the liquor for 3–4 times, the taste of alcohol is the strongest" (Mu and Bai, 2003).

By around 1500 BC, domestic horses were used in China. In the Yuan Dynasty, the use of mare's milk was the most common. Historic records from the *Yuan* Dynasty record that there were different levels based upon the mare's characteristics. *Khans, maharajas*, and other royalties used the milk from a white horse or a black horse. This milk was called *precious food*. Other colored mares also gave milk, which was drunk by the ordinary people. *Genghis Khan* specifically adopted 10,000 white horses and used the mares in the field. In periods of war, the Khan led 10,000 horse-riding soldiers and the accompanying tens of thousands of horses. During these series of war, when warriors got injured, they would drink mare's milk as nutritional supplements (Qin and Zhang, 1996). In Inner Mongolia and Xinjiang, the ethnic minorities had made and drunk *Koumiss* for hundreds of years. They accumulated a rich experience and understood how to apply the principles of the use of yeasts and the control of temperature, and how to adjust these to improve the production techniques of *Koumiss*. Over time, they continuously improved it (Zhao and Li, 2007).

2.6.2 Health Benefits of *Koumiss*

The function and nature of *Koumiss* are well known by many people. In 1858, Ostinkov opened a *Koumiss* therapy hospital in a town near Samaria. The Mongolians in China had noticed the special treatment effects of *Koumiss* early on. The Mongolian medical dictionary has recorded that "*Koumiss* tastes sour, sweet, bitter, helps digestion; strengthens the body, clears the esophagus, heals injuries, connects bones and strengthens the five sense organs. The bitterness will heal the blood heat, dissolve blood clots, reduce the obesity, and regenerate the skins."

It is known that *Koumiss* has the function of reducing blood pressure and dissolving blood clots, strengthening the kidneys, dispelling coolness, and enhancing gastric function and the mental system. In traditional Mongolian medicine, *Koumiss* is used as a medicine to treat certain diseases. On the basis of long-term medical practice and research, Mongolian medicine established *Koumiss* therapy to cure various diseases of the CV, digestive, and neurological systems, and to cure infectious diseases such as tuberculosis. It obtained good treatment effects. What is worth mentioning is that, from the pure pastoral area of prehistoric times until the present, Mongolians have eaten very simple foods and have not been able to eat fresh vegetables, but there has never been a reduction in their health or malnutrition. We think this has some link to drinking *Koumiss* (Ha et al., 2003).

In popular medicine today, *Koumiss* therapy still has value. It is widely applied in the treatment of various diseases. In 1980, the Xi Meng Mongolian Medical Research Hospital of Inner Mongolia started to use *Koumiss* in clinical practices, and it was discovered that it had good clinical effects in treating tuberculosis, emphysema, and chronic gastric diseases. In addition to the Inner Mongolian Zhong Meng Hospital and the XiLinGuoLe Mongolian Hospital, there has been a *Koumiss clinical center* specially built in Russia and Mongolia, especially for curing chronic diseases like CV diseases, digestive system diseases, neurological diseases, and tuberculosis (Zha, 1987). In the nationwide and worldwide science and technology archives, there are many relevant *Koumiss* medical research reports.

2.6.2.1 Treating Tuberculosis and Emphysema

Mongolian doctors use *Koumiss* to cure tuberculosis, and this has a long history resulting in a lot of clinical practical experience. The Ximeng Mongolian Medical Research Institute in Inner Mongolia uses *Koumiss* to cure tuberculosis every summer and autumn (Zha, 1987). It has obtained good clinical results. It discovered that *Mycobacterium tuberculosis* cannot survive in mare's milk or *Koumiss*. It attributed that to an antituberculosis element generated by the microorganisms in *Koumiss*. As described previously, Skordumova isolated *Saccharomyces lactis*, which is resistant to *M. tuberculosis* and other harmful microorganisms (Mu and Bai, 2003). Chahada used *Koumiss* in clinical practice to treat tuberculosis patients and has reported a 60–91% rate of recovery (through x-ray tests and tuberculosis tests; the indication of complete recovery is the disappearance of symptoms; the indication of effective treatment is the basic disappearance of the symptoms).

2.6.2.2 Clinical Effects of CV Diseases

Koumiss can enhance blood delusion and blood circulation. It can improve blood flow and reduce blood vessel pressure. It prevents the breakage of blood vessels and the formation of blood clots. It also has significant positive effects in treating high blood pressure. In addition, regular drinking of *Koumiss* will reduce the blood cholesterol content and control the formation of blood lipids (Zha et al., 1994).

2.6.2.3 Treatment of Digestive System Diseases

The microorganisms in *Koumiss* form a biological barrier on the walls of the stomach and the intestines. It prevents the growth of harmful bacteria with the micro-ecological environment. The antibacterial substances generated by the microorganisms can clear and kill the decayed bacteria. Also, it can modify the functions of the stomach and the intestines and enhance the digestive function. The large amount of fatty acid stimulates the stomach and the intestines to wriggle and increase the osmotic pressure and humidity of the feces, thus preventing restriction to transit (Liu and Ma, 2000).

2.6.2.4 Treatment of Neurological Diseases

Koumiss contains substantial amounts of V_{B1}, V_{B2}, V_C, and other rare items essential for the nervous system. Therefore, *Koumiss* can mend and adjust the functions of the neurological system. *Koumiss* can also improve blood circulation in the brain and the blood supply functions, thus curing various disorders of the neurological system, and of the digestive functions (Ha et al., 2003).

2.6.2.5 Treatment of Diabetes

Russian experts found that *Koumiss* can reduce the sugar content in a patient's blood and enhance insulin secretion, thereby adjusting sugar metabolism (Mu and Bai, 2003).

2.6.2.6 Treatment of Chloasma

Koumiss also has beautifying effects. Using *Koumiss* to wash the face can moisturize the skin, dispel dark skin and spots, smoothen the skin, and make the skin white and soft. Baodebili of the dermatological branch of the Inner Mongolia Hospital used *Koumiss* to make face creams and masks to cure chloasma (Bao et al., 1995). After the patients' use of *Koumiss*, the skin became soft, white, smooth, and flexible. It has good effects on the skin, with a total effectiveness of 95%.

2.6.2.7 Increase of Immunity

Fdeehck conducted research on the effects of *Koumiss* on the immune system. The results showed that it could considerably enhance the immune system of rats and chicklings against antigens. Liu and Ma (2000) also studied the immune functions of *Koumiss* in rats. The results show that fresh mare's milk can enhance the thymus index and the spleen index. It can also strengthen the functions of macrophages, and can increase the ratio of hemolysin in blood serum, and the erythrocyte rosette-forming rate of T-cells. Fresh mare's milk can increase the weight of the immune organs of rats and enhance their normal immune functions, control abnormal body fluid immune functions, and regulate cell immune abilities (Liu and Ma, 2000).

2.6.3 Chemical and Physical Properties of *Koumiss*

Among dairy animals, the only monogastric animal is the mare. Its secretion mechanism and other ecological functions and processes are different from those in multi-stomach animals such as cows, sheep, and camels. Mares can provide a special, high-quality milk. The physical properties and nutritional elements have their own special qualities (Huo et al., 2003).

Mare's milk is a white, thin liquid with a complex composition. It contains hundreds of chemical constituents. Its main components include water, protein, fat, lactose, minerals, vitamins, enzymes, immune bodies, pigments, and ash contents (Table 2.7). Compared to cow's and sheep's milk, mare's milk is rich in lactalbumin, peptone, amino acids, and essential fatty acids, and is relatively high in vitamins and minerals.

The nutritional characteristics of mare's milk and *Koumiss* are summarized as follows.

1. *Protein:* The protein content of mare's milk is lower than cow's and sheep's milk, on average 1.7–2.2%. There are differences in the protein content in different types of milk. Although mare's milk has a lower protein content than cow's milk, there are correct ratios of casein and lactalbumin contents, around 1:1, which is close to human milk.
2. *Fat:* Fat is the most important content of mare's milk, with the highest content of free fatty acids, especially essential fatty acids, which is four to five times that of cow's milk. The essential fatty acids needed by the human body (such as linoleic acid and linolenic acid) are relatively high in concentration in mare's milk. Mare's milk has a higher fatty acid content of

TABLE 2.7
Chemical Composition (%) of Human and Domestic Animal Milk

Chemical Composition	Human Milk	Mare Milk	Cow Milk	Goat Milk	Sheep Milk	Camel Milk
Total milk solids	12.6	11.0	12.5	13.2	17.9	14.6
Protein	1.5	2.09	3.3	3.8	5.8	3.5
Lactose	7.0	6.7	4.7	4.1	4.6	4.9
Fat	3.8	1.8	3.8	4.5	6.7	5.5
Ash content	0.28	0.30	0.7	0.79	0.8	0.7

Source: From Mu Z, Bai Y: *Journal of Inner Mongolia Agricultural University* 1:116–120, 2003. With permission.

C_{16} fats, around 20–35%. The fatty acid ratio of C_2 to C_4 is more than that of cow's milk.

3. *Lactose:* Mare's milk has the highest concentration of lactose and it is constant at 6.7%, close to that of human milk (7.4%), 1.5 times that which is found in cow's milk. The sugar in mare's milk is mainly lactose. The content of other sugars is low, which is conducive to *Koumiss* lactose fermentation.
4. *Minerals:* Mare's milk has many types of minerals beneficial to the human body such as calcium, phosphorus, magnesium, zinc, iron, copper, and manganese, although their contents are lower than in cow's and sheep's milk, at around 0.3%. Other minerals are in optimum ratios, as a proper Ca:P = 2:1 ratio, which is similar to human milk, and is conducive to absorption by the human body.
5. *Vitamins:* Mare's milk is rich in vitamins A, B, E, B_1, B_2, B_{12}, and pantothenic acid. Vitamin C is at the highest concentration, at 9.8–13.5 mg/100 mL, 5–10 times that of cow's milk and 2–3 times that of human milk. After fermentation, vitamin B_{12} and vitamin C contents are high in mare's milk, which is conducive to pharmacological action.
6. *Amino acids:* Mare's milk and *Koumiss* are similar to human milk. All of the amino acids essential to the human body can be found in mare's milk and in *Koumiss* (Mu and Bai, 2003).

2.6.4 Bioactivity of *Koumiss*

The microbial content of *Koumiss* is very high, with the main groups being lactic acid bacteria, acetic acid bacteria, leukonoid, and enterococcus (Li et al., 2002). The joint action of these creates the unique special taste, nutritional value, and health functions of *Koumiss*. Traditional *Koumiss* contains microbes that are different, due to the different fermentation and production methods used, and to seasonal changes. Also, *Koumiss* at different fermentation stages will have different microbial structures and ratios. In the first few hours of fermentation, lactobacilli and acetic acid bacteria are active with a low pH value. Later, microzymes grow quickly with rising alcohol content. As acidity and alcohol content rise, they inhibit the activity of microzymes and kill them. Microzyme activity can last for a long time. Table 2.8 contains the preliminary test results of microbes from 15 samples.

Koumiss is rich in those microbes that provide valuable resources for selecting probiotics. At present, the microbes isolated from *Koumiss*, especially probiotics, represent an excitingly hot research field. He et al. (2002, 2007) extracted *Enterococcus faecalis* from the *Koumiss* samples in the Xi Lin Guo Le Meng Region of Inner Mongolia. It was proved to have inhibitory effects toward the growth of *Listeria monocytogenes*. Zhang et al. (2006b) extracted acid-tolerant *Lactobacilli* and the highest resistance density to cholate in the medium was measured at 1.6 g/100 mL. *Lactobacillus* has certain cholate hydrolytic enzymes, which can cultivate sodium taurocholate in the medium and release free choleric acid. After holding at 37°C for 24 h, 49.91% of the cholesterol can be separated. These results showed that *Koumiss* can be used as a potential source of probiotics.

TABLE 2.8
Some Microbes Isolated from Koumiss

Species	Genus	Numbers
Lactic acid bacteria	*Lactobacillus*	12
	Streptococcus	5
	Leuconostoc	3
	Enterococcus	6
Acetic acid bacteria	*Acetobacter*	1
	Saccharomycodes	3
	Debaryomyces	3
	Kloeckera	2
	Brettanomyces	5
	Pichia	2
	Saccharomyces	5
	Zygosacchromyces	2

Source: From Mu Z, Bai Y: *Journal of Inner Mongolia Agricultural University* 1:116–120, 2003. With permission.

2.6.5 Distribution of Raw Materials for *Koumiss*

Mare's milk has very strong seasonal traits. Milk production is not large, and this inhibits the industrialization of *Koumiss* (Mu and Bai, 2003). To overcome this difficulty, modern biological techniques need to be used to introduce high-quality mare breeds, improved feeding conditions, mare's milk secretion quantity, and length of production.

Xinjiang is home to many mares. Traditional nomadic minority groups own the mares. Xinjiang's Kazakhstan mares have stronger galactophores, with a large milking capacity after years of nomadic milking. With sufficient nutrition available, adult mares can supply 6–7 kg of milk per day in addition to feeding young foals. The annual milking period can be up to 120 days. According to the statistics from 2005, Xinjiang now has around 960,000 horses. The annual mare's milk output can reach 360,000 tons. About 70% of mare's milk can be processed, and this has become the main income source for many people. Therefore, the Xinjiang regional development in *Koumiss* production has also facilitated regional economic development.

2.6.6 Processing Technology for Both Homemade and Industrial Production Scales

2.6.6.1 Homemade Koumiss

As early as 2500 years ago, the nomadic groups in Southeast Russia and the Scythian tribes were drinking mare's milk in the form of *Koumiss.* The famous Greek historian recorded in the year 5 BC that the Scythians drank and produced *Koumiss.* Li Qi wrote "drink mare's milk as alcohol, after constant stirring" (Qin and Zhang, 1996). The modern minorities in Kazakhstan, the Khalkhas groups, also made *Koumiss* as was done in ancient times. The method is described below. Place the mare's milk in a mare skin container with yeast; then place the container in an area with constant temperature to ferment; and finally stir with a specially designed wooden stick several times. After several days, the milk will be slightly sour with alcoholic smells; it is then clear and comfort-type *Koumiss*.

Mold culture is a kind of pasty milk after fermentation, and is used as the starter culture. Nomadic groups usually prepare yeast in the autumn. They select good-quality, unique flavored, and mildly fermented *Koumiss*. They select the yeast and preserve it by the following methods. One method is to put it in a deep well or cellar, and the other is to select cloth bags or boiled gauze and harrow to dry up. *Koumiss* is then stored under cool conditions. The key in making good *Koumiss* is to produce ethanol and lactic acid in the right ratio. Lactic acid bacteria are the easiest to grow at temperatures between 40°C and 45°C. The best temperature for enzyme activity is 20–25°C. The traditional production methods vary from season to season, and result in different flavors. The *Koumiss* produced in autumn is unique in taste. This is not only because of the good quality of mare's milk in autumn, but also because the low temperature is conducive to the production of enzymes (Qin and Zhang, 1996).

2.6.6.2 Industrial Production of Koumiss

Koumiss industrialization started late in China, around the 1920s. At present, Inner Mongolia and Xinjiang have several factories with limited scale production. Large amounts of *Koumiss* are produced in Yili of Xinjiang and are sold on the streets and in shops. For current *Koumiss* production, keeping the microbial strain viable and maintaining product freshness and quality are two key issues.

The present strains are mostly natural, but some are artificially grown. These are mainly enzymes and lactobacilli. The flavor, texture in the mouth, and quality are not as good as with natural starter cultures. Therefore, studying the yeasts and their coexisting relationship with microbes is one of the main directions for *Koumiss* development (Mu and Bai, 2003).

The industrial process for *Koumiss* production is shown in Figure 2.3. In the production process for *Koumiss*, sanitary conditions are a key control point and using clean raw milk and containers, selecting fresh milk, producing strains and inoculating, sterilizing, and critical control point (CPP) cleaning are important (Liu et al., 2003). The sanitary quality control measures are as follows:

1. Immunization of the young mares.
2. Clarification of the origin and composition of feed to ensure zero pollution.
3. Cleaning the surfaces of the mare's breasts, and the milking staff's hands. Because the density of lactose is high, maintaining sterile conditions is very important. The bacterial count in raw milk should be kept as low as possible to ensure sanitary quality. Before milking, the mare must be treated with measures to ensure suitable cleanliness, such as the use of ClO_2 solution to clean the milk container in order to effectively control bacterial concentration.
4. Raw milk testing: To ensure the mare's milk quality and feasibility measures, raw material testing indicators must at least include acidity and specific gravity. On this basis, there must be regular checks for microflora, coliforms, molds, and yeasts.

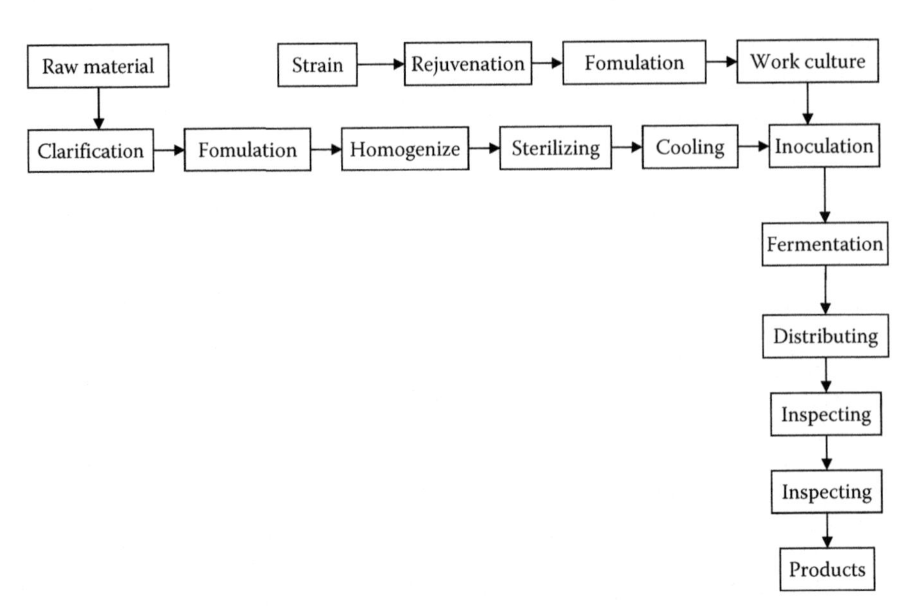

FIGURE 2.3 Industrial process for *Koumiss* production (Data from Liu S, Lin H: *Marine Sciences* 27:57–60, 2003.)

2.7 TRADITIONAL CHINESE FUNCTIONAL SEA CUCUMBER

The sea cucumber is an echinoderm of the class *Holothuroidea*, with an elongated body and leathery skin, and is found on the sea floor worldwide. It is so named for its cucumber-like shape and pebbly skin. In China's marine areas, there are 140 species, from *Bo Hai* Bay, *Liao Dong* Peninsula, to *Wei Zhou* Island, and the *Nan Sha* archipelagos. Among them, 20 species are edible, with 10 considered of better quality, as rare sea products. In the north of China, *Apostichopus japonicus* is the sole kind of edible sea cucumber. There are more species of sea cucumber in the south of China. Ten of them live in *Hai Nan* and *Xi Sha* archipelagos. *Stichopus variegatus* is the common edible sea cucumber (Cui et al., 2002).

The sea cucumber has a very rich nutrient content. The edible parts of dried sea cucumbers make up 93% of the total, supplying 1096 kJ of energy per 100 g. The typical nutrient contents of 100 g sea cucumber include water 18.9 g, protein 50.2 g, fat 4.8 g, carbohydrate 4.5 g, sodium 5.0 g, magnesium 1.0 g, potassium 0.4 g, selenium 0.15 g, niacin 1300 μg, cobalamin 2213 μg, vitamin A 39 μg, retinol 39 μg, and iron, manganese, zinc, lead, and phosphorus (Institute of Nutrition and Food Hygiene, Chinese Academy of Preventive Medicine, 2001). Many sea cucumbers contain toxins. The edible sea cucumber should be processed by washing away the toxins.

The Chinese began to use sea cucumbers more than 2000 years ago as a health food. It is listed in the *Yellow* Emperor's Canon of Internal Medicine, China's earliest medical work. In the *Food Record* of the Ming Dynasty, sea cucumbers were used for supplementing energy and for strengthening internal organs and the body. The *Herbal Outline Supplement* of the Qing Dynasty listed sea cucumbers as a health supplement.

Recent research has confirmed the medicinal value of sea cucumbers. *Apostichopus japonicus* and *Stichopus variegatus* have been shown to be beneficial in antioxidation (Mamelona et al., 2007), strengthening the immune system (Wang et al., 2007a), treating cancer (Wang et al., 2007a) and atherosclerosis (Wang et al., 2008; Zhang et al., 2007), and protecting the vascular endothelium (Wang et al., 2007b). The sea cucumber is an excellent source of nutrition for pregnant women, and is believed to preserve youth and beauty.

2.8 TRADITIONAL CHINESE FUNCTIONAL EDIBLE BIRD'S NEST

Edible bird's nests are built by the male swiftlet, after ingesting small fish and/or seaweeds during the breeding season, with gastric juice and quill coverts over a period of 35 days. They take the shape of a shallow cup stuck to the cave wall.

The nests are traditionally harvested from high up on cave walls. There is some risk to collectors who stand on bamboo scaffolding that is sometimes hundreds of feet tall and centuries old. This delicacy food was first recorded in Herbal Records, which states its properties as sweet, mild, aiding digestion, raising libido, improving the voice, alleviating asthma, and increasing concentration. In clinical practice, the nests cure coughing, hematemesis, malaria, choking, and nausea. They have been used widely by the general public as rare herbal medicines and delicacies (Kang et al., 2003).

Traditional Chinese medicine believes that edible bird's nests stimulate cell growth and strengthen human immunity against cancer as well as enhanced recovery after radiation. The numerous soluble proteins in edible bird's nests are functional components. Based on medical effects, these components are categorized into anti-inhibitor coagulant factors, virus neutralizers, epidermal growth factors, and mitogens (Lin et al., 2006b). The efficacies recorded in Chinese medicine include the following: blood clotting; enhancing lectin to facilitate the mitosis of lymphocytes; increasing immunity to prevent brain aging, and offsetting the effects of reactive oxygen species; and strengthening the heartbeat.

2.9 TRADITIONAL CHINESE FUNCTIONAL ROYAL JELLY

Royal jelly is a honey bee secretion that is used in the nutrition of the queen and larvae. During a season of May to June, a well-managed hive can produce approximately 500 g of royal jelly. The queen uses royal jelly as food for her whole life time. The result is a larger body size, stronger productive ability, and a greater weight of egg production in 24 h than her body weight (Hua, 1994).

Royal jelly contains 62.5–70% (w/w) water and 30–37.5% (w/w) dry substance (Tamura, 1985). The protein consists of two-thirds of albumin and one-third of globulin. The special element in royal jelly is 10-hydroxy-2-decylenic acid, which is the characteristic substance in royal jelly. People collect and sell royal jelly as a dietary supplement, claiming various health benefits because of components like B-complex vitamins such as pantothenic acid (vitamin B_5) and vitamin B_6 (pyridoxine). It can also be found in various beauty products. The annual global total consumption has reached 1 million tons. China has a large geographical area with mild climate and rich bee resources. Recently, there were 7 million bee colonies, each of which produced 0.2 tons of bee jelly every year; hence China is the largest bee-raising country in the world. The ethnic minorities of Yunnan Province of China have, for long, recognized the value of royal jelly, which is said to cure numerous diseases. Modern nutritional and medical research illustrates that royal jelly has several effects: antiaging, anticancer, immunity enhancement, blood circulation adjustment, and beautification (Cho, 1977).

With the increasing production of royal jelly products, medical and clinical research on royal jelly has also increased. Recent results include the following:

1. It cures hepatitis B (Zhu, 1998). It can have both antiviral and immunity adjusting effects. It can also enhance the human body and internal organs. Royal jelly contains proteins that can greatly improve the albumins and globulins, plus liver health. There are eight essential amino acids in royal jelly, which is the medicine essential for curing regular hepatitis B infection. Moreover, in royal jelly, there are numerous types of vitamins such as acetyl choline, which can prevent and cure liver diseases. Vitamin E can protect and maintain the normal fat structure and biomedical functions of cells. It can also inhibit recurrence after malignancy treatment, and assists in lung cancer treatment (Zhu, 1998).
2. Royal jelly's chemical composition is complex. It contains large amounts of biologically active substances and has wide medicinal effects. It also contains substances that can cure tumors.
3. It has treatment effects to cure diabetes. Royal jelly contains substances similar to insulin, which can reduce the blood sugar level. Royal jelly can be used by diabetic patients (Hua, 1994).
4. It can prevent aging (Zhang and Dai, 2000). This function illustrates two aspects: it contains superoxide dismutase, which is the main dispelling instrument against free radicals, and it has an immunity adjusting function.
5. Other effects include blood pressure reduction (Albert et al., 1999), antimetamorphosis (Oka et al., 2001), antiviral effects (Fujiwara et al., 1990), and sex hormone adjusting effects (Liang et al., 1999).

2.10 PRODUCT STANDARDS IN CHINA

Four standards are related to the production of *douchi* and *sufu*. These include two national standards (GB 2712-2003 "Hygienic standard for fermented bean product" and GB/T 5009.52 "Method for analysis of hygienic standard of fermented been products," issued by Chinese Health Ministry and Chinese Standardization Administration) and two commercial and industry standards (DB 51/T391-2006 "Technical standard for fermented soybeans" issued by Bureau of Quality and Technical Supervision of Sichuan Province, and SB/T 10170-2007 "Fermented bean curd" issued by Ministry of Commerce of the People's Republic of China).

The product standards concerning walnuts include SN/T 0880-2000 "Rules for the inspection of walnuts in shell for import and export," QB/T 1411-91 "Canned roasted and salted walnut flesh," QB/T 2301-1997 "Vegetable protein beverage, walnut milk," DB33/T 514-2004 "Rules for the inspection of hickory for export," SN/T 0881-2000 "Rules for the inspection of walnut meats for import and export," GB 10164-88 "Walnuts," GB 7907-87 "High yield of walnut and quality of nuts," GB/T 20398-2006 "Walnut quality grade," and QB/T 1410-91 "Canned amber sugar coated walnut flesh."

The product standards concerning *XiangGu* include SN/T 0632-1997 "Rules for inspection of dried *XiangGu* export," GH/T 1013-1998 "XiangGu," GB/T 18525.5-2001 "Code of food irradiation practice for the control of molds and insect disinfestation in desiccated *XiangGu*," GB 19087-2003 "Product of designations of origin or geographical indication—Qingyuan *XiangGu*," and NY/T 1061-2006 "Grades and specifications of *XiangGu*."

2.11 SUMMARY

The present-day paradigm for functional foods involving appreciation of the health-benefiting or ailment-preventative role of foods, as opposed to only the nutritive role, has been part of Chinese alimentology since 100 BC. With the advent of modern research, various health claims for Chinese traditional foods in preventing, managing, or delaying the premature onset of chronic disease later in life have been ascertained by identifying the functional constituents, elucidating the biochemical structures, and determining the mechanisms behind the physiological roles.

Various foods have been officially documented, most of which are less well known or appreciated outside the orient. Among them, the major ones include soybean products—*sufu* and douche, Chinese yellow wine, *Xiajiang* (fish sauce), walnut kernels, *XianGu*, Chinese royal jelly, Chinese traditional sea cucumber, and traditional edible bird's nests. There have been a reasonable number of research-based health and beneficial claims that have been made for some of the foods such as *Koumiss* going up to clinical trials level. The claims that have been made for various foods include, but are not limited to, antiviral, anti-inflammatory, antioxidant, liver-protecting, antifatigue, antidiabetes, antiobesity, hypocholesteremic, hypotensive, antimicrobial, and many other medicinal effects. As a result of the complex nature of these foods, although many have been studied, there is still a wide gap of knowledge regarding characterizing their beneficial effects.

The Chinese government, in the mean time, has set out specific guidelines and laws regarding the health claims and other various aspects pertaining to traditional Chinese foods. The government agencies mainly include the Chinese Health Ministry and Chinese Standardization Administration, Ministry of Commerce of the People's Republic of China, National Development and Reform Commission, and the Chinese Agriculture Ministry.

The beneficial effects of these foods are derived mainly from the special constitutive raw materials, enhanced through the special traditional Chinese processing technologies that have principally been passed on down through the generations from ancient times. These technologies function to develop the beneficial effects or to release bioactive substances from raw materials. The processing technologies mostly take advantage of natural conditions such as ambient temperature, humidity, and other product-specific aspects such as natural starter cultures to bring out the nutraceutically beneficial functions. For example, natural fermentations characterize products such as *dofu, sufu*, and fish sauce. Presently, processing ranges from homemade production technologies similar to the ancient Chinese technologies to small-scale and industrial production technologies that benefit from modern mechanization in many of their steps. There are regional variations within China in the processing of the various products, especially in terms of the starting raw materials. This mainly serves to match the variation in regional specific organoleptic preferences for the products. Within this, there lies the potential for the worldwide adoption of these products. Continuous research is under way presently on the improvement of the production and the quality of most of these products.

REFERENCES

Adlercreutz H, Mousavi Y, Clark J, Hockerstedt K, Hamalainen E, Wahala K, Makela T, Hase T: Dietary phytoestrogens and cancer: *In vitro* and *in vivo* studies. *Journal of Steroid Biochemistry and Molecular Biology* 41:331–337, 1992.

Al-Bachir M: Effect of gamma irradiation on fungal load, chemical and sensory characteristics of walnuts (*Juglans regia* L.). *Journal of Stored Products Research* 40:355–362, 2000.

Albert S, Bhattacharya D, Klaudiny J: The family of major royal jelly proteins and its evolution. *Journal of Molecular Evolution* 49:290–297, 1999.

Bai L: Analysis of some nutritive and medicinal components in Shiitake [*lentinus edodel* (Berk) Singer]. *Journal of Zhoukou Normal University* 24:101–102, 2007.

Bao D, Wu J, Ma F, Wang S, Hao B: The 44 samples of chloasma treatment with Koumiss facial mask. *The Chinese Journal of Dematovenereology* 12:110–110, 1995.

Chang S, Miles P: *Edible Mushroom and their Cultivation.* Boca Raton, FL, CRC Press, 1989.

Chen Q, Li LK, Wu Y: Progress of research on the chemical components and pharmaceutical action of walnut kernel. *Journal of Anhui University Natural Science Edition* 29:86–89, 2005.

Chihara G: *Medical Aspects of Lentinan Isolated from Lentinus edodes (berk) Sing*. The Chinese University Press, Hong Kong, 1993.

Chiraha G, Hamuro J, Maeda YY, Arai Y, Fukuoka F: Fractionation and purification of the polysaccharides with marked antitumor activity, especially Lentinan, from *Lentinus edodes*. *Cancer Research* 30:2776–2781, 1970.

Chihara G, Maeda Y, Hamuro J, Sasaki T, Fukuoka F: Inhibition of mouse sarcoma 180 by polysaccharide from *Lentinus edodes* (Berk) Sing. *Nature* 222:687–688, 1969.

Cho YT: Studies on royal jelly and abnormal cholesterol and triglycerides. *American Bee* 117:36–38, 1977.

Cui G, Zhao L: Names and species identifying of edible sea cucumber in China. *Cuisine Journal of Yangzhou University* 3:13–18, 2002.

Denter J, Rehm HJ, Bisping B: Changes in the contents of fat-soluble vitamins and provitamins during tempe fermentation. *International Journal of Food Microbiology* 45:129–134, 1998.

Dong X, Ning A, Cao J, Huang M: Progress of the research on *Lentinus edodes* and the pharmacological effect. *Journal of Dalian University* 26:63–68, 2005.

Duan Z, Xu H, Chen Y: The effects of prevention and repair of *Lentindus edodes* on genetic damage induced by radiation in mice. *Chinese Journal of Radiological Health* 6:196–198, 1997.

Fujiwara S, Imai J, Yaeshima T: A potent antibacterial protein in royal jelly. Purification and determination of the primary structure of royalisin. *Journal of Biological Chemistry* 265:11333–11337, 1990.

Fulushima M, Ohashi T, Fuliwara Y, Sonoyama K, Nakano M: Cholesterol-lowering effects of Maitake fibre, Shiitake fibre and Enokitake fibre in rats. *Experimental Biology and Medicine* 226:758–765, 2001.

Ha SAM, Mang L: Koumiss and its medical values. *China Journal of Chinese Materia Medica* 28:11–14, 2003.

Han B, Rombouts FM, Nout MJR: A Chinese fermented soybean food. *International Journal of Food Microbiology* 65:1–10, 2001.

He Y (ed): *Chinese XiangGu*, Shanghai Science and Technology Press, Shanghai, 1994.

He Y, Li S, Mu Z, Wang L, Zhao M, Shuang J, Wu J: Isolation, identification and anti-bacteria function of microorganisms from Koumiss. *Transactions of the Chinese Society of Agricultural Engineering* 18:91–95, 2002.

He Y, Wang L, Li S, Wu J, Tian J: Extraction and characteristics of antibacterial factor from *Enterococcus faecalis* in *Koumiss. Transactions of the Chinese Society of Agricultural Engineering* 23:264–267, 2007.

Hirabayashi M, Matsui T, Yano H: Fermentation of soybean flour with *Aspergillus usamii* improves availabilities of zinc and iron in rats. *Journal of Nutritional Science and Vitaminology* 44:877–886, 1998.

Hua M: Clinical observation of diabetes with royal jelly treatment. *Bee Journal* 7:5–6, 1994.

Huo Y, Rong HSS, Leng AMG: A study of Koumiss nutritional composition and active molecules. *Inner Mongolia Husbandry* 6:22–23, 2003.

Ishida H, Uesugi T, Hirai K, Toda T, Nukaya H, Yokotsuka K, Tsuji K: Preventive effects of the plant isoflavones, daidzin and genistin on bone loss in ovariectomized rats fed a calcium-deficient diet. *Biological and Pharmaceutical Bulletin* 21:62–66, 1998.

Ishimi Y, Miyaura C, Ohmura M, Onoe Y, Sato T, Uchiyama Y, Ito M, Wang X, Suda T, Ikegami S: Selective effects of genistein, a soybean isoflavone, on b-lymphopoiesis and bone loss caused by estrogen deficiency. *Endocrinology* 140:1893–1900, 1999.

Ji B, Li B: *Safe Production and Quality Control of Soybean Products*. Chemical Industry Press, Beijing, 2005.

Kang T, Li F, Hu, Y. *Yan Wo De Yan Jiu Jin Zhan. China Journal of Chinese Materia Medica* 28:1103–1105, 2003.

Kaplan NM: The dietary guideline for sodium: Should we shake it up? No! *The American Journal of Clinical Nutrition* 71:1020–1026, 2000.

Kasaoka S, Astuti M, Uehara M, Suzuki K, Goto S: Effect of Indonesian fermented soybean tempeh on iron bioavailability and lipid peroxidation in anemic rats. *Journal of Agricultual and Food Chemistry* 45:195–198, 1997.

Li J (ed): *The Principle and Method of Biochemical Experiments*. Peking University Press, Beijing, 1994.

Li L, Li Z, Yin L: *Soybean Processing and Application*. Chemical Industry Press, China, 2003.

Li S, Wu N: Studies on biological properties of lactic acid bacteria from Koumiss in Xilin Guole. *Journal of Inner Mongolia Agricultural University* 23:59–66, 2002.

Li W, Li J, Zhang H: Study on separating of undesirable constituents from *Lentinus edoes* and processing preserved *Lentinus edoes. Science and Technology of Food Industry* 4:41–43, 1998.

Li X, Chen L: *Practical Fruit Vegetables Refreshing Technology*. Scientific and Technical Documents Publishing House, Beijing, 2000.

Li Y: Research status and prospects of *Lentinula edous. Microbiology Letter* 32:149–152, 2005.

Liang M, Liu W, Liang X: Sex hormone effects of organic acid in RJ. *China Medicine Journal* 14:22–23, 1999.

Lin G, Xu X, Lian W: Antioxidant activity of edible mushroom polysaccharide extracts in vitro. *Journal of East China University of Science and Technology (Natural Science Edition)* 32:278–280, 2006a.

Lin J, Zhou H, Lai X. *Yan Wo Yan Jiu Gai Shu. Journal of Chinese Medicinal Materials* 29:85–90, 2006b.

Lin Z, Huang Y: Protective action of *Lentinus edodes* polysaccharide on experimental liver injury. *Journal of Peking University (Health Sciences)* 19:93–95, 1987.

Liu D, Qian JY, Bu M: Current research in anti-tumor activities of *Lentinan. Edible Fungi of China* 22:6–7, 2002.

Liu G, Ma L: Koumiss treatment of astriction. *China Folk Therapy* 8:27–28, 2000.

Liu H, Gao K, Liu L, Zhang H, Hou L, Gao J: HACCP application in the *Koumiss* production. *China Dairy Industry* 31:34–35, 2003.

Liu M: Development of walnut kernel health wine. *Liquor-making Science & Technology* 1:84–86, 2004.

Liu Q: The history and present status of the sufu study. *Chinese condiment* 2:9–27, 1988.

Liu S, Lin H: The preparation of shrimp brie with lower salinity by enzymolysis. *Marine Sciences* 27:57–60, 2003.

Luft FC: Salt and hypertension at the close of the millennium. *Wien Klin Wochenschr* 110:459–466, 1998.

Mamelona J, Pelletier E, Girard-Lalancette K, Legault J, Karboune S, Kermasha S: Quantification of phenolic contents and antioxidant capacity of Atlantic sea cucumber, *Cucumaria frondosa. Food Chemistry* 104(3):1040–1047, 2007.

Masatomo H, Naoto S, Tomotake N, Kazuo F, Kazuko T: Three kinds of antibacterial substances from *Lentinus edodes* (Berk.) Sing. (Shiitake, an edible mushroom). *International Journal of Antimicrobial Agents* 11:151–157, 1999.

McCarron DA: The dietary guideline for sodium: should we shake it up? Yes! *The American Journal of Clinical Nutrition* 71:1013–1019, 2000.

Mu Z, Bai Y: Koumiss. *Journal of Inner Mongolia Agricultural University* 1:116–120, 2003.

Ni L, Rao P, Wang Z: Isolation and characterization of bioactive peptides in sufu. *Journal of Zhejiang University (Agriculture and Life Science)* 21:93–97, 1997.

Nout MJR, Rombouts FM: Recent developments in tempe research. *Journal of Applied Bacteriology* 69:609–633, 1990.

Oka H, Emori Y, Kobayashi N, Hayashi, Y, Nomoto, K: Suppression of allergic reactions by royal jelly in association with the restoration of macrophage function and the improvement of Th1/Th2 cell responses. *International Immunopharmacology* 1:521–532, 2001.

Ou-Yang XN, Yu Z, Wang W, Zhang X: Trial study on an ti-inflammatory effects of Lentinan. *Academic Journal of People's Liberation Army Postgraduate Medical School* 27:56–57, 2006.

Pickering G: Salt intake and essential hypertension. *Cardiovascular Review Report* 1:13–17, 1980.

Prasad RNB: Walnuts and pecans, in *Encyclopaedia of Food Science and Technology*. Academic Press, London, 1993.

Qi C, Zhang W, Wang Y, Zhang G: The preparation of shrimp sauce by protease and its application. *Food Technology* 4:19–20, 2007.

Qin L, Zhang H: A study of Koumiss of Chinese minority groups. *Liquor-making Science & Technology* 78:72–75, 1996.

Ren H, Liu H, Endo H, Takagi Y, Hayashi T: Anti-mutagenic and anti-oxidative activities found in Chinese traditional soybean fermented products *furu. Food Chemistry* 95:71–76, 2006.

Samaranayaka AGP, John JA, Shahidi F. Antioxidant activity of English walnut (*Juglans regia* L.). *Journal of Food Lipids* 15(3): 384–397, 2008.

Sarkar PK, Morrison E, Tinggi U, Somerset SM, Craven GS: B-group vitamin and mineral contents of soybeans during kinema production. *Journal of the Science of Food and Agriculture* 78:498–502, 1998.

Shi M, Wen H: The processing technology of the traditional shrimp paste of the Dong nationality. *Fisheries Science and Technology Information* 33:143–144, 2006.

Sze-tao KWC, Sathe SK: Walnuts (*Juglans regia* L.): Proximate composition, protein solubility, protein amino acid composition and protein *in vitro* digestibility. *Journal of Science of Food and Agriculture* 80:1393–1401, 2000.

Tamura T: Royal jelly from the standpoint of clinical pharmacology. *Honeybee Science* 6:117–124, 1985.

Tao J, Xiao Q, Cai S, Chen N, Li D: Technique of baking and processing Xianggu mushroom. *Food Science and Technology* 4:23–25, 2003.

USDA, Agricultural Research Service. *Nuts, Walnuts, English. USDA National Nutrient Database for Standard Reference, Release 20 (2007).* U.S. Department of Agriculture, Agricultural Services, Washington, DC 2007 [cited 4/16 2008]. Available from http://www.nal.usda.gov/fnic/foodcomp/cgi-bin/list_nut_edit.pl

Wang J, Dong L, Dong P, Zhao Q, Xue C: Effects of two species of *Holothurian* on blood lipid regulation and vascular endothelium protection in experimental hyperlipidemia rats. *Chinese Journal of Marine Drugs* 26:10–13, 2007b.

Wang J, Gao S, Pang L, Zhao Q, Liu H, Xue C: Comparative investigation on the effects of triterpene glycosides from three kinds of sea cucumber on vascular endothelial cells. *Periodical of Ocean University of China* 38:221–224, 2008.

Wang J, Lu Z, Chi J, Wang W, Sua M, Kou W, Yu P, Yu L, Chen L, Zhu J, Chang J: Multicenter clinical trial of the serum lipid-lowering effects of a *Monascus purpureus* (red yeast) rice preparation from traditional Chinese medicine. *Current Therapeutic Research* 58:964–978, 1997.

Wang J, Wang Y, Zhao L, Feng L, Xue C: Effects of *Apostichopus japonicus* on antitumor and immune regulation in S_(180) bearing mice. *Periodical of Ocean University of China* 37:93–96, 2007a.

Wang L, Li L, Fan J, Qi: Antioxidative activity evaluation study on fermented tofu extracts by scavenging DPPH. *Food Science (China)* 25:169–172, 2004.

Wang L, Saito M, Tatsumi E, Li L: Antioxidative and angiotensin I-converting enzyme inhibitory activities of sufu (fermented tofu) extracts. *Japan Agricultural Research Quarterly* 37:129–132, 2003.

Wang R: Study on the Chinese preserved bean curd. *Chinese Condiment* 1:43–56, 2006.

Wei H, Wei L, Frenkel K, Bowen R, Barnes S: Inhibition of tumor promoter-induced hydrogen peroxide formation *in vitro* and *in vivo* by genistein. *Nutrition and Cancer* 20:1–12, 1993.

Wu J: Medical and health function of edible fungi and its improvement of food structure. *Edible Fungi of China* 22:9–11, 2002.

Wu X, Chen S: Development of the cultivation of Xianggu in China. *Journal of Zhejiang Forest Science and Technology* 22:14–18, 2002.

Xi R, Zhang R: *Chinese Fruit Tree Memory: Walnut Issue.* China Forestry Publishing House, Beijing, 1995.

Xu L, Pei Y, Xu S: The application of MAP in refreshing XiangGu. *North Horticulture* 129:31–33, 1999.

Xue H: The refreshing storage of XiangGu. *Edible Fungi of China* 26:51, 2007.

Yang J, Wu MC, Zhang SH, Liang GY: Study on the antifatigue effects of protein-band polysaccharide from *Lentinus edodes*. *Acta Nutrimenta Sinica* 4:350–353, 2001.

Yang J: The origin and development of China's China's bean curd. *Agricultural Archaeology* 1:217–226, 2004.

Yang Z, Li L, Liang S, Yang X: Determination of peptidase system of microorganisms in ripening of Chinese sufu. *Journal of South China University of Technology (Natural Science Edition)* 34:35–40, 2006.

Yin L, Li L, Li Z, Tatsumi E, Masayoshi S: Changes in isoflavone contents and composition of sufu (fermented tofu) during manufacturing. *Food Chemistry* 87:589–592, 2004.

Zha M, Liu Y, Jin Z: 50 clinical samples of Koumiss fat reduction and anti-clotting. *Inner Mongol Journal of Traditional Chinese Medicine* 13:16, 1994.

Zha M: *Kumiss Therapy*. Inner Mongolian People's Publishing House, Huhehaote, 1987.

Zhang F, Zhang S, Sun F, Liu Z, Chen X: Anti-virus effect of Lentinan. *Journal of Changchun University of Traditional Chinese Medicine* 22:11–12, 2006a.

Zhang H, Meng H, Wang J, Sun T, Xu J, Wang L, Yun Y, Wu R: Assessment of potential probiotic properties of *Lactobacillus casei* Zhang strain isolated from traditionally home-made Koumiss in Inner Mongolia of China. *China Dairy Industry* 34:4–10, 2006b.

Zhang J, Cai S, Yao C: Effects of mucopolysaccharide from *Stichopus variegatus* on the proliferation and apoptosis induced by fetal bovine serum in rat vascular smooth muscle cells. *Chinese Journal of Marine Drugs* 26:11–15, 2007.

Zhang J, Dai Q: Royal jelly frozen powder's effects to the immunity of rats. *Journal of Shanghai Tiedao University* 21:16–17, 2000.

Zhang R, Li L, Li J, Fu H, Rao P: A study of hypocholesterolemic effect of tofuru. *Journal of Fuzhou University (Natural sciences edition)* 26:132–135, 1998.

Zhang X, Li L, Li Z, Li X, Tatsumi E: Study on the changes of soybean isoflavone during the sufu making. *China Brewing* 6:17–20, 2002.

Zhao BJ, Gong YH: *Overview of Walnut Culture in China*. UCDAVIS 2006. Available from http://walnutresearch.ucdavis.edu/2006/2006_33.pdf.

Zhao H, Huang M (eds): *Aquatic Products Assay.* Science & Technology Press, Tianjin, 1986.

Zhao H, Li Y: Inner Mongolian national dairy overview. *Food and Nutrition in China* 10:49–50, 2007.

Zhou X, Xiang L, Chen J: The investigation of shrimp nutrition. *China Feed* 19:19–21, 2002.

Zhu C: Bai Chun Capsules: 64 treatment examples for malignancies. *TCM Research* 11: 23–24, 1998.

3 Traditional Indian Functional Foods

Krishnapura Srinivasan

CONTENTS

3.1 INTRODUCTION

The curative effect of food has been a traditionally established belief for many generations in India. The current view that food can have an expanded role that goes well beyond providing a source of nutrients truly applies to many traditional Indian foods. In fact, the traditional Indian diet is "functional" as it contains high amounts of dietary fiber (whole grains and vegetables), antioxidants (spices, fruits, and vegetables), and probiotics (curds and fermented batter products), which are wise choices for health promotion. Many Indian traditional foods impart beneficial effects on human physiology beyond providing adequate nutrition. The health benefits thus derived may range from ensuring normal physiological functions in the body such as improving gastrointestinal health, enhancing the immune system, weight management, and providing better skeletal health, among others, in order to reduce blood cholesterol, oxidative stress, the risk of cardiovascular diseases, inflammatory diseases, various types of cancers, and possible prevention of diabetes, and neurodegenerative diseases. A dietary ingredient that affects its host in a targeted manner so as to exert positive effects on health can be classified as a "functional" ingredient. The functional components present in Indian traditional foods may be chemical or biological in nature and play a key role in imparting beneficial physiological effects for improved health. Some of the ingredients that make Indian traditional foods functional include dietary fiber, vitamins and minerals, oligosaccharides, lignins, essential fatty acids, flavonoids, miscellaneous phytochemicals, and lactic acid bacterial cultures. These functional ingredients are abundantly available in foods such as fruits, vegetables, cereals, legumes, nuts, and milk and milk-based products.

3.2 HISTORY OF INDIAN FOOD CULTURE AND TRADITIONAL FOODS

Indian heritage foods are of considerable antiquity and not much is known about their origin. There are a number of regional heritage foods that evolved locally, depending on the availability of raw ingredients. Heritage foods in India are an integral part of Indian culture. Traditional foods started with the inception of tradition, which dates back to Aryan civilization (3000 BC) followed by Harappan (2000 BC), Vedic (1500 BC), and later the Hindu culture as influenced by other cultures, and Indian food habits followed the changing cultural patterns. With a history of 3000 years or more, the Indian civilization has given food a prominent place in the social and cultural lives of its people. India has had several philosophies and religions which grew from within, such as pre-Buddhist Hinduism, Buddhism, Jainism, and post-Buddhist Hinduism. These philosophies interacted with each other and made their impact felt on Indian traditional food cultures (Achaya, 1994).

The cereal grain barley was the major grain eaten by the Aryans, followed by *Apupa*, *Lajah*, *Soma* juice, and rice. Wheat was introduced during the Vedic period. Cattle were an integral part of the Vedic culture, and the literature before 800 BC is full of references to the milk of the cow and other cattle. Vedic literature also refers to curdling of milk with starter from an earlier run. Curds thus prepared were eaten with rice, barley, or *Soma* juice. Ghee (*Ghrita*) was prepared by melting down and desiccating butter and was considered a commodity of prestige. Ghee was also used in Vedic rituals (as offerings to God), for frying, and for dipping to add relish to other foods, and for mixing with *Soma* juice. According to the sage Sushruta (600 BC), the profounder of Ayurveda, the indigenous system of medicine in India, cow's milk had a stabilizing effect on body secretions, while the fat-rich buffalo's milk was more healing. It is also mentioned by Sushruta that cream of milk called *Santanika* had many beneficial effects on health. Curds were distinguished as sweet, slightly acidic, and strongly acidic for consumption regionwise. Ghee prepared and stored for 10–100 years in a vessel was called "Kumbha ghrita" and ghee stored for more than a century was termed "Maha ghrita." Such aged ghee preparations are of much value in the Ayurvedic system of medicine.

Diets were created by our ancestors originally to meet their survival needs. People of various Indian cultures gradually enriched them through long empirical experience using combinations of a variety of primary food materials, especially the locally available food grains and vegetables that nutritionally complement and supplement each other. This has contributed to better health protection, improvement of digestibility, resistance to health disorders, and increased human longevity. India has a heritage of many indigenous ethnic cultures, and thousands of delicious and functional diets have been developed over millennia. The foods of nearly 50 major Indian cultures and many minor cultures have created more than 5000 dietary preparations, which include many items for daily consumption to protect and sustain human health. People of these cultures have been brought together by several historic circumstances. One such occasion was the Empire of Ashoka (300–260 BC), which held almost the whole of India as one country and promoted Buddhism. Philosophies of both Buddhism and Jainism, which preached vegetarianism and reverence for all forms of life, had a significant impact on peoples' outlook on life and consequently on their foods. The Mughal Empire (1250–1650 AD) that ruled most of India for 400 years also brought many traditional cultures together and made them interact meaningfully. The British colonial rule (nineteenth to twentieth century) in greater India tremendously contributed to people with different ethnicities coming together and sharing their wider variety of heritage foods (Parpia, 2006).

Indian heritage foods, many of which are incidentally functional foods too, have developed over a long period and include cereal-based items such as rice or wheat specialties, meal adjuncts such as pickles, chutneys, papads, and similar items, medium of cooking—ghee, butter, or vegetable oils, a variety of fermented batter foods (steam cooked or lightly fried), milk and non-milk-based sweets, and an innumerable variety of snack foods. The traditional food pattern in India is comprised of fiber-rich menus, with moderate fat, selective carbohydrate sources, and curds. They cover the functional components, imparting wider health benefits, and such systematic food habits are an excellent preventive measure to ward off many diseases. Indian traditional meals, which are mainly based on plant products such as grains, vegetables, and fruits, are very rich in natural dietary fiber. Fiber-rich and low-fat traditional foods reduce the risk of coronary heart disease. Traditional plant foods specifically based on fruits and vegetables provide functional components such as β-carotene, vitamin C, vitamin E, folates, and antioxidant phytochemicals. Cereals, which are the staple of Indian traditional foods, provide thiamine, tocopherols, selective starches, and minerals that play a role in regulating metabolic functions in the body.

The traditional food habits of each specific region of India are primarily a component of its culture, and India's cultural diversity is reflected in the numerous traditional food preparations. Indian traditional dietary patterns have basically evolved from the combination of locally available crops. Every region of this vast country uses a different choice of ingredients with its own unique food. The cereals wheat and rice form the staple in Indian traditional food, followed by the coarse grains sorghum and finger millet. The whole meal flour from these grains is consumed traditionally in the form of roti and as dumpling, or muddle. Traditionally, a typical North Indian meal consists of unleavened breads, *chapati* or *paratha* made of whole wheat flour, and an assortment of side dishes such as soups, fried vegetables, curries, chutney, pickles, and curd (*dahi*). South Indian food, which is largely nongreasy, consists of cooked rice usually served with *sambar* (seasoned lentil broth), *rasam* (a thin soup), dry and curried vegetables, a curd preparation called *pachadi*, and curd. Vegetables in Indian dishes are generally stir-fried, steamed, braised, or curried to create various textures and flavors. Commonly consumed vegetables include leafy greens, radish, yam, beans, bamboo shoots, ladies finger, and cabbage. The use of pickles and chutneys is predominant in arid regions of India due to the low availability of fresh vegetables in these hot and dry regions.

South Indian breakfast items are most commonly pancakes made from a rice batter known as *dosa*, steamed rice cakes known as *idli*, deep-fried doughnuts made from a batter of lentils known as *vada*, rice pancakes known as *appam*, *upma* (cooked semolina seasoned in oil with mustard, pepper, cumin, and dry lentils), and *pongal* (a mash of rice and lentils boiled together and seasoned with ghee, cashew nuts, pepper, and cumin). Coconut is an important ingredient in South Indian food. Coconut milk and desiccated coconut are important flavorings in South Indian cuisines. The presence of coconut mellows out the hot curries and chutneys, and is used as a topping for vegetables.

To neutralize the pungency of red chilli and soothen the stomach, curd is used in a variety of South Indian dishes. Curd *sambhar*, *thambli* (fenugreek or other seeds in curd), fried black gram powder in butter milk with seasonings known as *uddinettu*, thin soup prepared from *Garcinia indica* known as *punar puli rasam*, thin soup prepared from cumin (*Cuminum cyminum*) and black pepper seed (*Piper nigrum*) powders known as *jeerige menasu rasam*, *kosambari* (salads from dehusked gram sprouts mixed with fresh coconut kernel, green chillies, and seasonings), and watery buttermilk garnished with ginger, asafetida, coriander leaves, curry leaves, and salt are among the traditional functional foods commonly consumed. Thus a perfect combination of protein from legumes and coconut, carbohydrates from rice, fat both visible and invisible from curry and fried savory items, vitamins and minerals from sprouted grams of *kosambari* (salads from sprouted legumes), and vitamins from curd and vegetables are obtained through this combination. The regular use of curd and watery buttermilk with accompaniments aid in digestion and provide considerable health benefits.

3.3 BASIS OF EVOLUTION OF TRADITIONAL FUNCTIONAL FOODS IN INDIA

Ancient India seems to have realized the importance of health and wellness much ahead of its time. The Indian dietary pattern and the traditional foods evolved are based on the indigenous Ayurvedic system of medicine, which professes natural ways of achieving physical and mental wellness. A balanced meal recommended by

Ayurveda takes into account the properties of the food (*gunas*), the characteristics of the individual (*dosha*), and the assimilation by the body (*sadhana*). Traditional Indian food formulations show ingenuity in the choice of ingredients and additives with critical attention to wholesome nutrition beyond taste. Indian cuisines have great aromas and in-depth taste profiles, which are derived from a complex combination of spices and preparation techniques. The well-balanced Indian meal contains all the six defined tastes, namely sweet, sour, salty, spicy, bitter, and astringent. Indian cooking principles go beyond the balancing of tastes, however. Every meal aims to achieve a good balance between these sensations to promote digestion and well-being. Side dishes and condiments contribute to the overall flavor and texture of an Indian meal. The hot, sour, and crunchy side dishes and condiments, whether chutneys, curries, or soups, enhance and provide balance to the overall flavor and texture of the main staple. Inclusion of natural antimicrobials and antioxidants in the form of spices and condiments also improves the shelf-life of prepared foods against spoilage. Ayurveda has elaborated the curative and therapeutic functions of herbs and spices. This is manifested in the commonality among heritage foods from different parts of India, in that almost all of them are rich in spices.

Traditional Indian food formulations show ingenuity in the choice of ingredients and additives with adequate attention to wholesome nutrition and tastes. Another feature of traditional Indian foods is the effective utilization of natural resources and minimization of waste. Historically, cooking techniques have been developed for the protection, storage, and preparation of diets, the ingredients of which mutually supplement and complement each other to provide nutritionally balanced and also delicious diets. Indian traditional foods are noteworthy not only for their food quality but also from the food safety perspective. Since some of them are steam-cooked, they are generally free from microorganisms. Others are boiled to the desired temperatures. As a general rule, they have to be consumed fresh and hot. Generally, Indian traditional foods, once cooked, are not preserved for future use.

Indian traditional foods can be classified into eight broad categories (Table 3.1): (1) processed grain products, (2) fermented foods, (3) dehydrated products, (4) pickles, chutneys, sauces, and relishes, (5) ground spice and spice mixtures, (6) fried food products, (7) dairy products, and (8) confections and sweets. The functionality of the majority of these categories of Indian traditional foods is delineated in the later paragraphs of this chapter.

3.4 TRADITIONAL FUNCTIONAL FOODS

Various Indian traditional foods with bioactive substances provide additional health benefits over and above the physiological roles of the nutrients present in such foods. Sprouting, malting, and fermentation are processes that enhance the functional properties of food and are widely used in the daily diet of Indians.

3.4.1 TRADITIONAL FOODS BASED ON WHOLE GRAIN CEREALS AND LEGUMES

The staples of Indian cuisine are rice, whole wheat flour, sorghum, finger millet, and a variety of pulses. Pulses are the main source of protein supplement for a large

TABLE 3.1
Categories of Typical Traditional Foods of India

Raw Material	Traditional Products	Food Category
Cereals and grains	Rice-based: parboiled rice, hand-pounded rice, flaked rice, puffed rice	Processed grain products
	Wheat-based products like *chapati*, *puri*, bread, *naan*, biscuit	
	Extruded products like rice noodles, vermicelli, snack items—*murukku*	
	Fermented products like *idli, dosa, vada*	Fermented foods
Coarse cereals	Puffed sorghum, maize, finger millet, pearl millet	Processed grain products
Legumes	Pulses (split legumes without husk), puffed legumes, sprouted legumes, legume flours (e.g., Bengal gram, soy), *papads*	Processed grain products
Gram flour, sugar/ jaggery	*Jilebi, laddu, chikki, Mysore pak*	Confections and sweets
Milk	*Peda, burfi, rasgolla, jamun, sandesh, kheer, halwa*	Confections and sweets
	khoa, rabri	Dehydrated dairy products
	Chhana, paneer	Coagulated dairy products
	Dahi (curd), butter milk, *lassi*, butter	Fermented dairy products
	Ghee, *malai, makkhan*	Fat-rich dairy products
Fruits/vegetables	Fruit leather, dried fruits, dehydrated vegetables	Dehydrated products
	Pickles, chutneys, murabbas, petha, candied fruits, amchur, and pickled vegetables	Pickles, chutneys, sauces
Spices and Condiments	Spice powders, spice mixes (e.g., *garam masala*, *sambar* powder, *rasam* powder)	Ground spices and spice mixes
Drinks and beverages	Neera, toddy, arrack, khanasari, rice beer, Indian beer, honey, vinegar, jaggery	Beverages (alcoholic/ nonalcoholic)

majority of the cereal-based ethnic diets because they are easy to cook and fit well into the traditional diets. A large variety of savory and sweet processed products are made from them. Each Indian ethnic culture has developed its own diets based on the variety of pulses that they grow. While pulses may be used whole, dehusked, or split, most of the grain legume-based or mixed grain preparations are made by using dehusked or split pulses. The most important pulses are red lentil (*masoor*), chickpea (Bengal gram), red gram (pigeon pea), black gram (*urad*), and green gram (*mung*). Some of the pulses, like chickpea and green gram, are also processed into flour. Pulses, commonly used in Indian cuisines, are fried, roasted, or boiled with spices and herbs for making fermented breads, soups, chutneys, snacks, purees, and sweets. Chickpea is used raw in chutneys, roasted whole for spicy snacks, and ground for sweets, or is used whole with vegetables. Black gram is popular in southern India, where it is fermented with rice and mixed with spices to make *dosa*, steamed *idli*, and snacks such as *vada* or papad. Red gram, which exhibits a thick and more gelatinous consistency, is combined with chickpea, spices, and red chilli for making the lentil broth *sambhar*. The use of whole grain cereals and legumes in these traditional foods ensures provision of the highly desirable dietary fiber and also the polyphenols and the micronutrients, vitamins, and minerals associated with the bran portion of the grain.

Unleavened breads from whole grain: Chapati and *paratha* are processed from whole wheat; hence they contain all the natural components (bran, endosperm, and germ) of wheat. *Chapati* is unleavened bread baked on a griddle while *paratha* is unleavened bread fried on a griddle. *Parathas* are rich in crude fiber (1.3–5.8%) and protein content (8.5–12.6%) and low in fat (7.5–12.8%). The assorted *paratha* formulations that contain soy protein isolate are supposed to contain all the essential amino acids. As these are made out of whole wheat, they provide the full complements of fiber, minerals, and polyphenols associated with the bran portion of the grain. Unleavened breads baked on a griddle made out of the coarse cereals sorghum and finger millet are also widely consumed by sections of the Indian population. The use of whole grain cereals in these traditional staple foods ensures provision of the highly desirable dietary fiber and also the polyphenols and micronutrient vitamins and minerals associated with the bran portion of the grain.

Finger millet dumpling: Finger millet dumpling is a common traditional food in southern India. Finger millet, although a minor cereal, has a major impact on health with functionalities of high dietary fiber and 20% resistance starch with a 30–40% slowly digestible starch fraction, a rich calcium content of 400 mg%, and an iron content of 17–20 mg%. Malted finger millet is traditionally consumed as a healthy beverage and is used in the preparation of weaning and geriatric foods.

Sprouted legumes as salads: Green gram and chickpea are commonly germinated prior to use in the preparation of specific traditional salad dishes, especially in southern India. Germination and malting have been found to enhance iron absorption due to elevated vitamin C content or reduced tannin or phytic acid content, or both (Tontisirin et al., 2002). These processes are known to activate endogenous phytases, which in turn hydrolyze phytate, rendering iron and zinc more available. During germination, endogenous phytase activity in cereals and legumes increases as a result of *de novo* synthesis and/or activation, resulting in reductions in the inositol phosphates (Lorenz, 1980; Chavan and Kadam, 1989; Reddy et al., 1989). Sprouting of legumes, green gram, chickpea, and finger millet is associated with significantly improved bioaccessibility of iron, which is due to a reduction in tannin content (Hemalatha et al., 2007; Prabhavathi and Rao, 1979). Studies *in vitro* on iron bioavailability have shown a twofold increase on germination and a five- to tenfold increase upon malting of the minor millets (De Maeyer et al., 1989).

Fermented batter foods from cereals and legumes (Figure 3.1): Fermented batter foods from cereals and legumes are the most common and nutritious Indian traditional breakfast items in the southern states of India. These include *dosa* (a pancake made from a fermented batter of rice and black gram (3:1), *idli* (steamed rice cakes made from a fermented batter of rice and black gram (2:1), and *vada* (a deep-fried doughnut made from a batter of lentils, usually black gram). Traditionally, for making these products, the mixtures of grains are soaked for 6–8 h and then ground. After grinding, the batter is allowed to undergo fermentation overnight, which makes use of the naturally occurring microorganism *Leuconostoc mesenteroides* present in

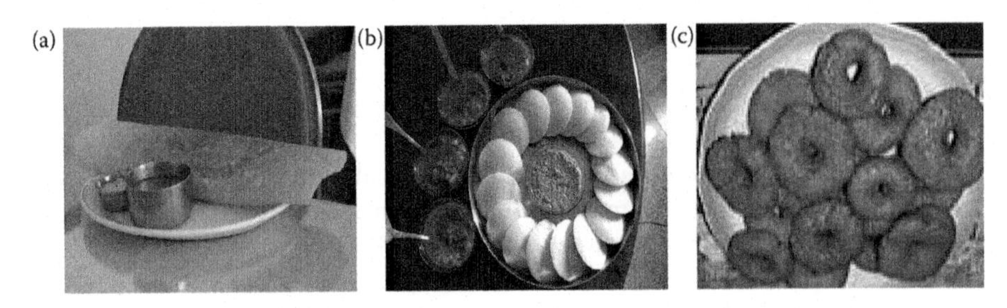

FIGURE 3.1 Fermented batter foods from grains: (a) *dosa*, (b) *idli*, (c) *vada*.

the batter. The microorganisms ferment the mixed batter and generate a unique flavor and texture. The fermented batter has an excellent foam-holding capacity to give the steamed/fried product an appetizing spongy texture. In the case of *idli*, the fermented batter is put into the cups of a special steaming vessel in the form of dumplings, after which it is consumed with a variety of garnishes (chutney) prepared from various natural flavorful ingredients such as coriander leaves, green chillies, and freshly ground coconut kernel, or with a thick soup (*sambhar*) made from legumes and vegetables flavored with spices. *Dosas* prepared by baking the batter with oil on a pan are often consumed with rolled-in or flavored vegetables.

These two fermented batter preparations are a rich source of good nutrients such as the essential amino acids methionine from rice and lysine from pulses as well as vitamins and minerals from the parboiled rice, vegetables, and nutrients generated by the fermentation of the batter. The products are also low in fat and rich in dietary fiber. Fermentation of the batter of cereal–pulse combinations in the preparation of *idli* and *dosa* is known to enhance the bioaccessibility of the micronutrients zinc and iron (Hemalatha et al., 2007). Fermentation of cereal–legume combinations of the *idli* and *dosa* batter significantly reduces both the phytate and tannins associated with the legumes. Food processing by fermentation is known to improve mineral bioavailability by reducing their absorption inhibitors such as the phytic acid present in the grains (Gibson and Hotz, 2001; Kaur and Kawatra, 2002). Besides reducing such factors, fermentation could also improve mineral bioavailability by virtue of the formation of organic acids, which form soluble ligands with the minerals, thereby preventing the formation of insoluble complexes with phytate (Tontisirin et al., 2002). A reduction in the pH by 1.5 units as a result of fermentation of the *idli* and *dosa* batters has been observed, which is attributable to the synthesis of organic acids during fermentation (Hemalatha et al., 2007). *Idli* can also be made using parboiled rice (which helps preserve vitamins, unlike regular rice) along with black gram, taken in the right combinations, and then fermented and steamed to deliver a profoundly nutritious, balanced, and tasty meal.

3.4.2 *Dahi* and Ghee: The Two Classical Milk-Based Traditional Health Foods of India (Figure 3.2)

Traditional Indian foods make use of dairy products such as curds (yogurt), cheese, buttermilk, and ghee. Ghee (clarified butter) and *dahi* (curd) are the two most

FIGURE 3.2 *Dahi* (curd) and ghee (dehydrated butter).

important age-old traditional milk products consumed in India. Ghee is a fat-rich dairy product whereas *dahi* is a fermented milk product. Both these milk products have higher keeping qualities than the other traditional dairy products. Several studies have been conducted on ghee and *dahi* by many investigators, and numerous claims have been made about their different nutritional and therapeutic values. Ghee is used as a cooking oil and for flavoring, especially in vegetarian cuisines. Fresh cheeses are added to vegetable curries or desserts, and are often mixed with sugar, milk, cardamom, and fruits for savory applications and desserts. In North India, yogurt and buttermilk are consumed as beverages. Yogurt provides consistency and flavor to Indian curries. Yogurt is also used in salads with cucumbers, onions, and tomatoes.

The gastrointestinal microflora in humans plays a key role in nutrition and health. A portion of the consumed food unabsorbed in the small intestine gets fermented in the large intestine. These foods, which are known as colonic foods, are metabolized through anaerobic fermentation by the gut microflora and the end products are usually short-chain fatty acids that are absorbed in the colon. Such foods are credited with protection from colon cancer, immune modulatory action, systemic effects on blood lipids, and a reduction of pathogens. Fermented milk products such as *dahi* are probiotics and are associated with positive effects such as reductions in cholesterol and triacylglycerols, protection against gastroenteritis, improved lactose intolerance, and stimulation of the immune system.

Dahi is traditionally consumed by Indians either as a part of their daily diet or as a refreshing beverage. It is characterized by its firm curd and delicate flavor. Investigations by various researchers have shown that regular consumption of *dahi* has many beneficial effects such as improvements in appetite and vitality, curing of dyspepsia, dysentery, and other intestinal disorders; removal of adverse effects of drugs, encouragement of thiamine synthesis, lowering the cholesterol level of blood, controlling cancer, easy digestion by lactose-intolerant persons, and so on. In addition, *dahi* has a high nutritive value due to the presence of all milk constituents and the various health-promoting lactic acid bacteria in it.

Dahi is considered as a functional food ingredient by virtue of its health-promoting probiotic effects. Probiotics fall within the realm of functional foods containing certain biologically active components that beneficially contribute to human health beyond basic nutrition. While earlier reports on the health-promoting effects of probiotics were largely limited to yogurt and other fermented dairy products containing *lactobacilli*, many reports since have shown that gut organisms such as bifidobacteria and other *lactobacilli* also beneficially affect the host through the development of resistance factors against diseases, the protective effect of the flora, and the production of microbial products. Several workers have corroborated the protective effect of gut flora and have shown that germ-free animals are more prone to diseases than their peers carrying a complete gut flora (Pollman et al., 1980; Saavendra, 1995). The documented beneficial effects of probiotics include their use in the treatment of various kinds of diarrhea, the alleviation of the gastrointestinal side effects of antibiotic treatment, the alleviation of lactose intolerance, relief during constipation, and the general balancing and stabilizing of the host's intestinal microbial integrity.

Ghee constitutes an important part of Indian life. Ghee is chiefly used in India as a part of the diet and as a cooking medium. It is valued for its pleasant cooked, caramelized flavor and granular texture. Ghee is made up mainly of fat, which gives energy to the body and forms an integral part of the body's cells. It helps to maintain the body's temperature. Recent studies have indicated that milk fat contains some anticarcinogenic substances such as conjugated linoleic acid, butyric acid, and vitamins, among others. Formerly, intake of ghee was presumed as one of the causes for the high incidence of heart ailments. However, paradoxically, later it was found that consumption of ghee has a hypocholesterolemic effect. Moreover, in the indigenous Ayurvedic medical treatments ghee is used for heart patients. Ghee contains the fat-soluble vitamins A, D, E, and K. The Indian Ayurvedic medical literature mentions various types of medicated ghee that can be used for the treatment of many diseases such as asthma, antiaging, cough, dermatitis, digestive problems, heart, hysteria, leprosy, leucoderma, and piles. Medicated ghee is used for either internal or external applications. Sometimes, various herbs are used along with ghee to enhance its therapeutic efficiency.

3.4.3 Traditional Food Adjuncts from Legumes and Spices

Food adjuncts include an assortment of items that are consumed as side dishes to staple foods. They add variety, spice, and crunch to the common menu with standard items. Traditional Indian food adjuncts may broadly be classified as pickles, chutneys, preserves, and dried vegetable products such as *sandige* and dry semiprocessed adjuncts such as *papad* (which are consumed after frying) (Table 3.2). Although consumed in small portions, adjuncts play an important role in nutrition and health. The wide range of ingredients used in the various types of adjuncts makes it difficult to generalize their nutritional importance.

Pickles are a good source of vitamin C, in spite of the fact that part of it is lost during processing and storage. Pickles and chutneys provide other valuable nutrients such as minerals, carotenoids, isoflavones, and antioxidants. Pickles and chutneys may thus

TABLE 3.2
Common Indian Traditional Food Adjuncts

Food Adjunct	Ingredients	Description of Preparation
Papads	Pulse flours (black gram/green gram) Starchy materials (rice flour/wheat starch/sorghum/minor millets/sago) Salt, spices, leavening agent	Pulse flour is made into a dough with spices; small dough balls are rolled and dried Starchy flours are mixed with water to a slurry; allowed to ferment; cooked and spread as thin sheets. The cooked slurry may also be extruded to a noodle-like structure, dried, and stored. Papads may be roasted or fried prior to consumption
Wadis	Legumes (green gram/black gram/Bengal gram) Rice, millets, vegetables (ash gourd, okhra, etc.)	Pulses are soaked and coarsely ground after draining excess water; mixed with shredded vegetables and allowed to ferment overnight; cooked and mixed with salt and spices; deposited as small masses or balls; dried. They are consumed after frying
Chutneys	Fruits, spices, jaggery, coconut, salt	The fruit pulp is mixed with jaggery, salt, spices, acidulants, and coconut, cooked to a viscous consistency, and seasoned
Dry *chutney* powders	Oil seeds (peanut, gingely, dry coconut) Legumes (Bengal gram/black gram) Salt, curry leaves, spices and condiments	Oil seeds and pulses are roasted and powdered with salt, spices, and condiments. Stable for 6 months. They are consumed as such or mixed with oil, and freshly cut vegetables
Pickles	Acidic fruits, salt, spices (especially chillies), condiments, optionally vegetable oil and vinegar	Acidic fruits are cut into small pieces; mixed with salt and spices fried in oil

extend the supply of nutrients from seasonal perishable items into the lean period. Mango pickle is the most widely consumed variety among pickles followed by lemon, emblica, swallow root, and mixed pickles. Chutneys usually accompany any traditional Indian meal. There are diverse varieties among chutneys; green chutneys are rich sources of nutraceuticals in the Indian meal. They can be served with assorted *parathas* to complement the nutritional requirement. Coriander chutney, mint chutney, and drumstick-curry leaf chutney are other chutneys whose ingredient herbs are good sources of vitamins and minerals. Chutneys generally contain pulses and oilseeds (e.g., black gram, chickpea, sesame seed, peanut, and coconut, among others) and are an excellent supplement to the cereal-based vegetarian staple diet by improving the protein quality. *Papads* are high-protein foods made from different pulses and certain cereals; they are flat, thin, round circular products that can be roasted or fried instantly. Legumes being the main ingredients, *papads* provide 7–15% protein. Legume-based *papad* and *wadi* add to the protein value of staple diet, and can act as a replacement for vegetables and legumes during the lean season. Fermentation in some *papad* and *wadi* products improves their digestibility and reduces antinutritional factors. Roasted *papad* is a healthy alternative to fried *papad*.

3.4.4 Indian Acidulant Fruits with Functional Properties (Figure 3.3)

Food acidulants such as lime, emblica, tamarind, kokum, and amchur are commonly used in traditional Indian culinary to impart a desirable sour taste to certain food preparations. Organic acids are known to promote the absorption of iron from plant foods (Gillooly et al., 1983). The food acidulants amchur and lime have been reported to significantly enhance the bioaccessibility (*in vitro* bioavailability) of zinc and iron from the food grains consumed in India (Hemalatha et al., 2005) and of β-carotene from green leafy or yellow-orange vegetables (Veda et al., 2008). This positive influence of food acidulants on the bioaccessibility of micronutrients from food grains or vegetables has been seen in both raw and cooked forms.

Emblica fruit (*Emblica officinalis*), commonly known as *amla* or Indian gooseberry, is one of the important subtropical fruits belonging to the family Euphorbaceae. Dried shreds of amla fruits are used as a food acidulant in Indian traditional foods. The fruits of *Emblica* are widely consumed raw, cooked, or pickled, but they are also principal constituents of many medicinal preparations in the indigenous system of

FIGURE 3.3 Food acidulants: (a) emblica fruit, (b) kokum, (c) amchur, and (d) tamarind.

medicine in India. Emblica occupies an important place in the preserve industry. Amla preserves, candy, squash, and burfi are some major traditional products of *amla*, which are widely used as health foods and as natural sources of vitamin C. Emblica fruits are highly nutritious and are very rich in vitamin C. These fruits have an ascorbic acid content of up to 950 mg/100 g, which is the second highest among all fruits, next only to Barbados cherry. The fruit contains considerable amounts of polyphenols that retard the oxidation of ascorbic acid. Emblica fruits are well known for their medicinal properties. These fruits, possessing rich antioxidant potency by virtue of both vitamin C and polyphenols, are used for curing chronic dysentery, bronchitis, diabetes, liver ailment, diarrhea, jaundice, and dyspepsia. Emblica fruits are valued highly among indigenous medicines in India and form the major ingredient in Ayurvedic preparations such as *Chavanprash* and *Trifala*.

Emblica fruit extract has been reported to have hypolipidemic, antidiabetic, and anti-inflammatory activity, and to inhibit retroviruses such as HIV-1, tumor development, and gastric ulcer (Sabu and Kuttan, 2002). Emblica fruit extract exhibits antioxidant properties, its aqueous extract being a potent inhibitor of lipid peroxidation and a scavenger of hydroxyl and superoxide radicals *in vitro* (Scartezzini and Speroni, 2000). Emblica fruit extract inhibits micronuclei formation, sister chromatid exchanges, clastogenesity, and mutagenesity induced by metals; it protects against radiations (Scartezzini and Speroni, 2000), inhibits clastogenesity of benzopyrene and cyclophosphamide (Sharma et al., 2000), and is gastroprotective (Al-Rehaily et al., 2002), cytoprotective, and immunomodulating (Sairam et al., 2002). Emblica fruits have been reported to protect against oxidative stress in ischemic reperfusion injury (Rajak et al., 2004), show antivenom capacity (Alam and Gomes, 2003), ameliorate hyperthyroidism and hepatic lipid peroxidation, display antiproliferative activity in breast cancer cell lines, show antitussive activity, and induce apoptosis in lymphoma ascites (Rajeshkumar et al., 2003).

The dried fruit rinds of kokum (*Garcinia indica*), commonly known as "Malabar tamarind," are liberally used in the coastal regions of India as a traditional food acidulant in culinary practices. The dark red fruit of *Garcinia indica* is valued for its nutritive value and outstanding medicinal properties. This fruit is known to reduce obesity and to beneficially regulate blood cholesterol levels. The antiobesity influence of kokum fruits is attributable to its organic acid constituent, hydroxycitric acid, present at a level of 22%. Hydroxycitric acid is a potent competitive inhibitor of ATP citrate lyase, the enzyme that catalyzes the cleavage of citrate to acetyl coenzyme A and oxaloacetate (Watson et al., 1969). As a consequence of this inhibition, a reduction in the rates of the *de novo* synthesis of fatty acids and cholesterol has been demonstrated in animal systems (Lowenstein, 1971; Sullivan et al., 1972). Kokum is also a source of pectin (6%) and fat (10%). Fresh fruits are cut into halves and the fleshy portion containing the seed is removed. The rind constitutes about 50–55% of the fruit and is generally sun-dried for future use. The kokum rind is used to make an attractive, red, pleasantly flavored extract for use as a beverage. Syrup from the fruit as traditionally prepared is popularly known as "Amrit kokum." Kokum seed is a good source of fat called "Kokum butter." The rind of fully ripe kokum contains 2–3% anthocyanin pigments and thus is a promising source of natural color.

Tamarind is the most common food acidulant used in southern India. The fruit rinds of tamarind (*Tamarindus indica*) have a fruity and sweet-and-sour taste and are used in sweetened drinks, curries, stews, or soups of South India. The food acidulants amchur and lime generally enhance the bioaccessibility of zinc and iron from the food grains (Hemalatha et al., 2005). This positive influence of acidulants on zinc bioaccessibility from food grains is seen in both the raw and cooked form. Tamarind is regarded as a refrigerant, digestive, carminative, and laxative. It is useful in diseases supposed to be caused by deranged bile (The Wealth of India, 1976). Tamarind has been shown to increase bile secretion with enhanced bile acid concentration (Sambaiah and Srinivasan, 1991).

3.4.5 Functional Oil Seeds

Vegetable oil is an important part of Indian culinary; mustard oil is most commonly used in northern India while sesame and coconut oils are used abundantly in the south. Most Indian curries are fried in vegetable oil. Sesame oil is an edible vegetable oil derived from sesame seeds. Besides being used as cooking oil in South India, it is often used as a flavor enhancer in Indian cuisine. Sesame seeds were one of the first crops processed for oil as well as one of the earliest condiments. Sesame oil is considered to be more stable than most vegetable oils due to the antioxidants contained in the oil. Sesame oil is least prone, among the cooking oils, to turn rancid. Because it has a very high boiling point, sesame oil retains its natural structure and does not break down even when heated to a very high temperature.

Sesame oil is a source of vitamin E, an antioxidant that protects low-density lipoproteins from oxidation. As with most plant-based condiments, sesame oil contains magnesium, copper, calcium, iron, zinc, and vitamin B_6. Copper provides relief for rheumatoid arthritis. Magnesium supports vascular and respiratory health. Calcium helps prevent colon cancer, osteoporosis, migraine, and postmenopausal syndrome. Zinc promotes bone health. It is suggested that due to the presence of high levels of polyunsaturated fatty acids in sesame oil, it may help to control blood pressure. Sesame oil is unique in that it has one of the highest concentrations of ω-6 fatty acids. At the same time, the oil contains two naturally occurring preservatives, sesamol and sesamin. The effect of this oil on blood pressure may be due to polyunsaturated fatty acids and the compound sesamin, a lignan present in sesame oil. There is evidence suggesting that both compounds reduce blood pressure in hypertensive rats and in humans (Matsumura et al., 1998; Miyawaki et al., 2009). Sesame lignans also inhibit the synthesis and absorption of cholesterol in these rats. Various constituents present in sesame oil have antioxidant and antidepressant properties. Hence its use may help fight senile changes and bring about a sense of well-being. It is suggested that regular topical application and/or consumption of sesame oil should mitigate the effects of anxiety, nerve and bone disorders, poor circulation, lowered immunity, and bowel problems. It is suggested that such use would also relieve lethargy, fatigue, and insomnia, while promoting strength and vitality, and enhancing blood circulation. There are claims that its use has relaxing properties, which eases pain and muscle spasm, such as sciatica, dysmenorrhea, colic, backache, and joint pain. There are claims similar to those for other therapeutic medicines that its having antioxidants explains the beliefs that it slows the aging process and promotes longevity. It is suggested that sesame oil could be used for the reduction of cholesterol levels (due to the presence of lignans, which are phytoestrogens), antibacterial effects, and even slowing down certain types of cancer (due to the antioxidant properties of the lignans).

Mustard seeds, a source of edible oil, are also a main ingredient of seasonings in Indian cuisine and a component of curry powders. Mustard possesses vermicidal, antihelminthic, and appetite-improving properties (Kirtikar and Basu, 1935). Mustard belongs to the cruciferous family, whose members include cabbage, broccoli, and cauliflower. All the above vegetable extracts have the property of inactivating the mutagenicity of food mutagens such as tryptophan pyrolysate. The active principle of mustard, namely dithiolthione, is also used as an antischistosomal drug. From epidemiological studies, it has been established that the regular consumption of cruciferous vegetables is associated with reduced cancer risk. Mustard seeds are rich in sulfur-containing compounds (dithiolthiones), which have a protective effect against liver toxicity induced by some chemicals and aflatoxins, potent toxic compounds present in fungal-contaminated peanuts. A concentration of 0.05% of dithiolthiones in the diet was found to stimulate the activity of protective enzymes. The mutagenic effects of mustard seed powder have been assessed in experimental animals treated with potent carcinogens. These experiments suggested that mustard, like turmeric, has excellent antimutagenic properties (NIN Annual Report, 1993–94).

3.4.6 Betel Leaves

Fresh green leaves of the betel vine (*Piper betle*), locally known as *paan*, are traditionally chewed, especially after meals, along with areca nut and lime, or with many other additional ingredients, mainly for mouth-freshening and digestive-stimulating effects. Betel leaf is aromatic and carminative. It is also an aphrodisiac and antiseptic. The habit of chewing betel leaves is claimed to be responsible for preventing osteoporosis among the economically weaker sections of the population in India.

3.4.7 Traditional Indian foods as Abundant Providers of Dietary Fiber

Traditional Indian cuisines are providers of liberal amounts of dietary fiber due to the extensive use of whole grains, vegetables, and fruits. For example, ash gourd fiber has a soluble fiber content of 22% while radish fiber has 16% and pea peels have 8–10%, with a total dietary fiber of 65–80%. Dietary fiber comprises a diverse group of compounds: insoluble cellulose, hemicelluloses, and lignins, and soluble gums and mucilages. These substances are exceedingly complex both chemically and morphologically and are resistant to hydrolysis by digestive enzymes in the human gut. The physiological roles of fiber in the diet are (1) filling the diet without adding calories, (2) increasing intestinal motility, (3) helping to reduce obesity, (4) preventing the absorption of cholesterol, (5) reducing the postprandial rise in blood glucose, (6) preventing diverticular diseases, and (7) softening of stools through absorption of water (preventing hemorrhoids) and promoting the growth of bifidobacteria in the gut.

A considerable body of evidence is available to show that dietary fiber improves glucose tolerance; soluble fiber has been especially effective in retarding postprandial glucose uptake in the intestine, thus reducing the insulin requirement. Soluble fiber has also been shown to enhance insulin receptor binding and improve glycemic response (i.e., increasing peripheral tissue insulin sensitivity). Ingestion of fiber suppresses energy intake by inducing satiety (by virtue of their bulking and viscosity-producing capabilities). Dietary fiber that forms viscous dispersions when hydrated affects every aspect of gastrointestinal function, gastric emptying, intestinal transit time, and absorption of digested products of fat and carbohydrates. Increased fecal bulk from dietary fiber is mainly due to the insoluble fiber fraction (cellulose, hemicellulose, and lignins), increased fecal water, and an increase in bacterial mass caused by soluble fiber (gums, mucilages, and pectins) fermentation. Dietary fiber has been shown to protect against colon cancer. Carcinogenic substances, either ingested as such or more likely produced by metabolic activation in the gut, are thought to induce malignant changes in mucosal cells. The protective effect of dietary fiber is thought to be due to its ability to increase stool bulk (dilution of toxic substances), increased transit duration (decreasing exposure duration), and altered fecal bacterial flora.

3.4.8 Traditional Indian Foods as Providers of Polyphenols

Traditional Indian cuisines also provide liberal amounts of polyphenols due to the extensive use of vegetables and fruits. Phenolic compounds are plant secondary metabolites, with a large variety of chemical structure. Phenols occurring in nature are of interest for many reasons, such as antioxidants, astringents, bitterness, browning reactions, color, and oxidation substrates, among others. They include simple phenols, hydroxycinnamates, and flavonoids. Phenols are responsible for the majority of the oxygen-utilizing capacity in most plant-derived products. With the exception of carotenes, the antioxidants in foods are phenolic compounds. Among those added to prevent oxidative rancidity in fats are the monophenols. Phenolic compounds have a wide range of biological properties. Of particular note are their anti-platelet aggregation property, anti-inflammatory potential, and antioxidant, antitumoral, and estrogenic activities, and hence they can potentially prevent coronary heart disease and cancer. Flavonoids are polyphenolic compounds that include flavonols, flavones, isoflavones, and anthocyanins, which have been suggested to play a dominant role in the prevention of cancer and heart diseases. Over 4000 flavonoids have been identified, many of which occur in fruits, vegetables, and beverages (tea, coffee, beer, wine, and fruit drinks). Epidemiological data indicate that high fruit and vegetable consumption has health benefits in the prevention of chronic diseases, including cardiovascular disease and certain types of cancer. Phytic acid present in legumes, oil seeds, and cereal bran is known to reduce blood glucose response to starchy foods. It is also known to lower blood cholesterol.

3.5 SPICES AS FUNCTIONAL FOOD ADJUNCTS WITH MULTIPLE HEALTH EFFECTS

Spices, which are used to enhance the flavor of a dish, form a vital part of Indian traditional food preparation. A correct blend of aromatic spices is crucial to every Indian cuisine. The most commonly used spices in Indian cuisine are black pepper,

FIGURE 3.4 Spices as components of *masala* powders (spice mixes).

chilli pepper, mustard seed, cumin, turmeric, fenugreek seed, ginger, coriander, asafetida, curry leaves, and garlic (Figure 3.4). Popular spice mixes are *garam masala* (which is usually a powder of five or more dried spices, commonly comprised of cardamom, cinnamon, and clove) and *sambar* masala powder (a popular spice mix in South India). The common use of curry leaves is typical of all Indian cuisine. In sweet dishes, cardamom, nutmeg, and saffron are used. The essential oils from spices and condiments enhance salivation and stimulate the digestion process. The abundance and variety of Asian spices and other flavorings creates a fresh taste found in no other cuisine. Spices in Indian cuisines create hot, sweet, sour, savory, and aromatic sensations all in one meal. Ginger, cumin, cassia, coriander, star anise, chilli peppers, coriander leaf, spearmint, turmeric, clove, and garlic are commonly used in Indian cooking. A few spices are pickled when fresh, such as ginger, mango ginger, and chilli peppers. Seasonings are a must for flavoring foods in India. Spice ingredients such as garlic, ginger, turmeric, chilli, and fenugreek seeds are known to contain functional constituents such as curcumin, capsaicin, flavonoids, and essential oils.

Besides contributing flavor, color, and aroma to the diet, spices have also long been recognized to possess physiological effects supposed to be beneficial to human health. They act as stimuli to the digestive system and relieve digestive disorders, and some are of antiseptic value. Their attributes such as tonic, carminative, stomachic, diuretic, and antispasmodic, largely empirical nevertheless efficacious, have earned them pharmacological applications in the indigenous systems of medicine in India and other countries (Table 3.3). With a long history of the use of spices and herbs dating back to 5000 years BC, and spices significantly contributing to human health by providing bioactives, they may be considered as one of the first

TABLE 3.3
Medicinal Properties of Spice Ingredients Recognized for a Long Time

Spice	Medicinal Properties
Turmeric (*Curcuma longa*)	Anti-inflammatory, diuretic, laxative, good for affections of the liver, jaundice, diseases of blood
Red pepper (*Capsicum annuum*)	Anti-inflammatory, for pain relief (rheumatism/neuralgia); useful in indigestion, rubefacient
Garlic (*Allium sativum*)	Antidyspeptic, antiflatulent, for ear infection, duodenal ulcers, as rubefacient in skin diseases
Onion (*Allium cepa*)	Diuretic, emmenagogue, expectorant, for bleeding piles
Fenugreek (*Trigonella Foenum-graecum*)	Diuretic, emmenagogue, emollient, useful in heart diseases
Cumin (*Cuminum cyminum*)	Antispasmodic, carminative, digestive stimulant
Coriander (*Coriandrum sativum*)	Antidyspeptic

ever recorded functional foods. Spices may also act synergistically to enhance the health-related properties of other food ingredients. Spices make foods palatable without salt, and hence may assist in meeting the recommended reduced daily intake of sodium. Similarly, they make foods palatable without fat, thus assisting in meeting the guidelines for healthy fat intake levels. During the last three decades, the beneficial effects of spices have been experimentally documented, which suggests that the use of these food adjuncts extends beyond taste and flavor (Srinivasan, 2005a, 2005b). The emerging research literature suggests that specific spices may confer unique health benefits. Although human studies are limited, considerable attention to this has been drawn because of the positive results from *in vitro* and *in vivo* animal studies.

3.5.1 Digestive Stimulant Action

The digestive stimulant action of spices is probably the most common experience. Several spices such as ginger, mint, ajowan, cumin, fennel, coriander, and garlic are common remedies used in traditional medicines or ingredients of pharmacological preparations to cure digestive disorders. The mechanism for the digestive stimulant action of spices has recently been understood through extensive animal studies (Platel and Srinivasan, 2004) (Figure 3.5). It has been shown that many commonly consumed spices (curcumin, capsaicin, ginger, fenugreek, mustard, cumin, coriander, ajowan, tamarind, and onion) stimulate bile acid production by the liver and its secretion into the bile (Sambaiah and Srinivasan, 1991; Platel and Srinivasan, 2000a). Bile acids play a major role in fat digestion and absorption. Several spices have also been shown to stimulate the activity of digestive enzymes from the pancreas, particularly lipase, and the terminal digestive enzymes from the small intestinal mucosa (Platel and Srinivasan, 1996, 2000b, 2001a). As a result of increased digestive capability, the spice-fed animals showed a reduced food transit time (Platel and Srinivasan, 2001b).

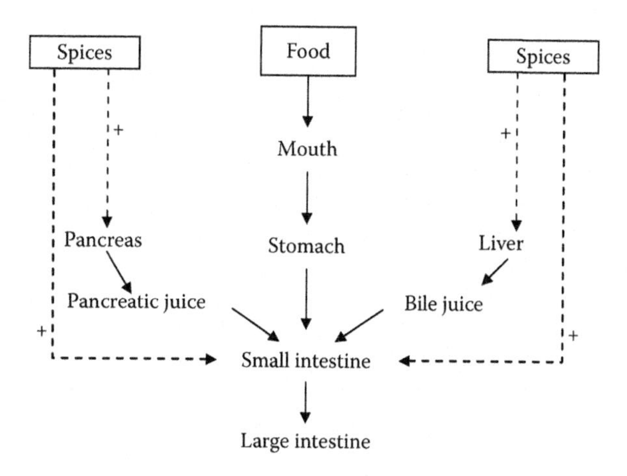

FIGURE 3.5 Digestive stimulant action of spice ingredients.

3.5.2 Antidiabetic Potential

Diet has been recognized as a corner stone in the management of diabetes mellitus. Among spices, the natural food adjuncts that have been evaluated in this context, fenugreek seeds, garlic and onion, and their sulfur compounds, turmeric and its yellow principle, curcumin, have been found to be effective in improving the glycemic status and glucose tolerance in diabetic animals and type 2 diabetic patients (Srinivasan, 2005c). Animal studies and clinical trials on antidiabetic properties of fenugreek and onion have been particularly extensive, while human studies are limited in the case of garlic and turmeric. Studies have unequivocally demonstrated the antidiabetic potential of fenugreek in both type 1 and type 2 diabetes. The addition of fenugreek seeds to the diets of diabetic patients or animals resulted in a fall in blood glucose and improvement in glucose tolerance (Sharma, 1986a; Sharma et al., 1996; Srinivasan, 2005c). The hypoglycemic effect is attributed to the fiber and gum, which constitute as much as 52% of the seeds. The fiber-rich fenugreek is believed to delay gastric emptying by direct interference with glucose absorption.

Garlic and onion are two other spices that have been widely tested for their antidiabetic potential. Both these spices have been shown to be hypoglycemic in different diabetic animal models and in limited human trials (Srinivasan, 2005c). The hypoglycemic potencies of garlic and onion are attributed to the disulfide compounds present in them, di(2-propenyl) disulfide and 2-propenylpropyl disulfide, respectively, which cause direct or indirect stimulation of insulin secretion by the pancreas (Kumudkumari et al., 1995; Augusti and Sheela, 1996). In addition, they may also have insulin-sparing action by protecting against sulfhydryl inactivation by reacting with endogenous thiol-containing molecules such as cysteine, glutathione, and serum albumin. Nephropathy is a common complication in chronic diabetes.

Turmeric is another spice that is claimed to possess beneficial hypoglycemic effects and to improve glucose tolerance in a limited number of studies (Tank et al., 1990). Nephropathy is a common complication in chronic diabetes. High blood cholesterol is an added risk factor that determines the rate of decline of kidney function in a diabetic situation. Dietary curcumin (of turmeric) and onion have been found to have a promising ameliorating influence on the severity of renal lesions in streptozotocin diabetic rats (Babu and Srinivasan, 1998, 1999). The hypocholesterolemic effect of these spices as well as their ability to lower lipid peroxidation under diabetic conditions is implicated in the amelioration of renal lesions. Capsaicin, the pungent principle of red pepper, has been shown to be useful in diabetic neuropathy (The Capsaicin Study Group, 1992).

3.5.3 Cardioprotective Effect

Specific spices exert cardioprotective influence through one or more of the following attributes: (1) hypolipidemic influence, (2) antithrombotic properties, (3) suppression of LDL oxidation, and (4) thermogenic influence (Srinivasan, 2008).

3.5.3.1 Hypolipidemic Influence

The importance of high blood cholesterol levels in relation to atherosclerosis and coronary heart disease is well known. Several spices consumed in India have been evaluated for their possible cholesterol-lowering effect in a variety of experimental situations with both animals and humans (Srinivasan et al., 2004). Garlic, onion, fenugreek, turmeric, and red pepper are found to be effective as hypocholesterolemic agents under various conditions of experimentally induced hypercholesterolemia or hyperlipidemia. Garlic, fenugreek, and onion are effective in humans with hyperlipidemic condition.

The consumption of garlic or garlic oil has been associated with a reduction in total cholesterol, low-density lipoprotein cholesterol, and triacylglycerol levels. There have been more than 25 clinical research publications concerning garlic and its preparations (Kleijnen et al., 1989). With the introduction of dehydrated garlic powder containing a standardized level of the parent sulfur compound alliin, effective clinical work could be undertaken with a relatively low and acceptable daily dosage of 300–900 mg (equivalent to 1 clove of garlic). Many clinical studies have indicated that consuming one clove of garlic (or equivalent) daily will have a cholesterol-lowering effect of up to 10% (Warshafsky et al., 1993; Gore and Dalen, 1994). This is consistent with a recent trial involving 780 patients taking 600–900 mg standardized garlic extract per day as a supplement that evidenced a modest 0.41 mmol/L decrease in serum cholesterol (Stevinson et al., 2000). Dietary supplementation with aged garlic extract showed better beneficial effects than fresh garlic on the lipid profile and blood pressure of moderately hypercholesterolemic subjects (Steiner et al., 1996). While garlic supplementation significantly decreased both total and LDL-cholesterol in hypercholesterolemic subjects, coadministration of garlic with fish oil had a more beneficial effect on serum lipid and lipoprotein concentrations by providing a combined lowering of total cholesterol, LDL-cholesterol, and triacylglycerol concentration as well as on the ratios of total cholesterol to HDL-cholesterol and LDL-cholesterol to HDL-cholesterol (Adler and Holub, 1997).

Fenugreek seeds were hypocholesterolemic in rats with hyperlipidemia induced by either high fat (Singhal et al., 1982) or a high cholesterol diet (Sharma, 1984, 1986b). Defatted fenugreek seed was effective in treating diabetic hypercholesterolemia in dogs (Valette et al., 1984) and in humans (Sharma, 1986b). The hypolipidemic effectiveness of turmeric and curcumin (Srimal, 1997), red pepper, and capsaicin (Suzuki and Iwai, 1984; Govindarajan and Satyanarayana, 1991; Surh and Lee, 1995) and of onion and garlic (Fenwick and Hanley, 1985; Carson, 1987; Jain and Apitz-Castro, 1994) has been periodically reviewed in recent years by different authors. The spice compounds curcumin and capsaicin have been associated with a reduction in LDL-cholesterol and an increase in HDL-cholesterol levels, but these results have been limited to animal studies.

3.5.3.2 Antithrombotic Properties

Besides the beneficial effect on serum lipid profiles (lowering of LDL-cholesterol and triglyceride levels), the antiplatelet aggregation and antiplatelet adhesion properties of several spices also contribute to cardiovascular protection. The spices or spice compounds thus far documented that have inhibitory effects on platelet aggregation are garlic, onion, curcumin, cuminaldehyde, eugenol, and zingerone. Garlic in particular exhibits antithrombotic and hypotensive properties, both of which also contribute to cardiovascular protection in addition to their hypolipidemic properties. Aged garlic extract (7.2 g) has been associated with anticlotting as well as modest reductions in blood pressure (an approximately 5.5% decrease in systolic blood pressure) (National Centre of Excellence, 2006). According to Lin (1994), the antiplatelet aggregation, antiplatelet adhesion, and antiproliferation properties of aged garlic extracts appear to contribute more to cardiovascular protection than the hypolipidemic properties.

3.5.3.3 Suppression of LDL Oxidation

The antioxidant properties of spices are of particular interest in view of the impact of oxidative modification of low-density lipoprotein (LDL)-cholesterol in the development of atherosclerosis. In recent years, a substantial body of evidence has indicated that free radicals contribute to cardiovascular disease. Oxidative modification of LDL is hypothesized to play a key role during the development of atherosclerosis. Since spices have high antioxidant concentrations that have the potential to inhibit the oxidation of LDL, the use of antioxidant spices is a promising proposition.

3.5.3.4 Thermogenic Influence

Obesity-related insulin resistance has emerged as a potent risk factor for cardiovascular disease. Dietary factors that affect satiety and thermogenesis could play an important role in determining the prevalence and severity of this problem. Among spices that may have a role to play in this regard, red pepper (or its pungent principle capsaicin) (Kawada et al., 1986) and garlic are promising, although more data are required to substantiate the benefits. The use of spices to displace fats and salt in the diet may reduce cardiovascular risk.

3.5.4 Antilithogenic Effect

A persistent lithogenic diet leads to cholesterol saturation in the bile, resulting in the formation of cholesterol crystals, that is, gallstones, in the gall bladder. The inhibitory effect of a curcuma mixture (Temoe Lawak Singer) on lithogenesis in rabbits has been reported (Beynen et al., 1987). Studies on the experimental induction of cholesterol gallstones in mice and hamsters by feeding a lithogenic diet have revealed that the incidence of gallstones is 40–50% lower when the animals are maintained on 0.5% curcumin or 0.015% capsaicin-containing diets (Hussain and Chandrasekhara, 1992, 1993). Animal studies have also revealed significant regression of preformed cholesterol gallstones by these spice principles in a 10-week feeding trial (Hussain and Chandrasekhara, 1994a). The antilithogenic potential of other known hypocholesterolemic spices (garlic, onion, and fenugreek seeds) has also been recently demonstrated in animal studies (Reddy and Srinivasan, 2009a, 2009b; Vidyashankar et al., 2009, 2010). The antilithogenicity of these spices is considered to be due to the lowering of cholesterol concentration and the enhancing of bile acid concentration, both of which contribute to lowering of the cholesterol saturation index and hence its crystallization. In addition to their ability to lower the cholesterol saturation index, the antilithogenecity of these spice principles may also be due to their influence on biliary proteins (Hussain and Chandrasekhara, 1994b).

3.5.5 Anti-Inflammatory Properties

Turmeric happens to be the earliest anti-inflammatory drug known in the indigenous system of medicine in India. Turmeric extract, curcuminoids, and volatile oil of turmeric have been found to be effective as anti-inflammatories in several studies involving mice, rats, rabbits, and pigeons. The efficacy of curcuminoids was also established in carrageenan-induced foot paw edema in mice and rats, and in cotton pellet granuloma pouch tests in rats (Srimal, 1997). Curcumin was considered to be advantageous over aspirin because it selectively inhibits the synthesis of the anti-inflammatory prostaglandin T_xA_2 without affecting the synthesis of the prostacyclin (PgI_2), which is an important factor preventing vascular thrombosis (Srivastava, 1986). Both *in vitro* and *in vivo* animal studies have documented the anti-inflammatory potential of the spice principles curcumin (of turmeric), capsaicin (of red pepper), and eugenol (of clove). Animal studies have revealed that curcumin and capsaicin also lower the incidence and severity of arthritis and delay the onset of adjuvant-induced arthritis. These spice principles also inhibited the formation of arachidonate metabolites (PgE_2, leukotrienes).

The anti-inflammatory effects of curcumin (400 mg) in patients undergoing surgery for hernia or hydrocele were found to be comparable to those of phenylbutazone (100 mg) (Satoskar et al., 1986). In rheumatoid arthritis patients, administration of curcumin (1.2 g/day) produced a significant improvement similar to phenylbutazone (Deodhar et al., 1980). Recently, capsaicin has received considerable attention as a pain reliever. In two trials with 70 and 21 patients with osteoarthritis and rheumatoid arthritis, topical application of creams containing 0.025% or 0.075% capsaicin was an effective and safe alternative to analgesics employed in systemic medications, which are often associated with potential side effects (Deal, 1991; McCarthy and McCarthy, 1991). Capsaicin has been suggested for the initial management of neuralgia consequent to herpes infections (Bernstein, 1989). There is also evidence for the benefit of ginger in ameliorating arthritic knee pain, although the effectiveness is lesser than that of ibuprofen. Ginger doses of 0.5–1.0 g per day have been found to be efficacious in osteoarthritis and rheumatoid arthritis. Experimental studies have shown that ginger constituents inhibit arachidonic acid metabolism, which is involved in the inflammation process (a key pathway in inflammation).

Natural anti-inflammatory compounds of spices (curcumin, capsaicin, gingerol) appear to operate by inhibiting one or more of the steps linking proinflammatory stimuli with cyclooxygenase activation, such as the blocking by curcumin of NFκB translocation into the nucleus. It has recently been shown that the natural anti-inflammatory compounds such as curcumin were as effective as indomethacin (a nonsteroidal anti-inflammatory drug) in inhibiting aberrant crypt foci in the rat.

3.5.6 Antimutagenicity and Anticancer Effects

Considerable attention has currently been paid to identifying naturally occurring chemopreventive substances capable of inhibiting, retarding, or reversing multi-stage carcinogenesis. A wide array of phenolic substances, some of those present in spices, have been reported to possess substantial anticarcinogenic activities (Milner, 1994; Coney et al., 1997; Guhr and LaChance, 1997). The majority of these naturally occurring phenolics possess antioxidative and anti-inflammatory properties, which appear to contribute to their chemopreventive or chemoprotective activity (Surh, 2002).

There are a number of *in vitro* and *in vivo* studies on rodents suggesting that spices may have a chemopreventive effect against the early initiating stages of cancer. Spices may act through several mechanisms to provide protection against cancer (Figure 3.6). Certain phytochemicals from spices have been shown to inhibit one or

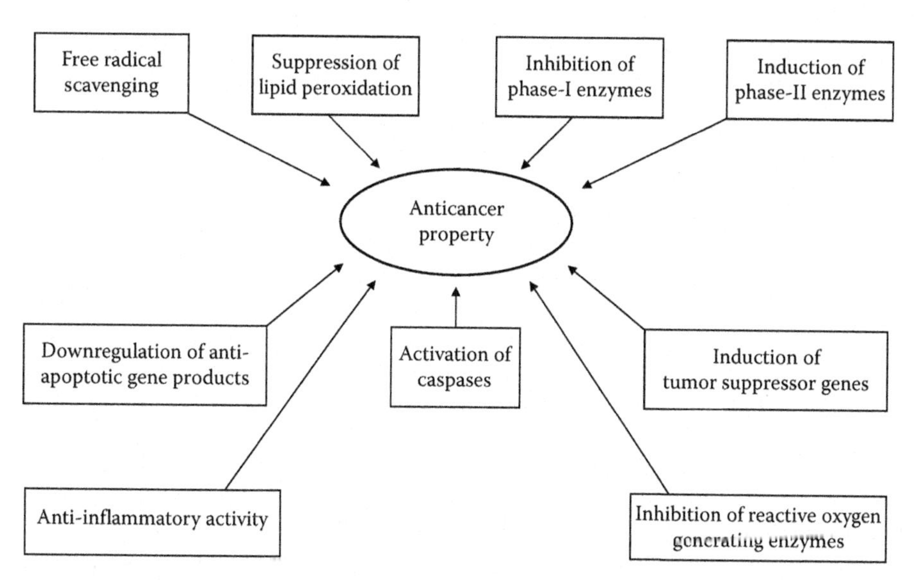

FIGURE 3.6 Mechanism of the anti-initiation, antipromotion, antiprogression, and anti-metastasis potential of cancer-preventive spice ingredients.

more of the stages of the cancer process (initiation, promotion, growth, and metastasis). Inhibition of phase I metabolic enzymes (involved in procarcinogen activation) and induction of phase II metabolic enzymes (involved in carcinogen deactivation) may account for the chemopreventive effects of spices. Spices may also protect against oxidative stress and inflammation, both of which are risk factors for cancer initiation and promotion (as well as other pathological conditions). Spices contain several natural lipid-soluble antioxidant biomolecules that may protect against the generation of genotoxic lipid peroxidation products.

Spices that have antioxidant properties can function as antimutagens. Since mutagenesis has a direct bearing on cancer initiation, antimutagenic spices can probably be anticarcinogenic. Turmeric or its bioactive compound curcumin, and garlic or its sulfur compounds have been shown to be antimutagenic in several experimental systems. Turmeric and curcumin were effective against benzo(α)pyrene and dimethyl benzanthracene in the Ames test (Nagabushan and Bhide, 1986). *In vivo* studies on experimental animals suggest that turmeric and curcumin inhibit the formation of mutagens. Mice and rats maintained on turmeric or curcumin-containing diet excreted lower levels of mutagenic metabolites as well as of carcinogens than the controls (Usha, 1994; Polasa et al., 1991). Turmeric and curcumin also inhibited the mutagenicity of cigarette and beedi smoke condensates as well as that of a tobacco-based dentifrice (Nagabushan et al., 1987). Further, curcumin was found to inhibit nitrosation of methylurea *in vitro* (Nagabushan et al., 1988). Studies on smokers revealed that administration of curcumin (1.5 g/day) for 30 days resulted in a significant reduction in the urinary excretion of mutagens (Polasa et al., 1992). Turmeric protected DNA against lipid peroxide-induced damage and against fuel smoke condensate-induced damage (Shalini and Srinivas, 1990). Eugenol (the flavour constituent of cloves) and mustard seeds (that contain dithiolthione) also produced antimutagenic effects by protecting the cells from damage to their DNA (NIN Annual Report, 1993–94). Dithiolthiones have been documented to have protective effects against liver toxicity induced by some chemicals and fungal aflatoxin (Ansher et al., 1986).

Spices (or their extracts and constituents) with known anticarcinogenic effects in animal models of cancer include turmeric, garlic, and ginger. Turmeric has been found to have chemopreventive effects against cancers of the skin, forestomach, liver, and colon, and oral cancer in mice. The anticancer potential of curcumin as evidenced by both preclinical and clinical studies has been exhaustively reviewed (Aggarwal et al., 2003). Animal studies involving experimental induction of tumors in specific tissues with potent carcinogens (such as benz(α)pyrene, 7,12-dimethylbenzanthracene, 3-methylcholanthrene, 12-*O*-tetradecanoylphorbol-13-acetate, and 1,2-dimethylhydrazine) have revealed significant reductions in the incidence of tumors by curcumin treatment. Several studies indicate that curcumin can suppress both tumor initiation and tumor promotion. Some of these studies, especially studies of skin tumorigenesis, have also employed topical application of curcumin (Aggarwal et al., 2003). It has been shown that the inhibition of arachidonic acid metabolism, modulation of cellular signal transduction pathways, and inhibition of hormones, growth factor, and oncogene activity are some of the mechanisms by which curcumin causes tumor suppression (Gescher et al., 1998). Chemopreventive activity of curcumin is observed when administered prior to, during, and after carcinogen treatment as well as when it is given only during the promotion or progression phase of colon carcinogenesis in rats (Kawamori et al., 1999). Curcumin is a powerful inhibitor of the proliferation of several tumor cells (Chuang et al., 2000a, 2000b; Dorai et al., 2001). With much evidence suggesting that curcumin can suppress tumor initiation, promotion, and metastasis, and with proven safety of its consumption (up to 10 g/day), curcumin offers enormous potential in the prevention and therapy of cancer (Aggarwal et al., 2003).

The cancer-preventive ability of garlic has been indicated by etiological studies wherein higher intakes of *Allium* products are associated with reduced risks of several types of cancer, especially stomach and colorectal (Fleischauer and Arab, 2001). Garlic is effective in the detoxification of carcinogens through its effects on phase I and phase II enzymes. The diallyl disulfide of garlic is an efficient inhibitor of the phase I enzyme cytochrome P-450 and significantly enhances a variety of phase II enzymes, including glutathione-*S*-transferase, quinone reductase, and UDP-glucuronyl transferase, which are responsible for the detoxification of carcinogens. Several mechanisms have been proposed to explain the cancer-preventive effects of garlic and its organosulfur compounds, as has been recently reviewed (Sengupta et al., 2004). These include inhibition of mutagenesis, modulation of enzyme activities thus suppressing bioactivation of carcinogen molecules, inhibition of carcinogen–DNA adduct formation, free radical scavenging, inhibitory effects on cell proliferation and tumor growth, and induction of apoptosis.

Pungent vanilloids, especially [6]-gingerol present in ginger (*Zingiber officinale*), have been found to possess potential chemopreventive activities. Prior topical applications of [6]-gingerol significantly suppressed the tumor promoter (phorbol ester)-stimulated skin inflammation initiated by 7,12-dimethylbenz [α] anthracene in mice (Surh et al., 1999). Reactive nitrogen species (RNS) such as nitric oxide (NO) have been proposed as being able to influence signal transduction and cause DNA damage, contributing to the carcinogenic processes. The pungent phenolic compound [6]-gingerol present in ginger has been shown to be a potent inhibitor of NO synthesis and also an effective protector against peroxynitrite-mediated damage in macrophages (Ippoushi et al., 2003). Dietary ginger constituents, galanals A and B, are potent apoptosis inducers in human T lymphoma cells (Miyoshi et al., 2003). Myristicin, a major volatile constituent of parsley, has been shown to strongly induce GSH-transferase in the liver and small intestinal mucosa of mice. This compound has been shown to lead to a 65% inhibition of tumor multiplicity in a rodent lung cancer model (Zheng et al., 1992a, 1992b).

3.5.7 Antioxidant Activity

The generation of reactive oxygen species and other free radicals during metabolism is a normal process that is ideally compensated for by an elaborate endogenous antioxidant defense system. Excessive free radical generation overbalancing the rate of their removal leads to oxidative stress. Oxidative damage has been implicated in the etiology of disease processes such as cardiovascular disease, inflammatory disease, cancer, neurodegenerative disease, and other degenerative diseases. Antioxidants are compounds that hinder the oxidative processes and thereby delay or suppress oxidative stress. There is a growing interest in the natural antioxidants found in herbs and spices. The bioactive compounds present in spices that possess potent antiatherogenic, anti-inflammatory, antimutagenic, and cancer-preventive activities are in fact antioxidants that have been experimentally shown to control cellular oxidative stress and thereby exert a beneficial role in preventing oxidative stress-mediated diseases.

Most of the health effects of spices on cancer, cardiovascular disease, inflammatory disease, and neurodegenerative disease may be mediated through their potent antioxidant effects. The antioxidant properties of spices are of particular interest in view of the impact of oxidative modifications of low-density lipoprotein cholesterol in the development of atherosclerosis. Suppression of oxidative stress and inflammation by spices is important in their cancer-preventive role, since both oxidative stress and inflammation are risk factors for cancer initiation and promotion (as well as other pathological conditions). Spices contain several natural antioxidant biomolecules: either water-soluble that can scavenge reactive oxygen species or lipid-soluble that may protect against the generation of genotoxic lipid peroxidation peroxides.

The antioxidative effects of curcumin, eugenol, capsaicin, piperine, gingerol, garlic, onion, and fenugreek have been experimentally evidenced (Srinivasan, 2010). The studies on this effect are exhaustive and experimental evidences are many in the case of curcumin of turmeric and eugenol of clove. Studies with several *in vitro* systems as well as *in vivo* animal studies have shown that the spice principles curcumin, eugenol, and capsaicin have beneficial antioxidant properties by quenching oxygen free radicals, by inhibiting the production of reactive oxygen radicals, and by enhancing antioxidant enzyme activities.

The antioxidant activities of spice compounds in mammalian systems involve one or more of the following: (1) free radical scavenging, (2) suppressing lipid peroxidation, (3) enhancing antioxidant molecules in tissues, (4) stimulating the activities of endogenous antioxidant enzymes, (5) inhibiting the activity of inducible nitric oxide synthase, (6) inhibiting LDL oxidation, and (6) inhibiting enzymes of arachidonate metabolism—5-lipoxygenase and 2-cyclooxygenase. By virtue of its antioxidant activity, curcumin has been documented to be anti-inflammatory, antimutagenic and cancer preventive, antiatherogenic and cardioprotective, hepatoprotective, neuroprotective, anticataractogenic, and an effective wound healant (Figure 3.7).

Thus, the multiple health beneficial attributes of these common food adjuncts include digestive stimulant action, cardioprotective potential, antilithogenic properties, protective effect on erythrocyte integrity, antidiabetic influence, anti-inflammatory properties, and cancer-preventive potential (Figure 3.8). The antioxidant and hypolipidemic properties of spices have far-reaching nutraceutical values. The antioxidant properties of the bioactive compounds present in spices are of particular interest in view of the impact of suppression of oxidative stress in the development of degenerative diseases such as cardiovascular disease, neurodegenerative disease, inflammatory disease, and cancer. In addition, by making the food attractive and palatable through flavor, aroma, and color, spices can reduce the need to use other less healthy ingredients such as salt, fat, or sugar. Spices thus deserve to be considered as a natural and necessary component of our daily nutrition, beyond their role in imparting taste and flavor to our food. It is presumed that the additive and

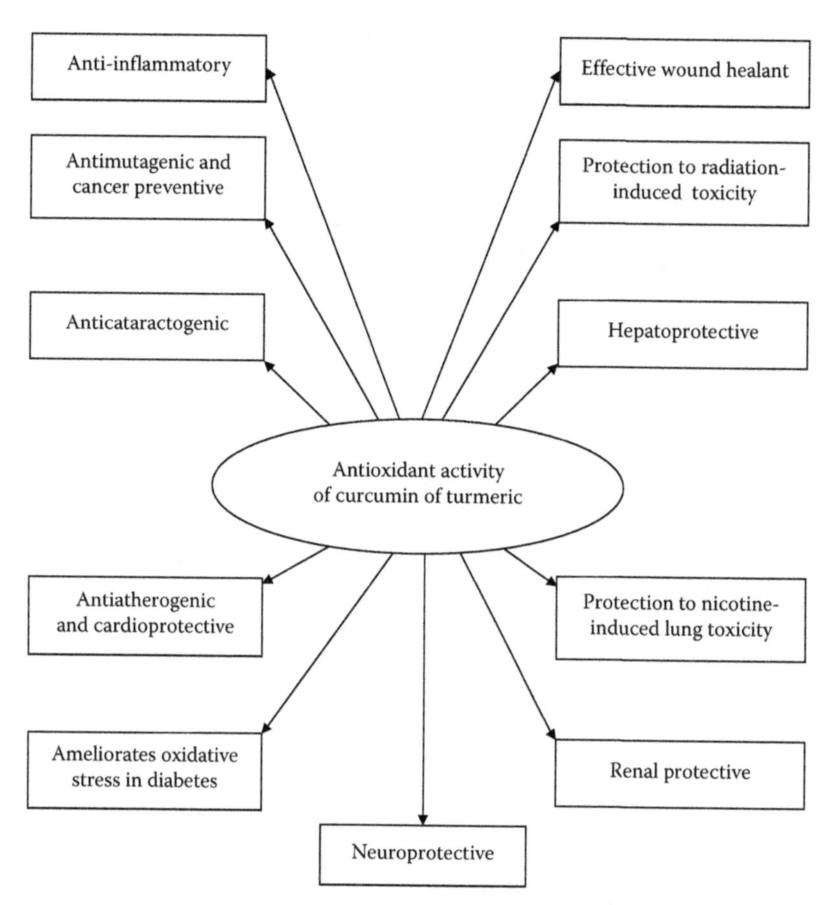

FIGURE 3.7 Health implications of the antioxidant property of spice ingredients.

synergistic effects of the complex mixture of phytochemicals present in vegetables, fruits, herbs, and spices are largely responsible for the health effects offered by those traditional Indian diets that are generally associated with lower incidence rates of some of the chronic diseases of aging, including cardiovascular disease and certain forms of cancer. The liberal consumption of spices is proved to be safe to derive their beneficial effects. Since each of the spices possesses more than one health beneficial property and there is also a possibility of synergy among them in their action when consumed in combination, a spiced diet is likely to make life not only more "spicy" but also more healthy.

3.6 SUMMARY

There is an abundance of scientific evidence which indicates that certain naturally occurring nonnutritive and some nutritive substances of spices, whole grain cereals and legumes, vegetables, fruits, sprouted grains, fermented grain products, and fermented milk products may prevent or reduce the risk of some chronic diseases

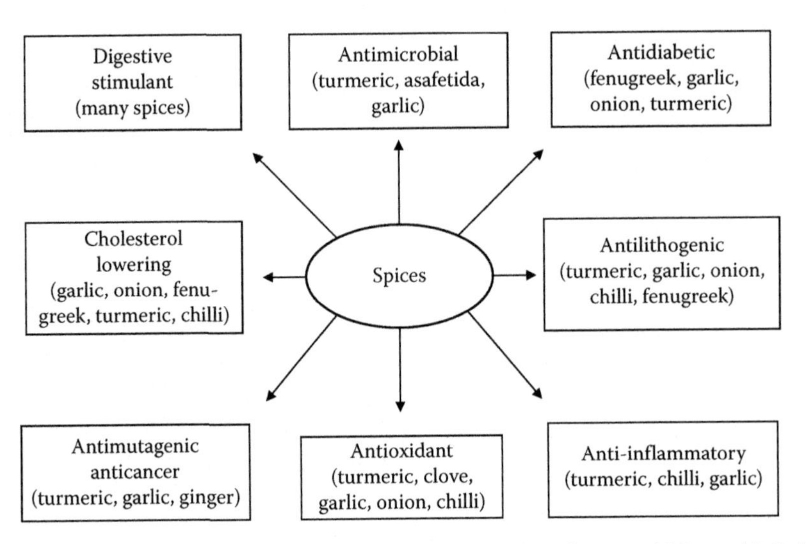

FIGURE 3.8 Summary of the multiple health effects of spices that are widely used in Indian traditional foods.

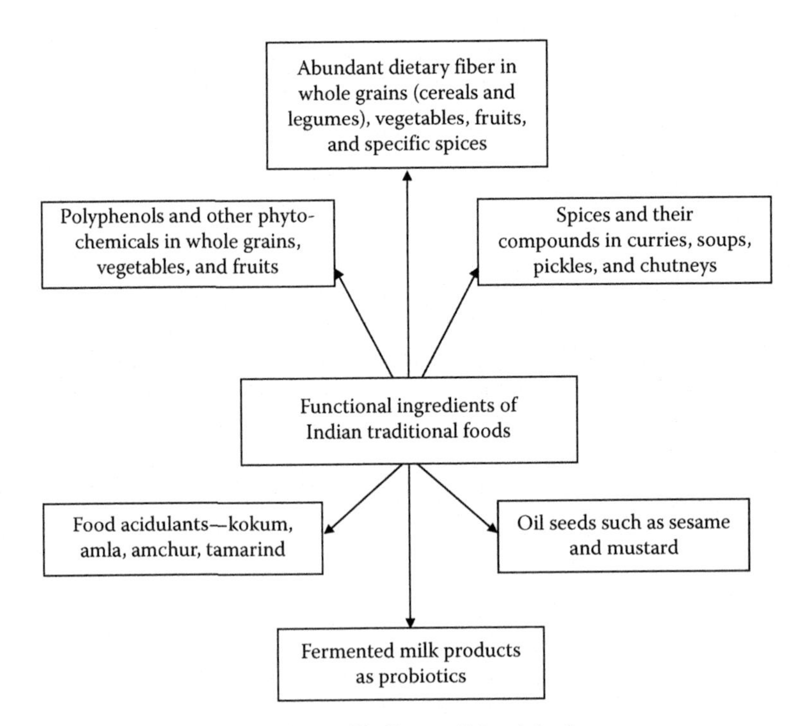

FIGURE 3.9 Functional ingredients of Indian traditional foods.

such as various cancers and cardiovascular disease. The health benefits of Indian heritage foods containing liberal quantities of these components may range from ensuring normal physiological functions in the body such as improving gastrointestinal health, enhancing the immune system, weight management, providing better skeletal health, and so on, to reduction of blood cholesterol, reduction of oxidative stress, reducing the risk of cardiovascular diseases, inflammatory diseases, and various types of cancers, and possible prevention of diabetes, neurodegenerative diseases, and so on. A perfect combination of protein from legumes, carbohydrates from rice, fat both visible and invisible from curry and fried savory items, vitamins and minerals from sprouted grains, and vitamins from curds and vegetables is obtained through typical Indian traditional meals (Figure 3.9).

REFERENCES

Achaya, K.T. 1994. *Indian Food: A Historical Companion*. Oxford University Press, Delhi.

Adler, A.J. and Holub, B.J. 1997. Effect of garlic and fish oil supplementation on serum lipid and lipoprotein concentration in hypercholesterolemic men. *Am. J. Clin. Nutr.* 65: 445–450, 1997.

Aggarwal, B.B., Kumar, A., and Bharti, A.C. 2003. Anticancer potential of curcumin: Preclinical and clinical studies. *Anticancer Res.* 23: 363–398.

Alam, M.I. and Gomes, A. 2003. Snake venom neutralization by Indian medicinal plants (*Vitex negundo* and *Emblica officinalis*) fruit extract. *J. Ethnopharmacol.* 86: 75–80.

Al-Rehaily, A.J., Al-Howiriny, T.A., Al-Sohaibani, M.O., and Rafatullah, S. 2002. Gastroprotective effects of amla (*Emblica officinalis*) on *in vivo* test models in rats. *Phytomedicine* 9: 515–522.

Ansher, S.S., Dolan, P., and Bueding, E. 1986. Biochemical effects of dithiolthiones. *Food Chem. Toxicol.* 24: 405–415.

Augusti, K.T. and Sheela, C.G. 1996. Antiperoxide effect of *S*-allyl cysteine sulfoxide, an insulin secretagogue in diabetic rats. *Experientia* 52: 115–119.

Babu, P.S. and Srinivasan, K. 1998. Amelioration of renal lesions associated with diabetes by dietary curcumin in experimental rats. *Mol. Cell. Biochem.* 181: 87–96.

Babu, P.S. and Srinivasan, K. 1999. Renal lesions in streptozotocin induced diabetic rats maintained on onion and capsaicin containing diets. *J. Nutr. Biochem.* 10: 477–483.

Bernstein, J.E. 1989. Treatment of chronic post-herpetic neuralgia with topical capsaicin. *Am. J. Dermatol.* 21: 265–270.

Beynen, A.C., Visser, J.J., and Schouten, J.A. 1987. Inhibitory effects on lithogenesis by ingestion of a curcuma mixture (Temoe Lawak Singer). *J. Food Sci. Tech.* 24: 253–256.

Carson, J.F. 1987. Chemistry and biological properties of onion and garlic. *Food Rev. Int.* 3: 71–103.

Chavan J.K. and Kadam, S.S. 1989. Nutritional improvement of cereals by fermentation. *Crit. Rev. Food Sci. Nutr.* 28: 349–400.

Chuang, S.E., Cheng, A.L., Lin, J.K., and Kuo, M.L. 2000a. Inhibition by curcumin of diethylnitrosamine-induced hepatic hyperplasia, inflammation, cellular gene products, and cell cycle related proteins in rats. *Food Chem. Toxicol.* 38: 991–995.

Chuang, S.E., Kuo, M.L., Hsu, C.H., et al. 2000b. Curcumin-containing diet inhibits diethylnitrosamine induced murine hepatocarcinogenesis. *Carcinogenesis* 21: 331–335.

Coney, A.H., Lou, Y.R., Xie, J.G., et al. 1997. Some perspectives on dietary inhibition of carcinogenesis : Studies with curcumin and tea. *Proc. Soc. Exp. Biol. Med.* 216: 234–245.

Deal, C.L. 1991. Effect of topical capsaicin: A double blind trial. *Clin. Therap.* 13: 383–395.

De Maeyer, B.M., Dallman, P., Gurncy, J.M., Hallberg, L., Sood, S.K., and Srikantia, S.G. 1989. *Preventing and Controlling Iron Deficiency Anaemia through Primary Health Care—A Guide for Health Administrators and Programme Managers.* Geneva: WHO.

Deodhar, S.D., Sethi, R., and Srimal, R.C. 1980. Preliminary studies on anti-rheumatic activity of curcumin. *Indian J. Med. Res.* 71: 632–634.

Dorai, T., Cao, Y.C., Dorai, B., Buttyan, R., and Katz, A.E. 2001. Therapeutic potential of curcumin in human prostate cancer. III. Curcumin inhibits proliferation, induces apoptosis, and inhibits angiogenesis of LNCaP prostate cancer cells *in vivo. Prostate*, 47: 293–303.

Fenwick, G.R. and Hanley, A.B. 1985. The genus *Allium*: Part-3, *Crit. Rev. Food Sci. Nutr.* 23: 1–73.

Fleischauer, A.T. and Arab, L. 2001. Garlic and cancer: A critical review of the epidemiologic literature. *J. Nutr.* 131:1032S–40S.

Gescher, A, Pastorino, U., Plummer, S.M., and Manson, M.M. 1998. Suppression of tumour development by substances derived from the diet—Mechanism and clinical implications. *Br. J. Clin. Pharmacol.* 45: 1–12.

Gibson, R.S. and Hotz, C. 2001. Dietary diversification/modification strategies to enhance micronutrient content and bioavailability of diets in developing countries. *Br. J. Nutr.* 85 (Suppl 2): S159–S166.

Gillooly, M., Bothwell, T.H., Torrance, J.D., et al. 1983. The effect of organic acids, phytate and polyphenols on the absorption of iron from vegetables. *Br. J. Nutr.* 49: 331–342.

Gore, J.M. and Dalen, J.E. 1994. Cardiovascular disease. *JAMA* 271: 1660–1661.

Govindarajan, V.S. and Satyanarayana, M.N. 1991. Capsicum: Production, technology, chemistry & quality; Impact on physiology, nutrition & metabolism, structure, pungency, pain and desensitization sequences. *Crit. Rev. Food Sci. Nutr.* 29: 435–474.

Guhr, G. and LaChance, P.A. 1997. Role of phytochemicals in chronic disease prevention. In: La Chance, P.A., ed. *Nutraceuticals: Designer Foods—III. Garlic, Soy and Licorice.* Trumbull, Connecticut, Food & Nutrition Press Inc, 311–364.

Hemalatha, S., Platel, K., and Srinivasan, K. 2005. Influence of food acidulants on bioaccessibility of zinc and iron from selected food grains. *Mol. Nutr. Food Res.* 49: 950–956.

Hemalatha, S., Platel, K., and Srinivasan, K. 2007. Influence of germination and fermentation on bioaccessibility of zinc and iron from cereals and pulses. *Eur. J. Clin. Nutr.* 61: 342–348.

Hussain, M.S. and Chandrasekhara, N. 1992. Influence of curcumin on cholesterol gall stone induction in mice. *Indian J. Med. Res.* 96: 288–291.

Hussain, M.S. and Chandrasekhara, N. 1993. Influence of curcumin and capsaicin on cholesterol gall stone induction in hamsters and mice. *Nutr. Res.* 13: 349–357.

Hussain, M.S. and Chandrasekhara, N. 1994a. Effect of curcumin and capsaicin on the regression of pre-established cholesterol gallstones in mice. *Nutr. Res.* 14: 1561–1574.

Hussain, M.S. and Chandrasekhara, N. 1994b. Biliary proteins from hepatic bile of rats fed curcumin or capsaicin inhibit cholesterol crystal nucleation in supersaturated model bile. *Indian J. Biochem. Biophys.* 31: 407–412.

Ippoushi, K., Azuma, K., Ito, H., Horie, H., and Higashio, H. 2003. [6]-Gingerol inhibits nitric oxide synthesis in activated J774.1 mouse macrophages and prevents peroxynitrite-induced oxidation & nitration reactions. *Life Sci.* 73: 3427–3437.

Jain, M.K. and Apitz-Castro, R. 1994. Garlic: A matter for heart. In: Charalambouis, G. ed. *Spices, Herbs and Edible Fungi.* New York: Elsevier, pp. 311–364.

Kaur, M. and Kawatra, B.L. 2002. Effect of domestic processing on zinc availability from rice bean (*Vigna umbellata*) diets. *Plant Food Hum. Nutr.* 57: 307–318.

Kawada, T., Hagihara, K., and Iwai, K. 1986. Effect of capsaicin on lipid metabolism in rats fed a high fat diet. *J. Nutr.* 116: 1272–1278.

Kawamori, T., Lubet, R., Steele, V.E., et al. 1999. Chemopreventive effect of curcumin, a naturally occurring anti-inflammatory agent, during the promotion/progression stages of colon cancer. *Cancer Res.* 59: 597–601.

Kirtikar, K.R. and Basu, B.D. 1935. *Indian Medicinal Plants.* Lalit Mohan Basu, Allahabad, India, p. 168, 2149.

Kleijnen, J., Knipschild, P., and Terriet, G. 1989. Garlic, onion and cardiovascular risk factors. *Br. J. Clin. Pharmacol.* 28: 535–544.

Kumudkumari, Mathew, B.C., and Augusti, K.T. 1995. Anti-diabetic and hypolipidemic effects of *S*-methyl cysteine sulfoxide isolated from *Allium cepa. Indian J. Biochem. Biophys.* 32: 49–54.

Lin, R.I. 1994. Phytochemicals and antioxidants. In: Goldberg, I., ed., *Functional Foods, Designer Foods, Pharma Foods, Nutraceuticals.* Chapman & Hall, London, pp. 393–450.

Lorenz, K. 1980. Cereal sprouts: Composition, nutritive value, food applications. *CRC Crit. Rev. Food Sci. Nutr.* 13: 353–385.

Lowenstein, J.M. 1971. Effect of (–)-hydroxycitrate on fatty acid synthesis by rat liver in vivo. *J. Biol. Chem.* 246: 629–632.

Matsumura, Y., Kita, S., Tanida, Y., Taguchi, Y., Morimoto, S., Akimoto, K., and Tanaka, T. 1998. Antihypertensive effect of sesamin. III. Protection against development and maintenance of hypertension in stroke-prone spontaneously hypertensive rats. *Biol. Pharm. Bull.* 21: 469–473.

McCarthy, G.M. and McCarthy, D.J. 1991.. Effect of topical capsaicin in the therapy of painful osteoarthritis of the hand. *J. Rheumatol.* 19: 604–607.

Milner, J.A. 1994. Reducing the risk of cancer. In: Goldberg, I., ed., *Functional Foods, Designer Foods, Pharma Foods, Nutraceuticals.* Chapman & Hall, London, pp. 39–70.

Miyawaki, T., Aono, H., Toyoda-Ono, Y., Maeda, H., Kiso, Y., and Moriyama K. 2009. Antihypertensive effects of sesamin in humans. *J. Nutr. Sci. Vitaminol.* (*Tokyo*) 55: 87–91.

Miyoshi, N., Nakamura, Y., Ueda, Y., Abe, M., Ozawa, Y., Uchida, K., and Osawa, T. 2003. Dietary ginger constituents, galanals A and B, are potent apoptosis inducers in Human T lymphoma Jurkat cells. *Cancer Lett.* 199: 113–119.

Nagabushan, M., Amonkar, A.J., and Bhide, S.V. 1987. *In vitro* anti-mutagenicity of curcumin against environmental mutagenesis. *Food Chem. Toxicol.* 25: 545–547.

Nagabushan, M. and Bhide, S.V. 1986. Non-mutagenicity of curcumin and its anti-mutagenic action versus chilli and capsaicin. *Nutr. Cancer* 8: 201–205.

Nagabushan, M., Nair, U.J., Amonkar, A.J., D'Souza, A.V., and Bhide, S.V. 1988. Curcumins as inhibitors of nitrosation *in vitro. Mutation Res.* 202: 163–169.

National Centre of Excellence 2006. Health benefits of herbs and spices: The past, the present, the future. *Med. J. Aus.* 185: S1–S24.

NIN Annual Report—1993–94. National Institute of Nutrition, Hyderabad, India.

Parpia, H.A.B. 2006. Heritage foods: Challenges and opportunities. *Indian Food Industry*, 25 (6): 13–15.

Platel, K. and Srinivasan, K. 1996. Influence of dietary spices or their active principles on digestive enzymes of small intestinal mucosa in rats. *Int. J. Food Sci. Nutr.* 47: 55–59.

Platel, K. and Srinivasan, K. 2000a. Stimulatory influence of select spices on bile secretion in rats. *Nutr. Res.* 20: 1493–1503.

Platel, K. and Srinivasan, K. 2000b. Influence of dietary spices or their active principles on pancreatic digestive enzymes in albino rats. *Nahrung* 44: 42–46.

Platel, K. and Srinivasan, K. 2001a. A study of the digestive stimulant action of select spices in experimental rats. *J. Food Sci. Technol.* 38: 358–361.

Platel, K. and Srinivasan, K. 2001b. Studies on the influence of dietary spices on food transit time in experimental rats. *Nutr. Res.* 21: 1309–1314.

Platel, K. and Srinivasan, K. 2004. Digestive stimulant action of spices: A myth or reality? *Indian J. Med. Res.* 119: 167–179.

Polasa, K., Raghuram, T.C., Krishna, T.P., and Krishnaswamy, K. 1992. Effect of turmeric on urinary mutagens in smokers. *Mutagenesis* 7: 107–109.

Polasa, K., Sesikaran, B, Krishna, T.P., and Krishnaswamy, K. 1991. *Curcuma longa* induced reduction in urinary mutagens. *Food Chem. Toxicol.* 29: 699–706.

Pollman, D.S., Danielson, D.M., Wren, W.B., Peo, E.R., and Shahani, K.M. 1980. Influence of *Lactobacillus acidophilus* inoculum on gnotobiotic and conventional pigs. *J. Animal Sci.* 51: 629–637.

Prabhavathi, T. and Rao, B.S.N. 1979. Effects of domestic preparation of cereals and legumes on ionizable iron. *J. Sci. Food Agric.* 30: 597–602.

Rajak, S., Banerjee, S.K., Sood, S., Dinda, K.A., Gupta, Y.K., and Maulik, S.K. 2004. *Emblica officinalis* causes myocardial adaptation and protects against oxidative stress in ischaemic reperfusion injury in rats. *Phytother. Res.* 18: 54–60.

Rajeshkumar, N.V., Pillai, M.R., and Kuttan, R. 2003. Induction of apoptosis in mouse and human carcinoma cell lines by *Emblica officinalis* polyphenols and its effect on chemical carcinogenesis. *J. Exp. Clin. Cancer Res.* 22: 201–212.

Reddy, N.R., Pierson, M.D., Sathe, S.K., and Salunke, D.K. 1989. *Phytates in Cereals and Legumes*, Boca Raton, FL: CRC Press, pp. 68–72.

Reddy, R.L.R. and Srinivasan, K. 2009a. Dietary fenugreek seed regresses preestablished cholesterol gallstones in mice. *Can. J. Physiol. Pharmacol.* 87: 684–693.

Reddy, R.L.R. and Srinivasan, K. 2009b. Fenugreek seeds reduce atherogenic diet induced cholesterol gallstone formation in experimental mice. *Can. J. Physiol. Pharmacol.* 87: 933–943.

Saavendra, J.M. 1995. Microbes to fight microbes: A not so novel approach to controlling diarrhoeal disease. *J. Pediatr. Gastroenteol.* 21: 125–129.

Sabu, M.C. and Kuttan, R. 2002. Antidiabetic activity of medicinal plants and its relationship with their antioxidant property. *J. Ethnopharmacol.* 81: 155–160.

Sairam, M., Neetu, D., Yogesh, B., et al. 2002. Cytoprotective and immunomodulating properties of amla (*Emblica officinalis*) on lymphocytes: An *in vitro* study. *J. Ethnopharmacol.* 81: 5–10.

Sambaiah, K. and Srinivasan, K. 1991. Secretion and composition of bile in rats fed diets containing spices. *J. Food Sci. Technol.* 28: 35–38.

Satoskar, R.R., Shah, S.J., and Shenoy, S.G. 1986. Evaluation of anti-inflammatory property of curcumin (diferuloylmethane) in patients with post-operative inflammation. *Int. J. Clin. Pharmacol. Therap. Toxicol.* 24: 651–654.

Scartezzini, P. and Speroni, E. 2000. Review on some plants of Indian traditional medicine with antioxidant activity. *J. Ethnopharmacol.* 71: 23–43.

Sengupta, A., Ghosh, S., and Bhattacharjee, S. 2004. *Allium* vegetables in cancer prevention: An overview. *Asia Pac. J. Cancer Prev.* 5: 237–245.

Shalini, V.K. and Srinivas, L. 1990. Fuel smoke condensate induced DNA damage in human lymphocytes and protection by turmeric. *Mol. Cell. Biochem.* 95: 21–30.

Sharma, N., Trikha, M., and Raisuddin, S. 2000. Inhibitory effect of *Emblica officinalis* on the *in vivo* clastogenecity of benzopyrene and cyclophosphamide in mice. *Human Exp. Toxicol.* 19: 377–384.

Sharma, R.D. 1984. Hypocholesterolemic activity of fenugreek (*T. foenum-graecum*): An experimental study in rats. *Nutr. Rep. Int.* 30: 221–231.

Sharma, R.D. (1986a) Effect of fenugreek seeds and leaves on blood glucose and serum insulin responses in human subjects. *Nutr. Res.* 6: 1353–1364.

Sharma, R.D. (1986b) An evaluation of hypocholesterolemic factor in fenugreek seeds (*T. foenum-graecum*). *Nutr. Rep. Int.* 33: 669–677.
Sharma, R.D., Sarkar, A., Hazra, D.K., et al. 1996. Use of fenugreek seed powder in the management of NIDDM. *Nutr. Res.* 16: 1331–1339.
Singhal, P.C., Gupta, R.K., and Joshi, L.D. 1982. Hypocholesterolemic effect of *T. foenum-graecum. Nutr. Rep. Int.* 33: 669–677.
Srimal, R.C. 1997. Turmeric: A brief review of medicinal properties. *Fitoterapia* 68: 483–490.
Srinivasan, K., Sambaiah, K. and Chandrasekhara, N. 2004. Spices as beneficial hypolipidemic food adjuncts: A review. *Food Rev. Int.* 20: 187–220.
Srinivasan, K. 2005a. Role of spices beyond food flavouring: Nutraceuticals with multiple health effects. *Food Rev. Int.* 21: 167–188.
Srinivasan, K. 2005b. Spices as influencers of body metabolism: An overview of three decades of research. *Food Res. Int.* 38: 77–86.
Srinivasan, K. 2005c. Plant foods in the management of diabetes mellitus: Spices as potential antidiabetic agents. *Int. J. Food Sci. Nutr*. 56: 399–414.
Srinivasan, K. 2008. Reason to season: Spices as functional food adjuncts with multiple health effects (A Technical Review). *Indian Food Industry* 27 (5): 36–47.
Srinivasan, K. 2010. Antioxidant potential of spices and their active constituents. *Crit. Rev. Food Sci. Nutr.* 50: in press.
Srivastava, V. 1986. Effect of curcumin on platelet aggregation and vascular prostacyclin synthesis. *Arznei Forch* 36: 715–717.
Steiner, M., Khan, A.H., Holbert, D., and Lin, R.I.S. 1996. A double-blinded crossover study in moderately hypercholesterolemic men that compared the effect of aged garlic extracts and placebo administration on blood lipids. *Am. J. Clin. Nutr.* 64: 866–870.
Stevinson, C., Pittler, M.H., and Ernst, E. 2000. Garlic for treating hypercholesterolemia. A meta-analysis of randomized clinical trials. *Ann. Intern. Med.* 133: 420–429.
Sullivan, A.C., Hamilton, J.G., Miller, O.N., and Wheatly, V.R. 1972. Inhibition of lipogenesis in rat liver by (-) hydroxycitrate. *Arch. Biochem. Biophys.* 150: 183–190.
Surh, Y.J. 2002. Anti-tumor promoting potential of selected spice ingredients with antioxidative and anti-inflammatory activities: A short review. *Food Chem. Toxicol.* 40: 1091–1097.
Surh, Y.J. and Lee, S.S. 1995. Capsaicin—A double-edged sword: Toxicity, metabolism and chemopreventive potential. *Life Sci.* 56: 1845–1855.
Surh, Y.J., Park, K.K., Chun, K.S., Lee, L.J., Lee, E., and Lee, S.S. 1999. Anti-tumor-promoting activities of selected pungent phenolic substances present in ginger. *J. Environ. Pathol. Toxicol. Oncol.* 18: 131–139.
Suzuki, T. and Iwai, K. 1984. Constituents of red pepper species: Chemistry, biochemistry, pharmacology and food science of the pungent principle of *Capsicum* species. In: Brossi, A., ed., *The Alkaloids—Chemistry & Pharmacology*; Vol. 23, New York: Academic Press, pp. 227–299.
Tank, R., Sharma, R, Sharma, T., and Dixit, V.P. 1990. Anti-diabetic activity of *Curcuma longa* in Alloxan induced diabetic rats. *Indian Drugs* 27: 587–589.
The Capsaicin Study Group 1992. Effect of treatment with capsaicin on daily activities of patients with painful diabetic neuropathy. *Diabetic Care* 15: 159–165.
The Wealth of India. 1976. Council of Scientific and Industrial Research, New Delhi, 10: 114.
Tontisirin, K., Mantel, G., and Battacharjee, L. 2002. Food based strategies to meet the challenges of micronutrient malnutrition in the developing world. *Proc. Nutr. Soc.* 61: 243–250.
Usha, K. 1994. The possible mode of action of cancer chemopreventive spice—turmeric. *J. Am. Col. Nutr.* 13: 519–521.
Valette, G., Sauvaire, Y., Baccon, J.C., and Ribes, G. 1984. Hypocholesterolemic effect of fenugreek seeds in dogs. *Atherosclerosis* 50: 105–111.

Veda, S., Platel, K., and Srinivasan, K. 2008. Influence of food acidulants and antioxidant spices on the bioaccessibility of β-carotene from selected vegetables. *J. Agric. Food Chem.* 56: 8714–8719.

Vidyashankar, S., Sambaiah, K., and Srinivasan, K. 2009. Dietary garlic and onion reduce the incidence of atherogenic diet induced cholesterol gallstone in experimental mice. *Br. J. Nutr.* 101: 1621–1629.

Vidyashankar, S., Sambaiah, K., and Srinivasan, K. 2010. Regression of preestablished cholesterol gallstones by dietary garlic and onion in experimental mice. Metabolism, 10.1016/j.metabol.2009.12.032; 2010; 59: in press.

Warshafsky, S., Kamer, R.S., and Sivak, S.L. 1993. Effect of garlic on total serum cholesterol: a meta-analysis. *Ann. Intern. Med.* 119: 599–605.

Watson, J.A., Fang, M., and Lownstein, J.M. 1969. Tricarballylate and hydroxycitrate: substrate and inhibitor of ATP: citrate oxaloacetate lyase. *Arch. Biochem. Biophys.* 135: 209–217.

Zheng, G., Kenney, P.M., and Lam, L.K. 1992a. Myristicin: A potential cancer chemopreventive agent from parsley leaf oil. *J. Agric. Food Chem.* 40, 107–110.

Zheng, G., Kenney, P.M., Zhang, J., and Lam, L.K. 1992b. Inhibition of benzo[α]pyrene induced tumorigenesis by myristicin, a volatile aroma constituent of parsley leaf oil. *Carcinogenesis* 13: 1921–1923.

4 Some Biological Functions of Carotenoids in Japanese Food

Takashi Maoka and Hideo Etoh

CONTENTS

4.1 CAROTENOID IN JAPANESE FOODS

Carotenoids exhibit red, orange, and yellow colors, and are distributed in microorganisms, plants, and animals. They are tetraterpene pigments consisting of eight isoprenoid units having 40 carbon skeletons. Their structures commonly consist of polyene chains with nine conjugated double bonds near the center of the molecule and end groups at both sides of the polyene chain. Carotenoids are divided into two groups, that is, carotenes and xanthophylls. Carotenes contain carbon and hydrogen atoms. In addition to these two elements, xanthophylls contain oxygen atoms, which can be represented by hydroxyl, carbonyl, carboxylic, epoxide, and/or lactone groups.

Some xanthophylls exist as fatty acid esters or protein complexes. More than 700 types of carotenoids have been reported in nature (Britton et al., 2004).

Photosynthetic organisms are able to synthesize carotenoids *de novo*. On the other hand, animals cannot synthesize carotenoids. Therefore, animals need to obtain carotenoids from their food. A part of the ingested carotenoids get converted in their bodies through metabolic reactions such as *cis–trans* isomerization, oxidation, reduction, translocation of double bonds, and oxidative cleavage of double bonds (Schweigert, 1998).

Some Japanese foods contain various carotenoids. For example, many common vegetables contain β-carotene, α-carotene, lutein, zeaxanthin, antheraxanthin, violaxanthin, and neoxanthin; tomato contains lycopene; paprika has capsanthin and capsorubin; seaweed is a source of fucoxanthin; salmon, shrimp, and crab contain astaxanthin; shellfish and tunicates possess halocynthiaxanthin and mytiloxanthin; Satuma mandarin fruits contain β-cryptoxanthin and zeaxanthin. These are the typical carotenoids found in Japanese foods. The structures of some of the carotenoids commonly found in Japanese foods are shown in Figure 4.1.

4.2 ABSORPTION, METABOLISM, AND DISTRIBUTION OF CAROTENOIDS IN HUMANS

About 50 kinds of carotenoids are found in common foods, and among them, about 20 types ingested from foods are found in the blood (Khachik et al., 1991, 1997a, 1997b; Wada et al., 2007). Among them, β-carotene, α-carotene, lycopene, β-cryptoxanthin, lutein, and zeaxanthin have been found to be the major components and consist of more than 90% of the total carotenoids (Khachik et al., 1991, 1997a, 1997b; Wada et al., 2007). Capsanthin and capsorubin are also found in the blood when foods containing paprika are ingested (Etoh et al., 2000; Oshima et al., 1997). On the other hand, epoxycarotenoids such as violaxanthin, antheraxanthin, and neoxanthin, which are widely distributed in vegetables, are not found in blood. The reason might be due to the degradation of epoxycarotenoids by the gastric juice (Asai et al., 2004b). Fucoxanthin, a major carotenoid in seaweed, is also not found in blood. However, Asai et al. (2004a) demonstrated that fucoxanthin was converted into amarouciaxanthin A via fucoxanthinol and accumulated in mice and human hepatoma cell HepG2.

Carotenoids ingested from the diet are absorbed by the small intestine. Xanthophyll esters are hydrolyzed by lipase or esterase and then get absorbed. A part of provitamin A carotenoids, such as β-carotene, gets converted into retinal in a mucous of the small intestine by β-carotene-15,15′-dioxygenase. Absorbed carotenoids are incorporated into chylomicrons and then transported to the liver and various organs through the blood. All three major lipoproteins, very low-density lipoprotein (VLDL), low-density lipoprotein (LDL), and high-density lipoprotein (HDL), are involved in the transport of carotenoids. Carotenes are primarily associated with LDL. Carotenoids in the liver are released by VLDL.

Carotenoids can be found in several human organs, such as the liver, adrenal gland, ovaries, skin, eyes, lungs, testes, prostate, and blood serum. The distribution

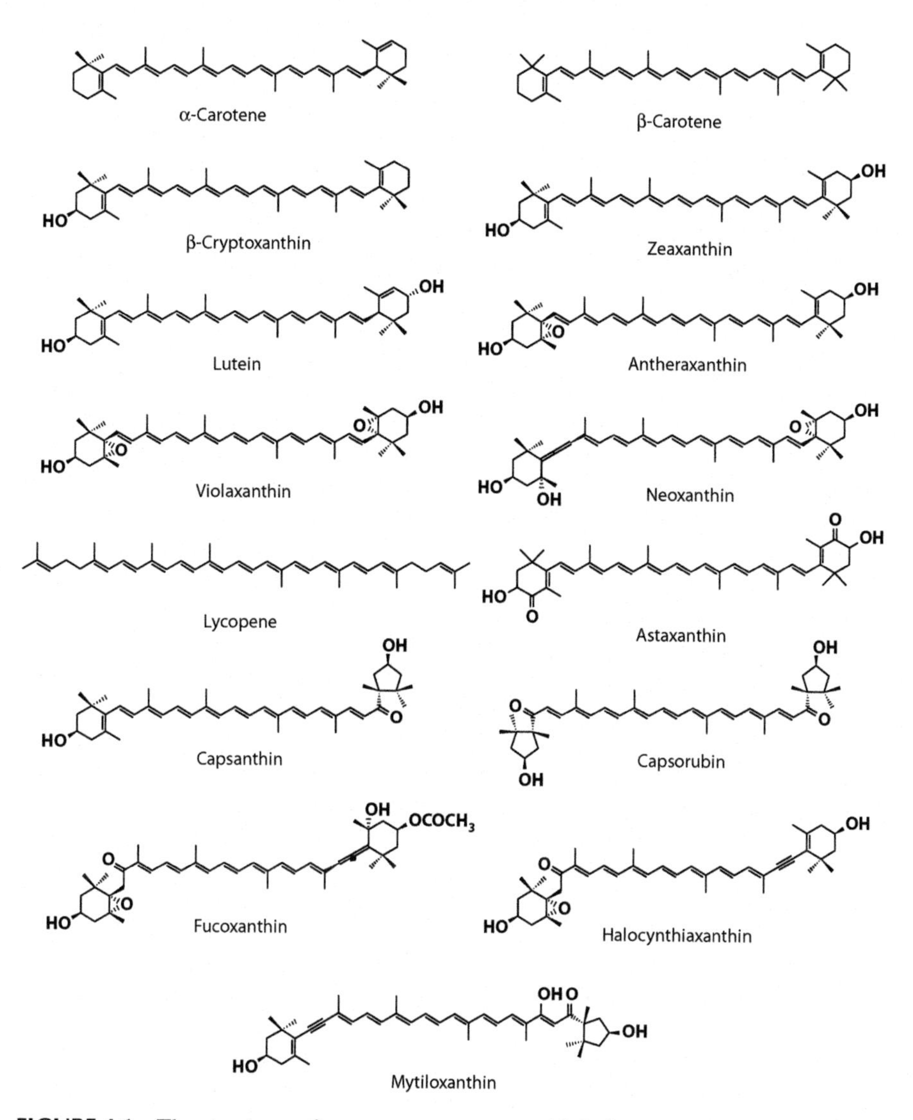

FIGURE 4.1 The structures of some common carotenoids in Japanese foods.

of carotenoids in human organs indicates specificity. Lutein and zeaxanthin are found in the surface of the skin and subcutaneous tissue in an esterified form and act as a UV absorber and quencher of singlet oxygen (Wingerath et al., 1998). Recently, xanthophylls such as β-cryptoxanthin, lutein, and zeaxanthin have been found in the brain (Craft et al., 2004). Lutein, zeaxanthin, and their oxidative metabolite, 3′-*O*-dehydrolutein, were present in the macula of the retina to protect it from photooxidation (Khachik et al., 1997a, 1997b). Lycopene accumulates in the prostate (van Breemen et al., 2002).

4.3 PROVITAMIN A ACTIVITY OF CAROTENOIDS

The best-known biological function of carotenoids such as β-carotene is their role as pro-vitamin A. Carotenoids having unsubstituted β-end groups, such as β-carotene, α-carotene, and β-cryptoxanthin, show pro-vitamin A activity. They are transformed into retinal by β-carotene-15,15′-dioxygenase. Retinal gets reduced to retinol (vitamin A) and then is distributed to various organs. It acts as a chromophore in visual pigments. Retinoic acid plays an important role in reproduction and differentiation in the cells (Schweigert, 1998).

4.4 ANTIOXIDATIVE ACTIVITY OF CAROTENOIDS

Carotenoids have excellent quenching activity for singlet oxygen and lipid peroxidation. Foote and Denny (1968) were the first to demonstrate that β-carotene has a strong quenching activity for singlet oxygen. Since then, it has been shown that several natural carotenoids with more than 11 conjugated double bonds show excellent quenching activity for singlet oxygen. Among them, astaxanthin, lycopene, and capsorubin exhibited the strongest quenching activities for singlet oxygen (Hirayama et al., 1994; Miki, 1991; Shimizu et al., 1996).

The mechanism for quenching of singlet oxygen is a physical reaction. Carotenoids take up thermal energy from singlet oxygen and release this energy by polyene vibration. Singlet oxygen is always generated during a photosynthetic reaction. Therefore, carotenoids are essential compounds in plants and photosynthetic organisms. Singlet oxygen is also generated in humans during phagocytolysis of macrophages and neutrophiles and causes some inflammation and arteriosclerosis. Oshima et al. (1999) reported on a carotenoid-rich, LDL that was prepared by humans ingesting long-term supplementation with tomato juice and inhibited the formation of singlet oxygen-mediated lipid peroxidation better than the LDL devoid of carotenoid. Furthermore, it was reported that supplementation with astaxanthin could increase the antioxidative activity of human LDL (Iwamoto et al., 2000). Therefore, it can be strongly suggested that carotenoids suppress oxidative damage induced by singlet oxygen, in humans.

Active oxygen and free radicals induce a chain reaction of lipid peroxidation. Burton and Ingold (1984) found that carotenoids showed excellent antioxidative activities against lipid peroxidation at low oxygen pressure. Oxygen pressure in tissues and cells is 40 and 1 mm Hg, respectively. Therefore, it was assumed that carotenoids might act as excellent antioxidants in human tissues and cells. The antioxidative activity of carotenoids is influenced by the number of conjugated double bonds. Carotenoids having long conjugated double bond systems showed excellent inhibitory effects on lipid peroxidation. It was reported that astaxanthin, canthaxanthin, capsanthin, and capsorubin having conjugated carbonyl groups showed the strongest quenching activities for lipid peroxidation, among the naturally occurring carotenoids (Lim et al., 1992; Miki, 1991; Terao, 1989).

Until now, it has been assumed that carotenoids do not directly scavenge superoxide and hydroxyl radicals (Trevithick-Sutton et al., 2006). However, recent studies have shown that β-carotene, lycopene, zeaxanthin, lutein (Trevithick-Sutton et al.,

2006), fucoxanthin, fucoxanthinol, and halocynthiaxanthin (Sachindra et al., 2008) could scavenge superoxide and hydroxyl radicals. Furthermore, some carotenoids could inhibit the generation of superoxide or nitric oxide (NO) radicals in cells. Thus, halocynthiaxanthin, capsanthin, and capsorubin were found to suppress the generation of superoxide and/or NO in stimulated leucocytes (Murakami et al., 2000). Therefore, carotenoids possess potency for inhibiting cancer, cardiac disease, inflammation, and arteriosclerosis caused by oxidative damage.

The molecular mechanisms for the reaction of active oxygen and free radicals with carotenoids have been discovered. Carotenoids take up oxygen from active oxygen into their conjugated double bond system to form apocarotenals, apocarotenones, and epoxy carotenoids (Kim et al., 2001). Recently, Yoshioka et al. (2006) reported that astaxanthin and β-carotene can react with peroxynitrite to form nitrocarotenoids.

4.5 ANTICARCINOGENIC ACTIVITIES AND CANCER-PREVENTIVE EFFECTS OF THE CAROTENOIDS

Peto et al. (1981) reported that dietary β-carotene can reduce human cancer rates. Since then, β-carotene has received much attention for its cancer-preventive effects, and several human intervention studies have been done using β-carotene. However, the results from some of the trials were not consistent. In the study carried out at Linxian, China, β-carotene proved to reduce the cancer incidence (Blot et al., 1993). However, in the α-tocopherol, β-carotene (ATBC) cancer prevention study and in the β-carotene and retinal efficiency trial (CARET), β-carotene supplements were shown to increase the risk of lung cancer among smokers (The α-tocopherol β-carotene cancer prevention study group, 1994; The ATBC cancer prevention study group, 1994; Ziegler, 1989).

On the other hand, numerous epidemiological studies have demonstrated that intake of greenish-yellow vegetables and fruits, which contain various carotenoids, are associated with a reduced risk of cancer. Furthermore, various carotenoids besides β-carotene have been found in human plasma and tissues. The Epstein–Barr virus (EBV) activation inhibition test has been widely used for the estimation of the anticancer-promoting activities of natural products. Tsushima et al. (1995) examined the anticancer-promoting activities of 50 kinds of carotenoids found in vegetables, fruits, and seafood. All 50 carotenoids investigated showed inhibitory effects on TPA-induced expression of the early antigen of the EBV. β-Cryptoxanthin suppressed activation of EBV early antigen induced by TPA at the highest potency among the carotenoids tested. Furthermore, lactucaxanthin, lutein, lycopene, fucoxanthin, and capsanthin showed stronger activities than β-carotene. These results suggested that there are many carotenoids characterized by anticancer or cancer-preventive effects, apart from β-carotene.

Nishino et al. (1999, 2002) studied the anticarcinogenesis activity of some natural carotenoids using two-stage mouse carcinogenesis models and revealed that β-carotene, α-carotene, lycopene, β-cryptoxanthin, lutein, and zeaxanthin, which are very widely distributed in vegetables, have inhibitory effects on carcinogenesis.

Among them, α-carotene showed a higher activity than β-carotene in suppressing tumorigenesis in the skin, lungs, liver, and colon. Lycopene and β-cryptoxanthin also strongly inhibited mouse carcinogenesis compared to β-carotene. Furthermore, marine carotenoids such as fucoxanthin, halocynthiaxanthin, astaxanthin, and peridinin also inhibited mouse carcinogenesis (Nishino et al., 2002, 2006; Tanaka et al., 1994). Paprika carotenoids, capsanthin, and casorubin also exhibited anticarcinogenesis activity (Maoka et al., 2001; Narisawa et al., 2000). Therefore, it has been made clear that various natural carotenoids are valuable in cancer prevention.

Furthermore, Nishino et al. (2002) demonstrated that a combination of carotenoids (mixture of β-carotene, α-carotene, lutein, lycopene, β-cryptoxanthin, and zeaxanthin) showed much stronger inhibitory effects against carcinogenesis compared to single compound use of carotenoids in various models for carcinogenesis in rodents. Clinical trials have also revealed that the administration of natural multicarotenoids (mixtures of β-carotene, α-carotene, lutein, and lycopene) and α-tocopherol resulted in significant suppression of hepatoma development in a hepatitis virus-induced cirrhosis patient (Nishino et al., 2002; Nishino, 2008).

4.5.1 Molecular Mechanism for Cancer Prevention

In order to reveal the molecular mechanism for growth inhibition of cancer cells, the effects of natural carotenoids on the expressions of genes and protein, and on cell cycle progression, have been investigated.

Gap junctional intracellular communications are known to be involved in carcinogenesis and the expression of connexins is often reduced in cancer cells (Yamasaki, 1990). Bertram showed that carotenoids and their metabolites enhance the expression of gap-junction communication proteins. This is an important mechanism exhibited by carotenoids for chemoprevention at the cellular level (Bertram, 2004). Zhang et al. (1991, 1992) reported that carotenoids enhanced the expression of connexins and gap-junction communication in murine fibroblast cells. Carotenoids have also been reported to cause cell cycle arrest. α-Carotene (Murakoshi et al., 1989), fucoxanthin (Das et al., 2005; Satomi and Nishino, 2007), and phytoene and phytofluene (Satomi and Nishino, 2004) have been proven to induce gastrointestinal (G1)-arrest in the process of the cell cycle. Furthermore, carotenoids induced apoptosis and differentiation. It has been reported that β-carotene, lycopene, lutein, and canthaxanthin induced apoptosis in human cancer cells (Nara et al., 2001; Oalozza et al., 1998; Palozza et al., 2001; Sumantran et al., 2000). Growth-inhibitory effects of carotenoids on human cancer cell lines have been investigated. Marine carotenoids, halocynthiaxanthin, and fucoxanthin were found to inhibit the cell growth of HL-60 human leukemia cells and induced apoptosis in HL-60 cells. These activities were higher than that of β-carotene (Hosokawa et al., 2004; Konishi et al., 2006). The inhibitory effects of carotenoids present in foodstuffs have been evaluated on the basis of the growth of the human prostate cancer cell lines, PC-3, DU 145, and LNCap by Kotake-Nara et al. (2001, 2005). Among the carotenoids evaluated, neoxanthin and fucoxanthin, having allenic groups, showed remarkable reduction in the growth of prostate cancer cells. These allenic carotenoids also showed antiproliferative effects on cultured cells MRC-5, HUC-Fm, B16, Caco-2, HCT116, and PC-3. The tumor necrosis factor-related apoptosis-inducing ligand (TRAIL) is a promising candidate for cancer therapy due to its ability to induce apoptosis. Recently, Yoshida et al. (2007) reported that halocynthiaxanthin and peridinin sensitized the cancer cells to TRAIL-induced apoptosis. The paprika carotenoid, capsorubin, remarkably inhibited the growth of the human lung A549 and prostate cancer cells. Furthermore, capsorubin was observed to be taken up into the nucleus of A549 cells and reduced the expression of NFκB, which is an important transcription regulation protein of cancer genes, and the induction of G1 arrest on cell cycles (Maoka et al., 2008). Angiogenesis, the processes of new blood vessel growth, is involved in many physiological and pathological situations. The antiangiogenesis therapy has been established as a strategy for cancer prevention. Fucoxanthin and fucoxanthinol have been observed to significantly suppress angiogenesis. These results imply that fucoxanthin, having antiangiogenic activity, might be useful in preventing angiogenesis-related diseases (Sugawara et al., 2006).

4.5.2 Recent Results of Cohort and Epidemiological Studies

β-Cryptoxanthin is especially rich in Satuma mandarin (*Citrus unshiu* Marc.) fruit, abundantly found in Japan. Cohort studies revealed that β-cryptoxanthin could be associated with a reduced risk of cancer (Mannisto et al., 2004; Yuan et al., 2001). Furthermore, epidemiological studies have shown that intake of lycopene reduced the risk of prostate cancer (Gann et al., 1999; Giovannucci, 1999).

4.6 CAROTENOIDS IN OCULAR HEALTH

There is increasing evidence that carotenoids play an important role in the prevention and treatment of age-related macular degeneration (AMD), cataracts, and other blinding disorders. Lutein, zeaxanthin, and their oxidative metabolites such as 3′-*O*-dehydrolutein have been detected in ocular tissues. Lutein and zeaxanthin have been found to protect against the development or progression of AMD and other eye diseases (Mozaffarieh et al., 2003; Semba and Dagnelie, 2003). One of the underlying hypotheses for the positive role of carotenoids in AMD and cataracts has been based on the ability of these carotenoids to act as antioxidants and to protect the human retina from photooxidation. Carotenoids in human retinal pigments of epithelium might be provided by both mechanisms of antioxidation and light screening. Carotenoids in the human ciliary body are more likely to act as antioxidants in this tissue, while in the iris and the lens, these pigments might also play a role in filtering out phototoxic short-wavelength visible light. Therefore, lutein is used as a supplement for AMD.

Astaxanthin was also found to decrease UVB-induced lipid peroxidation of human lens epithelia cells and stress signaling, representing a stronger effect than that of α-tocopherol (Chitchumroonchokchai et al., 2004). Astaxanthin has also been found to improve the accommodation amplitude to visual displays of terminal workers (Nagaki et al., 2002).

4.7 PREVENTION OF DIABETES AND DIABETIC COMPLICATIONS BY CAROTENOIDS

Oxidative stress is implicated as an important mechanism by which diabetes causes nephropathy. Naito et al. (2004), reported that astaxanthin can reduce diabetic nephropathy in diabetic db/db mice, a rodent model for type II diabetes. The astaxanthin- treated mice showed a lower level of blood glucose compared to the non-treated group. The mesangial area was observed to be significantly ameliorated in the treated group. Astaxanthin also showed prevention of the β-cell function of insulin secretion, decreased the higher levels of blood glucose in the diabetic animals (Uchiyama et al., 2002), and protected mesangial cells from hyperglycemia-induced oxidative signaling (Manabe et al., 2007). Furthermore, epidemiological studies have revealed that the intake of Satuma mandarin (*Citrus unshiu* Marc.) decreased the risk of diabetes (Sugiura et al., 2002). This effect might be due to the antioxidative activity of β-cryptoxanthin (Montonen et al., 2004; Sugiura et al., 2002).

4.8 CAROTENOIDS FOR SKIN COLOR AND CONDITIONING

The skin is exposed to both endogenous and environmental prooxidant agents, leading to reactive oxygen species (ROS) generation and possibly oxidative stress, which has been suggested to have an affect on membrane lipids and protein. Astaxanthin has been found to inhibit UVA-induced DNA damage in human skin (fiber) blastoma, human melanocyte, and human intestinal Caco-2 cells (Lyons and O'Brien, 2002). Furthermore, dietary astaxanthin significantly improved the fine lines/wrinkles and elasticity according to dermatologist's assessment. Additionally, an instrumental assessment suggested that the moisture content was improved after 6 weeks when compared to the baseline initial values (Yamashita, 2006).

A citrus carotenoid, β-cryptoxanthin, exhibited prevention of the melanogenesis dose dependency on mouse melanoma and human tyrosinase, and further repressed melanin production (Takayanagi et al., 2006).

4.9 ANTIOBESITY EFFECTS OF CAROTENOIDS

Obesity is defined as an accumulation of body fat. The accumulation of fat, especially around the internal organs, is a major risk factor causing many types of diseases. When the fat cells differentiate their accumulation of the fat into and around the cells, the cells secrete various bioactive compounds called adipocytokines, and these adipocytokines induce various diseases such as diabetes, hypertension, and cardiovascular disease.

Maeda et al. (2005) reported on the antiobesity effect of fucoxanthin through the mitochondria uncoupling protein and gene expression in white adipose tissues in rats and mice. Astaxanthin also showed antiobesity effect. Astaxanthin-administered mice exhibited lower body weights, lower volumes of fat tissue, lower liver masses, lower liver triacylglycerol, lower blood serum triacylglycerol, and lower total cholesterol than the control mice (Ikeuchi et al., 2007).

4.10 ANTI-INFLAMMATORY EFFECTS OF CAROTENOIDS

Several marine carotenoids including fucoxanthin and astaxanthin have been shown to possess anti-inflammatory potency. Fucoxanthin exhibits a dose-dependent antiocular inflammatory effect on endotoxins-induced uveitis (Tsukui et al., 2007). Astaxanthin was also found to reduce the production of NO, the activity of inducible nitric oxide synthase (NOS), and the production of PEG2 and TNF-α in RAW264.7 cells in a dose-dependent manner (Goureau et al., 1995; Ohgami et al., 2003).

4.11 ANTIFATIGUE EFFECTS FOR SKELETAL MUSCLES BY CAROTENOIDS

Free radicals play an important role as mediators of skeletal muscle damage and inflammation after strenuous exercise. Astaxanthin was shown to reduce exercise-induced damage in mouse skeletal muscle and heart, including an associated neutrophil infiltration that induces further damage (Aoi et al., 2003).

4.12 SUMMARY

In this review, we have summarized the various recent studies on physiological functions of carotenoids. Apart from the above studies, there are other reports available mentioning an enhancement of immunity by carotenoids. In addition, there is a recent report suggesting that β-cryptoxanthin promotes the uptake of calcium by the osseous tissue (Yamaguchi and Uchiyama, 2003). In this way, carotenoids are very useful components for improving and maintaining human health. Many of the physiological functions in the research summarized in this chapter have only been elucidated at the cellular level and at an experimental animal level, but the elucidation of the carotenoid functions in humans can be expected in the future.

REFERENCES

Aoi, W., Y. Naito, K. Sakuma, M. Kuchide, H. Tokuda, T. Moaka, S. Toyokuni, S. Oka, M. Yasuda, and T. Yoshikawa. 2003. Astaxanthin limits exercise-induced skeletal and cardiac muscle damage in mice. *Antioxid. Redox Signal* 5:139–144.

Asai, T., Sugawara, H. Ono, and A. Nagao. 2004a. Biotransformation of fucoxanthinol into amarouciaxanthin A in mice and HepG2 cells: Formation and cytotoxicity of fucoxanthin metabolites. *Drug Metab. Dispos*. 32:205–211.

Asai, A., M. Terasaki, and A. Nagao. 2004b. An epoxide-fyranoid rearrangement of spinachi neoxanthin occurs in the gastrointestinal tract of mice and *in vitro*: Formation and cytostatic activity on neochrome stereoisomer. *J. Nutr.* 134:2237–2243.

Bertram, S. 2004. Induction of connexin 43 by carotenoids: Functional consequence. *Arch. Biochem. Biophys*. 430:120–126.

Blot, J. W., J., Y. Li, R. R. Taylor, W. Guo, S. Dawsey, G. Q. Wang, C. S. Yang et al. 1993. Nutrition intervention trial in Linxian, China: Supplementation with specific vitamins/mineral combinations, cancer incidence, and disease-specific mortality in the general population. *J. Natl. Cancer Inst*. 85:1483–1492.

van Breemen, R. B., X. Xu, M. A. Viana, L. Chen, M. Stacewicz-Sapuntzaleis, C. Duncan, P. E. Bowen, and R. Sharifi. 2002. Liquid chromatography–mass spectrometry of *cis*- and all-*trans*-lycopene in human serum and prostate tissue after dietary supplementation with tomato sauce. *J. Agric. Food Chem.* 50:2214–2219.

Britton, G., S. Liaaen-Jensen, and H. Pfander. 2004. *Carotenoids Hand Book.* Birkhäuser, Basel, Switzerland.

Burton, G. W. and K. U. Ingold. 1984. Beta-carotene an unusual type of lipid antioxidant. *Science* 224:569–573.

Chitchumroonchokchai, C., J. Bomser, J. Glamm, and M. Failla. 2004. Xanthophyll and alpha-tocopherol decrease UVB-induced lipid peroxidation and stress signaling in human lens epithelial cells. *J. Nutr.* 134:3225–3232.

Craft, N. E., T. B. Haltema, K. M. Garnett, K. A. Fitch, and C. K. Dorey. 2004. Carotenoid, tocopherol, and retinal concentrations in elderly human brain. *J. Nutr. Health Aging* 8:156–162.

Das, S. K., T. Hashimoto, K. Shimizu, T. Yoshida, T. Sakai, Y. Sowa, A. Komoto, and K. Kanazawa. 2005. Fucoxanthin induces cell cycle arrest at G0/G1 phase in human colon carcinoma cells through up-regulation of p21WAF1/Cip1. *Biochim. Biophys. Acta* 1726:328–335.

Etoh, H., Y. Utsunomiya, A. Komori, Y. Nurakami, S. Ohshima, and T. Inakuma. 2000. Carotenoids in human blood plasma after ingesting paprika juice. *Biosci. Biotech. Biochem.* 64:1096–1098.

Foote, C. S. and R. W. Denny. 1968. Chemistry of singlet oxygen. VII, Quenching by β-carotene. *J. Am. Chem. Soc.* 90:6233–6235.

Gann, P. H., J. Ma, E. Giovannucci, W. Willett, F. M. Sacks, C. H. Hennekens, and M. J. Stampfer. 1999. Lower prostate cancer risk in men with elevated plasma lycopene levels: Results of a prospective analysis. *Cancer Res.* 59:1225–1230.

Giovannucci, E. 1999. Tomatoes, tomato-based products, lycopene and cancer: Review of epidemiological literature. *J. Natl. Cancer Inst.* 91:317–331.

Goureau, O., J. Bellot, B. Thillaye, Y. Courtois, and Y. de Kozak. 1995. Increased nitric oxide production in endotoxin-induced uveitis. Reduction of uveitis by an inhibitor of nitric oxide synthase. *J. Immunol.* 154:6518–6523.

Hirayama, O., K. Nakamura, S. Hamada, and Y. Kobayashi. 1994. Singlet oxygen quenching ability of naturally occurring carotenoids. *Lipids* 29:149–150.

Hosokawa, M., M. Kuso, H. Maeda, H. Kohno, T. Tanaka, and K. Myyashita. 2004. Fucoxanthin induced apoptosis and enhances the antiproliferative effect of the PPARγ ligand troglitazone, on colon cancer cells. *Biochem. Biophys. Acta* 1675:113–119.

Ikeuchi, M., T. Koyama, J. Takahashi, and K. Yazawa. 2007. Effects of astaxanthin in a obese mice fed a high fat diet. *Biosci. Biotechnol. Biochem.* 71:893–899.

Iwamoto, T., K. Hosoda, R. Hirano, H. Kurata, A. Matsumoto, W. Miki, M. Kamiyama, H. Itakura, S. Yamamoto, and K. Kondo. 2000. Inhibition of low-density lipoprotein oxidation by astaxanthin. *J. Atheroscler. Thromb.* 7:216–222.

Khachik, F., G. R. Beecher, M. B. Goli, and W. R. Lusby. 1991. Separation, identification, and quantification of carotenoids in fruits, vegetable, and human plasma by high performance liquid chromatography. *Pure Appl. Chem.* 63:71–80.

Khachik, F., P. Bernstein, and D. L. Garvalho. 1997a. Identification of lutein and zeaxanthin oxidation products in human and monkey retinas. *Invest. Ophthalmol. Visual Sci.* 38:1802–1811.

Khachik, F., C. J. Spangler, J. C. Smith Jr., L. M. Canfield, A. Steck, and H. Pfander. 1997b. Identification, quantification, and relative concentration of carotenoids and their metabolites in human milk and serum. *Anal. Chem.* 69:1873–1881.

Kim, S. J., E. Nara, H. Kobayashi, J. Terao, and A. Nagao. 2001. Formation of cleavage products by autoxidation of lycopene. *Lipids* 36:191–199.

Konishi, I., M. Hosokawa, T. Sashima, H. Kobayashi, and K. Miyashita. 2006. Halocynthiaxanthin and fucoxanthin from *Halocynthia roretzi* induce apoptosis in human leukemia, breast and colon cancer cells. *Comp. Biochem. Physiol. Part C.* 142:53–59.

Kotake-Nara, E., M. Kushiro, Zhang H. T. Sugawara, K. Miyashita, and A. Nagao. 2001. Carotenoids affect proliferation of human prostate cancer cells. *J. Nutr.* 131: 3303–3306.

Kotake-Nara, E., T. Sugiwora, and A. Nagao. 2005. Antiproliferative effect on neoxanthin and fucoxanthin on cultured cells. *Fish. Sci.* 71:459–461.

Lim, B. P., A. Nagao, J. Terao, K. Tanaka, T. Suzuki, and T. Tanaka. 1992. Antioxidant activity of xanthophylls on peroxy radical-mediated phospholipid peroxidation. *Biochem. Biophys. Acta* 1126:178–184.

Lyons, N. and N. O'Brien. 2002. Modulatory effects of an algal extract containing astaxanthin on UVA-irradiated cell in culture. *J. Dermatol. Sci.* 30:73–84.

Maeda, H., M. Hosokawa, T. Sashima, K. Funayama, and K. Miyashita. 2005. Fucoxanthin from edible seaweed, *Undaria pinnatifida*, shows antiobesity effect through UPCi expression in white adipose tissues. *Biochem. Biophys. Commun.* 332:392–397.

Manabe, E., O. Hanada, Y. Naito, K. Mizushima, S. Akagiri, S. Adachi, T. Takagi, S. Kokura, T. Maoka, and T. Yoshikawa. 2007. Astaxanthin protects mesangial cells from hyperglycemia-induced oxidative signaling, *J. Cell Biochem.* Published online, October 22, 2007.

Mannisto, S., S. A. Smith-Warner, D. Spiegelman, D. Albanes, K. Anderson, P. A. van den Brandt, J. R. Cerhan et al. 2004. Dietary carotenoids and risk of lung cancer in a pooled analysis of seven cohort studies. *Cancer Epidemiol. Biomarkers Prev.* 13:40–48.

Maoka, T., K. Mochida, F. Enjo, H. Tokuda, and H. Nishino. 2008. Growth inhibitory effects of paprika carotenoids, capsorubin from human cancer cells. *J. Clin. Biochem. Nutr.* 43 Suppl: 281–284.

Maoka, T., K. Mochida, M. Kozuka, M. Ito, Y. Fujiwara, K. Hashimoto, F. Enjo et al. 2001. Cancer chemopreventive activity of carotenoids in the fruits of red paprika *Capsicum annuum* L. *Cancer Lett.* 172:103–109.

Miki, W. 1991. Biological functions and activities of animal carotenoids. *Pure Appl. Chem.* 63:141–146.

Montonen, J. P., P. Knekt, R. Jarvinen, and A. Reunanen. 2004. Dietary antioxidant intake and risk of type 2 diabetes. *Diabetes Care.* 27:362–366.

Mozaffarieh, M., S. Sacu, and A. Wedrich. 2003. The role of the carotenoids, lutein and zeaxanthin, in protecting against age-related macular degeneration: A review based on controversial evidence. *Nutr. J.* 2:20.

Murakami, A., A. Nakashima, T. Koshiba, T. Maoka, H. Nishino, M. Yano, T. Sumida, K. Kim, K. Koshimizu, and H. Ohigashi. 2000. Modifying effects of carotenoids on superoxide generation from stimulated leukocytes. *Cancer Lett.* 149:115–123.

Murakoshi, M., J. Takayasu, O. Kimura, E. Kohmura, H. Nishino, A. Iwashima, J. Okuzumi et al. 1989. Inhibitory effects of alpha-carotene on proliferation of the human neuroblastoma cell line GOTO. *J. Natl. Cancer Inst.* 81:1649–1652.

Nagaki, Y., S. Hayasaka, T. Yamada, Y. Hayasaka, M. Sanada, and T. Unomi. 2002. Effects of astaxanthin on accommodation, critical flicker fusion, and pattern visual evoked potential in visual display terminal workers. *J. Trad. Med.* 19:170–73.

Naito, Y., K. Uchiyama, W. Aoi, G. Hasegawa, N. Nakamura, N. Yoshida, T. Maoka, J. Yakahashi, and Y. Yoshikawa. 2004. Prevention of diabetic nephropathy by treatment with astaxanthin in diabetic db/db mice, *Bio. Factors* 20, 49–59.

Nara, E., H. Hayashi, M. Kotake, K. Miyashita, and A. Nagao. 2001. Acyclic carotenoids and their oxidation mixtures inhibit the growth of HL-60 human promyelocytic leukemia cells. *Nutr. Cancer* 39:273–283.

Narisawa, T., Y. Fukaura, M. Hasebe, S. Nomura, S. Oshima, and T. Inakuma. 2000. Prevention of *N*-methylnitrosourea-induced colon carcinogenesis in rats by oxygenated carotenoid capsanthin and capsanthin-rich paprika juice. *Proc. Soc. Exp. Biol. Med.* 224:116–122.

Nishino, H. 2008. Cancer prevention by carotenoids. *J. Clin. Biochem. Nutr.* 43 Suppl: 105–107.

Nishino, H., M. Murakoshi, T. Ii, M. Takemura, M. Kuchide, M. Kanazawa, Y. M. Mou et al. 2002. Carotenoids in cancer chemoprevention. *Cancer Metastasis Rev.* 21:257–264.

Nishino, H., H. Tokuda, M. Masuda, X. Y. Mou, Y. Ohsaka, Y. Satomi, M. Murakoshi, M, Yano, and J. Jinno. 2006. Cancer prevention by natural carotenoids. *Carotenoid Sci.* 10:54–58.

Nishino, H., H. Tokuda, Y. Satomi, M. Masuda, P. Bu, M. Onozuka, S. Yamaguchi et al. 1999. Cancer prevention by carotenoids. *Pure Appl. Chem.* 71:2273–2278.

Oalozza, P., N. Maggiano, G. Calviello, P. Lanza, E. Piccioni, F. O. Ranelleti, and G. M. Bartoli. 1998. Canthaxanthin induces apoptosis in human cancer cell lines. *Carcinogenesis* 19:373–376.

Ohgami, K., K. Shiratori, S. Kotake, T. Nishida, N. Mizuki, K. Yazawa, and S. Ohno. 2003. Effects of astaxanthin on lipopolysaccharide-induced inflammation *in vitro* and *in vivo. Invest Ophthalmol. Vis. Sci.* 44:2694–2701.

Oshima, S., F. Ojima, H. Sakamato, Y. Ishiguro, and J. Terao. 1999. Supplementation with carotenoids inhibits singlet oxygen-mediated oxidation of human plasma low-density lipoprotein. *J. Agric. Food Chem.* 44:2306–2309.

Oshima, S., H. Sakamoto, Y. Ishiguro, and J. Terao. 1997. Accumulation and clearance of capsanthin in blood plasma after the ingestion of paprika juice in men. *J. Nutr.* 127:1475–1479.

Palozza, P., G. Calviello, S. Serini, N. Maggiano, P. Lanza, F. O. Ranelletti, and G. M. Bartoli. 2001. β-Carotene at high concentrations induces apoptosis by enhancing oxy-radical production in human adenocarcinoma cells. *Free Radic. Biol. Med.* 30:1000–1007.

Peto, R., R. Doll, J. D. Buckley, and M. B. Sporn. 1981. Can dietary beta-carotene materially reduce human cancer rate? *Nature* 290:201–208.

Sachindra, N. M., E. Sato, H. Maeda, M. Hosokawa, Y. Niwano, M. Kohno, and K. Miyashita. 2008. Radical scavenging and singlet oxygen quenching activity of marine carotenoid fucoxanthin and its metabolites. *J. Agric. Food Chem.* 55:8516–8522.

Satomi, Y. and H. Nishino. 2004. Carotenoids induce the expression of connexin 32 and G1 arrest in human hepatocellular carcinoma cells. *Carotenoid Sci.* 7:46–48.

Satomi, Y. and H. Nishino. 2007. Fucoxanthin, a natural carotenoid, induces G1 arrest and GADD45 gene expression in human, cancer cells. *In Vivo* 21:305–309.

Schweigert, F. J. 1998. Metabolism of carotenoids in mammals. In *Carotenoids Volume 3: Biosynthesis and Metabolism*, ed. G. Britton, S. Liaaen-Jensen, and H. Pfander, 249–284. Birkhäuser, Basel, Switzerland.

Semba, R. D. and G. Dagnelie. 2003. Are lutein and zeaxanthin conditionally essential nutrients for eye health? *Med. Hypoth.* 61:465–472.

Shimizu, N., M. Goto, and W. Miki. 1996. Carotenoids as singlet oxygen quenchers in marine organisms. *Fisheries Sci.* 62:134–137.

Sugawara, T., K. Matsubara, R. Akagi, M. Mori, and T. Hirata. 2006. Antiangiogenic activity of brown algae fucoxanthin and its deacylated product, fucoxanthinol. *J. Agric. Food Chem.* 54:9805–9810.

Sugiura, M., H. Matsumoto, and M. Yano. 2002. Cross-sectional analysis of Satuma mandarin (*Citrus unshiu* Marc.) consumption and health status based on a self-administered questionnaires. *J. Health Sci.* 48:366–369.

Sumantran, V. N., R. Zhang, D. S. Lee, and M. S. Wichia. 2000. Differential regulation of apoptosis in normal versus transformed mammary epithelium by lutein and retinoic acid. *Cancer Epidemiol. Biomark. Prev.* 9:257–263.

Takayanagi, K., R. Nakamura, and K. Mukai. 2006. Mechanism of melanogenesis repression by β-cryptoxanthin. *Carotenoid Sci.* 10:85–90.

Tanaka, T., Y. Morishita, M. Suzui, T. Kojima, A. Okumura, and H. Mori. 1994. Chemoprevention of mouse urinary bladder carcinogenesis by the naturally occurring carotenoid astaxanthin. *Carcinogenesis* 15:15–19.

Terao, J. 1989. Antioxidative activity of beta-carotene-related carotenoids in solution. *Lipids* 24:659–661.

The α-tocopherol β-carotene cancer prevention study group. 1994. The effects of vitamin E and beta carotene on the incidence of lung cancer and other cancers in male smokers. *New Engl. J. Med.* 330:1029–1035.

The ATBC cancer prevention study group. 1994. The alpha-tocopherol, beta-carotene lung cancer prevention study: Design, methods, participant characteristics, and compliance. *Ann. Epidemiol.* 4:1–9.

Tsukui, T., K. Konno, M. Hosokawa, H. Maeda, T. Sashima, and K. Miyashita. 2007. Fucoxanthin and fucoxanthinol enhance the amount of docosahexanoic acid in liver of KKAy obese/diabetic mice. *J. Agric. Food Chem.* 55:5025–5029.

Tsushima, M., T. Maoka, M. Katsuyama, M. Kozuka, T. Matsuno, H. Tokuda, H. Nishino, and A. Iwashima. 1995. Inhibitory effect of natural carotenoids on Epstein–Barr virus activation activity of a tumor promoter in Raji cells. A screening study for antitumor promoters. *Biol. Pharm. Bull.* 18:227–233.

Trevithick-Sutton, C. C., C. S. Foote, M. Collinus, and J. R. Trevithick, 2006. The retinal carotenoids zeaxanthin and lutein scavenge superoxide and hydroxyl radicals: A chemiluminescence and ESR study. *Mol. Vision* 12:1127–1135.

Uchiyama, K., Y. Naito, G. Hasegawa, N. Nakamura, J. Takahashi, and T. Yoshikawa. 2002. Astaxanthin protects beta-cell against glucose toxicity in diabetic db/db mice. *Redox. Rep.* 7:290–293.

Wada, T., S. Ito, Y. Yuasa, and T. Maoka. 2007. Analysis of carotenoids in human serum by high performance liquid chromatography (in Japanese). *Food Clin. Nutr.* 2:15–26.

Wingerath, T., H. Sies, and W. Stahl. 1998. Xanthophyll esters in human skin. *Arch. Biochem. Biophys.* 355:271–274.

Yamaguchi, M. and S. Uchiyama. 2003. Effect of carotenoid on calcium content and alkaline phosphates activity in rat femoral tissues *in vitro*: The unique anabolic effect of β-cryptoxanthin. *Biol. Pharm. Bull.* 26:1189–1191.

Yamasaki, H. 1990. Gap junctional communication and carcinogenesis. *Carcinogenesis* 11:1051–1058.

Yamashita, E. 2006. The effects of a dietary supplement containing astaxanthin on skin condition. *Carotenoid Sci.* 10:91–95.

Yoshida, T., T. Maoka, S. K. Das, K. Kanazawa, M. Horinaka, M. Wakada, Y. Satomi, H. Nishino, and T. Sakai. 2007. Halocynthiaxanthin and priedinin sensitize colon cancer cell lines to tumor neurosis factor-related apoptosis-inducing ligand. *Mol. Cancer Res.* 5:615–625.

Yoshioka, R., T. Hayakawa, K. Ishizuka, A. Kulkarni, Y. Terada, T. Moaka, and H. Etoh. 2006. Nitration reaction of astaxanthin and β-carotene by peroxynitrite. *Tetrahedron Lett.* 47:3637–3640.

Yuan, J. M., R. K. Ross, X. D. Chu, Y. T. Gao, and M. C. Yu. 2001. Prediagnostic levels of serum beta-cryptoxanthin and retinal predicts smoking-related lung cancer risk in Shanghai, China. *Cancer Epidemiol. Biomarkers Prev.* 10:767–773.

Zhang, L. X., R. V. Cooney, and J. S. Bertram. 1991. Carotenoids enhance gap junctional communication and inhibit lipid peroxidation in C3H/10T1/2 cells: relationship to their cancer chemopreventive action. *Carcinogenesis* 12:2109–2114.

Zhang, L. X., R. V. Conney, and J. S. Bertram. 1992. Carotenoids up regulated connexin 43 gene expression independent of their provitamin A or antioxidant properties. *Cancer. Res.* 52:5707–5712.

Ziegler, R. G. 1989. A review of epidemiologic evidence that carotenoids reduce risk of cancer. *J. Nutr.* 119:116–122.

5 Traditional Chinese Medicated Diets

John Shi, Yueming Jiang, Xingqian Ye, Sophia Jun Xue, and Yukio Kakuda

CONTENTS

5.1 INTRODUCTION

The Chinese are proud of their heritage and the ingenuity of their early scientific and cultural developments. One of their most remarkable contributions to civilization undoubtedly is the abundance of information on and descriptions of the uses of natural substances, plants, chemicals, and animals in treating illnesses. Health and "healing" foods have a long history in Chinese culture, where food and medicine are considered to be equally important in preventing and curing diseases. China has long had the concepts "food can treat illness and build up life force" and "food and medicine come from the same source" (Li, 1578; Beijing Traditional Chinese Medical College and Hospital, 1981; Huang, 1993). Because Western medicine is sometimes limited in effectiveness against chronic illnesses, help is being sought in traditional concepts of preventive medicine, using medicinal herbs to restore balance to the body rather than turning to pharmaceuticals. Chinese medicated diets have a long history as one application of traditional Chinese medicine. Traditional Chinese medicine and diet both originate from practice and experience in daily life. Dietetic therapy is always considered as the first choice for most chronic diseases rather than chemical medicine (Lee, 1990; Chen, 1997; Ma, 2002).

Traditional Chinese medicine practitioners consider that foods and medicine are based on the same basic theories, and have the same uses. Chinese medicated diets are foods that have unique effects on human physiology and health such as maintaining and improving health status, to prevent and help in treating diseases or to facilitate rehabilitation. Many unique traditional Chinese medicated diets or foods were developed by combining food materials with medicinal herbs. Western medicine and Chinese traditional medicine vary in practice, theory, and medication choices. Western medicine uses pure natural chemicals or synthetic compounds aimed at a single target, while traditional Chinese medicine uses processed crude multicomponent natural products in various combinations and formulations, aimed at multiple targets to treat a totality of different symptoms (Shen, 1999).

Chinese medicated diets (or medicinal cuisine) are also popular in Korea, Japan, and Southeastern areas such as Vietnam and other countries. Based on traditional Chinese herbal medicine practice, it combines strictly processed traditional Chinese medicine with traditional culinary materials to produce delicious foods with health-restoring qualities. In China, people contend that a food tonic is much better than a medicine tonic in fortifying one's health. Chinese medicated diets have been one of the important branches and unique applications of traditional Chinese medicine during its evolution over the past 2000 years.

5.2 HISTORY OF CHINESE MEDICATED DIETS

Chinese medicated diets have a long history and are an important composite part of the theoretical system of traditional Chinese medicine. The ancient legend *Shennong Chang Bai Cao* (*Shennong* tastes a hundred grasses) from about 2800 BC shows that early in remote antiquity the Chinese nation began to explore the function of foods and medicaments, hence the saying "Traditional Chinese medicine and diet both originate from the practice and experience in daily life" (Li, 1578; Huang, 1993). In the Zhou Dynasty (11–256 BC), royal doctors were divided into four kinds, one of which was a dietetic doctor, who was in charge of the emperor's health care and health preservation, and was responsible for preparing diets for the emperor. In the book *Huangdi Nei Jing* (*Yellow Emperor's Internal Classic*), a medical classic in traditional Chinese medicine that appeared in approximately 100 BC, several medicated diet prescriptions were recorded. The book *Shennong Bencao Jing* (*Shennong Emperor's Classic of Materia Medica*) was published in approximately the Qin and Han periods (221 BC to 8 AD) and is the earliest monograph on materials for medicated diets, from which many sorts of medicinal herbs and foods such as Chinese date (*Fructus Ziziphi Jujubae*), sesame seed (*Semen Sesami*), Chinese yam (*Rhizoma Dioscoreae*), grape (*Vitis*), walnut kernel (*Semen Fuglandis*), lily bulb (*Bulbus Lilii*), fresh ginger (*Rhizoma Zingiberis Recens*), and Job's tears seed (*Semen Coicis*), among others, were recorded. In the book *Shanghan Zabing Lun* (*Treatise on Febrile and Miscellaneous Diseases*) written by Zhang Zhongjing, a famous medical doctor in the Eastern Han Dynasty (25–220 AD), some medicated diet recipes such as "Soup of Chinese Angelica root," "Fresh ginger and mutton" (*Danggui Shengjiang Yangrou Tang*), "Decoction of Pig-skin" (*Zhufu Tang*), and the like were recorded for their important values in treating diseases (Wu, 1982; Wang, 1983; Zhou, 1986; Huang, 1993).

Li Chi and Sun Shiang, well-known Chinese medical doctors in the Tang Dynasty (618–907 AD), listed and discussed dietetic treatments and medicated diet prescriptions for senile health care and health preservation in their famous books *Beiji Qianjin Yaofang* (*Essential Prescriptions Worth a Thousand Pieces of Gold for*

Emergencies) and *Qianjin Yifang* (*Additions to the Prescriptions Worth a Thousand Pieces of Gold*). According to these history books, up to the period of the Sui and Tang Dynasties (581–907 AD), more than 60 kinds of books on dietetic treatments had been published.

Writing on nutrition also flourished in the Tang Dynasty. Meng Seng, a prominent physician, wrote the three-volume "Diet *Bencao*," which was preserved in the west part of China. The book "Diet *Bencao*" (*Dietotherapy of Materia Medica*) by Meng Song covers many important methods of identifying varieties and characteristics of plants that are nutritious and able to improve health and life. This book also describes in detail the preparation of medicated diets and diseases due to malnutrition or deficiency of certain food ingredients. The book had a great influence on later generations. It is the earliest monograph on dietetic treatments (Szechuan Medical College, 1978; Ling et al., 1984).

In the Song Dynasty (960–1279 AD), Wang Huaiyin discussed medicated diet treatments for many diseases. The book *Helping the Old Preserve Health and Your Kith and Kin Prolong Lives* by Chen Zhi is an early monograph on gerontology in China. In this book, Chen Zhi emphasized that "dietetic therapy should go first for any senile diseases, and then be followed by medicine if they are not cured" (Li, 1578; Hsu and Peacher, 1976; Ho, 1993, Zhang, 1997).

In the book *Principles of Correct Diet*, a monograph on medicated diet by Hu Sihui, a royal doctor in the Yuan Dynasty (1271–1368 AD), large quantities of medicated diet prescriptions were recorded. Some questions such as diet contraindications for pregnancy, for wet nurses, and for drinking, among others, were also discussed in the book.

In the Ming Dynasty (1368–1644 AD), the most important work on Chinese medicine written in this period was the "Ben Cao Kong Mu," recognized worldwide as an outstanding work. The author was Li Shizhen (1518–1593 AD). By studying hard and reading every book he could obtain, Li Shizhen became a well-known physician and his exact diagnoses and treatments were widely regarded. Li Shizhen made a significant improvement in the classification of medicinal herbs listed in his book. He categorized them according to biological terminology. Li Shizhen divided the total numbers of medicinal herbal materials into 16 groups and more in subgroups such as "water," "fire," "earth," "gold-mineral," "plants," "vegetables," "fruits," "wood," and "animals," among others. He also made diet prescriptions on medicated gruel and medicated wine. In "Eight Essays on Life Preservation," a monograph on preserving health in the Ming Dynasty, many medicated diets on health preservation and health care were recorded.

Monographs on medicated diet treatments in the Qing Dynasty (1644–1911 AD), with over 300 species belonging to seven phyla of medicated food and drinks, were introduced in the book *Recipes of Suixiju* by Wang Shixiong. In the book *Analysis of Food and Drinks for Treatment of Diseases* by Zhang Mu, more medicated foods were recommended. In the *Cookbook of Suiyuan*, cooking principles and methods were introduced. In the book *Common Sayings for Senile Health Preservation* by Cao Tingdong, about 100 medicated gruel prescriptions for geriatric treatments were listed (Quinn, 1972; Wu, 1982; Wang, 1983; She-Yue, 1984; Zhou, 1986; Yin, 1987; Lang, 1988).

5.3 CHARACTERISTICS OF CHINESE MEDICATED DIETS

Chinese medicated diets are not a simple mixture of food and medicinal herbs, but special diets prepared from Chinese medicinal materials such as herbs and food as well as other condiments under the theoretical guidance of diet preparation based on differentiation of symptoms and signs of traditional Chinese medicine. It addresses not only the efficacy of medicine but also delicate foods, and has the function to prevent and cure diseases, build up one's health, and prolong one's life. Chinese medicated diets have combination effects in preventing and curing diseases.

Medicated diets can usually be prepared from the combination of Chinese medicinal materials and food according to certain prescriptions by processing and cooking. Natural food materials for dietetic Chinese diets include cereals, fruits, nuts, vegetables, seasonings, birds and animals, aquatic products, and so on. Based on the different preparation processes, the forms of the medicated diets can be as medicated teas, wines, gruels, honey extracts, cakes, pancakes, soups, or an entire meal.

The body's internal balance and immune system are intimately connected. Moreover, nutrition strongly affects the immune system's capabilities. An unbalanced diet impairs the effectiveness of immune cells, making the body more vulnerable to bacterial and viral invasions and to chronic ailments. Traditional Chinese medicine doctors usually first make an overall analysis of the patient's physical and health condition, the nature of his illness, the season that he got ill, geographical information, and so on, to form a judgment on the type of syndrome, and then make a decision based on corresponding principles for dietetic therapy and select suitable medicated diets. By the principle of "laying stress on the whole, selecting medicated diets on the basis of differential diagnosis," the prescription of medicated diets is to stress as a whole, selecting proper medicated diets (Hsu and Peacher, 1976; Zhang, 1990; Lu, 1991; Ho, 1993; Zhang, 1997).

Although medicated diets usually show mild effects on diseases, they have a positive effect on their prevention and cure without any side effects. Therefore, many individuals use medicated diets to treat chronic diseases (as senior people) or to build up their health and prevent diseases as healthy people, with outstanding results. This is one of the characteristics by which medicated diets are different from treatments provided by Western-style chemical medicines.

Most of the medicinal herbal materials used in medicated diets are both edible and medicinal, and retain features of food such as color, sweet-smelling nature, and natural flavor. Tasty medicated diets are made by mixing herbal material with food and carefully cooking them. Medicated diets are also convenient to ingest, for both children and adults.

5.4 FORMULA AND PREPARATION OF CHINESE MEDICATED DIETS

The formulas for Chinese medicated diets are composed of several medicinal herbs at suitable dosage, some food materials selected on the basis of definite diagnosis, and a therapy chosen by means of syndrome differentiation. All these mixtures are cooked into a delicate diet. Mixtures of medicinal herbal and food materials have cooperative effects. On the basis of interaction of herbs, it can increase curative effects through additive and synergistic effects, and decrease toxic and side effects through mutual restraint and antagonism. The design of a formula for Chinese medicated diets is guided by traditional Chinese medicine principles based on syndrome differentiation. Some herbs are chosen to constitute them in a certain pattern of forming the formula of Chinese medicated diets, so that they can complement each other to reduce toxicity and side effects and to increase the efficacy of herbs. A formula for Chinese medicated diets can be modified according to the status of disease, constitution, sex, and age, so it can be flexible and suitably accurate to treat complicated and changeable diseases.

In the formula for a Chinese medicated diet, each herbal material is termed as "principal herb," "assistant herb," "adjuvant herb," or "dispatcher herb" (Li, 1578; Huang, 1993; Lu, 1994). Their meanings are given below:

"Principal herb" material. Herbs that are used to produce leading effects that dominate the disease and control the main symptom of a disease are termed "principal herb" or "imperial herb" material. These herbs are also called "monarch herbs" in traditional Chinese medicine books. They are the chief herbal materials (main ingredients) of a formula.

"Assistant herb" material. Herb materials that help to increase the treatment effects of principal herbs are termed "assistant herbs." They are also called "minister herbs" in traditional Chinese medicine books.

"Adjuvant herb" material. Herb materials that are opposite and complementary to "assistant herbs" (minister herbs) are called "adjuvant herbs" in traditional Chinese medicine books. "Adjuvant herbs" can be divided into three types. The first one assists "principal herbs" and "assistant herbs" in strengthening their therapeutic effects or by treating the accompanying symptoms. The second one restrains or eliminates drastic actions or side effects of "principal herbs" and "assistant herbs," and the toxic effects of some herbs. The third type deals with possible vomiting due to some potent effect. It possesses properties opposite to "principal herbs" in compatibility, but creates supplementing effects in the treatment of diseases and controls the main symptoms of a disease (Zhang et al., 1988; Lu, 1994).

"Dispatcher herb" material. Herbs that follow the instructions of "minister herbs" are known as "dispatcher herbs." "Dispatcher herbs" can be divided into two types. The first one is known as herbs for meridian guiding, leading other herbs in the formula to the affected site. The second one is called "mediating herbs," and serves to coordinate all effects of various ingredients in the formulas of Chinese medicated diets.

"Principal herb" material is absolutely necessary in a formula of Chinese medicated diets, but whether or not "assistant herbs," "adjuvant herbs," or "dispatcher herbs" are included depends on the disease state and the herbal properties, and on the respective importance of effect, potency, and dosage in a formula. For example, the "principal herbs" or "assistant herbs" in some formulas have determination actions comprised of the effects from "adjuvant herbs" or "dispatcher herbs." Under such conditions, the "assistant herb" or "adjuvant herb" material need not be used in the formula for Chinese medicated diets. It is difficult to distinguish "principal herbs," "assistant herbs," "adjuvant herbs," and "dispatcher herbs" in a complex formula of

Chinese medicated diets with many medicinal ingredients. In a mixture of "principal herbs," "assistant herbs," "adjuvant herbs," and "dispatcher herbs" in any given formula, "principal herbs" are always selected as the main ingredient in large dosages with great potency, while "assistant herbs" and "adjuvant herbs" are usually used as medicinal ingredients in smaller dosages (Zhang, 1990, 1997; Lu, 1991, Ho, 1993).

The dosage of each ingredient in the formula for a Chinese medicated diet is based on the disease, patient, and symptom conditions such as the mild, serious, chronic, or emergent states of illness, the age, sex, diathesis, occupation, and birth place of the patient, the season in which the disease occurs, and the weather change, among others, so as to match the illness and enhance its therapeutic effects. Therefore, the modification of formulas of Chinese medicated diets is a normal practice for Chinese medicine doctors, adjusting for an individual patient's condition, including modification of herb varieties, of dose, and of preparation and cooking processes. Modification of herb varieties includes selecting corresponding "assistant herbs" and "adjuvant herbs" in accordance with variations in the accompanying symptoms presented without any change in the main symptoms and the "principal herbs" chosen to meet the conditions of the illness. Modification of dose is an increase or decrease in the dose quantity of herbs in the same formula of Chinese medicated diets, which will change the functionality of the formula. Modification of preparation and cooking procedures is also important because the same formula with different preparation and cooking procedures will change the therapeutic effects (Zhang, 1990; Ho, 1993; Su, 1993; Weng and Chen, 1996).

Formulas for Chinese medicated diets have been classified by disease and syndrome, formula structure, effects, causes of the disease, viscera, main formula, clinic disciplines, and preparation methods. According to Cheng Wuji of the Jin Dynasty in his book *Forward of Concise Expositions on Cold Attack*, all formulas can be classified into seven categories: "heavy," "light," "mild," "urgent," "odd," "even," and "complex."

The "heavy" formula of Chinese medicated diets includes more varieties of herbal material and a large dosage of herbal material. The "heavy" formula of Chinese medicated diets is capable of treating severe disease. The "light" formula of Chinese medicated diets is used to treat mild disease cases. The "mild" formula of Chinese medicated diets has a moderate action for a debilitating case with a chronic disease. The "odd" formula in Chinese medicated diets usually has a prescription containing one single herbal material ingredient or odd-numbered ingredient mixtures. The "even" formula of Chinese medicated diets has a prescription with two herbal ingredients or even-numbered ingredient mixtures. The "complex" formula of Chinese medicated diets has a combination of "odd" and "even" formula recipes. The formulas of Chinese medicated diets can also be classified into 22 categories, including formulas for tonification, exterior-relieving, emetic, interior-purging both diaphoretics and purgative mediation, activating energy, promoting blood circulation, dispelling wind, cold-dispelling, summer heat clearing, dampness-dispelling, dryness-moistening, fire-purging, phlegm-resolving, digestive retention-relieving, astringency, expelling or antihelminthic, eye-sight improving, treating external carbuncles, and menstruation-regulating, among others, according to the effects on the disease and its symptoms. This classification is moderate and applicable. Modern textbooks of Chinese herb formulas are mostly based on this kind of classification (Beijing Medical College Chinese Herbal Medicine Research Group, 1959; Ho, 1993; Su, 1993; Zhang, 1997).

5.5 BASIC THEORIES OF CHINESE MEDICATED DIETS

5.5.1 Yin–Yang Theory

For thousands of years, traditional Chinese medicine has made use of a myriad of herbs with great success. The basic theories of traditional Chinese medicine arose from the yin–yang concept. Nearly all symptomatic diagnoses are based on the philosophy of the "yin" and "yang," the two forces that, traditional Chinese medicine believes, control the working of the universe (Yu, 1999; Williams, 1995).

The yin represents the feminine side of nature, encompassing darkness, tranquility, depth, cold, and wetness. The earth, the moon, and water are all yin elements. The yang represents the masculine principle, encompassing light, activity, height, heat, and dryness. The heaven, the sun, and fire are yang elements. By the opposition of yin–yang, all things and phenomena in the natural world contain the two opposite components, for example the heaven and the earth, and cold and hot. The yin and yang not only oppose but also contain each other; without the other, neither can exist. For instance, there would be no earth without heaven, and vice versa. Without cold, there would be no heat, and vice versa. This relationship of coexistence is known as interdependence. So the yin and yang oppose each other and yet depend on each other. Although the yin is commonly interpreted to be a negative force and the yang represents a positive force, the philosophy of the yin and yang does not parallel Western philosophy's dualism between good and evil. Instead, the two are complementary. Thus, Chinese medical treatments attempt to reach a balance between the yin and yang in a patient's body (Huang, 1979; You, 1996; Zhang, 1997).

Traditional Chinese medicine considers people's health condition as a balance between the yin and yang inside the body. Traditional Chinese medicine attempts to consider a person's body as a whole, and uses external phenomena to figure out what is going on inside the human body. The main principle of traditional Chinese medicine treatments is to establish a holistic balance of such forces in the human body and to promote health. Some specific traditional Chinese medicine treatments are based on specific diagnostic observations and systematic principles that are originally from yin–yang theory. Traditional Chinese medicine doctors believe that all diseases are due to an imbalance of yin and yang in a person's body. If a patient has a fever, for example, extra yang in the human body, a yin (cooling) medicine herbal material should be used to reach a yin–yang balance. If the patient has a yin (cooling) problem, a yang (warming) medicine herbal material must be used (Lu, 1991; Ho, 1993; You, 1996). Chinese philosophy focuses mainly on developing universal harmony by using a balanced approach. According to traditional Chinese medicine, the human body is an integrated whole; hence treatment is based on the concept of an integral human body and is decided through differentiation of symptoms and signs.

5.5.2 Food Materials in Traditional Chinese Medicine by Properties

How does the concept of yin and yang relate to food? A basic adherence to this philosophy can be found in any Chinese dish, from stir-fried beef with broccoli to sweet and sour pork. There is always a balance in color, flavor, and texture. However, belief in the importance of following the principles of yin and yang in the diet extends further. Associated with the yin and yang theory, the terms "coolness" and "heat" are conditions that are generally believed to stem from types of foods that affect the human body in different ways. "Heat" refers to a condition in which the body consumes a large amount of energy, thus generating heat. Certain foods are thought to have yin or "cooling" properties, while others have warm, yang properties. The challenge is to consume a diet that contains a healthy balance between the two. When treating illnesses, an oriental physician will frequently advise dietary changes in order to restore a healthy balance between the yin and yang in the body. For example, let's say you're suffering from heartburn, caused by consuming too many spicy (yang) foods. Instead of antacids, you're likely to take home a prescription for herbal teas to restore the yin forces. Similarly, coughs or flu are more likely to be treated with dietary changes with "hot" (yang) food rather than antibiotics or cough medicines.

"Hot" foods include nearly all fried foods, red meats, and vegetables and fruits with high fat content such as avocado and the durian fruits of Southeast Asia. However, fruits such as lychee and mangoes are also considered to be "hot," despite their refreshing nature. "Cooling" foods include most fruits and vegetables, and many herbs. For a yin–yang balance in a patient's body, excessive "hot" in the human body requires treatments with "cooling" herbs, while excessive "coolness" in the human body requires treatments with "heating" herbs. Imperial traditional Chinese medicine doctors emphasized the "attainment of nourishment by selecting appropriate food in a somewhat philosophical way" (Ho, 1993). Accordingly, "appropriate food" means both a moderate intake of food and variety in the diet.

In the book *The Yellow Emperor's Internal Classic*, various sources in the diet were recommended for healthy intake. According to the book, some cereals and legumes such as rice, sesame seeds, soybeans, wheat, and millet would provide nourishment; five fruits (date, plum, chestnut, apricot, and peach) would produce complementarity; five animals (beef, dog meat, pork, mutton, and chicken) would give advantage; and five vegetables (marrow, chives, bean spouts, shallot, and onion) were for supplementarity (Ho, 1993). Such a concept of "a balanced and complete diet" in traditional Chinese medicine can be compared to theories of balanced dietary energy and nutrients in modern nutritional science.

In many cases, Chinese medicinal herbs, *Panax ginseng* for example, may contain dozens of different constituents. Thus, their actions are so diverse and multitudinous that we classify them in one category (Liu and Zhung, 1998; Yun and Choi, 1998; Lee et al., 1999; Wu and Zhong, 1999). In view of the numerous actions of some herbs, they have multiple therapeutic uses. Some of the principal ingredients produce a similar pharmacodynamic effect although they are structurally different. A possible synergistic effect suggests that the herb itself might be more potent than a purified single compound.

5.5.2.1 Food Materials and Medicinal Herbs with "Cooling" and "Cold" Properties

According to the yin–yang theory, traditional Chinese medicine considers the following food material and medicinal herbs to have "cooling" and "cold" effects on the human body, based on the yin property (Ling et al., 1984; Ho, 1993; Zhang, 1993; Weng and Chen, 1996; Zhang, 1997; Zhang et al., 1998).

Cereals: Barley, millet, buckwheat, Job's tears, mung bean, south wheat, proso millet, hyacinth bean, black bean, garden pea, red bean, and green bean.

Fruits: Apple, pear, orange, banana, litchi, pineapple, green orange, sweet orange, fig, strawberry, crab apple, mango, carambola, mume, watermelon, loquat, kumquat, lemon, musk melon, olive, kiwi, persimmon, sugarcane, and hawthorn.

Nuts and seeds: Almond, walnut, chestnut, pine nut, watermelon seed, pumpkin seed, black sesame, white sesame, and bitter apricot nut.

Vegetables: Arrowhead, celery, spinach, lettuce, turnip (white), lotus seed, lotus root, lily bulb, bamboo shoots, eggplant, tomato, winter melon, cucumber, sponge gourd, water chestnut, leather vegetable, greens, radish, asparagus, wax gourd, towel gourd, spinach, mung bean sprouts, purslane, bitter vegetable, bitter green, bitter gourd, black fungus, white fungus, and yam.

Meats: Antler, rabbit meat, frog meat, duck meat, donkey meat, pork liver, pork skin, pork kidney, goose blood, horse flesh, chicken liver, goat liver, duck blood, duck egg, egg white, and pigeon egg.

Aquatic products: Oyster, kelp, laver (seaweed), fragrant salmon, frog, snail, laurel fish, rubber fish, glass fish, sea turtle, squid fish, octopus, scallop, loach, turtle, abalone, black fish, field snail, clam, red laver, star jelly, river crab, sea crab, and carp.

Herbals: Acanthanccous, lucerne, bamboo leaf, bitter orange, coix seed, eclipta, fragrant solomon seal rhizome, honeysuckle flower, Indian bread, white peony root, mulberry leaf, isatis leaf, subprostrate, sophora root, dansui root, knocia root, genkwas flower, glossy privet fruit, indigo wood root, kudzu vine root, rhubarb root, mirabitite, pseudostellaria root, glehnia root, ophiopogon root, gold thread, dendrobium, rehmannia root, wild rice stem, chrysanthemum flower, American ginseng, duckweed, cicada molting, polyporus, platycodon, gastrodia tuber, spiny jujube, cimicifuga, bupleurum, capillaris, gromwell, Mongolian dandelion, moutan, alisma, fangji, gentian, fritillaria, uncaria, leonurus, *Redix rehmania*, scrophularia, anemarrhena, bezoar, purslane, philodendron, forsythia, dandelion, gardenia fruit, wormwood, cynanchum root, prunella, morning glory, isatis root, thichosanthes, gypwum, gentian, coptis, antelope horn, tea, and coffee.

5.5.2.2 Food Materials and Medicinal Herbs with "Warm" and "Hot" Properties

According to the yin–yang theory, traditional Chinese medicine considers that the following food material and medicinal herbs have "warm" and "hot" effects on the human body based on the yang property (Weng and Chen, 1996).

Vegetables: Ginger, onion, fennel green, garlic, chives, potato, pumpkin, and leek.

Fruits: Coriander, plum, peach, pineapple, red bayberry, cherry, longan, apricot, Chinese date, and pomegranate.

Cereals: Glutinous rice.

Meats: Chicken, dog meat, deer meat, mutton, and turkey.

Aquatic products: Grass carp, trout, sea slug, silver carp, and sea cucumber.

Herbals: Angelica, *Astragalus membranaceus*, cassia seed, cassia bark, copper leaf, dogwood fruit, Korea ginseng, pepper, pilose, curculigo rhizome, rhizome of rehmannia, mustard green, epimedium, chili, and schisandra fruit.

5.5.2.3 Food Materials and Medicinal Herbs with "Neutral" Properties

The concept of "neutral" and tonic foods further supports the yin–yang principle in traditional Chinese medicine. Foods that are between "cold" and "hot" are considered "neutral," while foods that are used for strengthening and nourishment are "tonic." Usually "cold" food is cooked or consumed together with some "hot" foods for the purpose of neutralization. According to the yin–yang theory, traditional Chinese medicine considers the following foods and medicinal herbs to have "neutral" effects on the human body, based on the yin–yang property (Ling et al., 1984; Ho, 1993; Weng and Chen, 1996).

Herbals: Dangshan, *Plastrum testudinis*, rhizome, licorice root, poria, and wild juba seed.

Vegetables: Rice, pea, carrot, daylily, olive, wheat, red bean, mushroom, potato, yellow croaker, corn, cabbage, soybean, cauliflower, and peanut.

Ingredients: Oolong tea, jasmine tea, honey, and white sugar.

Fruits: Grape, cherry, and apple.

Meats: Beef, quail meat, goose meat, quail egg, pork, pigeon meat, spring chicken, and donkey-hide gelatin.

Aquatic products: Eel and jellyfish.

5.6 CHINESE MEDICATED DIET PREPARATION AND SOME CLASSIC FORMULAS

Chinese medicated diets can be prepared, according to the doctor's advice, from edible Chinese herbs and foods by processing and culinary skills such as stewing, braising, simmering, steaming, boiling, cooking in water, stir-frying, roasting, fricasseeing, and deep-frying, among others. To cook medicinal foods, one has a large variety of fine materials to choose from, and each material has its own unique flavor. Generally, processed herbal materials are more commonly used in order to avoid strong odors. However, individuals of different physical status need to select different herbs. The selection of herbs will depend on each individual's condition of health, according to the doctor's prescription.

In the preparation of Chinese medicated diets, the "cooking" procedure is a very important are, often to cook a mixture of Chinese medicinal herbs and food materials by boiling for a certain time. Some popular methods are making gruel (also called congee), soup, or meal through direct cooking or decoction, and then cooking the mixture of herbal and food materials together. In the direct cooking process, the mixture of medicinal herbs and foods along with sufficient water is cooked directly. In the decoction procedure, the herbal material is decocted first with sufficient water, then the solution is separated from the cooked residue, and the decoction solution is mixed with the food and cooked further to obtain the final diet foods.

The selected herbs are soaked in an appropriate amount of water, and then cooked together with selected food materials. After the cooking procedure, health-promoting components from Chinese medicated diets are particularly easily absorbed, quickly effective, and easily modified to meet the needs of the diseases. Most people can prepare Chinese medicated diets in their home. For effective absorption, the preparation technology for Chinese medicated diets is related to several factors (Weng, 1999; Wu and Zhong, 1999; Yeung, 1985).

5.6.1 Selection of Materials

The medicinal herbs and food materials used for medicated diets should be chosen carefully. The chosen items must be clean, pure, and dustless, and none of them can be mildewed or rotten. For example, Chinese dates are selected as big, purplish red, with plenty of flesh, smooth, and not moth-eaten; otherwise, they are of poor quality and should not be used. Wolfberry fruit (*Frutus Lycii*) is another example; large and soft fruits with plenty of flesh, few seeds, and red color should be selected.

5.6.2 Cooking Container

A cooker made of silverware is the best. A cooker made of chinaware is also good. Tin or iron cooking pots are forbidden because some components from the medicinal herbs may produce sediment, lower solubility, or chemical reactions.

5.6.3 Water

The water should be pure and clean. Well water or spring water is often chosen. Sometimes rice-wine (yellow wine) is used as a cooking solution; distilled water can also be used.

5.6.4 Fire

The fire for cooking the mixture of medicinal herbs and food material is of two types: soft fire and strong fire. A soft fire makes the temperature rise slowly and the mixture to boil gently. A strong fire is used to make the cooking temperature rise quickly and make the mixture boil vigorously. The strong fire is usually used first, and the soft fire is applied after boiling.

5.6.5 Cooking Technology

Cooking procedure must be taken into consideration to prepare good medicated diets. Besides the color, the fragrance, taste, and shape that the food is commonly cooked in dish. When medicated diets are being cooked, attention must be paid to retaining their nutritional and effective constituents as much as possible, in order to bring into full play their functions for treatment and health care. The main purpose of cooking medicated diets with the proper condiments is to maintain the special properties of medicinal herbs so that their nature and flavor might be closely combined. As a result, medicated diets have good color, fragrance, taste, and shape, thus arousing the appetite; also, the treatment of the disease and health care are brought into play. Some edible medicinal herbs with no unpleasant flavor (or their fine powder) can be cooked together with the food. If some herbs have distinctly improper flavors, they can be wrapped in gauze and cooked along with the food, and the property of the herbs can go into the foods or the soup. The residue must be strained out before consumption.

The herbs are first put into a pot. Water is added until the herbs are submerged. This is soaked for 30 min. Then the herb–food mixture material is placed in the cooker and the cooking is started. In this way, the effective components of herbs can be extracted. First, a strong fire is used until the mixture boils, and then a soft fire is applied. Meanwhile, the pot should be covered with a lid so as to minimize the loss of volatile components and their therapeutic effects. Tonic herbs with strong flavors are usually cooked with a soft fire so that their effective components may be released completely.

Some minerals or shell material such as tortoise plastron, turtle shell, red ochre, shell of abalone, oyster shell, dragon's bone, magnetite, and gypsum are so hard that their effective components cannot be extracted easily. They should first be crushed and then boiled for 10–20 min before other herbs are placed in the cooker. Some crude herbs contaminated with sand and mud, glutinous rice roots, some herbs with small weight but large bulk such as reed rhizome, cogongrass rhizome, prunella spike, and bamboo shavings should be decocted and separated from the precipitate first in order to obtain a clear solution; then the solution can be mixed with food material for further cooking. Some aromatic herbs containing volatile oils should be cooked with the all herb–food mixture by a soft fire in order to prevent the volatilization of their active components. Some herbs such as red hollysite, talc, or inula flower should be wrapped in a thin cloth in the herb–food mixture to avoid causing a turbid solution. Some of the gluey, viscous, easily dissolved herbs, such as donkey-hide gelatin, antler gelatin, honey, and malt sugar, should be cooked and melted separately and then blended with the herb–food mixture after the mixture has finished cooking.

5.6.6 Types of Chinese Medicated Diets

There are many types of Chinese medicated diets. Most of them are prepared as meal, cake, soup, congee, dumpling soup, and sweets, among others.

5.6.6.1 Medicated Cakes

This is a cooked food prepared by first grinding edible medicinal herbs into a powder, then mixing the powder with rice, wheat, or bean flour, or a proper amount of white sugar and cooking oil, and finally making the cakes with the mixture by either steaming or baking in a pan or in an oven, or by frying. Take an "eight-ingredient cake" for example. It can strengthen the stomach of patients suffering from chronic gastritis of cold or insufficiency type, or a peptic ulcer.

5.6.6.2 Cooked Dishes

There are hundreds of medicated diets (including varieties of meat and vegetable dishes) that are curative and can be used for health care. They are prepared by cooking chicken, duck, fish, or vegetables along with herbs and condiments. They can be stewed, braised, simmered, steamed, boiled, cooked in water, stir-fried, roasted, fricasseed, or deep-fried. For example, "Baby Pigeon Stewed with Ginseng, Astragalus Root and Cordyceps" can be used to treat bronchial asthma at the remission stage and "Turtle Stewed with Chinese Angelica Root and Wolfberry Fruit" can be used to treat early cirrhosis. "Baby Pigeon Stewed with Gouqi (Medlar) and Huangqi (Membranous Milk Vetch)," "Pork Simmered with Lotus Seed and Lily," and "Pig's Kidney Stewed with Eucommia Bark" are typical medicine cuisines used to treat the most common chronic diseases.

5.6.6.3 Honey Extract

Honey extract, also known as soft extract or decoction extract, is a thick half-liquid prepared by first decocting edible medicine herbs in water and, after enriching the liquid, mixing in honey or sucrose. For example, the beverage "Flavored Extract of Fritillary Bulb and Pear" is applicable to patients with bronchial asthma.

5.7 SOME CHINESE MEDICATED DIETS WITH A SPECIFIC PURPOSE

5.7.1 Prevention of Aging

According to traditional Chinese medicine, prevention of the aging process involves nourishing yin, to facilitating blood circulation, and eliminating excessive yang (Figure 5.1). Chinese medicated diets that have functionality in prevention of the aging process are usually selected food materials and medicinal herbs from the following natural sources.

5.7.1.1 Food Materials

These include Chinese yam, lotus seed, lotus root, tremella, pork skin, donkey-hide gelatin, beef, deer tail, wood ear, quail meat, quail egg, hawthorn, lily bulb, green tea, konjac, celery, pear, coriander, black sesame, honey, and mulberry fruits.

5.7.1.2 Medicinal Herbal Materials

These include American ginseng, Korean ginseng, cordyceps, poria, pilose antler, antler glue, gastrodia tuber, chrysanthemum, orange peel, acanthopanax bark, Solomon seal rhizome, roxburgh rose, fleece flower root, and barbary wolfberry fruits.

5.7.1.3 Some Classic Menus

These include the following:

1. Korean red ginseng (20 g), whole chicken (1000 g), dried abalone (60 g), sea cucumber (200 g), fish stomach (45 g), lean Chinese ham (150 g), dried scallops (45 g), and some salt and water, cooked into a dish.

FIGURE 5.1 Cordyceps (Chinese caterpillar fungus), a famous fungus in medicated diets for preventing aging.

2. Ginseng (2 g), ginger (6 g), walnut (6 g), Chinese onion (20 g), and rice (300 g) are put into a pig's stomach, and some salt and water, and then cooked into a meal.
3. Turtle (1000 g), ginseng (5 g), Chinese onion (10 g), ginger (15 g), rice wine (20 mL), salt, water, and pepper cooked into a dish.
4. Stewed pig's feet (500 g), chestnuts (200 g), and some salt and water, cooked into a dish.

The meal pictures shown in Figure 5.2 (whole chicken and ginseng cooked for a meal), Figure 5.3 (shrimps and cordyceps (Chinese caterpillar fungus), and Chinese dates cooked for a meal), and Figure 5.4 (chicken meal, fresh ginseng, barbary wolfberry fruits, and Chinese dates cooked for a meal) are all for the prevention of aging.

5.7.2 Immune Regulations

According to traditional Chinese medicine, Chinese medicated diets that have functionality in improving immune regulation processes are usually selected food materials and medicinal herbs from the following natural sources.

5.7.2.1 Food Materials

These include walnut, Chinese date, royal jelly, and mulberry fruits.

5.7.2.2 Medicinal Herbal Materials

These include American ginseng, Korean ginseng, acanthopanax, barbary wolfberry fruits, astragalus, Chinese caterpillar fungus, gingko leaf, Chinese angelia, and Ganoderma.

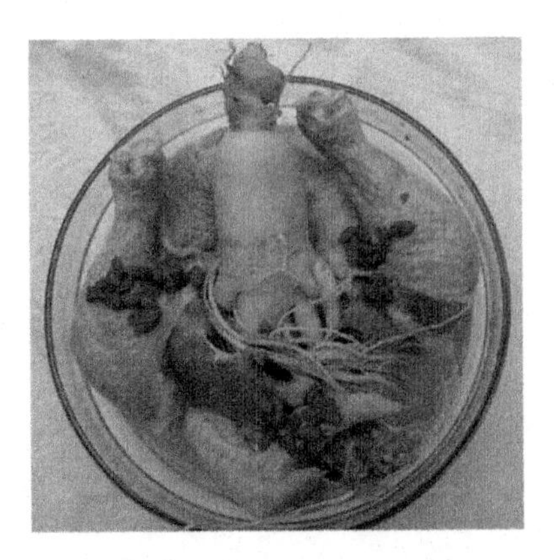

FIGURE 5.2 Whole chicken and ginseng cooked for a meal.

5.7.2.3 Some Classic Menus

These include the following:

1. Chinese yam (50 g), logan (15 g), lychee (20 g), magnolia vine fruit (5 g), rice (50 g), crystal sugar (10 g), and some water, cooked into a soup.
2. Turtle (1000 g), Chinese caterpillar fungus (10 g), Chinese date (30 g), ginger (5 g), Chinese onion (5 g), rice wine (50 mL), and some water, cooked into a meal.
3. Duck meat (200 g), sea cucumber (50 g), and some salt and water, cooked into a dish.

A picture of a meal with whole chicken, Chinese dates, barbary wolfberry fruits, and onion cooked for improving immune regulations is shown in Figure 5.5.

FIGURE 5.3 Shrimps and cordyceps (Chinese caterpillar fungus), and Chinese dates cooked for a meal.

FIGURE 5.4 Chicken meal, fresh ginseng, barbary wolfberry fruits, and Chinese dates cooked for a meal.

FIGURE 5.5 Whole chicken, Chinese dates, barbary wolfberry fruits, and onion cooked for a meal.

5.7.3 Reduction of Cholesterol

According to traditional Chinese medicine, Chinese medicated diets that have functionality in reduction of cholesterol are usually selections of foods and medicinal herbs from the following natural sources.

5.7.3.1 Food Materials

These include soybean, hawthorn fruits, peach seed, fish oil, corn oil, buckwheat, pumpkin, pig pancreas, and mulberry fruits.

5.7.3.2 Medicinal Herbal Materials

These include chrysanthemum flower, spine date seed, flax seed, safflower seed, licorice root, dried tangerine peel, barbary wolfberry fruit, five-leaf gynosttemma, Chinese yam, Cocosporia, spirulina, and germinated barley.

5.7.3.3 Some Classic Menus

These include the following:

1. Barbary wolfberry fruits (5 g), white wood ear (10 g), Chinese date (10 g), rice (100 g), and water, cooked into a congee.
2. Mung bean (50 g), mature pumpkin (50 g), salt (2 g), and some water, cooked into a soup.
3. Hawthorn (15 g), *Rhizoma polygonati* (30 g), rice (100 g), and water, cooked into a congee.
4. Fresh reed rhizome (30 g), brown rice (50 g), and water, cooked into a congee.

5.7.4 Cosmetological Purposes

According to traditional Chinese medicine, to maintain and improve the appearance of people is to improve the luster of their face skin, to limit and reduce the skin's wrinkling process, to remove dark pigments in the face skin, to promote blood circulation, and to increase moisture on the facial surface, which is to eliminate extra yang. Chinese medicated diets that have functionality for cosmetological purposes are usually selected with food and medicinal herbal materials from the following natural sources.

5.7.4.1 Food Materials

These include Chinese yam, sesame seed, honey, sea cucumber, turtle, mutton, milk, cherry, lychee, pine nut, coix seed, bean sprouts, Chinese toon, lotus flower, and barbary wolfberry fruit.

5.7.4.2 Medicinal Herbal Materials

These include astragalus, Solomon seal rhizome, fingered citron, *Artemisia apiacea*, and stewed sea cucumber.

5.7.4.3 Some Classic Menus

These include the following:

1. Green cabbage (200 g), shrimp (100 g), barbary wolfberry fruits (10 g), and some salt, plus water, cooked into a dish.
2. Lotus seed (15 g), white wood ear (25 g), crystal sugar (50 g), and water, all cooked into a soup.
3. Lily bulb (50 g), platycladi seed (10 g), honey (10 g), and water, cooked into a soup.
4. Lotus seed (15 g), lily bulb (15 g), rice (100 g), crystal sugar (10 g), and water, cooked into a soup.
5. Turtle (300 g), *Artemisia apiacea* (20 g), and some salt and water, cooked into a dish.
6. Bean sprouts (100 g), portulaca (20 g), and some salt and water, cooked into a dish.

5.7.5 Controlling Body Weight

According to traditional Chinese medicine, controlling the body weight process is to strengthen the yang function and to remove excessive body fat. Chinese medicated diets that have functionality in controlling body weight processes are usually selected food materials and medicinal herbs from the following natural sources.

5.7.5.1 Food Materials

These include winter melon, Chinese date, water caltrop, lotus root, lotus seed, oats, mung bean sprouts, watermelon, cucumber, onion, turnip, hot pepper, old hen meat, and mutton.

5.7.5.2 Medicinal Herbal Materials

These include astragalus root, poria peel, winter melon peel, corn stigma, lotus leaf, Chinese cassia tree bark, and nutmeg.

5.7.5.3 Some Classic Menus

These include the following:

1. Astragalus (20 g) and winter melon (300 g), with some salt and water, cooked into a soup.
2. Mutton (100 g), hot pepper (20 g), onion (50 g), and some salt and water, cooked into a dish.
3. Bean sprouts (200 g), lotus leaf (20 g), lotus seed (20 g), lotus root (100 g), and some salt and water, cooked into a soup.

A picture of a meal with Chinese dates, lotus roots, and pork rib cooked for controlling body weight is shown in Figure 5.6.

FIGURE 5.6 Chinese dates, lotus roots, and pork rib cooked for a meal.

5.8 TREATING HYPERTENSION

According to traditional Chinese medicine, reduction of hypertension is to reduce yang and to nourish yin. Chinese medicated diets that have functionality in treating hypertension are usually food and herbs selected from the following natural sources.

5.8.1 Food Materials

These include Chinese date, celery, sponge gourd, tomato, wood ear, hawthorn, green tea, mulberry leaf, and barbary wolfberry fruit.

5.8.2 Medicinal Herbal Materials

These include gastrodia tuber, chrysanthemum flower, prunella spike, dogbane leaf, Solomon seal rhizome, fingered citron, and corn stigma.

5.8.3 Some Classic Menus

These include the following:

1. Hawthorn (200 g), cassia seed, and water, cooked into a soup
2. Hawthorn (200 g), wood ear (50 g), prunella (10 g), and water, cooked into a soup.
3. Fish head (300 g), gastrodia tuber (100 g), and some salt and water, cooked into a soup.
4. Tremella (15 g), Jew's ear (15 g), crystal sugar (10 g), and water, cooked into a soup.

5.9 TREATING DIABETES

According to traditional Chinese medicine, the treatment of diabetes is to eliminate excessive yang, nourish yin, and strengthen internal organ health. Chinese medicated diets that have functionality in treating diabetic conditions are usually selected food materials and medicinal herbal materials from the following natural sources.

5.9.1 Food Materials

These include bitter melon, pumpkin, Chinese yam, west lake greens, spinach, black plum, mulberry fruit, green tea, gluten, wood ear, quail egg, and duck egg.

5.9.2 Medicinal Herbal Materials

These include coco sporia, five-leaf gynosttemma, hawthorn, rehmannia root, ophiopogon root, barbary wolfberry fruit, Solomon seal rhizome, and rhizome of wind weed.

5.9.3 Some Classic Menus

These include the following:

1. Crucian carp (200 g), green tea (20 g), and some salt and water, cooked into a soup.
2. Silver carp (300 g), Solomon seal rhizome (50 g), and some salt and water, cooked into a soup.
3. Duck egg (100 g), green tea (20 g), chrysanthemum (5 g), and some salt and water, cooked into a dish.
4. Spinach (200 g), egg white (100 g), honeysuckle flower (5 g), and some salt and water, cooked into a dish.
5. Gluten (50 g), barbary wolfberry fruit (15 g), honeysuckle flower (5 g), and water, cooked into a soup.

5.10 NEW DEVELOPMENTS

Medicated diets have been developed in a great assortment on the basis of traditional processes. Today, products of salutary foods and drinks produced on the basis of achievements in scientific research and having the effect of curing diseases are a variety of sorts, and vary in characteristics. There are medicated foods suitable for patients suffering from diabetes, obesity, and angiocardiopathy, or for prolonging the life of the aged. There are some health-promoting foods and drinks suitable for athletes, actors, actresses, and miners; there are also health-promoting foods or medicated diets suitable for the promotion of children's health and growth. Chinese medicated diets have expanded internationally. Some medicated cans, health-promoting drinks, and medicated wine made from traditional Chinese medicine are being sold in international markets. Medicated diet dining halls have been set up in some countries. Chinese medicated diets will make contributions to the health of people all over the world.

REFERENCES

Beijing Traditional Chinese Medical College and Hospital, Ed. *A Collection of Herbal Prescriptions*. Beijing Traditional Chinese Medical College Publisher, Beijing, 1978–1981.

Beijing Medical College Chinese Herbal Medicine Research Group. 1959. *J. Beijing Medical College,* **1**: 104.

Chen, W.C. 1997. *Chinese Herb Cooking for Health.* Chin-Chin Publishing Co., Ltd., Taipei, Taiwan.

Hsu H.Y. and Peacher, W.G. 1976. *Chinese Herb Medicine and Therapy*. Nashville, TN: Aurora Publishers Inc.

Huang, J. 1979. *Discuss of Ways in Traditional Chinese Medicine*. Hunan People Publishing House.

Huang, K.C. 1993. *The Pharmacology of Chinese Herbs*. Boca Raton, FL: CRC Press.

Ho, Z.C. 1993. Principles of diet therapy in ancient Chinese medicine. *Asia Pacific Journal of Clinical Nutrition*, **2**: 91–95.

Lang, F.N., Ed. 1988. *Chinese Basic Herbs and Prescriptions*. Beijing: People's Health Publisher.

Lee, Y.H. 1990. *Home-Cook Nutritional Recipes*. Hong Kong: Haibin Publishing.

Lee, S.J., Sung, J.H., Lee, S.J., et al. 1999. Antitumor activity of a novel ginseng saponin metabolite in human pulmonary adenocarcinoma cells resistant to cisplatin. *Cancer Letters*, **144**: 39–43.

Li, S.C. 1578. *The Chinese Pharmacopoeia* (new version published in 1991). Beijing: People's Hygiene Publisher,

Ling, Iqi, Zhong, C., and Yao, Y.J. 1984. *Chinese Herbal Studies*. Shangahi: Shanghai Science Technology Publisher

Liu, S. and Zhong, J.J. 1998. Phosphate effect on production of ginseng saponin and polysaccharide by cell suspension cultures of *Panax ginseng* and *Panax quinquefolium*. *Process Biochemistry,* **33**: 69–74.

Lu, H.C. 1991. *Chinese Foods for Longevity*. Taipei: Yuan-Lion Publishing Co.

Ma, B.L. Ed. 2002. *"Yin-Yang" Balance and Health Care*. Beijing: People's Military Medical Publisher.

Quinn, J. F. ed. 1972. *Medicine and Public Health in the People's Republic of China*. U.S. Department of Health, Education and Welfare Publisher.

She, Y. ed. *Chinese Pharmaceutical History*. Beijing: People's Health Publisher, 1984.

Su, W.L. 1993. *Oriental Herbal Cook Book for Good Health*. New York: Shun An Tong Corp.

Shen, J.H.F. 1999. *Chinese Medicine*. Web site: http://www.infinite.org/shen/cmbook/index.html

Szechuan Medical College, Ed. 1978. *Pharmacy of Chinese Herbs*. Beijing: People's Republic Health Publisher.

Wang, Y.S. Ed. 1983. *The Pharmacology of Chinese Herbs and Their Uses*. Beijing: People's Public Health Publisher.

Weng, W.J. 1999. *Menu Collection of Chinese Medicated Diets*. Beijing: People's Public Health Publisher.

Weng, W.J. and Chen, J.S. 1996. The eastern perspective functional foods based on traditional Chinese medicine. *Nutrition Reviews*, **54**(11): 11–16.

Williamas, T. 1995. *Chinese Medicine*. Rockport, MA: Element.

Wu, D.X. 1996. *Review on Healthy Liquors in China*. Shanghai: Publishing House of Shanghai Science and Technology, 1996, pp. 30–60.

Wu, J. and Zhong, J.J. 1999. Production of ginseng and its bioactive components in plant cell culture: current technological and applied aspects. *Journal of Biotechnology,* **68**: 89–99.

Wu, P.J. 1982. *The Pharmacology of Chinese Herbs*. Beijing: People's Public Health Publisher.
Yin, Gi-ye. 1987. *Chinese Medical History*. Jiangsi: Jiangsi Science and Technology Publisher.
Yeung, H. 1985. *Handbook of Chinese Herbs and Formulas*, Vol. 1. Institute of Chinese Medicine, Los Angeles, CA.
You, J. 1996. *Preliminary Explore of Yin-Yang*. China Overseas Chinese Publishing House.
Yun, T.K. and Choi, S.Y. 1998. Non-organ specific cancer prevention of ginseng: A prospective study in Korea. *International Journal of Epidemiology*, **27**: 359–364.
Zhang, E. 1997. *Basic Theory of Traditional Chinese Medicine*. Publishing House of Shanghai College of Traditional Chinese Medicine.
Zhang, E.Q. 1990. *Chinese Medicated Diet*. Shangai: Publishing House of Shanghai College of Traditional Chinese Medicine.
Zhang, W., Jia, W., Li, S., Zhang, J., Qu, Y., and Xu, X. 1988. *Chinese Medicated Diet*. Shanghai: Publishing House of Shanghai College of Traditional Chinese Medicine.
Zhou, J. H. 1986. *Chinese Herbs Pharmacology*. Shanghai: Shanghai Science Technology Publisher.

6 Functional Foods and Men's Health

A. Venket Rao and Amir Al-Weshahy

CONTENTS

6.1 INTRODUCTION

Chronic diseases including cancer, coronary heart disease, diabetes, and osteoporosis are the major cause of mortality all over the world. Many factors contribute to the risk of such chronic diseases including genetics, age, and environmental factors. Among the environmental factors, diet has emerged as one of the major contributing factors. High-calorie, high-fat, high-salt and -sugar, and low-fiber diets are typically associated with the high risk of chronic diseases. Based on scientific data, dietary guidelines now recommend increased consumption of plant-based nutrients for the prevention of chronic diseases. Although both males and females are equally at risk, certain diseases such as prostate disorders, testicular problems, and impotency related to the sperm, are unique to men. The incidence of certain other diseases tends to be higher in men due to their lifestyle habits and advancing age. Genetic makeup, environmental factors, and diet also influence the incidence and patterns of human diseases in different regions of the world. Typically, infectious diseases were the main cause of mortality among the Asian, South American, and African population compared to the more industrialized Western and European population where chronic, metabolic diseases are more prevalent. However, such differences in human disease patterns are declining due to changes in lifestyle and dietary habits.

The roles of dietary components, particularly the role of nutrients such as vitamins and minerals in health and disease have long been established. However, one of the most significant recent developments is the study of the minor components of food substances termed "phytochemical" in human health. Traditionally, these phytochemicals were considered as "antinutrients." Recent studies have shown that many of the phytochemicals may in fact be essential for maintaining good health and the prevention of several diseases. They are increasingly being referred to as "phytonutrients." Examples of some phytochemicals of recent interest include: carotenoids (β-carotene and lycopene), terpenoids (saponins), phytosterols, polyphenols (flavonoids and isoflavones), indoles, isothiocyanates, lignans, phytates, and fiber (soluble, insoluble, gelling, and viscous). In recognition of the importance of these compounds in human health, a new field of study has emerged referred to as "nutraceuticals" and "functional ingredients." Foods containing such "functional" ingredients are now referred to as "functional foods." Table 6.1 shows examples of some functional components in foods, their sources, and potential health benefits. By definition, functional foods are those that go beyond providing the essential nutrients and contain biologically active compounds, that is, the functional ingredients that lower the risk of or prevent specific human diseases.

TABLE 6.1
Examples of Functional Ingredients, Sources, and Potential Benefits

Class/Components	Source	Potential Benefits
Carotenoids		
β-Carotene	Carrots and various fruits	Neutralizes free radicals that may damage cells; bolsters cellular antioxidant defenses
Lutein, Zeaxanthin	Kale, collards, spinach, corn, eggs, citrus	May contribute to maintenance of healthy vision
Lycopene	Tomatoes and processed tomato products	May contribute to maintenance of prostate health
Polyphenolic Compounds		
Anthocyanins	Berries, cherries, red grapes	Bolster cellular antioxidant defenses; may contribute to maintenance of brain functions
Flavanols-Catechines, Epicatechines, procyanidines	Tea, cocoa, chocolate, apples, grapes	May contribute to maintenance of heart health
Flavanones	Citrus	Neutralize free radicals that may damage cells; bolster cellular antioxidant defenses
Flavonols	Onion, apples, tea, broccoli	Neutralize free radicals that may damage cells; bolster cellular antioxidant defenses

TABLE 6.1 (continued)
Examples of Functional Ingredients, Sources, and Potential Benefits

Class/Components	Source	Potential Benefits
Proanthocyanidins	Cranberries, cocoa, apples, strawberries, grapes, wine, peanuts, cinnamon	May contribute to maintenance of urinary tract health and heart health
Caffeic acid Ferulic acid	Apples, pears, citrus fruits, some vegetables	May bolster cellular antioxidant defenses; may contribute to maintenance of healthy vision and heart health
Isothiocyanates		
Sulforaphane	Cauliflower, broccoli, broccoli sprout, cabbage, kale, horseradish	May enhance detoxification of undesirable compounds and bolster cellular antioxidants defenses
Sulfides/Thiols		
Diallyl sulfide	Garlic, onions, leeks, scallions	May enhance detoxification of undesirable compounds; may contribute to maintenance of heart health and healthy immune function
Allyl Methyl Trisulfide		
Dithiolthiones	Cruciferous vegetables (broccoli, cabbage, bok choy, collards)	May contribute to maintenance of healthy immune function
Phytoestrogens		
Isoflavones (Daidezein, Genistein)	Soybeans and soy-based foods	May contribute to maintenance of bone health, healthy brain, and immune function
Lignans	Flax, rye, some vegetables	May contribute to maintenance of heart health and healthy immune function
Plant Sterols and Stanols		
Free sterols and stanols	Corn, soy, wheat, wood oils, fortified foods and vegetables	May reduce risk of coronary heart disease
Stanol/sterol esters	Fortified table spreads, stanol esters dietary supplements	May reduce risk of coronary heart disease
Probiotics		
Yeast, Lactobacilli, Bifidobacteria, and other specific strains of beneficial bacteria	Certain yogurts and other cultured dairy and nondairy applications	May improve gastrointestinal health and systemic immunity; benefits are strain-specific
Prebiotics		
Inulin, Fructo-oligosaccharides, polydextrose	Whole grains, onions, some fruits garlic, honey, leeks, fortified foods and beverages	May improve gastrointestinal health; may improve calcium absorption

Source: Adapted from International Food Information Council Foundation. 2007. *Functional Foods.* With permission.

This chapter reviews specific aspects of gender-specific health issues and the role of functional foods in the management of such diseases.

6.2 MEN'S GENDER-SPECIFIC HEALTH ISSUES

In many cases, men's health problems may overlap with that of the females, ranging from chronic conditions such as diabetes, heart disease, and high blood pressure, certain lung diseases, and neurodegenerative diseases such as Alzheimer's, to wellness issues such as exercise and fitness, nutrition, weight control, and emotional health. Other specific health problems, specific to men, include prostate health, testicular disorders, infertility, and other disorders related to low levels of testosterone. Work-related stress has also been identified as a risk factor contributing to the mental health. Lifestyle factors such as smoking, drinking, and physical exercise as well as seeking medical health are all contributing factors to men's health problems. With advancing age, men's health issues become compounded as a result of altered nutrition, compromised immune system, changing endocrine status, advancing neuronal damage as well as alterations in the gut microflora. Examples of some health problems faced by men are shown in Table 6.2.

TABLE 6.2
Common Health Problems Associated with Men

Unique to Men	
• Prostate health	
• Testicular disorders	
• Infertility	
• Disorders related to low levels of testosterone	
Other Diseases	
• Heart diseases	• Kidney diseases
• Hepatitis	• Alzheimer
• Arthritis	• Pulmonary diseases
• Stroke	• Hypertension
• Orthopedic	• Cancer (lung, colorectal, non-Hodgkin lymphoma, leukemia, melanoma, bladder)
• Fibromyalgia	• Influenza/pneumonia
• Parkinson's	
Other Conditions	
• Acne	• Alcohol
• Allergies	• Back pain
• Dental health	• Depression
• Headaches/Migraines	• Heartburns
• Hernia	• Skin care
• Sleep disorders	• Stress
• Smoking	• Influenza/pneumonia
• Suicide	• Unintentional injuries

6.3 NUTRITIONAL REQUIREMENTS OF ADULT MEN

Nutritional requirements for men vary considerably from that of females and also through their growth cycle. Table 6.3 shows some examples of the recommended nutrient requirements for adult males and females. It is essential that foods provide all the recommended nutrients daily and failure to meet these requirements may result in several health problems.

The information provided in Table 6.3 pertains merely to some important essential nutrients. However, health professionals are now recommending intake of the phytochemical nutrients, the functional ingredients, to complete the list of healthy sources of food, the functional foods.

6.4 CHARACTERISTICS OF FUNCTIONAL FOODS TO MEET MEN'S HEALTH AND NUTRITIONAL NEEDS

To understand the beneficial role of functional foods with respect to human health, it is important to know the mechanism and causation of common human diseases. It is only through such understanding that appropriate functional foods can be developed to meet specific health requirements. Some common mechanisms of chronic diseases include: immune-related, hormone-mediated, oxidative stress, genetic, and metabolic (Phase I and II reactions). Recognition of the important role of antioxidants in the prevention of human diseases has led to extensive research directed toward the formulation of functional foods targeted at reducing the risk of oxidative stress-related human diseases. Similarly, other phytochemicals having the ability to influence other risk factors such as high cholesterol, carcinogenic metabolites, hormones, and antigens are also being studied. The role of specific functional ingredients in men's health-related disorders will be reviewed in the following section.

6.5 FUNCTIONAL INGREDIENTS AND THEIR ROLE IN DISEASES RELATED TO MEN

6.5.1 ANTIOXIDANTS

Oxidative stress induced by the formation of reactive oxygen species (ROS) as a result of normal metabolic activity, lifestyle factors, and diet are now recognized as playing an important role in the causation of many human diseases. Figure 6.1 shows the role of ROS and antioxidants in chronic diseases. Oxidation of low-density lipoprotein (LDL) cholesterol is considered as an important early step leading to increased risk of coronary heart disease. Similarly, oxidation of DNA can increase the risk of cancers. Antioxidants, by virtue of their ability to interact with ROS, can mitigate their damaging effect and reduce the risk of chronic diseases. As a result, antioxidants have been the focus of recent research in the management of human diseases. The origin of antioxidants can be: endogenous, which include antioxidant enzymes (superoxide dismutase, glutathione peroxidase, and catalase) and glutathione; and dietary, that include vitamins (A, C, and E), minerals (iron and selenium), and phytochemicals (polyphenols and carotenoids). Although vitamins and minerals have been

TABLE 6.3
Dietary Reference Intakes (DRIs): Estimated Average Requirements for Adults

Daily Requirements (Male)

Life Span	CHO (g/d)	Protein (g/d)	Vit A (µg/d)	Vit C (mg/d)	Vit E (mg/d)	Thiamin (mg/d)	Riboflavin (mg/d)	Niacin (mg/d)	Vit B6 (mg/d)	Folate (µg/d)	Vit B12 (µg/d)	Copper (µg/d)	Iodine (µg/d)	Iron (mg/d)	Magnesium (mg/d)	Molybdenum (µg/d)	Phosphorus (mg/d)	Selenium (µg/d)	Zinc (mg/d)
9–13	100	27	445	39	9	0.7	0.8	9	0.8	250	1.5	540	73	5.9	200	26	1.055	35	7
14–18	100	44	630	63	12	1	1.1	12	1.1	330	2	685	95	7.7	340	33	1.055	45	8.5
19–30	100	46	625	75	12	1	1.1	12	1.1	320	2	700	95	6	330	34	580	45	9.4
31–50	100	46	625	75	12	1	1.1	12	1.1	320	2	700	95	6	350	34	580	45	9.4
51–70	100	46	625	75	12	1	1.1	12	1.4	320	2	700	95	6	350	34	580	45	9.4
>70	100	46	625	75	12	1	1.1	12	1.4	320	2	700	95	6	350	34	580	45	9.4

Daily Requirements (Female)

Life Span	CHO (g/d)	Protein (g/d)	Vit A (µg/d)	Vit C (mg/d)	Vit E (mg/d)	Thiamin (mg/d)	Riboflavin (mg/d)	Niacin (mg/d)	Vit B6 (mg/d)	Folate (µg/d)	Vit B12 (µg/d)	Copper (µg/d)	Iodine (µg/d)	Iron (mg/d)	Magnesium (mg/d)	Molybdenum (µg/d)	Phosphorus (mg/d)	Selenium (µg/d)	Zinc (mg/d)
9–13	100	28	420	39	9	0.7	0.8	9	0.8	250	1.5	540	73	5.7	200	26	1.055	35	7
14–18	100	38	485	56	12	0.9	0.9	11	1	330	2	685	95	7.9	200	33	1.055	45	7.3
19–30	100	38	500	60	12	0.9	0.9	11	1.1	320	2	700	95	8.1	255	34	580	45	6.8
31–50	100	38	500	60	12	0.9	0.9	11	1.1	320	2	700	95	8.1	265	34	580	45	6.8
51–70	100	38	500	60	12	0.9	0.9	11	1.3	320	2	700	95	5	265	34	580	45	6.8
>70	100	38	500	60	12	0.9	0.9	11	1.3	320	2	700	95	5	265	34	580	45	6.8

Source: Adapted from Institute of Medicine. 2000. *Dietary Reference Intakes: Applications in Dietary Assessment*. The National Academies Press. With permission.

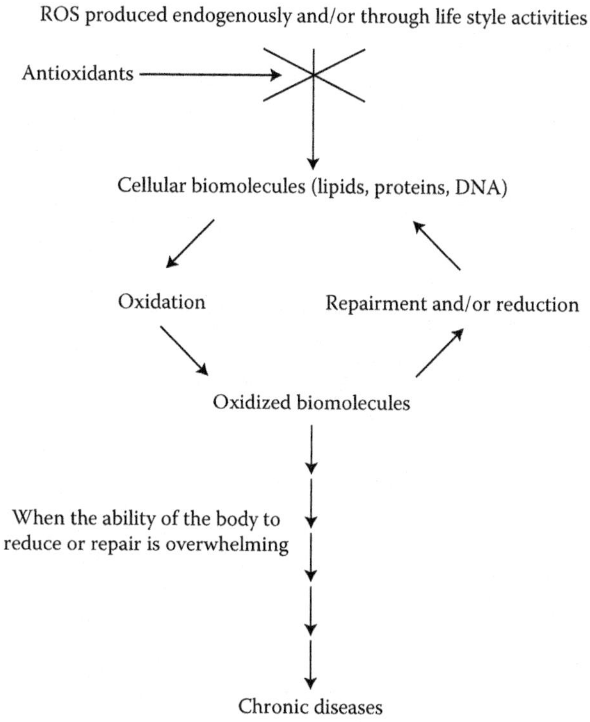

ROS produced endogenously and/or through life style activities

Antioxidants

Cellular biomolecules (lipids, proteins, DNA)

Oxidation

Repairment and/or reduction

Oxidized biomolecules

When the ability of the body to reduce or repair is overwhelming

Chronic diseases

FIGURE 6.1 Oxidative stress, antioxidants, and chronic diseases.

studied extensively in the past, the antioxidant properties of phytochemicals are an area of intense research in recent years.

6.5.1.1 Carotenoids

Carotenoids are a family of pigmented compounds that are synthesized by plants and microorganisms, but not humans. Diet becomes an important source of the required carotenoids. Fruits and vegetables constitute the major sources of carotenoids in human diet (Mangels et al., 1993; Agarwal and Rao, 2000; Johnson, 2002). Table 6.4 shows typical food sources and amounts of some common carotenoids. Based on epidemiological studies, a positive link is suggested between higher dietary intake and tissue concentrations of carotenoids and the risk of chronic diseases (Agarwal and Rao, 2000; Johnson, 2002; Elliott, 2005). β-Carotene and lycopene were shown to be inversely related to the risk of certain cancers (Johnson, 2002; Ribaya-Mercado and Blumberg, 2004). Of all the cancers, the role of lycopene in the prevention of prostate cancer has been studied the most. A publication in 1995 was the first to demonstrate an inverse relationship between the consumption of tomatoes and the risk of prostate cancer (Giovannucci et al., 1995). Lycopene was suggested as being the beneficial compound present in tomatoes. A follow-up meta-analysis of 72 different studies in 1999 showed that lycopene intake as well as serum lycopene levels were inversely related to several cancers including the prostate (Giovannucci, 1999).

TABLE 6.4
Major Carotenoids, Sources, and Amounts

Carotenoid	Food Source	Amount (μg/100 g)
β-Carotene	Apricot, dried	17.600
	Carrot, cooked	9.771
	Spinach, cooked	5.300
	Green collard	5.400
	Cantaloupe	3.000
	Beet green	2.560
	Broccoli, cooked	1.300
	Tomato, raw	520
α-Carotene	Carrot, cooked	3.723
Lycopene	Tomato, raw	3.100
	Tomato juice	10.000
	Tomato paste	36.500
	Tomato ketchup	12.390
	Tomato sauce	13.060
β-Cryptoxanthin	Tangerine	1.060
	Papaya	470
Lutein	Spinach, cooked	12.475
	Green collard	16.300
	Beet, green	7.700
	Broccoli, cooked	1.839
	Green peas, cooked	1.690

Source: From Rao, A. V. and Rao, L. G. 2007. *Pharmacol Res* **55**, 207–216. With permission.

Several other studies since then demonstrated that with increased intake of lycopene and serum levels of lycopene the risk of cancers were reduced significantly (Kucuk et al., 2001; Giovannucci et al., 2002; Rao and Rao, 2004; Rao et al., 2006). A study (Rao et al., 1999) undertaken to investigate the status of oxidative stress and antioxidants in prostate cancer patients showed significant differences in the levels of serum carotenoids, biomarkers of oxidation, and prostate-specific antigen (PSA) levels in these subjects. Although there were no differences in the levels of β-carotene, lutein, cryptoxanthin, vitamins E and A, between prostate cancer patients and their controls, levels of lycopene were significantly lower in the cancer patients. As expected, the PSA levels were significantly elevated in the cancer patients who also had higher levels of lipid and protein oxidation, indicating higher levels of oxidative stress in prostate cancer patients. In the same study, the serum PSA levels were shown to be inversely related to serum lycopene (Rao et al., 1999). Other carotenoids did not show similar inverse relationship. Other studies have also shown a decrease in the levels of PSA as well as the growth of prostate cancer in newly diagnose prostate cancer patients receiving 15 mg of lycopene daily for 3 weeks prior to radical prostatectomy (Kucuk et al., 2001; Bowen et al., 2002; Kucuk and Wood, 2002; Heath et al., 2006). In another study, when tomato sauce was used as a source of lycopene, providing 30 mg lycopene/day for 3 weeks preceding prostatectomy in men diagnosed with prostate cancer, the serum and prostate lycopene levels were elevated significantly (Bowen et al., 2002). The oxidative damage to DNA was reduced and the serum PSA levels declined significantly by 20% with lycopene treatment. Although small in number, these observations raise the possibility that lycopene may be involved not only in the prevention of cancers, but may also play a role in the treatment of the disease.

Carotenoids in general and lycopene in particular have been also been studied for their role in the prevention of coronary heart disease (Arab and Steck, 2000; Rissanen et al., 2000; Rissanen, 2006). The strongest population-based evidence comes from a multicenter case-control study (EURAMIC) that evaluated the relationship between adipose tissue antioxidant status and acute myocardial infarction (Kohlmeier and Hastings, 1995; Kohlmeier et al., 1997). Subjects that included 662 cases and 717 controls were recruited from 10 different European countries. Results of this study showed a dose-response relationship between adipose tissue lycopene and the risk of myocardial infarction. Another study that compared the Lithuanian and Swedish populations showed lower lycopene levels to be associated with increased risk and mortality from coronary heart disease (CHD) (Kritenson et al., 1997). Serum cholesterol level has traditionally been used as a biomarker for the risk of CHD. Oxidation of the circulating low-density lipoprotein (LDL) that carries cholesterol into the blood stream to oxidized LDL (LDL_{ox}) is also thought to play a key role in the pathogenesis of arteriosclerosis, which is the underlying disorder leading to heart attacks and ischemic strokes (Parthasarathy et al., 1992; Witztum, 1994; Heller et al., 1998). Lycopene was also shown to significantly reduce the levels of LDL_{ox} in subjects consuming tomato sauce, tomato juice, and lycopene oleoresin capsules as sources of lycopene (Agarwal and Rao, 1998). In another small study, lycopene was shown to reduce serum total cholesterol levels, thereby lowering the risk of cardiovascular diseases (CVD) (Fuhramn et al., 1997).

Male infertility, a common reproductive disorder, is now being associated with oxidative damage of the sperm, leading to the loss of its quality and functionality. Significant levels of ROS are detectable in the semen of up to 25% of infertile men, whereas fertile men do not produce detectable levels of ROS in their semen (Iwasaki and Gagnon, 1992; Zini et al., 1993). A number of studies have reported the beneficial effects of vitamins C and E as well as other antioxidants, including taurine (Alvarez and Storey, 1983), L-carnitine (Moncada et al., 2002), coenzyme Q10 (Alleva et al., 1997; Lewin and Lavon, 1997), and glutathione (Lenzi et al., 1998) on sperm quality. Researchers are in the process of investigating the role of lycopene in protecting the sperm from oxidative damage. Men with antibody-mediated infertility were found to have lower semen lycopene levels than fertile controls (Palan and Naz, 1996). In another study, infertile men consumed a daily dose of capsules containing 8 mg lycopene, for 12 months after which a significant increase in the serum lycopene concentration and improvements in sperm motility, sperm motility index, sperm morphology, and functional sperm concentration was observed. Lycopene treatment also resulted in 36% successful pregnancies. Other studies are now in progress and their results will further advance our knowledge of the beneficial role of lycopene in reducing the incidence of male infertility.

6.5.1.2 Polyphenols

Polyphenols are a specific group of secondary plant metabolites derived mainly from phenylalanine or tyrosine (Shahidi and Naczk, 2004) and occur naturally in medicinal plants, spices, vegetables, fruits, grains, pulses, and other seeds. Table 6.5 gives an account of the main classes of polyphenols found in foods, their sources, and amounts.

TABLE 6.5
Polyphenols in Foods, Sources, and Amounts

Polyphenolic Compounds	Food Source	Amount (mg/100 g Edible Portion or mg/100 mL)
	Celery	3.5
	Onions	15.4–38.8
	Fennel	16–50
	Hot peppers	2.3–20.3
	Cherry tomatoes	1.2–9.4
	Spinach	3.5
Quercetin	Lettuce	23.1
Myricetin	Buckwheat	2.55
Kaempferol	Apples	2.54
	Apricots	1.2
	Grapes	1.10–42
	Plums	3.01
	Berries	55
	Black currant juice	2.69
	Tea	2.07
	Black cocoa powder	20.3
(–)-Epicatechin	Apples	9
(+)-Catechin	Apricots	11
(–)-Epigallocatechin gallate	Grapes	17.6
	Peaches	2.3
	Nectarines	2.75
	Pears	3.34
	Plums	6.19
	Raisins	3.68
	Berries	1.1–18.7
	Tea	33–132
	Chocolate	53.5
	Hot peppers	5.4
	Celery hearts	22.6
Apigenin	Fresh parsley	303

TABLE 6.5 (continued)
Polyphenols in Foods, Sources, and Amounts

Polyphenolic Compounds	Food Source	Amount (mg/100 g Edible Portion or mg/100 mL)
Luteolin	Oregano	4.5
	Rosemary	4
	Thyme	56
	Dry parsley	13.51
	Lemon	49.9
Naringenin	Lemon juice	18.3
Hesperetin	Lime juice	11.5
Eriodictyol	Orange juice	15
	Tangerine juice	10.8
	Peppermint	20
	Orange	43.9
	Grapefruit	54.5
Cyanidin	Berries	25–749
Malvidin	Black grape	30–750
Delphinidin	Cherries	35–450
Pelargonidin	Plums	0.20–25

Source: Modified from Lotito, S. B. and Frei, B. 2006. *Free Radic Biol Med* **41**, 1727–1746.

There is well-documented evidence in support of the role of polyphenols in the prevention of chronic diseases. In the case of cancer, polyphenols, acting as antioxidants, trap free radicals and prevent DNA oxidation and uncontrolled proliferation of the initiated cells (Thompson, 2004). They can also prevent DNA oxidation by chelating oxidizing metal ions and by reactivating endogenous antioxidant enzymes (Kanatt et al., 2005). The ability of polyphenols to inhibit prostate cancer has been shown in animal models (Yang et al., 2008). Heber (2008) described the effect of polyphenols and their metabolites on prostate cancer in human patients as well as mice implanted with cancer cell lines. Consumption of pomegranate as a source of polyphenols reduced the expression of prostate cancer gene by twofold. The low incidence of prostate cancer among Asians (Sim and Cheng, 2005) is associated with high consumption of green tea and herbs that are considered as the primary source of polyphenols (Siddiqui et al., 2007). Epidemiological studies have shown a positive correlation between the consumption of fruits and vegetables, as major sources of polyphenolic compound, and reduction in the incidence of CVD and the risk of stroke (Arts et al., 2001; Arts and Hollman, 2005; Walter et al., 2008). As antioxidants, they are also able to inhibit atherogenesis by inhibiting oxidation of LDL cholesterol and its accumulation in macrophages (Zock and Katan, 1998). Seven prospective studies reported in the literature also showed protective effects of polyphenolic compounds to fatal or non-fatal coronary arterial disease (CAD) with a reduction of up to 65% in the risk of mortality. In a large cohort study of 35.000

male U.S. health professionals, suggestion of a reduction in coronary mortality rates with high intake of polyphenols was found only among men with a previous history of CAD (Rimm et al., 1996).

The protective role of polyphenolic compounds also extends to the brain and cognitive system. Since the brain contains the highest amount of unsaturated fatty acids, high consumption of oxygen and lack of an antioxidant system (Mattson et al., 2002), it is particularly vulnerable to oxidative damage leading to degenerative diseases such as the Alzheimer's disease. Polyphenolic compounds, as well as other dietary antioxidants including vitamins E, C, and carotenoids, are able to protect the brain and neurons from free radical damage and prevent the incidence of neuronal diseases (Deschamps et al., 2001). In a population-based prospective study of 1836 Japanese Americans in King County, Washington, who were dementia-free at baseline, subjects who drank fruit and vegetable juices at least thrice a week, compared to those who drank less often than once a week, the probable hazard ratio for Alzheimer's disease was significantly reduced (Dai et al., 2006). The protective role of polyphenols as antioxidants against brain damage is also supported by animal laboratory experiments (Simonyi et al., 2005).

Similar to the carotenoid antioxidants, polyphenols can also improve male infertility problems. Several studies have shown that peroxidase-positive leucocytes in semen (mostly polymorphonuclear leucocytes and macrophages) produce large amounts of ROS (Saleh et al., 2002; Agarwal et al., 2003). Excessive ROS levels disrupt the human sperm function by inducing peroxidative damage to the unsaturated fatty acids within the sperm plasma membrane, diminishing motility and leading to incompetence of sperm–oocyte fusion, leading to a decline in sperm counts consequently resulting in male infertility and deterioration of semen quality (Agarwal et al., 2003; Kullisaar et al., 2007). Polyphenols by virtue of their ability to react with ROS have been shown to reduce the oxidative damage to the sperm and improve its functionality.

6.5.2 Phytosterols

Plant sterols (phytosterols) are fat-like substances that occur naturally in many fruits and vegetables. They are chemically related and structurally similar to cholesterol. The most common phytosterols are β-sitosterol, campesterol, and stigmasterol. Of the plant stanols, which are the saturated derivatives of sterols, sitostanol is the most common. Current dietary guidelines recommend increased consumption of phytosterol- and sterol-containing plant foods to control serum cholesterol levels and to reduce the risk of coronary heart disease. Plant sterols and stanols decrease total and LDL cholesterol levels by reducing dietary and biliary cholesterol absorption via the displacement of cholesterol from micelles resulting in reduced cholesterol solubility in the intestine (Van Horn et al., 2008).

A large body of clinical research supports the use of plant sterols for various health benefits. On an average, a reduction of 10% in LDL cholesterol levels has been found in a recently published meta-analysis of the clinical research investigating the effects of phytosterols and stanols on lipid concentrations in familial hypercholesterolemia (FH) subjects (Moruisi et al., 2006; Martikainen et al., 2007). The authors concluded that phytosterols and stanols can provide effective adjunct treatment strategies for lowering cholesterol in FH patients. They suggest the use of plant stanol ester-containing spreads as part of a daily diet replacing regular spreads, for the prevention of CHD. Another recently published study also shows the role of sterols in reducing serum cholesterol levels (Silbernagel et al., 2008). Jones et al. (1999) examined the effects of sitostanol-containing phytosterols on plasma lipid and phytosterols concentrations and *de novo* cholesterol synthesis rate in the context of a controlled diet. Thirty-two hypercholesterolemic men were fed either a diet of prepared foods alone or a diet containing 1.7 g phytosterols/day for 30 days in a parallel study design. The LDL cholesterol decreased significantly with improvement in the lipid profile in those subjects consuming the phytosterols. In another study, 67 subjects were randomized to one of the two dietary treatments for a period of 6 weeks: a cocoa flavanol-enriched snack bar containing 1.5 g phytosterol or a control product containing no phytosterols. Consumption of only the phytosterol-enriched snack bars for 6 weeks was associated with significant reductions in plasma and LDL cholesterol, and the ratio of total to high-density lipoprotein (HDL) cholesterol (Polagruto et al., 2006).

6.5.3 Phytoestrogens

Phytoestrogens are a group of plant compounds that are structurally similar to endogenous sex hormones (Wiseman, 2000). Due to this similarity, they can act as agonists or antagonist of sex hormones and influence hormone-sensitive cancers (Cade et al., 2001). The main sources for phytoestrogens are soybean and its products, flaxseed, thyme, and turmeric (Liu et al., 2001). Figure 6.2 shows the classes of phytoestrogens and their metabolites in the body.

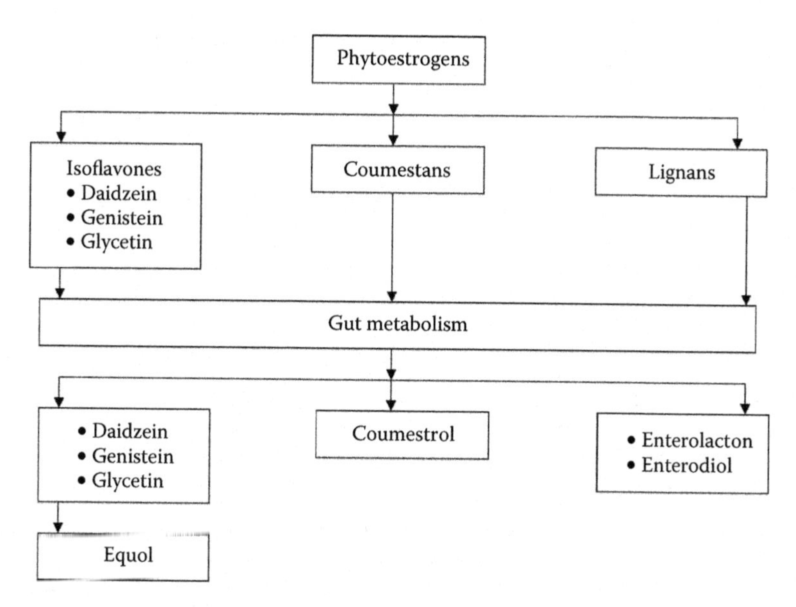

FIGURE 6.2 Phytoestrogens present in foods and their metabolites.

Cell culture, animal, and human studies have all provided evidence to the role of phytoestrogens in the prevention of prostate cancer. Genistein, equol, and enterolactone, have all been shown to inhibit the growth of the androgen-sensitive LNCaP cells, and reduce both intracellular and extracellular PSA concentrations. In another study, genistein and its precursor biochanin-A at relatively high concentrations inhibited the growth of prostate cancer cells (Peterson and Barnes, 1993). In a recent animal study, dietary genistein was shown to improve survival and inhibit progression to advanced prostate cancer in transgenic adenocarcinoma of the mouse prostate (TRAMP) associated with reduced expression of osteopontin, a specific extracellular protein that develops the metastatic process (Mentor-Marcel et al., 2001). In addition, genistein was demonstrated to be an effective inhibitor of angiogenesis and to decrease PSA levels and prevent metastatic diseases in male Lobund–Wistar rats (Schleicher et al., 1999). Enterolactone has also been shown to inhibit the growth and development of prostate cancer by triggering apoptosis of these cells (Chen et al., 2007). Population based case-control studies in Scotland (Heald et al., 2007) and Japan (Kurahashi et al., 2007) showed that consuming soy-based diets containing lignans and isoflavones significantly decreased the risk of prostate cancer. A 16-year-long prospective health study showed that men who consumed more than one glass of soy milk per day had a 70% lower risk of prostate cancer (Jacobsen et al., 1998; Han et al., 2008). In another study, when diets containing fermented soybean and placebo diets were fed to patients with prostate cancer, men classified into the fermented soybean group had slower rise in their PSA compared to the placebo group (Barken et al., 1999). When male subjects over the age of 65 years were treated with 60 mg of isoflavones from red clover per day for a year, a significant decrease in the total PSA levels by more than 30% was observed (Engelhardt and Riedl, 2008). Based on this observation, the author suggests red clover and other sources of phytoestrogens as a potent and tolerated remedy for prostate cancer. In addition to the protective effect of phytoestrogens toward prostate cancer, they have also been shown to be beneficial toward lower urinary tract symptoms (LUTS) of cancer in a cohort of elderly Chinese men (Wong et al., 2007).

Phytoestrogens can also have an important cardioprotective effect mediated through several mechanisms that include reduction in blood total cholesterol, LDL cholesterol, and triglyceride levels (Vanharanta et al., 1999); quenchers of free radicals (Wiseman, 1996), inhibition of cell adhesion, altering the growth factor activity, inhibiting cell proliferation involved in atherosclerotic lesion formation (Anthony et al., 1996), and the contraction of smooth muscles (Figtree et al., 2000). In the Zutphen Elderly Study, a prospective cohort study in which 570 men aged 64–84 years were followed for 15 years, an inverse relationship between the consumption of plant lignins and the incidence of CHD, CVD, cancer, and all causes of mortality was observed (Milder et al., 2006). In another study, when daidzein tablets were consumed by senile patients with CHD for 6 weeks, levels of serum inflammatory factors were reduced significantly supporting the suggestion that isoflavones have anti-inflammatory effects in patients with coronary atherosclerosis (Liu et al., 2006). In a study with middle-aged men from Eastern Finland, an inverse relationship between serum enterolactone levels and the incidence of acute coronary events was shown (Vanharanta et al., 1999). Similar results were also reported in the Kuopio Ischaemic Heart Disease Risk Factor Study (Vanharanta et al., 2003).

6.5.4 Phytochemicals Influencing Phase I and II Metabolic Reactions

One of the body's primary self-defense mechanisms is the conversion and neutralization of xenobiotic compounds including carcinogens that enter through the diet and the environment, into soluble and safe by-products that can be eliminated via urine or the bile. This process of detoxification involves both Phase I and Phase II enzymes. In general, Phase I reactions result in the formation of active metabolites that are then acted upon by Phase II enzymes. A proper balance of the two phases is therefore essential for the detoxification process. Certain phytochemicals have the ability to influence both these phases and thereby play an important role in the prevention of chronic diseases and maintaining good health (Akbaraly et al., 2008). Among the phytochemicals, phenolic compounds, carotenoids, and trace elements such as zinc and selenium have received great deal of recent interest. Rubiolo et al. (2008) demonstrated the ability of resveratrol, a polyphenolic compound present in grapes, grape juice and red wine, to enhance Phase II detoxifying enzymes against oxidative stress in rat liver. Chemopreventive potential of the black tea polyphenols, Polyphenol-B and BTF-35, were tested in a hamster model for 18 weeks (Vidjaya Letchoumy et al., 2008). Phase I and Phase II xenobiotic-metabolizing enzymes and the DNA oxidation product 8-hydroxy-deoxyguanosine (8-OH-dG) in the buccal pouch and liver were used as bio-markers of chemoprevention. Administration of the polyphenols significantly decreased tumor incidence, oxidative DNA damage, Phase I enzyme activities as well as expression of CYP1A1 and CYP1B1 isoforms, while enhancing Phase II enzyme activities. In another study, the effect of the bifunctional inducer flavone on the expression of Phase I and II biotransformation enzymes was examined in an animal model. Enzymatic activity and mRNA levels of cytochrome P450 mono-oxygenase (CYP) isoforms (CYP1A1, CYP1A2, and CYP2B1/2) and glutathione-*S*-transferase (GST) isoforms (GSTA, GSTM, and GSTP), were used as indices for the changes in expression. A significant increase in the Phase I and II enzyme expression was observed in response to the diet containing flavone (Rudolf et al., 2008). Another member of the polyphenolic group, the flavonoids, was also shown to affect cytochrome P450 (CYP) enzymes involved in the activation of procarcinogens and Phase II enzymes responsible for the detoxification of carcinogens (Moon et al., 2006).

Epidemiological and dietary studies have also shown an association between the high dietary intakes of cruciferous vegetables and decreased prostate cancer risk. Studies have shown that indole-3-carbinol (I3C), a common phytochemical in the cruciferous vegetables, and its *in vivo* dimeric product 3,3′-diindolylmethane (DIM) upregulate the expression of Phase I and Phase II enzymes, suggesting increased capacity for detoxification and inhibition of carcinogens (Sarkar and Li, 2004). Similarly, epidemiological studies have also implicated low dietary and serum levels of retinol with an increased risk for the development of human prostate cancer. In a recent report, dietary fenretinide [*N*-[(4-hydroxyphenyl)] retinamide], a synthetic retinoid with low toxicity, decreased the incidence of experimentally induced prostate cancer. Fenretinide is currently being evaluated in Phase I and Phase II clinical trials as an agent for both the treatment and chemoprevention of human prostate cancer (Slawin et al., 1993).

Research on the potential health benefits of soy isoflavones and other polyphenols contained in red wine, green and black tea, and dark chocolate, along with recent clinical trials and studies in animal models and cultured endothelial cells provide important and novel insights into the mechanisms by which dietary polyphenols afford protection against oxidative stress (Mann et al., 2007). It is suggested that nitric oxide and reactive oxygen radicals may mediate dietary polyphenol-induced activation of Nrf2, which in turn triggers an antioxidant response element (ARE)-driven transcription of Phase II detoxifying and antioxidant defense enzymes in vascular cells.

6.5.5 Probiotics and Prebiotics

Among the most promising targets for functional ingredients and functional foods are the gastrointestinal functions, particularly the microflora of the large intestine. The large intestine is the most heavily colonized part of the gastrointestinal tract with several hundred species of bacteria, which play a significant role in influencing both the health and disease of the host. With advancing age, a significant alteration in the composition of intestinal microorganisms takes place resulting in the decline in the numbers of beneficial bacteria. With the decline of the beneficial bacteria, the incidence of infectious and metabolic diseases increases. It is therefore considered essential to maintain a predominance of beneficial bacteria to prevent human diseases. In recognition of the important role of bacteria in human health, new concepts of probiotics and prebiotics are being utilized increasingly as effective tools in the management of health.

A probiotic is broadly defined as a viable microbial dietary supplement that beneficially affects the host through its effects on the intestinal tract. More specifically, probiotic is a live microbial food ingredient that is beneficial to the host's health (Fuller, 1989). For the probiotics to be beneficial they must be safe, present in high numbers, be viable when consumed, have antibiotic resistance, and be able to colonize in the intestinal epithelium (Havenaar et al., 1992; Salminen et al., 1998). The most widely used probiotic bacteria are *lactobacilli* and *bifidobacteria*. However, other species like Gram-positive *cocci*, *bacilli*, *yeasts*, and *Escherichia coli* can also be beneficial to the host (Holzapfel and Schillinger, 2002). Health benefits of the probiotics include inhibition of the pathogens (Bruno and Shah, 2002), stimulation of the immune system (Cross, 2002), reduction in the levels of cholesterol and the risk of cardiovascular disease (Naruszewicz et al., 2002; Pereira and Gibson, 2002), and reduction of the risk of cancer (Rao, 1999; Reddy, 1999; Burns and Rowland, 2000). Other benefits include reduction of lactose intolerance in infants (Marteau et al., 2001; Ouwehand et al., 2002), improvement of calcium absorption (Fairweather-Tait and Johnson, 1999), action as vector to vitamins and insulin (Gorbach, 2002). They also function as carriers of mucosal vaccines to prevent the spread of the HIV virus, and as natural antibiotic to avoid multidrug resistance problems (Alvarez-Olmos and Oberhelman, 2001). Some of the health benefits of probiotic bacteria are shown in Figure 6.3. Popular functional foods that contain probiotic bacteria include yoghurt and other fermented dairy products.

Prebiotic, on the other hand, is defined as a "non-digestible food ingredient that beneficially affects the host by selectively stimulating the growth and/or activity of

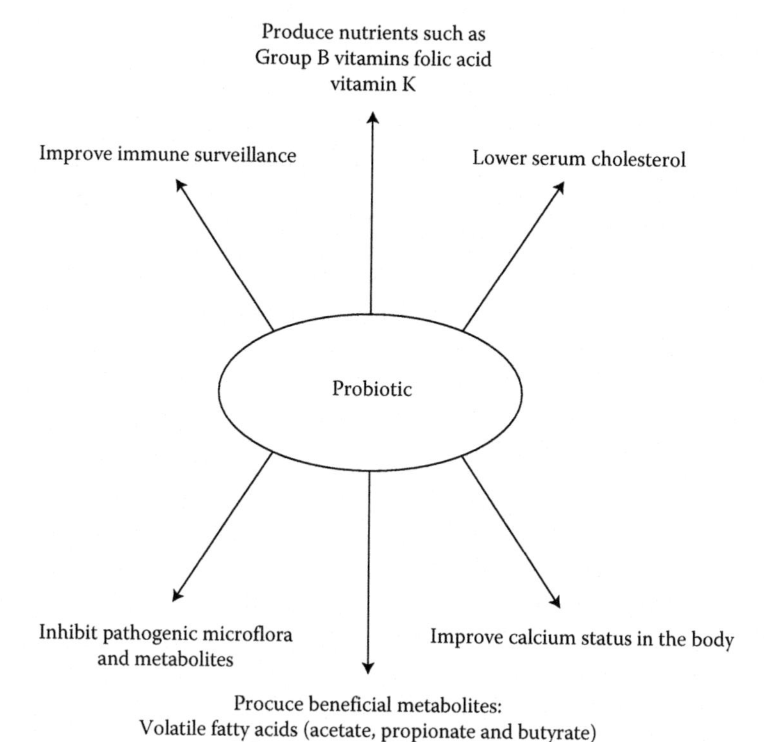

FIGURE 6.3 Beneficial properties of probiotics.

one or a limited number of bacteria in the colon, and thus improves host health" (Gibson and Roberfroid, 1995). For a food ingredient to be considered as prebiotic, it must be able to survive the gastrointestinal transit environment, promote the growth and/or metabolic activity of the beneficial bacteria present in the intestinal tract, alter the colonic microflora in favor of a composition more favorable to the individual's health, and must induce luminal or systemic effects influencing a beneficial change in the health of the host organism (Gibson and Roberfroid, 1995). Although many different nutrients reach the colon, only a few meet the above criteria. Inulin-type fructans, which include native inulin, enzymatically hydrolyzed inulin or oligofructose, and synthetic fructo-oligosaccharides, are the ones that have been studied (Roberfroid and Delzenne, 1998; Roberfroid et al., 1998). Prebiotic functional ingredients are used in fermented dairy products and in other foods such as ice cream, soups, and sauces as thickeners.

6.5.6 Other Functional Ingredients of Recent Interest

To meet the nutritional and health needs of men as they advance in age, several functional ingredients in addition to the ones discussed above have been identified. They include: omega-3 fatty acids for triglyceride reduction and lowering the risk of heart attack and strokes; calcium and vitamin D in the management of osteoporosis and other bone disorders; and insoluble, soluble, viscous, and gelling fibers for laxation and control of metabolic disorders. Undoubtedly, many other functional ingredients are also being investigated as nutritional and health supplements.

6.6 FUTURE DIRECTIONS AND CONCLUSIONS

At present, there is convincing scientific evidence to suggest an important relationship between diet and health. Although similar features characterize human diseases around the globe, epidemiological studies have also shown differences based on regions, gender, and age. Health disorders such as prostate disorders, testicular problems, and impotency related to the sperm are specific to men. Other health disorders though in common with women occur at a higher frequency in men. Traditionally, foods are looked upon as sources of nutrients that influence health outcomes. However, more recently the concept of functional foods is being considered as a viable alternate to compliment the use of supplements and medication in the prevention of human diseases. Newer functional ingredients are constantly being identified and evaluated through systematic studies for their role in human health. Rising to the challenge, healthy industry representatives are introducing new and innovative products in the form of supplements, beverages, and foods, to the market. It is unfortunate however, that the very popularity of such functional foods has also given rise to unregistered and nonclinically tested food products that be claimed as functional foods. The lack of stringent regulations requiring science-based evidence in support of the health claims has contributed to consumer confusion. However, government regulatory agencies are beginning to take an initiative in many countries to establish dedicated departments that deal with the efficacy and regulation of such functional ingredients and foods. It is in the best interest of all stakeholders including the manufacturers and the consumers to establish safety and uniform standards of identity for functional ingredients, supplements, foods, and beverages. Combined with a healthy lifestyle practices and good diet, functional foods will certainly have an important role to play in the management of men's health.

REFERENCES

Agarwal, A., Saleh, R. A., and Bedaiwy, M. A. 2003. Role of reactive oxygen species in the pathophysiology of human reproduction. *Fertil Steril* **79**, 829–843.

Agarwal, S. and Rao, A. V. 1998. Tomato lycopene and low density lipoprotein oxidation: A human dietary intervention study. *Lipids* **33**, 981–984.

Agarwal, S. and Rao, A. V. 2000. Carotenoids and chronic diseases. *Drug Metabol Drug Interact* **17**, 189–210.

Akbaraly, T. N., Favier, A., and Berr, C. 2008. Total plasma carotenoids and mortality in the elderly: Results of the Epidemiology of Vascular Ageing (EVA) study. *British Journal of Nutrition* **101**, 86–92.

Alleva, R., Sacararmucci, A., Mantero, F., Bompadre, S., Leoni, L., and Littarru, G. P. 1997. The protective role of ubiquinol-10 against formation of lipid hydroperoxides in human seminal fluid. *Mol Aspects Med* **18**, S221–S228.

Alvarez, J. G. and Storey, B. T. 1983. Taurine, hypotaurine, epinephrine and albumin inhibit lipid peroxidation in rabbit spermatozoa and protect against loss of motility. *Biol Reprod* **29**, 548–555.

Alvarez-Olmos, M. I. and Oberhelman, R. A. 2001. Probiotic agents and infectious diseases: A modern perspective on a traditional therapy. *Clin Infect Dis* **32**, 1567–1576. doi:CID000780 [pii].

Anthony, M. S., Clarkson, T. B., Hughes, C. L., Jr., Morgan, T. M., and Burke, G. L. 1996. Soybean isoflavones improve cardiovascular risk factors without affecting the reproductive system of peripubertal rhesus monkeys. *J Nutr* **126**, 43–50.

Arab, L. and Steck, S. 2000. Lycopene and cardiovascular disease. *Am J Clin Nutr* **71**, 1691S–1695S; discussion 1696S–1697S.

Arts, I. C. and Hollman, P. C. 2005. Polyphenols and disease risk in epidemiologic studies. *Am J Clin Nutr* **81**, 317S–325S.

Arts, I. C., Hollman, P. C., Feskens, E. J., Bueno de Mesquita, H. B., and Kromhout, D. 2001. Catechin intake and associated dietary and lifestyle factors in a representative sample of Dutch men and women. *Eur J Clin Nutr* **55**, 76–81.

Barken, I., Eliaz, I., Baranov, I. E., and Geller, J. 1999. The effect of two soy preparation on prostate specific antigen levels in patients with prostate cancer and the correlation of prostate-specific antigen changes with plasma genistein. In: *Third International Symposium on the Role of Soy in Preventing and Treating Chronic Disease*. Washington, DC, USA.

Bowen, P., Chen, L., Stacewicz-Sapuntzakis, M., Duncan, C., Sharifi, R., Ghosh, L., Kim, H. S., Christov-Tzelkov, K., and van Breemen, R. 2002. Tomato sauce supplementation and prostate cancer: Lycopene accumulation and modulation of biomarkers of carcinogenesis. *Exp Biol Med* **227**, 886–893.

Bruno, F. A. and Shah, N. P. 2002. Inhibition of pathogenic and putrefactive microorganisms by *Bifidobacterium* sp. *Milchwissenschaft* **57**, 617–621.

Burns, A. J. and Rowland, I. R. 2000. Anti-carcinogenicity of probiotics and prebiotics. *Curr Issues Intest Microbiol* **1**, 13–24.

Cade, J., Burley, V., and Kirk, S. 2001. Phytoesterogen and health. In: *Food and Nutritional Supplements: Their Role in Health and Disease* (eds.) J. K. Ransley, J. K. Donnelly, and N. W. Read, pp. 141–155. Berlin, Germany: Springer-Verlag.

Chen, L. H., Fang, J., Li, H., Demark-Wahnefried, W., and Lin, X. 2007. Enterolactone induces apoptosis in human prostate carcinoma LNCaP cells via a mitochondrial-mediated, caspase-dependent pathway. *Mol Cancer Ther* **6**, 2581–2590.

Cross, M. L. 2002. Microbes versus microbes: Immune signals generated by probiotic lactobacilli and their role in protection against microbial pathogens. *FEMS Immunol Med Microbiol* **34**, 245–253. doi:S0928824402003772 [pii].

Dai, Q., Borenstein, A. R., Wu, Y., Jackson, J. C., and Larson, E. B. 2006. Fruit and vegetable juices and Alzheimer's disease: The Kame Project. *Am J Med* **119**, 751–759.

Deschamps, V., Barberger-Gateau, P., Peuchant, E., and Orgogozo, J. M. 2001. Nutritional factors in cerebral aging and dementia: Epidemiological arguments for a role of oxidative stress. *Neuroepidemiology* **20**, 7–15.

Elliott, R. 2005. Mechanisms of genomic and non-genomic actions of carotenoids. *Biochim Biophys Acta* **1740**, 147–154.

Engelhardt, P. F. and Riedl, C. R. 2008. Effects of one-year treatment with isoflavone extract from red clover on prostate, liver function, sexual function, and quality of life in men with elevated PSA levels and negative prostate biopsy findings. *Urology* **71**, 185–190.

Fairweather-Tait, S. J. and Johnson, I. T. 1999. Bioavailability of minerals. In: *Colonic Microbiota, Nutrition and Health* (eds.) G. R. Gibson and M. B. Roberfroid. Dordrecht, the Netherlands: Kluwer Academic Press.

Figtree, G. A., Griffiths, H., Lu, Y., Webb, C. M., MacLeod, K., and Collins, P. 2000. Plant-derived estrogens relax coronary arteries *in vivo* by a calcium antagonistic mechanism. *J Am Coll Cardiol* **35**, 1977–1985.

Fuhramn, B., Elis, A., and Aviram, M. 1997. Hypocholesterolemic effect of lycopene and b-carotene is related to suppression of cholesterol synthesis and augmentation of LDL receptor activity in macrophage. *Biochem Biophys Res Commun* **233**, 658–662.

Fuller, R. 1989. Probiotics in man and animals. *J Appl Bacteriol* **66**, 365–378.

Gibson, G. R. and Roberfroid, M. B. 1995. Dietary modulation of the human colonic microbiota: Introducing the concept of prebiotics. *J Nutr* **125**, 1401–1412.

Giovannucci, E. 1999. Tomatoes, tomato-based products, lycopene, and cancer: Review of the epidemiologic literature. *J Natl Cancer Inst* **91**, 317–331.

Giovannucci, E., Ascherio, A., Rimm, E. B., Stampfer, M. J., Colditz, G. A., and Willett, W. C. 1995. Intake of carotenoids and retinol in relation to risk of prostate cancer. *J Natl Cancer Inst* **87**, 1767–1776.

Giovannucci, E., Rimm, E. B., Liu, Y., Stampfer, M. J., and Willett, W. C. 2002. A prospective study of tomato products, lycopene, and prostate cancer risk. *J Natl Cancer Inst* **94**, 391–398.

Gorbach, S. L. 2002. Probiotics in the third millennium. *Dig Liver Dis* **34** Suppl 2, S2–S7.

Han, H. Y., Wang, X. H., Wang, N. L., Ling, M. T., Wong, Y. C., and Yao, X. S. 2008. Lignans isolated from *Campylotropis hirtella* (Franch.) Schindl. decreased prostate specific antigen and androgen receptor expression in LNCaP cells. *J Agric Food Chem* **56**, 6928–6935. doi:10.1021/jf800476r [doi].

Havenaar, R., Brink, B. T., and Veld, H. J. 1992. Selection of strains for probiotic use. In: *Probiotics: The Scientific Basis* (ed.) R. Fuller. Boundary Row, London: Chapman & Hall.

Heald, C. L., Ritchie, M. R., Bolton-Smith, C., Morton, M. S., and Alexander, F. E. 2007. Phyto-oestrogens and risk of prostate cancer in Scottish men. *Br J Nutr* **98**, 388–396.

Heath, E., Seren, S., Sahin, K., and Kucuk, O. 2006. The role of tomato lycopene in the treatment of prostate cancer. In: *Tomatoes, Lycopene and Human Health* (ed.) A. V. Rao, pp. 127–140. Scotland: Caledonian Science Press.

Heber, D. 2008. Multitargeted therapy of cancer by ellagitannins. *Cancer Lett* **269**, 262–268.

Heller, F. R., Descamps, O., and Hondekijn, J. C. 1998. LDL oxidation: Therapeutic perspectives. *Atherosclerosis* **137** Suppl, S25–S31.

Holzapfel, H. W., and Schillinger, W. 2002. Introduction to pre- and probiotics. *Food Res Int* **35**, 109–116.

Institute of Medicine. 2000. *Dietary Reference Intakes: Applications in Dietary Assessment.* Washington, DC: The National Academies Press.

International Food Information Council Foundation. 2007. *Functional Foods.* http://www.foodinsight.org/OpenSearchEngine/tabid/85/Default.aspx?xsq=functional+foods

Iwasaki, A. and Gagnon, C. 1992. Formation of reactive oxygen species in spermatozoa of infertile patients. *Fertil Steril* **57**, 409–416.

Jacobsen, B. K., Knutsen, S. F., and Fraser, G. E. 1998. Does high soy milk intake reduce prostate cancer incidence? The Adventist Health Study (United States). *Cancer Causes Control* **9**, 553–557.

Johnson, E. J. 2002. The role of carotenoids in human health. *Nutr Clin Care* **5**, 56–65.

Jones, P. J., Ntanios, F. Y., Raeini-Sarjaz, M., and Vanstone, C. A. 1999. Cholesterol-lowering efficacy of a sitostanol-containing phytosterol mixture with a prudent diet in hyperlipidemic men. *Am J Clin Nutr* **69**, 1144–1150.

Kanatt, S., Chander, R., Radhakrishna, P., and Sharma, A. 2005. Potato peel extract: A natural antioxidant for retarding lipid peroxidation in radiation processed lamb meat. *J Agric Food Chem* **53**, 1499–1504.

Kohlmeier, L. and Hastings, S. B. 1995. Epidemiologic evidence of a role of carotenoids in cardiovascular disease prevention. *Am J Clin Nutr* **62**, 1370S–1376S.

Kohlmeier, L., Kark, J. D., Gomez-Garcia, E., Martin, B. C., Steck, S. E., Kardinaal, A. F. M., Ringstad, J. et al. 1997. Lycopene and myocaedial infarction risk in the EURAMIC study. *Am J Epidemiol* **146**, 618–626.

Kritenson, M., Zieden, B., Kucinskiene, Z., Abaravicius, A., Razinkovienë, L., Elinder, L. S., Bergdahl, B., Elwing, B., Calkauskas, H., and Olsson, A. G. 1997. Antioxidant state and mortality from coronary heart disease in Lithuanian and Swedish men: Concomitant cross sectional study of men aged 50. *Br Med J* **314**, 629–633.

Kucuk, O., Sarkar, F. H., and Sakr, W. 2001. Phase II randomized clinic trial of lycopene supplementation before radical prostatectomy. *Cancer Epidemiol Biomarkers Prev* **10**, 861–868.

Kucuk, O. and Wood, D. P. J. 2002. Response of hormone refractory prostate cancer to lycopene. *J Urol* **167**, 651.

Kullisaar, T., Turk, S., Punab, M., Korrovits, P., Kisand, K., Rehema, A., Zilmer, K., Zilmer, M., and Mandar, R. 2007. Oxidative stress in leucocytospermic prostatitis patients: Preliminary results. *Andrologia* **40**, 161–172.

Kurahashi, N., Iwasaki, M., Sasazuki, S., Otani, T., Inoue, M., and Tsugane, S. 2007. Soy product and isoflavone consumption in relation to prostate cancer in Japanese men. *Cancer Epidemiol Biomarkers Prev* **16**, 538–545.

Lenzi, A., Gandini, L., and Picardo, M. 1998. A rationale for glutathione therapy. Debate on: Is antioxidant therapy a promising strategy to improve human reproduction. *Hum Reprod* **13**, 1419–1424.

Lewin, A. and Lavon, H. 1997. The effect of coenzyme Q10 on sperm motility and function. *Mol Aspects Med* **18**, S213–S219.

Liu, J., Burdette, J. E., Xu, H., Gu, C., van Breemen, R. B., Bhat, K. P., Booth, N. et al. 2001. Evaluation of estrogenic activity of plant extracts for the potential treatment of menopausal symptoms. *J Agric Food Chem* **49**, 2472–2479.

Liu, P., Zhao, Y. X., and Zhang, Y. 2006. Clinical observation of daidzein intervention on serum inflammatory factors in senile patients with coronary heart disease. *Zhongguo Zhong Xi Yi Jie He Za Zhi* **26**, 42–45.

Lotito, S. B. and Frei, B. 2006. Consumption of flavonoid-rich foods and increased plasma antioxidant capacity in humans: Cause, consequence, or epiphenomenon? *Free Radic Biol Med* **41**, 1727–1746.

Mangels, A. R., Holden, J. M., Beecher, G. R., Forman, M. R., and Lanza, E. 1993. Carotenoid contents of fruits and vegetables: An evaluation of analytical data. *J Am Diet Assoc* **93**, 284–296.

Mann, G. E., Rowlands, D. J., Li, F. Y., de Winter, P., and Siow, R. C. 2007. Activation of endothelial nitric oxide synthase by dietary isoflavones: Role of NO in Nrf2-mediated antioxidant gene expression. *Cardiovasc Res* **75**, 261–274.

Marteau, P. R., de Vrese, M., Cellier, C. J., and Schrezenmeir, J. 2001. Protection from gastrointestinal diseases with the use of probiotics. *Am J Clin Nutr* **73**, 430S–436S.

Martikainen, J. A., Ottelin, A. M., Kiviniemi, V., and Gylling, H. 2007. Plant stanol esters are potentially cost-effective in the prevention of coronary heart disease in men: Bayesian modelling approach. *Eur J Cardiovasc Prev Rehabil* **14**, 265–272.

Mattson, M. P., Chan, S. L., and Duan, W. 2002. Modification of brain aging and neurodegenerative disorders by genes, diet, and behavior. *Physiol Rev* **82**, 637–672.

Mentor-Marcel, R., Lamartiniere, C. A., Eltoum, I. E., Greenberg, N. M., and Elgavish, A. 2001. Genistein in the diet reduces the incidence of poorly differentiated prostatic adenocarcinoma in transgenic mice (TRAMP). *Cancer Res* **61**, 6777–6782.

Milder, I. E., Feskens, E. J., Arts, I. C., Bueno-de-Mesquita, H. B., Hollman, P. C., and Kromhout, D. 2006. Intakes of 4 dietary lignans and cause-specific and all-cause mortality in the Zutphen Elderly Study. *Am J Clin Nutr* **84**, 400–405.

Moncada, M. L., Vicari, E., Cimino, C., Calogero, A. E., Mongioi, A., and D'Agata, R. 2002. Effect of acetylcarnitine in oligoasthenospermic patients. *Acta Eur Fertil* **23**, 221–224.

Moon, Y. J., Wang, X., and Morris, M. E. 2006. Dietary flavonoids: effects on xenobiotic and carcinogen metabolism. *Toxicol In Vitro* **20**, 187–210.

Moruisi, K. G., Oosthuizen, W., and Opperman, A. M. 2006. Phytosterols/stanols lower cholesterol concentrations in familial hypercholesterolemic subjects: A systematic review with meta-analysis. *J Am Coll Nutr* **25**, 41–48. doi:25/1/41 [pii].

Naruszewicz, M., Johansson, M. L., Zapolska-Downar, D., and Bukowska, H. 2002. Effect of *Lactobacillus plantarum* 299v on cardiovascular disease risk factors in smokers. *Am J Clin Nutr* **76**, 1249–1255.

Ouwehand, A. C., Salminen, S., and Isolauri, E. 2002. Probiotics: An overview of beneficial effects. *Antonie Van Leeuwenhoek* **82**, 279–289.

Palan, P. and Naz, R. 1996. Changes in various antioxidant levels in human seminal plasma related to immunofertility. *Arch Androl* **36**, 139–143.

Parthasarathy, S., Steinberg, D., and Witztum, J. L. 1992. The role of oxidized low-density lipoproteins in pathogenesis of atherosclerosis. *Annu Rev Med* **43**, 219–225.

Pereira, D. I. and Gibson, G. R. 2002. Effects of consumption of probiotics and prebiotics on serum lipid levels in humans. *Crit Rev Biochem Mol Biol* **37**, 259–281.

Peterson, G. and Barnes, S. 1993. Genistein and biochanin-A inhibit the growth of human prostate cancer cells but not epidermal growth factor receptor tyrosine autophosphorylation. *Prostate* **22**, 335–345.

Polagruto, J. A., Wang-Polagruto, J. F., Braun, M. M., Lee, L., Kwik-Uribe, C., and Keen, C. L. 2006. Cocoa flavanol-enriched snack bars containing phytosterols effectively lower total and low-density lipoprotein cholesterol levels. *J Am Diet Assoc* **106**, 1804–1813.

Rao, A. V. 1999. Large bowel cancer and colonic foods. In: *Colonic Microbiota, Nutrition and Health* (eds.) G. R. Gibson and B. M. Roberfroid. London: Kluwer Academic Publishers.

Rao, A. V., Fleshner, N., and Agarwal, S. 1999. Serum and tissue lycopene and biomarkers of oxidation in prostate cancer patients: A case-control study. *Nutr Cancer* **33**, 159–164.

Rao, A. V. and Rao, L. G. 2004. Lycopene and human health. *Curr Top Nutraceut Res* **2**, 127–136.

Rao, A. V. and Rao, L. G. 2007. Carotenoids and human health. *Pharmacol Res* **55**, 207–216.

Rao, A. V., Ray, M. R., and Rao, L. G. 2006. Lycopene. *Adv Food Nutr Res* **51**, 99–164.

Reddy, B. S. 1999. Possible mechanisms by which pro- and prebiotics influence colon carcinogenesis and tumor growth. *J Nutr* **129**, 1478S–1482S.

Ribaya-Mercado, J. D. and Blumberg, J. B. 2004. Lutein and zeaxanthin and their potential roles in disease prevention. *J Am Coll Nutr* **23**, 567S–587S.

Rimm, E. B., Katan, M. B., Ascherio, A., Stampfer, M. J., and Willett, W. C. 1996. Relation between intake of flavonoids and risk for coronary heart disease in male health professionals. *Ann Intern Med* **125**, 384–389.

Rissanen, T. 2006. Lycopene and cardiovascular disease. In: *Tomatoes, Lycopene and Human Health* (ed.) A. V. Rao, pp. 141–152. Scotland: Caledonian Science Press.

Rissanen, T., Voutilainen, S., Nyyssonen, K., Salonen, R., and Salonen, J. T. 2000. Low plasma lycopene concentration is associated with increased intima-media thickness of the carotid artery wall. *Arterioscler Thromb Vasc Biol* **20**, 2677–2681.

Roberfroid, M. B. and Delzenne, N. 1998. Dietary fructans. *Annu Rev Nutr* **18**, 117–143.

Roberfroid, M. B., Van Loo, J. A. E., and Gibson, G. R. 1998. The bifidogenic nature of chicory inulin and its hydrolysis products. *J Nutr* **128**, 11–19.

Rubiolo, J. A., Mithieux, G., and Vega, F. V. 2008. Resveratrol protects primary rat hepatocytes against oxidative stress damage: Activation of the Nrf2 transcription factor and augmented activities of antioxidant enzymes. *Eur J Pharmacol* **591**, 66–72.

Rudolf, J. L., Bauerly, K. A., Tchaparian, E., Rucker, R. B., and Mitchell, A. E. 2008. The influence of diet composition on phase I and II biotransformation enzyme induction. *Arch Toxicol*. doi:10.1007/s00204–008–0310–1 [doi].

Saleh, R. A., Agarwal, A., Kandirali, E., Sharma, R. K., Thomas, A. J. J., and Nada, E. A. 2002. Leukocytospermia is associated with increased reactive oxygen species production by human spermatozoa. *Fertil Steril* **78**, 1215–1224.

Salminen, S., Bouley, C., Boutron-Ruault, M. C., Cummings, J. H., Franck, A., Gibson, G. R., Isolauri, E., Moreau, M. C., Roberfroid, M., and Rowland, I. 1998. Functional food science and gastrointestinal physiology and function. *Br J Nutr* **80** Suppl 1, S147–S171.

Sarkar, F. H. and Li, Y. 2004. Indole-3-carbinol and prostate cancer. *J Nutr* **134**, 3493S– 3498S. doi:134/12/3493S [pii].

Schleicher, R. L., Lamartiniere, C. A., Zheng, M., and Zhang, M. 1999. The inhibitory effect of genistein on the growth and metastasis of a transplantable rat accessory sex gland carcinoma. *Cancer Lett* **136**, 195–201.

Shahidi, F. and Naczk, M. 2004. *Phenolics in Food and Nutraceuticals.* Boca Raton, FL: CRC Press.

Siddiqui, I. A., Saleem, M., Adhami, V. M., Asim, M., and Mukhtar, H. 2007. Tea beverage in chemoprevention and chemotherapy of prostate cancer. *Acta Pharmacol Sin* **28**, 1392–1408.

Silbernagel, G., Fauler, G., Renner, W., Landl, E. M., Hoffmann, M. M., Winkelmann, B. R., Boehm, B. O., and März, W. 2008. The relationships of cholesterol metabolism and plasma plant sterols with the severity of coronary artery disease (The Ludwigshafen Risk and Cardiovascular Health Study). *J Lipid Res.* **50**, 334–341.

Sim, H. G. and Cheng, C. W. 2005. Changing demography of prostate cancer in Asia. *Eur J Cancer* **41**, 834–845.

Simonyi, A., Wang, Q., Miller, R. L., Yusof, M., Shelat, P. B., Sun, A. Y., and Sun, G. Y. 2005. Polyphenols in cerebral ischemia: Novel targets for neuroprotection. *Mol Neurobiol* **31**, 135–147.

Slawin, K., Kadmon, D., Park, S. H., Scardino, P. T., Anzano, M., Sporn, M. B., and Thompson, T. C. 1993. Dietary fenretinide, a synthetic retinoid, decreases the tumor incidence and the tumor mass of ras + myc-induced carcinomas in the mouse prostate reconstitution model system. *Cancer Res* **53**, 4461–4465.

Thompson, H. J. 2004. DNA oxidation products, antioxidant status, and cancer prevention. *J Nutr* **134**, 3186S–3187S.

Van Horn, L., McCoin, M., Kris-Etherton, P. M., Burke, F., Carson, J. A., Champagne, C. M., Karmally, W., and Sikand, G. 2008. The evidence for dietary prevention and treatment of cardiovascular disease. *J Am Diet Assoc* **108**, 287–331.

Vanharanta, M., Voutilainen, S., Lakka, T. A., van der Lee, M., Adlercreutz, H., and Salonen, J. T. 1999. Risk of acute coronary events according to serum concentrations of enterolactone: A prospective population-based case-control study. *Lancet* **354**, 2112–2115.

Vanharanta, M., Voutilainen, S., Rissanen, T. H., Adlercreutz, H., and Salonen, J. T. 2003. Risk of cardiovascular disease-related and all-cause death according to serum concentrations of enterolactone: Kuopio Ischaemic Heart Disease Risk Factor Study. *Arch Intern Med* **163**, 1099–1104.

Vidjaya Letchoumy, P., Chandra Mohan, K. V., Stegeman, J. J., Gelboin, H. V., Hara, Y., and Nagini, S. 2008. Pretreatment with black tea polyphenols modulates xenobiotic-metabolizing enzymes in an experimental oral carcinogenesis model. *Oncol Res* **17**, 75–85.

Walter, A., Etienne-Selloum, N., Sarr, M., Kane, M. O., Beretz, A., and Schini-Kerth, V. B. 2008. Angiotensin II induces the vascular expression of VEGF and MMP-2 *in vivo*: Preventive effect of red wine polyphenols. *J Vasc Res* **45**, 386–394.

Wiseman, H. 1996. Role of dietary phytoestrogens in the protection against cancer and heart disease. *Biochem Soc Trans* **24**, 795–800.

Wiseman, H. 2000. The therapeutic potential of phytoestrogens. *Expert Opin Investig Drugs* **9**, 1829–1840.

Witztum, J. L. 1994. The oxidation hypothesis of artherosclerosis. *Lancet* **344**, 793–796.

Wong, S. Y., Lau, W. W., Leung, P. C., Leung, J. C., and Woo, J. 2007. The association between isoflavone and lower urinary tract symptoms in elderly men. *Br J Nutr* **98**, 1237–1242.

Yang, C. S., Ju, J., Lu, G., Xiao, H., Hao, X., Sang, S., and Lambert, J. D. 2008. Cancer prevention by tea and tea polyphenols. *Asia Pac J Clin Nutr* **17** Suppl 245–248.

Zini, A., de Lamirande, E. and Gagnon, C. 1993. Reactive oxygen species in semen of infertile patients: Levels of superoxide dismutase and catalase-like activities in seminal plasma and spermatozoa. *Int J Androl* **16**, 183–188.

Zock, P. L. and Katan, M. B. 1998. Diet, LDL oxidation, and coronary artery disease. *Am J Clin Nutr* **68**, 759–760.

7 Therapeutic Potential of Ginseng for the Prevention and Treatment of Neurological Disorders

Jung-Hee Jang and Young-Joon Surh

CONTENTS

7.1 ORIGIN AND GENERAL MEDICINAL USE OF GINSENG

Ginseng, the root of the plant *Panax* (C.A. Meyer Araliaceae), has been used as an important component of herbal prescriptions in the eastern countries such as Korea, China, and Japan for more than 2000 years and now represents one of the most popular natural remedies with numerous applications. The genus "Panax," named by the Russian botanist C.A. Meyer, was derived from the Greek word *panacea*, representing "cure-all and longevity." The term "ginseng" derived from the Chinese word *rensheng*, which means "human" as ginseng roots resemble the shape of the human body (Radad et al., 2006). Since its introduction into western cultures, ginseng has occupied a permanent and prominent position in the best-seller list of herbal medicine worldwide.

Ginseng products are empirically and traditionally used as a general tonic and a psychic energizer to provide resistance to a wide range of physical, chemical, and biological stresses. It is regarded as an adaptogen, a substance which protects the body from harmful and deleterious factors by strengthening the normal physiological functions and improves the restoration of homeostasis (Nocerino et al., 2000). Therefore, ginseng is believed to enhance vitality, health, and longevity. Besides these tonic and adaptogenic effects, ginseng exhibits a variety of pharmacological activities including rejuvenating, antistress, and antifatigue properties (Kitts and Hu, 2000; Yue et al., 2007). Moreover, studies have shown the beneficial effects of ginseng against diverse pathological conditions such as cardiovascular diseases, cancer, diabetes mellitus, allergy, insomnia, gastritis, and hepatotoxicity (Kitts and Hu, 2000; Radad et al., 2006; Yue et al., 2007; Xiang et al., 2008). Ginseng and its active ingredients have been shown not only to potentiate physical functions, but also to make a positive contribution to amelioration of neurodegenerative diseases. These compounds have been reported to exhibit a number of actions in the central nervous system (CNS), such as anticonvulsant or antipsychotic effects and improvement of memory (Gillis, 1997; Radad et al., 2006; Rausch et al., 2006; Nah et al., 2007). Although ginseng exhibits multiple neuropharmacological properties in both *in vitro* and *in vivo* experimental systems, underlying molecular mechanisms are, in general, not fully defined. Accumulating data have revealed the antioxidant, anti-inflammatory, antiapoptotic, and immune-modulatory activities as plausible mechanistic basis for ginseng-mediated health beneficial effects (Radad et al., 2006). This chapter highlights the biological effects of ginseng in the CNS, especially in the context of ginseng as a potentially useful therapeutic choice for the prevention and/or treatment of neurological disorders.

7.2 PHYSICAL AND CHEMICAL PROPERTIES OF GINSENG

The species of the genus *Panax* include *Panax ginseng* C.A. Meyer (Korea), *Panax quinquefolius* (North America), *Panax notoginseng* (China), *Panax japonicus* (Japan), *Panax pseudoginseng* (eastern Himalayas), and *Panax vietnamensis* (Vietnam), depending on the cultivation area. Ginseng roots are harvested when the plant is 3- to 6-year old, and then the roots are subjected to air dry (white ginseng) or steam dry (red ginseng). Red ginseng is produced by steaming raw ginseng at 98–100°C for 2–3 h. Red ginseng has been considered to possess more pronounced biological activities than white ginseng, which may result from changes in the chemical constituents during the steaming process. Recently, several methods have been developed to enhance the yield of red ginseng-specific ingredients by steaming ginseng at a temperature higher than that adopted for the conventional preparation of red ginseng. These heat-processed ginsengs exhibited distinct pharmacological activities, such as vasorelaxation, antioxidant and antitumor effects (Keum et al., 2000; Park et al., 2002; Kang et al., 2006). In particular, the radical scavenging activities of the heat-processed ginseng were stronger than those of white ginseng (Kang et al., 2006).

In general, the ginseng root consists of two major ingredients: saponin and nonsaponin fractions. Ginseng saponins, more specifically ginsenosides, are derivatives of triterpene dammarane and have a common basic structure, characterized by a steroid-like four-ring system with carbohydrate moieties attached. Ginsenosides are the principal active ingredients of ginseng, and more than 40 different ginsenosides have been isolated and identified. Based on the position, the number and the type of different carbohydrate moieties C-3 and C-6, ginsenosides are mainly classified into three major categories: 20(*S*)-protopanaxadiol (e.g., Rb_1, Rb_2, Rb_3, Rc, Rd, Rg_3, Rh_2, Rs_1), 20(*S*)-protopanaxatriol (e.g., Re, Rf, Rg_1, Rg_2, Rh_1), and oleanolic acid groups (e.g., Ro) (Radad et al., 2006; Nah et al., 2007; Choi, 2008). The ginsenoside content of ginseng varies depending on the plant species, age and size, part, growing and harvesting conditions, preservation methods, and extraction protocols. Moreover, pharmacological actions of the panaxadiol-type ginsenosides have been attributed to their metabolites (Hasegawa et al., 1996; Hasegawa, 2004). For example, protopanaxadiol-type ginsenodies are transformed to compound K and protopanaxatriol-type ginsenosides are converted into ginsenoside Rh_1 by intestinal bacteria when taken orally (Hasegawa, 2004). The nonsaponin fraction includes essential oils, polyacetylenes, alkaloids, phytoesterols, phenolics, polysaccharides, peptides, fatty acids, and trace minerals. The structures of many native ginsenosides have been summarized in Table 7.1.

7.3 NEUROPROTECTIVE EFFECTS OF GINSENG

Recently, herbal medicine has received much attention and is recommended as a natural alternative to maintain or improve mental as well as physical health. An increasing number of studies have described the beneficial effects of ginseng and its principal ginsenosides on several types of neurodegenerative diseases and neuropathological conditions. The neuropharmacological effects of ginseng extracts as well as their active ingredients are described below.

7.3.1 Alzheimer's Disease

Alzheimer's disease (AD) is a representative neurodegenerative disease, which accounts for most dementia diagnosed in the individuals after the age of 60. The two neuropathological hallmarks of AD are extracellular deposition of senile plaques and intracellular formation of neurofibrillary tangles in the brain (Citron, 2004). β-Amyloid peptide (Aβ) produced from amyloid precursor protein (APP) via proteolytic cleavage by sequential actions of β- and γ-secretases forms the major constituent of the plaque core and plays a critical role in neuronal damage and cell death implicated in AD (Citron, 2004).

Ginsenoside Rg_1 effectively rescued Chinese hamster ovary (CHO) cells from the cytotoxicity induced by excessive endogenous $A\beta_{1-42}$ (Wei et al., 2008). In this study, to overproduce $A\beta_{1-42}$ peptide, CHO cells were stably transfected with APP751 and mutant presenilin 1 (PS1). Chen et al. compared the effects of several ginsenosides on Aβ production in CHO cells stably transfected with human βAPP 695 wt and found that Rg_3 was the most potent in terms of reducing $A\beta_{1-40}$ and

TABLE 7.1
Representative Ginsenosides and Their Chemical Structures

Groups	Ginsenosides	R1	R2	R3
PPD	Ra_1	*glc* (2-1) *glc*	H	*glc* (6-1) *arap* (4-1) *xyl*
	Rb_1	*glc* (2-1) *glc*	H	*glc* (6-1) *glc*
	Rb_2	*glc* (2-1) *glc*	H	*glc* (6-1) *arap*
	Rb_3	*glc* (2-1) *glc*	H	*glc* (6-1) *xyl*
	Rc	*glc* (2-1) *glc*	H	*glc* (6-1) *araf*
	Rd	*glc* (2-1) *glc*	H	*glc*
	Rg_3	*glc* (2-1) *glc*	H	H
	Rh_2	*glc*	H	H
	Rs_1	*glc* (2-1) *glc*	H	*glc* (6-1) *arap*
PPT	Re	H	O-*glc* (2-1) *rha*	*glc*
	Rf	H	O-*glc* (2-1) *glc*	H
	Rg_1	H	O-*glc*	*glc*
	Rg_2	H	O-*glc* (2-1) *rha*	H
	Rh_1	H	O-*glc*	H

Abbreviations: PPD, 20(*S*)-protopanaxadiol; PPT, 20(*S*)-protopanaxatriol; *glc*: β-D-glucopyranosyl, *arap*: α_L-arabinopyranosyl, *araf*: α_L-arabinofuranosyl, *xyl*: β_D-xylopyranosyl, *rha*: α_L-rhamanopyranosyl.

$A\beta_{1-42}$ accumulation (Chen et al., 2006). Furthermore, studies in the Tg2576 mouse model revealed that even a single oral administration of ginsenoside Re, Rg_1, or Rg_3, resulted in a significant decrease in the amount of Aβ (Chen et al., 2006). Exogenous $A\beta_{25-35}$-induced cytotoxicity was significantly reduced by Re treatment (Ji et al., 2006). In a mouse model of AD, which employed i.c.v. injection of $A\beta_{25-35}$, impaired spatial memory was recovered by oral administration of Rb_1, or its bioactive metabolite, compound K (Tohda et al., 2004). Rb_1 and compound K had axonal extension activity in degenerated neurons and improved memory as well as synaptic loss caused by $A\beta_{25-35}$ (Tohda et al., 2004). After the treatment with Rb_1 and compound K, mice did better on the hidden platform on successive trials in the water-maze test. In Sprague-Dawley (SD) rats, pretreatment of ginseng saponins minimized the impairment of memory caused by the i.c.v injection of $A\beta_{25-35}$ and attenuated the $A\beta_{25-35}$-induced decrease in the hippocampal acetylcholine (Ach) release (Wang et al., 2006a).

Ach is a very important neurotransmitter in the brain, a relatively low concentration of which is usually associated with learning and memory impairments (Musial et al., 2007). Some ginsenosides facilitated neurotransmission by increasing cholinergic metabolism in the CNS. Rg_1 and Rb_1 elevated the level of Ach in the CNS, which might be due to enhancement of choline acetyltransferase (ChAT) activity and inhibition of acetylcholine esterase (AchE) (Cheng et al., 2005). The abnormal functions of ChAT, a key enzyme in the Ach synthesis, and AchE, a critical enzyme in the degradation of Ach, correlated with the severity of dementia and the gradual development of the neuropathological manifestations in AD (Cheng et al., 2005). Rb_1 has been considered to partially prevent the experimentally induced memory deficits by modulating the Ach metabolism in CNS (Benishin et al., 1991). Rb_1 treatment facilitated the release of Ach from hippocampal slices, which was associated with increased uptake of choline into the central cholinergic nerve endings (Benishin et al., 1991; Lee et al., 2001). In the forebrain, Rb_1 upregulated the mRNA expression of ChAT and tropomyosin-related kinase A (TrkA), a receptor that binds with high affinity to the neurotrophin ligand, nerve growth factor (NGF) (Salim et al., 1997). Along with the loss of cholinergic neurons, muscarinic acetylcholine receptors (mAchRs), a member of the G-protein-coupled receptor family in the cerebral cortex and hippocampus, are functionally inactive in AD, and recent drug development for AD treatment focuses on identifying muscarinic receptor agonists (Fisher et al., 2003). Panaxynol concentration dependently upregulated the mRNA expression of M1 muscarinic receptor via the cAMP-mediated pathway in CHOm1 cells (Hao et al., 2005).

7.3.2 Cognitive Impairment and Memory Deficit

Scopolamine is a nonselective muscarinic receptor antagonist, which has been widely used to induce learning and memory deficit in animals as the experimental models of amnesia. The extracts from the root of American ginseng (Sloley et al., 1999) as well as Korean red ginseng (Jin et al., 1999), panaxynol (Yamazaki et al., 2001) and ginsenosides Rg_1 and Re (Yamaguchi et al., 1997) reversed the scopolamine-induced learning and memory dysfunction by improving cholinergic activity. The American ginseng extract, HT-1001, increased choline uptake in synaptosomal preparations without affecting accumulation of other aminergic neurotransmitters such as norepinephrine, dopamine, 5-HT (serotonin), 3,4-dihydroxyphenylacetic acid, and 5-hydroxyindoleacetic acid (Sloley et al., 1999). The protective effects of Rg_1 and Re on scopolamine-induced impaired memory performance were related to the increased activity of ChAT in the medial septum, which may stimulate cholinergic neuronal terminals to release Ach, thereby ameliorating the memory deficits caused by scopolamine (Yamaguchi et al., 1997). Oral administration of ginseng extracts improved the 8-arm radial maze performance disrupted by scopolamine in a dose-related manner (Nitta et al., 1995a). The memory impairment induced by oral administration of ethanol or intraperitoneal injection of scopolamine was significantly restored by ginsenosides $Rg_3(S)$ and Rg_5/Rk_1 (Bao et al., 2005).

Aging is one of the major risk factors for cognitive dysfunction and memory deficit. Subchronic treatment of the ginseng extract (8 g/kg/day, p.o. for 12–33 days) attenuated the age-related impairment of spatial cognitive learning performance in the radial maze task (Nitta et al., 1995b). In the passive avoidance test, the *Panax ginseng* (30 mg/kg/day, p.o. for 13 days) treatment considerably prolonged the latency in 27-month-old rats when compared with untreated control animals (Jaenicke et al., 1991). However, in the open field, the treated rats exhibited neither an altered locomotion or exploration nor a behavioral change, which could explain the improved avoidance reaction (Jaenicke et al., 1991). In addition, the nonsaponin fraction of red ginseng also significantly ameliorated impairments in place-navigation learning in the aged rats as assessed by the place-learning task and augmented long-term potentiation (LTP) in the mossy fiber-CA3 synapses (Kurimoto et al., 2004). LTP has been considered as an important index of cognitive activities in cellular and synaptic levels and proposed as a neural model for memory storage (Wang and Zhang, 2001). LTP in the hippocampus is the key form of longlasting synaptic plasticity and connects the behavior of learning and memory with the plasticity of neurons. Rg_1 was found to improve the basic synaptic transmission and increased the amplitude of LTP induced by high-frequency stimulation in anesthetized rats (Wang and Zhang, 2001).

7.3.3 Parkinsons's Disease

Parkinson's disease (PD) is currently regarded as the second common degenerative disorder of the aged brain after AD. PD is characterized by the tetrad of tremor at rest, slowness of voluntary movements, rigidity, and postural instability. These neurophysiological changes are caused by the loss of dopamine (DA)-producing neurons in the ventral mid-brain with cell bodies in the substantia nigra pars compacta (SNpc) that project to the striatum (nigrostriatal pathway) (Bóve et al., 2005; Moore et al., 2005). Among the neurotoxins generally used in experimental models of PD, 1-methyl-4-phenyl-1,2,3,6-tetrahydropyridine (MPTP), rotenone, 6-hydroxydopamine (6-OHDA) and paraquat (Bóve et al., 2005) have remained the most popular tools to produce selective neuronal death.

Prolonged oral administration of the ginseng extract significantly protected against neurotoxicity induced by MPTP and its toxic metabolite 1-methyl-4-phenylpyridinium (MPP^+) in rodent models of PD (Van Kampen et al., 2003). MPP^+, the active metabolite of MPTP, is highly selective and toxic for nigrostriatal dopaminergic neurons. MPTP and its metabolite inhibit NADH-dehydrogenase (Complex I) of the mitochondrial respiratory chain, which leads to increased production of free radicals and impaired production of ATP, thereby causing oxidative damage to dopaminegic neurons (Bóve et al., 2005). Ginseng treatment effectively attenuated MPP^+-induced neuronal damage and tyrosine hydroxylase (TH)-positive neuronal loss in SNpc (Van Kampen et al., 2003). In this study, striatal dopamine transporter was significantly preserved following the ginseng treatment. Pretreatments of C57-BL mice with Rg_1 protected against MPTP-induced neuronal loss in substantia nigra (Chen et al., 2005). MPTP decreased the number of TH- and Nissl-positive neurons in substantia nigra, which were restored by pretreatment with Rg_1 (Chen et al., 2002). In an *in vitro* model, aqueous extracts of ginseng radix and Rg_1 ameliorated MPP^+-induced cytotoxicity in PC12 (Kim et al.,

2003) and SH-SY5Y (Chen et al., 2003a) cells, respectively. Ginsenosides Rb_1 and Rg_1 enhanced the survival of dopaminergic cells in primary culture, promoted their neurological growth and attenuated the degenerative morphological changes after exposure to MPP^+ (Radad et al., 2004a).

Rotenone is the most potent member of the rotenoids extracted from *Leguminosae* plants and has been used as an insecticide. Rotenone is known to impair oxidative phosphorylation in mitochondria by specific and irreversible inhibition of the mitochondria complex (Bóve et al., 2005). Co-treatment of Rg_1 with rotenone increased the survival of cells comprising substantia nigra in a primary culture (Leung et al., 2007). As determined by the lactate dehydrogenase cytotoxicity assay and flow cytometry, co-treatment of Rg_1 increased the LC_{50} value of rotenone in the whole neuronal culture as well as in the TH-immunopositive dopaminergic neurons (Leung et al., 2007). Dopamine can be oxidized to generate semiquinone and quinone, with concomitant generation of reactive oxygen species (ROS), which are responsible for neuronal cell death (Miyazaki and Asanuma, 2008). Administration of dopamine directly into the brain leads to neuronal death in experimental animal models *in vivo*. Pretreatment of Rg_1 attenuated dopamine-induced apoptosis in PC12 cells by blunting oxidative stress (Chen et al., 2003b).

7.3.4 Ischemic Brain Injuries

Vascular dementia is the second most common cause of dementia in the elderly population after AD. Vascular dementia is defined as a loss of cognitive functions resulting from ischemic or hemorrhagic brain lesions due to cerebrovascular diseases. Ischemic injury is mainly caused by the interruption of blood flow, hypoxia, ATP depletion and subsequent reoxygenation of the brain in ischemia–reperfusion (Mehta et al., 2007). Experimental models of stroke have been developed in animals to mimic the neuropathological condition of human cerebral ischemia. The focal ischemia is induced by transient or permanent occlusion of the middle cerebral artery, which is the result of blood vessel occlusion due to thrombosis or embolism and affects the portion of the brain depending upon the vessel occluded (Millikan, 1992; Shah et al., 2005). On the other hand, global ischemia is induced by bilateral occlusion of carotid and vertebral arteries, which is similar to the events caused by the systemic decrease in blood flow and affects the entire brain (Millikan, 1992; Shah et al., 2005).

Korean ginseng tea exhibited neuroprotective and cerebroprotective properties against global as well as focal ischemia in rats (Shah et al., 2005). The administration of red ginseng powder and its active constituent Rb_1 before transient forebrain ischemia in gerbils increased response latency, as determined by the passive avoidance test, and rescued a significant number of ischemic hippocampal CA1 pyramidal neurons (Wen et al., 1996). In addition, Siberian ginseng (Bu et al., 2005) and Rg_2 (Zhang et al., 2008) reduced the infarct volume of focal cerebral ischemia induced by transient middle cerebral artery occlusion in SD rats. The intracerebroventricular infusion of ginsenoside Rb_1 after transient forebrain ischemia prevented disability in the passive avoidance task and rescued hippocampal CA1 pyramidal cells from delayed neuronal death (Lim et al., 1997). In addition, postischemic intravenous infusion of Rb_1 attenuated cerebrocortical infarct formation, place navigation disability and secondary thalamic degeneration in stroke-prone spontaneously hypertensive rats with permanent occlusion of the unilateral middle cerebral artery distal to the striate branches (Zhang et al., 2006). Orally administered ginsenoside Rh_2, a metabolite of Rg_3 formed by intestinal bacteria, substantially protected against ischemia–reperfusion brain injury induced by occlusion of the middle cerebral artery in SD rats (Park et al., 2004). The combined treatment of *Panax notoginseng* saponins with icariin, a major flavonoid component of *Epimedium pubescens*, was found to be protective against learning and memory impairment caused by transient cerebral ischemia–reperfusion injuries in SD rats (Zheng et al., 2008). This combination therapy significantly improved the rat passive avoidance task in step-down latency and spatial cognition in the 8-arm radial maze concomitant with an improvement of blood viscosity.

7.3.5 Other Neuronal Injuries

Glutamate plays a dominant role in central excitatory neurotransmission and is known to be a potent neurotoxin when present in excess at synapses. The excitotoxicity of glutamate has been shown to contribute to neuronal degeneration in acute as well as chronic disease conditions, such as stroke, epilepsy, trauma, hypoxia, and hyperglycemia (Lipton and Rosenberg, 1994; Kim et al., 1998; Radad et al., 2004b). Glutamate-mediated toxicity has been attributed to a massive influx of calcium through the NMDA receptor, which leads to mitochondrial depolarization and increased production of ROS. Rb_1 and Rg_3 protected cultured cortical neurons from the deleterious effects of glutamate (Kim et al., 1998). Rb_1 and Rg_1 increased the survival of primary cultures from embryonic mouse mesencephala and promoted lengths and numbers of the neurites in surviving dopaminergic cells after glutamate toxicity, which may be mediated by reducing Ca^{2+} overinflux into the mitochondria and increasing energy production by the mitochondria (Radad et al., 2004b). In line with this notion, ginsenoside Rg_2 effectively rescued PC12 cells from glutamate-induced neural injury and decreased Ca^{2+} influx (Li et al., 2007).

Kainic acid is a glutamate analogue with excitotoxic properties and used as a model agent for the study of excytotoxicity. In ICR mice, ginseng total saponin inhibited the kainic acid-induced lethal toxicity in a dose-dependent manner (Lee et al., 2003). Ginsenosides abrogated kainic acid-induced neuronal damage in CA1 and CA3 regions of rat hippocampus (Lee et al., 2002). In particular, Rb_1 and Rg_1 protected neurons in the spinal cord culture from neurotoxicity caused by glutamate, kainic acid, and hydrogen peroxide (Liao et al., 2002).

3-Nitropropionic acid (3-NP), a compound found in crops contaminated with fungi, is a reversible succinate dehydrogenase inhibitor and causes cellular ATP depletion and selective striatal degeneration similar to Huntington's disease (Miller and Zaborszky, 1997). Ginseng total saponins protected against systemic 3-NP-induced lesions in rat striatum, improved behavioral impairments such as dystonia, limb paralysis, and other abnormal motor abilities, and consequently extended survival (Kim et al., 2005). Pretreatment of the Rb-enriched extract or individual ginsenosides Rb_1, Rb_3, or Rd significantly reduced the 3-NP-induced motor impairment and cell loss in the striatum and completely prevented mortality (Lian et al.,

2005). In addition, ginseng total saponins protected hippocampal neurons in the experimentally induced traumatic brain injury (Ji et al., 2005). Thus, ginseng total saponin decreased hippocampal neuronal loss in the CA2 and CA3 regions, cortical contusion volume, and neurological dysfunction caused by a controlled cortical impact injury (Ji et al., 2005). The controlled cortical impact model can produce graded, reproducible, and quantifiable brain injury in rats. Intravenous infusion of dihydroginsenoside Rb_1, a stable chemical derivative of ginsenoside Rb_1 ameliorated morphological damage of the spinal cord, motor deficit, and behavioral abnormalities after spinal cord injury (Sakanaka et al., 2007). The diverse neuropharmacological effects of ginseng extracts and ginsenosides are summarized in Table 7.2.

7.4 POTENTIAL NEUROPROTECTIVE MECHANISMS OF GINSENG

Although the exact molecular mechanisms of neurological damages caused under diverse neuropathological conditions have yet to be elucidated, extensive experimental and clinical studies have provided evidence for the possible involvement of oxidative stress, proinflammatory responses, excitotoxicity, and apoptotic cell death (Kajta, 2004; Halliwell, 2006; Zipp and Aktas, 2006). The following sections deal with the antiapoptotic, antioxidative, and anti-inflammatory effects of ginseng in relation to its potential for the prevention and/or treatment of various types of neurological disorders.

7.4.1 ANTIAPOPTOSIS

Apoptosis or programmed cell death is normally required for maintenance of the cellular homeostasis and an essential physiological process during development, differentiation, and aging. However, defective regulation of apoptosis is implicated in the etiology of cancer, aging, autoimmune diseases, and neurodegenerative disorders. Apoptotic cell death is frequently observed under neuropathological conditions (Kajta, 2004; Ribe et al., 2008).

Apoptosis is characterized by distinct morphological and biochemical changes, including cell shrinkage, chromatin condensation, internucleosomal DNA fragmentation into 180–200 base pairs and formation of apoptotic bodies (Cummings et al., 1997). Pretreatment with Rg_1 inhibited the $A\beta_{1-42}$-induced apoptosis in mutant PS1 M146L cells as determined by annexin V/propidium iodine and TUNEL staining (Wei et al., 2008). Compared with the MPTP treatment alone, pretreatment with Rg_1 increased the number of Nissl- and TH-positive neurons and decreased the proportion of TUNEL-positive cells in the SNpc (Chen et al., 2002). In addition, ginseng radix exerts a significant neuroprotective effect against MPP^+-induced apoptosis in PC12 cells (Kim et al., 2003). In a primate model of thromboembolic stroke induced by occlusion of the middle cerebral artery, Rb_1 decreased the ischemic peri-infarct area by suppressing the activity of astrocytes, thereby preventing a neuronal cell death caused by ischemic stroke (Yoshikawa et al., 2008).

At the molecular level, apoptosis is mediated by the activation of the caspase (cysteine-dependent aspartate-directed protease) cascade. Activation of caspase-3, the most extensively investigated member of the caspase family, is a key step in the

TABLE 7.2
Protective Effects of Ginseng Extracts and Individual Ginsenosides on Experimentally Induced Neuronal Injury

Neuropathological Condition	Neuronal Damage/ Neurotoxicant	Ginseng/Ginsenoside	Experimental Model	References
Alzheimer's disease	$A\beta_{25-35}$	Rb_1, compound K	ddY mice	Tohda et al. (2004)
		Ginseng saponin	SD rats	Wang et al. (2006a)
		Rb_1	hippocampal slice	Lee et al. (2001)
		Re	PC12 cells	Ji et al. (2006)
	$A\beta_{1-42}$	Rg_1	CHO cells	Wei et al. (2008)
		Rg_3	BV-2 cells	Joo et al. (2008)
	$A\beta_{1-40}$, $A\beta_{1-42}$	Re, Rg_1, Rg_3	CHO cells	Chen et al. (2006)
		Re, Rg_1, Rg_3	Tg2576 mice	Chen et al. (2006)
Cognitive Impairment/Memory Deficit	Scopolamine	American ginseng	SD rats	Sloley et al. (1999)
		Korean red ginseng	ICR mice	Jin et al. (1999)
		Panaxynol	ddY mice	Yamazaki et al. (2001)
		Rg_1, Re	SD rats	Yamaguchi et al. (1997)
		Ginseng extract	Wistar rats	Nitta et al. (1995a)
	Aging	Ginseng extract	Fisher 344 rats	Nitta et al. (1995b)
		Ginseng extract	Wistar rats	Jaenicke et al. (1991)
		Nonsaponin fraction	Fisher 344 rats	Kurimoto et al. (2004)
Parkinson's disease	MTP^+	Aqueous extract	PC12 cells	Kim et al. (2003)
		Rg_1	SH-SY5Y cells	Chen et al. (2003a)
		Rb_1, Rg_1	Primary culture	Radad et al. (2004a)
		Ginseng extract	SD rats	Van Kampen et al. (2003)
	MPTP	Ginseng extract	C57B16 mice	Van Kampen et al. (2003)
		Rg_1	C57-BL mice	Chen et al. (2002, 2005)

	Rotenone	Rg_1	Primary culture	Leung et al. (2007)
	Dopamine	Rg_1	PC12 cells	Chen et al. (2003b)
Ischemic brain injuries	Global ischemia	Ginseng tea	Swiss albino rats	Shah et al. (2005)
	Focal ischemia	Ginseng tea	Swiss albino rats	Shah et al. (2005)
		Rb_1, Ginseng powder	Mongolian gerbils	Wen et al. (1996)
		Siberian Ginseng	SD rats	Bu et al. (2005)
		Rg_2	SD rats	Zhang et al. (2008)
		Rb_1	Mongolian gerbils	Lim et al. (1997)
		Rb_1	SHR-SD rats	Zhang et al. (2006)
		Rh_2	SD rats	Park et al. (2004)
Other neurotoxicants-induced neuronal damage	Glutamate	Rb_1, Rg_3	Primary culture	Kim et al. (1998)
		Rb_1, Rg_1	Primary culture	Radad et al. (2004b)
		Rg_2	PC12 cells	Li et al. (2007)
	Kainic acid	Ginseng saponin	ICR mice	Lee et al. (2003)
		Ginsenosides	SD rats	Lee et al. (2002)
	Glutamate, H_2O_2, Kainic acid	Rb_1, Rg_1	Primary culture	Liao et al. (2002)
	3-NP	Ginseng saponin	SD rats	Kim et al. (2005)
		Rb extract, Rb_1, Rb_3, Rd	SD rats	Lian et al. (2005)
	Traumaric injury	Ginseng extract	SD rats	Ji et al. (2005)
	Spinal cord injury	DihydroRb_1	Wistar rats	Sakanaka et al. (2007)

execution of apoptosis and its inhibition blocks apoptotic cell death (Brown and Borutaite, 2008; Wong and Puthalakath, 2008). Cytosolic cytochrome *c* has previously been demonstrated to bind apoptosis protease activating factor-1 (Apaf-1) and to subsequently trigger the sequential activation of caspase-9 and caspase-3 (Brown and Borutaite, 2008; Wong and Puthalakath, 2008). Rg_1 pretreatment reduced the levels of $A\beta_{1-42}$ and activation of caspase-3 in mutant PS1 M146L cells (Wei et al., 2008). Pretreatment of PC12 cells with ginsenoside Rg_1 markedly suppressed the dopamine-induced release of mitochondrial cytochrome *c* into the cytosol and eventually inhibited caspase-3 activation (Chen et al., 2003b). Rotenone-induced mitochondrial apoptotic signaling as evidenced by dissipation of mitochondria membrane potential and cytochrome *c* release was attenuated by Rg_1 (Leung et al., 2007). Moreover, Rg_2 inhibited glutamate-induced apoptotic cell death by downregulating the expression of proapoptotic factor calpain II and caspase-3 (Li et al., 2007).

Mitochondrial membrane permeability to cytochrome *c* is primarily regulated by Bcl-2 family proteins, some of which are antiapoptotic (e.g., Bcl-2, Bcl-X_L, and Bcl-w), while others are proapoptotic (e.g., Bax, Bad, Bcl-X_S, Bid, and Bak). The relative ratio of pro- to antiapoptotic members of Bcl-2 protein family is crucial for determining neuronal survival or death (Wong and Puthalakath, 2008). Rg_2 increased the ratio of Bcl-2 to Bax and decreased expression of the tumor suppressor p53 (Zhang et al., 2008). p53 exerts proapoptotic actions following cellular damage and stress in various types of cells and causes cell cycle arrest. Ginsenoside Rg_1 reduced the levels of cleaved caspase-3 and proapoptotic Bax, but elevated the expression of antiapoptotic Bcl-2 and Bcl-X_L during the MPTP-induced apoptosis (Chen et al., 2002). In addition, Rg_1 stimulated the phosphorylation of the proapoptotic protein Bad through activation of the PI3K/Akt pathway (Leung et al., 2007). An aqueous extract of ginseng radix inhibited MPP^+-induced cytotoxicity, morphological changes, and apoptotic cell death through inhibition of p53 and Bax and activation of caspase-3 in PC12 cells (Kim et al., 2003). Ginsenoside Rb_1 stimulated neuronal Bcl-X_L protein expression and prevented neuronal apoptosis *in vitro*, and intravenous infusion of Rb_1 in the middle cerebral artery occluded SHR-SP mice increased mRNA and protein levels of Bcl-X_L (Zhang et al., 2006). Moreover, intravenous infusion of dihydroginsenoside Rb_1 prevented spinal cord injury and ischemic damage via upregulation of Bcl-X_L and vascular endothelial growth factor (VEGF), a potent angiogenic and neurotrophic factor (Sakanaka et al., 2007).

Another key regulatory factor involved in the apoptotic process is c-Jun *N*-terminal kinase (JNK). JNK increases phosphorylation and expression of c-Jun, a major subunit of the AP-1 transcription factor family, which mediates an immediate-early gene response following cellular exposure to external stimuli. AP-1 is also involved in the increased expression of proapoptotic Bax (Lei et al., 2002). JNK is speculated to phosphorylate other proteins such as Bcl-2 and Bcl-X_L and inhibit their antiapoptotic potential (Harada and Grant, 2003). MPP^+-induced apoptosis in SH-SY5Y cells via activation of JNK and caspase-3 was attenuated by Rg_1 (Chen et al., 2003a). Rg_1 inhibited the phosphorylation of JNK and c-Jun following the MPTP treatment in C57-BL mice (Chen et al., 2005). The intracellular signaling molecules involved in regulating cell survival and death that can be potential targets for antiapoptotic effects of ginseng, are depicted in Figure 7.1.

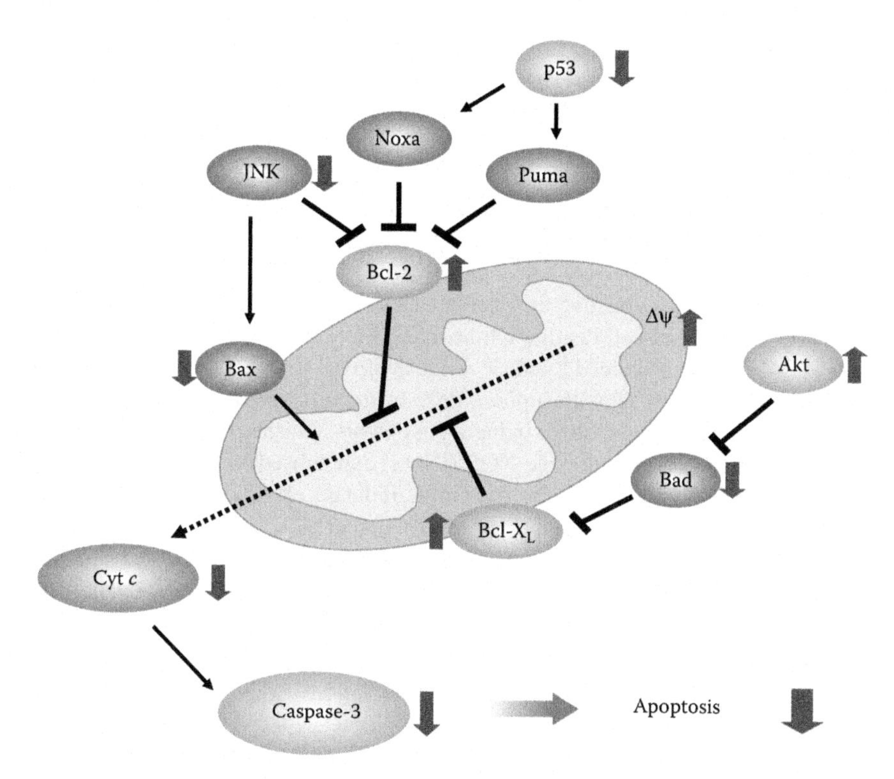

FIGURE 7.1 Intracellular signaling molecules as potential targets for the antiapoptotic effects of ginseng and its bioactive ingredients. The arrows indicate the up- or downregulation of the target molecules.

7.4.2 Antioxidant

Oxidative stress is one of the common players in the apoptotic cascades, leading to cell death in neurodegenerative diseases (Contestabile, 2001). Numerous neurotoxicants can induce apoptosis directly or indirectly by increasing the generation of ROS including hydrogen peroxide (H_2O_2), superoxide anion ($O_2^{\bullet -}$), and hydroxyl radical (•OH). Particularly, ROS accumulation leads to critical damages to cellular macromolecules, which include oxidation of proteins, lipid peroxidation, and oxidative DNA damage. The brain is particularly vulnerable to oxidative stress because of a relatively high metabolic rate, oxygen consumption, and fatty acid levels.

Ginsenosides could protect astrocytes and spinal cord neurons from oxidative stress generated by hydrogen peroxide, a representative ROS (Liao et al., 2002; López et al., 2007). Rb_1 rescued hippocampal CA1 neurons from lethal damage caused by ischemia, the hydroxyl radical-promoting agent $FeSO_4$, and the Fenton reaction system containing *p*-nitrosodimethylaniline, which is likely to be attributable to its hydroxyl radial-scavenging activity (Lim et al., 1997). The North American ginseng extract exhibited antioxidant activity by chelating transition meal ions as well as by scavenging free radicals (Kitts et al., 2000). Heat-processed ginseng showed more potent superoxide anion-, peroxynitrite-, and hydroxyl radical-scavenging activities than white ginseng (Kang et al., 2006). During heat processing, the content of maltol was remarkably increased in a temperature-dependent manner, which confers strong hydroxyl radical scavenging activity. American ginseng with steam-heat treatment exhibited higher levels of radical scavenging and antioxidant enzyme activities (Kim et al., 2007). The protective effect of Rg_1 against MPP^+-induced SH-SY5Y cell death was mediated by suppression of intracellular ROS accumulation (Chen et al., 2003a). Furthermore, Rg_1 attenuated dopamine-induced apoptosis through amelioration of oxidative stress in PC12 cells (Chen et al., 2003b).

In addition to ROS, reactive nitrogen species (RNS), such as nitric oxide (NO), play a role in neuronal cell death (Achike and Kwan, 2003; Calabrese et al., 2004). NO is endogenously produced by nitric oxide synthase (NOS). Under certain conditions, NO can rapidly react with superoxide to form peroxynitrite ($ONOO^-$), a highly reactive nitrating agent capable of inducing apoptosis. Ginsenoside Rg_1 pretreatment reduced protein levels of inducible NOS (iNOS) and subsequent production of NO. Rg_1 pretreatment could reduce the dopamine-induced elevation of NO generation and attenuate the neurotoxicity of dopamine (Chen et al., 2003b). Ginsenosides Rb_1 and Rg_3 inhibited the overproduction of NO associated with glutamate neurotoxicity, decreased the formation of malondialdehyde, a compound produced during lipid peroxidation, and diminished the influx of calcium (Kim et al., 1998).

To counteract excess ROS and/or RNS, our body operates cellular defense systems, which consist of antioxidants and antioxidant enzymes that lower intracellular levels of ROS and/or RNS or repair oxidative cellular damage (Jang and Surh, 2003; Dringen, 2005). Antioxidant enzymes include superoxide dismutase (SOD), catalase (CAT), and glutathione peroxidase (GPx). These antioxidant enzymes function in concert to inactivate ROS and/or RNS and thus rescue cells from oxidative/nitrosative damage (Jang and Surh, 2003; Dringen, 2005). SOD converts superoxide anion into hydrogen peroxide, which is in turn decomposed to water and oxygen by CAT. GPx detoxifies hydrogen peroxide in the presence of GSH, producing water and oxidized glutathione (GSSG), which is recycled to GSH by glutathione reductase (GR). GSH, a ubiquitous tripeptide thiol, is considered as a central component of the intracellular antioxidant defense system and acts to directly detoxify ROS/RNS (Dringen and Hirrlinger, 2003).

There is a considerable body of evidence to support that ginseng has important roles in maintaining the cellular redox balance by fortifying cellular antioxidant functions. Korean ginseng tea exhibited marked protection against experimentally induced ischemia in rats. Global and focal ischemia elevated lipid peroxidation and depleted protective antioxidant enzymes such as GR, CAT, GST, GPx, and SOD, which were effectively restored by the Korean ginseng tea treatment (Shah et al., 2005). A mixture of ginsenosides significantly improved SOD and GPx activity and reduced lipid peroxidation in cerebral ischemia/reperfusion in Wistar rats (Zhou et al., 2006). Ginsenosides decreased H_2O_2-induced cytotoxicity as well as ROS formation in primary cultured astrocytes mainly through the activation of antioxidant enzymes such as CAT, SOD, GPx, and GR (Kitts et al., 2000). In glutamate-treated cortical cultures, Rb_1 and Rg_3 significantly prevented a decrease in the SOD activity whereas the CAT or GPx activity was not much affected by these ginsenosides (Kim et al., 1998). Rg_1 prevented GSH reduction and SOD activation in substantia nigra

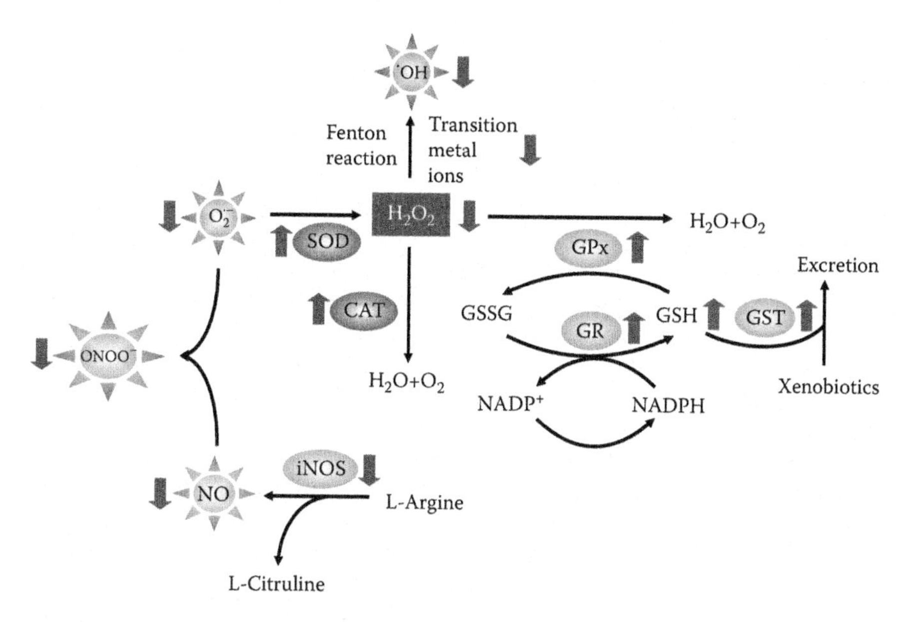

FIGURE 7.2 Proposed molecular mechanisms responsible for the antioxidant effects of ginseng and ginsenosides. The arrow denotes effects of ginseng.

following MPTP treatment in C57-BL mice (Chen et al., 2005). Although total saponin and panaxatriol did not activate the transcription of Cu/Zn-SOD (SOD1) and CAT genes but panaxadiol including Rb_2 served as a major inducer of the aforementioned antioxidant genes (Chang et al., 1999). SOD1 gene was greatly activated by ginsenoside Rb_2 through the activation of the transcription factor AP2 (Kim et al., 1996). Mutations of the AP2 binding sites in the heterologous promoter and natural context systems abolished the transcriptional activation of SOD1 by ginsenoside Rb_2 (Kim et al., 1996). Figure 7.2 illustrates representative entities mediating oxidative or nitrosative stress and antioxidant enzymes responsible for inactivation/elimination of ROS or RNS.

7.4.3 Anti-Inflammation

Proinflammatory cascades occurring in the CNS are closely related to pathways leading to neuronal cell death in many types of neurodegenerative conditions such as AD, PD, and ischemia (Zipp and Aktas, 2006). In particular, chronic inflammation in the brain is harmful because of the possible vulnerability of neurons to the inherent damage caused by inflammatory processes. Suppressing inflammation is now regarded as an important preventive or therapeutic approach in the management of chronic neurodegenerative diseases, as corroborated by the observation that long-term use of nonsteroidal anti-inflammatory drugs (NSAIDs) provides some protection against the development and progression of AD and PD (Lleo et al., 2007).

A variety of inflammatory modulators including complement proteins, cell adhesion molecules, and inflammatory cytokines are formed and secreted by both microglia and astrocytes in the brain. These factors propagate and maintain neuroinflammation via multiple mechanisms, including the activation of proinflammatory enzymes such as cyclooxygenases (COX) and NOS (Zipp and Aktas, 2006). COX converts arachidonic acid to prostaglandin H_2. Another prominent inflammatory mediator, NO is endogenously produced by NOS. Cytokines are major effectors of the neuroinflammatory cascade and play an important role in neural cell response to infection and brain injury (Zipp and Aktas, 2006). The expression of genes involved in the inflammatory response is regulated both transcriptionally and posttranscriptionally. The released cytokines react with their receptors leading to activation of protein kinase cascades and the pathways, resulting in the activation of transcription factors such as NF-κB. In the nucleus, NF-κB regulates the transcription of many genes implicated in inflammatory and immune responses (Dutta et al., 2006). These genes include those encoding COX-2, iNOS, tumor necrosis factor-α (TNF-α), interleukin-1β (IL-1β), and cell adhesion molecules.

Panax notoginseng inhibited lipopolysaccharide (LPS)-induced production of TNF-α and IL-6 in RAW264.7 cells (Rhule et al., 2006). Furthermore, mRNA expression of COX-2 and IL-1β was attenuated in LPS-stimulated cells following exposure to notoginseng (Rhule et al., 2006). RAW264.7 cells represent an immortalized murine cell line that has been extensively used to evaluate the fate and the function of monocytes and macrophages *in vitro*. Compound K (Park et al., 2005) and BST204 (Seo et al., 2005a, 2005b), a fermented ginseng extract, markedly inhibited the production of NO and prostaglandin E_2 in LPS-treated RAW264.7 cells. Compound K also reduced the protein expression of the iNOS and COX-2 and suppressed the activation of NF-κB, an upstream regulator of these proinflammatory enzymes (Park et al., 2005). In line with this notion, 20(*S*)-protopanaxatriol inhibited iNOS and COX-2 expression via the inactivation of NF-κB in LPS-stimulated RAW264.7 cells (Oh et al., 2004).

Ginseng extracts and ginsenosides have been shown to exhibit anti-inflammatory properties in neuronal cells as well as murine macropahges. The water extract of Siberian ginseng inhibited ischemia-induced inflammation and microglia activation as revealed by reduced expression of COX-2 and OX-42 (Bu et al., 2005). Rg_3 suppressed the $A\beta_{1\text{-}42}$-induced mass production of the proinflammatory repertoire of microglial cells via downregulation of NF-κB DNA binding. Rg_3 effectively reduced expression of inflammatory cytokines such as TNF-α and IL-1β in $A\beta_{1\text{–}42}$-treated cells, and inhibited the binding of NF-κB to its DNA consensus sequences, and significantly reduced iNOS levels in activated microglia (Joo et al., 2008). Ginsan, a polysaccharide extracted from ginseng, inhibited the p38 MAPK-NF-κB pathway and decreased the production of proinflammatory cytokines (Ahn et al., 2006). Rh_2 and compound K significantly reduced TNF-α-induced expression of intercellular adhesion molecule-1 in human astroglial cells (Choi et al., 2007). The inhibitory effects of Rh_2 and compound K are likely to be mediated by the suppression of IκBα and subsequent phosphorylation and degradation of IκBα which are responsible for the activation of NF-κB (Choi et al., 2007). Furthermore, ginsenoside Rh_2 inhibited LPS/IFN-γ-induced production of NO, which was correlated with the decreased protein and mRNA expression of iNOS in BV-2 microglial cells (Bae et al., 2006). Additionally, Rh_2 decreased the expression of COX-2, TNF-α, and

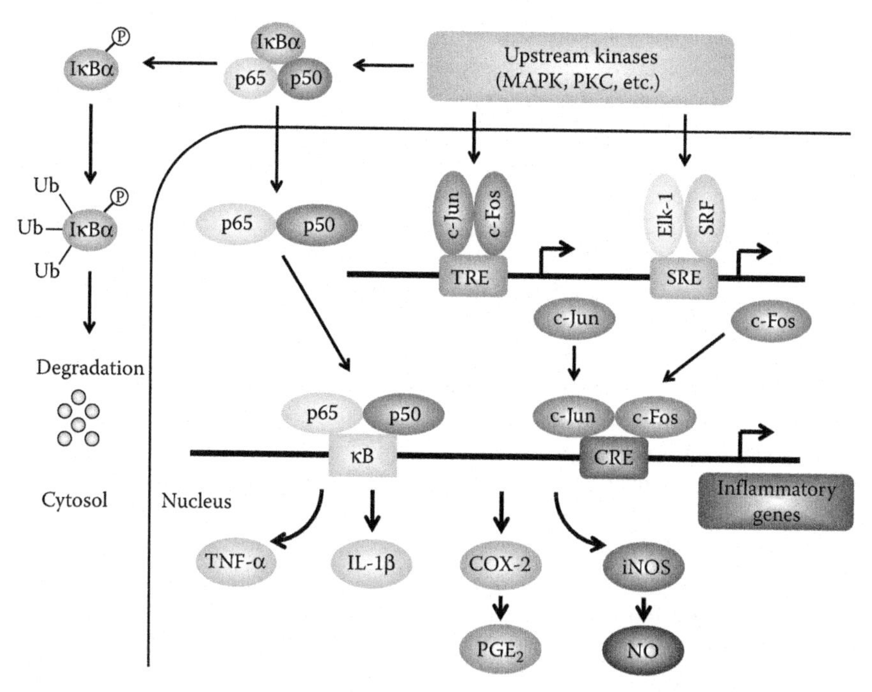

FIGURE 7.3 Intracellular signaling molecules as potential targets for the anti-inflammatory effects of ginseng and its bioactive ingredients.

IL-1β under the same experimental conditions. The anti-inflammatory effect of Rh_2 appears to depend on the AP-1 and protein kinase A (PKA) (Bae et al., 2006). However, the anti-inflammatory property of Rg_3 against LPS/IFN-γ-activated BV-2 cells was less prominent than that of its metabolite Rh_2. The anti-inflammatory effects and underlying mechanisms of ginseng are illustrated in Figure 7.3.

7.4.4 Other Mechanisms

In the CNS, abnormalities in intracellular Ca^{2+} homeostasis systems are involved in neurological disorders, such as excitatory neurotoxicity, epilepsy, AD, PD, and ischemia. Particularly, the excess Ca^{2+} influx by glutamate receptor activation causes oxidative stress, which leads to neuronal degeneration. Ginsenosides Rb_1, Rg_1, and Rg_2 attenuated glutamate-induced neuronal damage by reducing Ca^{2+} overinflux (Radad et al., 2004b; Li et al., 2007). Furthermore, ginseng total saponin, Rf and Rg_3 inhibited $[Ca^{2+}]_i$ overload and neuronal cell death (Kim et al., 2008). However, Rg_1 activated the extracellular signal-regulated kinase 1/2 (ERK1/2) pathway by increasing the intracellular calcium ion flux and phosphorylating CaMKIIa, which eventually led to the activation of cAMP response element binding protein (CREB) in PC12 cells (Hu et al., 2008). CREB was shown to be necessary for neurogenesis including cell proliferation and survival. The activation of the ERK pathway may contribute to the effects of Rg_1 on LTP induction and maintenance. NMDA receptor is a ligand-gated ion channel, which is permeable to Ca^{2+}. The activation of NMDA receptor is essential for synaptic plasticity such as LTP and long-term depression, which are thought to be involved in memory acquisition and learning. Prolonged intracerebroventricular infusion of Rg_1 increased the mRNA expression of NMDA receptor subunit NR1 in the caudate putamen, hippocampus, and granule layer of the cerebellum (Kim et al., 2000). On the other hand, the NMDA receptor plays a crucial role in glutamate-induced neurotoxicity via intracellular Ca^{2+} overload, and the NMDA receptor antagonists could attenuate neuronal cell death. In accordance with this notion, Rg_3 protected against NMDA-induced excitotoxicity in primary hippocampal neurons by competing for interaction with glycine-binding site of NMDA receptor (Kim et al., 2004).

Ginsenosides also exert their pharmacological activities in the CNS by modulating the availability of neurotransmitters. The plasticity of the synaptic function is regarded as one of the most important mechanisms underlying the process of learning and memory. Synapsins are abundant phosphoproteins essential for regulating neurotransmitter release (Fiumara et al., 2004). Ginsenoside Rb_1 promotes release of neurotransmitters by increasing the phosphorylation of synapsins via the cAMP-dependent PKA pathway (Xue et al., 2006). Compound K restored the $A\beta_{25-35}$-reduced levels of phosphorylated NF-H, an axonal marker and synaptophysin, a synaptic marker in the cerebral cortex and the hippocampus (Tohda et al., 2004). Compound K had axonal extension activity in degenerated neurons, and improved the memory disorder and synaptic loss induced by $A\beta_{25-35}$ through the teneurin-2-PI3-kinase cascade (Tohda et al., 2004). Rb_1 and Rg_1 enhanced spatial learning ability in the Morris water-maze task by increasing the hippocampal synaptic density without changing the plasticity of individual synapses in electrophysiological recordings (Mook-Jung et al., 2001). The hippocampus, but not the frontal cortex, of treated mice contained higher density of a synaptic marker protein synaptophysin, compared to control mice (Mook-Jung et al., 2001).

The proliferative ability of neural progenitor cells was reduced in association with age- or disease-related cognitive decline. Therefore, neurogenesis is regarded as a potential therapeutic target for neurodegenerative disorders (Kelleher-Andersson, 2004). Ginseng administration increased the number of BrdU-positive cells (neurogenesis) of the dentate gyrus and enhanced performance in contextual fear conditioning, which is hippocampus- and amygdala-dependent learning (Qiao et al., 2005). Ginsenoside, especially Rg_1, might regulate the proliferation and/or surviving ability of the neural progenitor class (Shen and Zhang, 2004). Rg_1 exerts partial neurotrophic and neuroprotective effects by increasing the mRNA expression of nerve growth factor (NGF) in the hippocampus (Shen and Zhang, 2004). Rb_1 also enhanced the stimulatory effect of NGF on neurite outgrowth (Nishiyama et al., 1994). In addition, Rg_5 serves as a neurogenic molecule, which can induce neuronal differentiation in epidermal growth factor-responsive neurosphere stem cell cultures (Liu et al., 2007). Panaxynol could morphologically promote neurite outgrowth in PC12D cells and expression of microtubule-associated phosphoprotein (MAP1B) via c-AMP- and MAPK-dependent mechanisms (Wang et al., 2006b). MAP1B is highly expressed in developing neurons especially in the extension of axons and dendrites (Gordon-Weeks and Fischer, 2000).

7.5 CONCLUDING REMARKS

Ginseng consumption is one of the most unique traditional herbal remedies with a long history of use and worldwide popularity. Ginseng has diverse biological functions. Various ginsenosides, the major well-defined constituents of ginseng, exhibit a vast variety of pharmacological activities in the CNS including memory improvement, neuroprotection, and neurogenesis. Recently, specific mechanisms underlying neuropharmacological effects of ginseng and its ingredients have been unraveled at the molecular and cellular levels. However, ginsenosides and other bioactive constituents of ginseng exhibit different efficacies, and a single ginsenoside initiates multiple actions in the same tissue, which makes the overall pharmacology of ginseng complicated. Therefore, further research is necessary to identify specific molecular targets for neuroprotection and to elucidate neurospecific components of ginseng. Additional studies should also be done to verify clinical efficacy and safety of the commercially available formulated ginseng products as well as individual ginsenosides. Elucidation of molecular mechanisms of ginseng and individual ginsenosides will provide new strategies for the prevention and/or treatment of neurodegenerative diseases and neurological disorders.

ACKNOWLEDGMENTS

This study was supported by the Biogreen 21 Program (no. 20070301-034-02) and Korea Ginseng Corporation.

REFERENCES

Achike, F. I. and Kwan, C. Y. 2003. Nitric oxide, human diseases and the herbal products that affect the nitric oxide signalling pathway. *Clin. Exp. Pharmacol. Physiol.* 30: 605–15.

Ahn, J. Y., Choi, I. S., Shim, J. Y., Yun, E. K., Yun, Y. S., Jeong, G., and Song, J. Y. 2006. The immunomodulator ginsan induces resistance to experimental sepsis by inhibiting Toll-like receptor-mediated inflammatory signals. *Eur. J. Immunol.* 36:37–45.

Bae, E. A., Kim, E. J., Park, J. S., Kim, H. S., Ryu, J. H., and Kim, D. H. 2006. Ginsenosides Rg_3 and Rh_2 inhibit the activation of AP-1 and protein kinase A pathway in lipopolysaccharide/interferon-gamma-stimulated BV-2 microglial cells. *Planta Med.* 72: 627–33.

Bao, H. Y., Zhang, J., Yeo, S. J., Myung, C. S., Kim, H. M., Kim, J. M., Park, J. H., Cho, J., and Kang, J. S. 2005. Memory enhancing and neuroprotective effects of selected ginsenosides. *Arch. Pharm. Res.* 28:335–42.

Benishin, C. G., Lee, R., Wang, L. C., and Liu, H. J. 1991. Effects of ginsenoside Rb_1 on central cholinergic metabolism. *Pharmacology* 42:223–9.

Bóve, J., Prou, D., Perier, C., and Przedborski, S. 2005. Toxin-induced models of Parkinson's disease. *NeuroRx.* 2:484–94.

Brown, G. C. and Borutaite, V. 2008. Regulation of apoptosis by the redox state of cytochrome c. *Biochim. Biophys. Acta* 1777:877–81.

Bu, Y., Jin, Z. H., Park, S. Y., Baek, S., Rho, S., Ha, N., Park, S. K., Kim, H., and Kim, S. Y. 2005. Siberian ginseng reduces infarct volume in transient focal cerebral ischaemia in Sprague–Dawley rats. *Phytother. Res.* 19:167–9.

Calabrese, V., Boyd-Kimball, D., Scapagnini, G., and Butterfield, D. A. 2004. Links nitric oxide and cellular stress response in brain aging and neurodegenerative disorders: The role of vitagenes. *In Vivo* 18:245–67.

Chang, M. S., Lee, S. G., and Rho, H. M. 1999. Transcriptional activation of Cu/Zn superoxide dismutase and catalase genes by panaxadiol ginsenosides extracted from *Panax ginseng*. *Phytother. Res.* 13:641–4.

Chen, X. C., Chen, Y., Zhu, Y. G., Fang, F., and Chen, L. M. 2002. Protective effect of ginsenoside Rg_1 against MPTP-induced apoptosis in mouse substantia nigra neurons. *Acta Pharmacol. Sin.* 23:829–34.

Chen, F., Eckman, E. A., and Eckman, C. B. 2006. Reductions in levels of the Alzheimer's amyloid beta peptide after oral administration of ginsenosides. *FASEB J.* 20: 1269–71.

Chen, X. C., Fang, F., Zhu, Y. G., Chen, L. M., Zhou, Y. C., and Chen, Y. 2003a. Protective effect of ginsenoside Rg_1 on MPP^+-induced apoptosis in SH-SY5Y cells. *J. Neural. Transm.* 110:835–45.

Chen, X. C., Zhou, Y. C., Chen, Y., Zhu, Y. G., Fang, F., and Chen, L. M. 2005. Ginsenoside Rg_1 reduces MPTP-induced substantia nigra neuron loss by suppressing oxidative stress. *Acta Pharmacol. Sin.* 26:56–62.

Chen, X. C., Zhu, Y. G., Zhu, L. A., Huang, C., Chen, Y., Chen, L. M., Fang, F., Zhou, Y. C., and Zhao, C. H. 2003b. Ginsenoside Rg_1 attenuates dopamine-induced apoptosis in PC12 cells by suppressing oxidative stress. *Eur. J. Pharmacol.* 473:1–7.

Cheng, Y., Shen, L. H., and Zhang, J. T. 2005. Anti-amnestic and anti-aging effects of ginsenoside Rg_1 and Rb_1 and its mechanism of action. *Acta Pharmacol. Sin.* 26:143–9.

Choi, K. T. 2008. Botanical characteristics, pharmacological effects and medicinal components of Korean Panax ginseng C A Meyer. *Acta Pharmacol.* 29:1109–18.

Choi, K., Kim, M., Ryu, J., and Choi, C. 2007. Ginsenosides compound K and Rh_2 inhibit tumor necrosis factor-α-induced activation of the NF-κB and JNK pathways in human astroglial cells. *Neurosci Lett.* 421:37–41.

Citron, M. 2004. Strategies for disease modification in Alzheimer's disease. *Nat. Rev. Neurosci.* 5:677–85.

Contestabile, A. 2001. Oxidative stress in neurodegeneration: Mechanisms and therapeutic perspectives. *Curr. Top. Med. Chem.* 1:553–68.

Cummings, M. C., Winterford, C. M. and Walker, N. I. 1997. Apoptosis. *Am. J. Surg. Pathol.* 21:88–101.

Dringen, R. 2005. Oxidative and antioxidative potential of brain microglial cells. *Antioxid. Redox Signal.* 7:1223–33.

Dringen, R. and Hirrlinger, J. 2003. Glutathione pathways in the brain. *Biol. Chem.* 384:505–16.

Dutta, J., Fan, Y., Gupta, N., Fan, G., and Gélinas, C. 2006. Current insights into the regulation of programmed cell death by NF-κB. *Oncogene* 25:6800–16.

Fisher, A., Pittel, Z., Haring, R., Bar-Ner, N., Kliger-Spatz, M., Natan, N., Egozi, I., Sonego, H., Marcovitch, I., and Brandeis, R. 2003. M1 muscarinic agonists can modulate some of the hallmarks in Alzheimer's disease: Implications in future therapy. *J. Mol. Neurosci.* 20:349–56.

Fiumara, F., Giovedì, S., Menegon, A., Milanese, C., Merlo, D., Montarolo, P. G., Valtorta, F., Benfenati, F., and Ghirardi, M. 2004. Phosphorylation by cAMP-dependent protein kinase is essential for synapsin-induced enhancement of neurotransmitter release in invertebrate neurons. *J. Cell. Sci.* 117:5145–54.

Gillis, C. N. 1997. Panax ginseng pharmacology: A nitric oxide link? *Biochem. Pharmacol.* 54:1–8.

Gordon-Weeks, P. R. and Fischer, I. 2000. MAP1B expression and microtubule stability in growing and regenerating axons. *Microsc. Res. Tech.* 48:63–74.

Halliwell, B. 2006. Oxidative stress and neurodegeneration: where are we now? *J. Neurochem.* 97:1634–58.

Hao, W., Xing-Jun, W., Yong-Yao, C., Liang, Z., Yang, L., and Hong-Zhuan, C. 2005. Up-regulation of M1 muscarinic receptors expressed in CHOm1 cells by panaxynol via cAMP pathway. *Neurosci. Lett.* 383:121–6.

Harada, H. and Grant, S. 2003. Apoptosis regulators. *Rev. Clin. Exp. Hematol.* 7:117–38.

Hasegawa, H. 2004. Proof of the mysterious efficacy of ginseng: Basic and clinical trials: metabolic activation of ginsenoside: deglycosylation by intestinal bacteria and esterification with fatty acid. *J. Pharmacol. Sci.* 95:153–7.

Hasegawa, H., Sung, J. H., Matsumiya, S., and Uchiyama, M. 1996. Main ginseng saponin metabolites formed by intestinal bacteria. *Planta Med.* 62:453–7.

Hu, J. F., Xue, W., Ning, N., Yuan, Y. H., Zhang, J. T., and Chen, N. H. 2008. Ginsenoside Rg_1 activated CaMKIIα mediated extracellular signal-regulated kinase/mitogen activated protein kinase signaling pathway. *Acta Pharmacol. Sin.* 29:1119–26.

Jaenicke, B., Kim, E. J., Ahn, J. W., and Lee, H. S. 1991. Effect of *Panax ginseng* extract on passive avoidance retention in old rats. *Arch Pharm. Res.* 14:25–9.

Jang, J. H. and Surh, Y. J. 2003. Potentiation of cellular antioxidant capacity by Bcl-2: Implications for its antiapoptotic function. *Biochem. Pharmacol.* 66:1371–9.

Ji, Z. N., Dong, T. T., Ye, W. C., Choi, R. C., Lo, C. K., and Tsim, K. W. 2006. Ginsenoside Re attenuate beta-amyloid and serum-free induced neurotoxicity in PC12 cells. *J. Ethnopharmacol.* 107:48–52.

Ji, Y. C., Kim, Y. B., Park, S. W., Hwang, S. N., Min, B. K., Hong, H. J., Kwon, J. T., and Suk, J. S. 2005. Neuroprotective effect of ginseng total saponins in experimental traumatic brain injury. *J. Korean Med. Sci.* 20:291–6.

Jin, S. H., Park, J. K., Nam, K. Y., Park, S. N., and Jung, N. P. 1999. Korean red ginseng saponins with low ratios of protopanaxadiol and protopanaxatriol saponin improve scopolamine-induced learning disability and spatial working memory in mice. *J. Ethnopharmacol.* 66:123–9.

Joo, S. S., Yoo, Y. M., Ahn, B. W., Nam, S. Y., Kim, Y. B., Hwang, K. W., and Lee, D. I. 2008. Prevention of inflammation-mediated neurotoxicity by Rg_3 and its role in microglial activation. *Biol. Pharm. Bull.* 31:1392–6.

Kajta, M. 2004. Apoptosis in the central nervous system: Mechanisms and protective strategies. *Pol. J. Pharmacol.* 56:689–700.

Kang, K. S., Kim, H. Y., Pyo, J. S., and Yokozawa, T. 2006. Increase in the free radical scavenging activity of ginseng by heat-processing. *Biol. Pharm. Bull.* 29:750–4.

Kelleher-Andersson, J. 2004. Neurogenesis as a potential therapeutic strategy for neurodegenerative disorders. *J. Alzheimers Dis.* 6:S19–25.

Keum, Y. S., Park, K. K., Lee, J. M., Chun, K. S., Park, J. H., Lee, S. K., Kwon, H., and Surh, Y. J. 2000. Antioxidant and anti-tumor promoting activities of the methanol extract of heat-processed ginseng. *Cancer Lett.* 150:41–8.

Kim, H. S., Hwang, S. L., and Oh, S. 2000. Ginsenoside Rc and Rg_1 differentially modulate NMDA receptor subunit mRNA levels after intracerebroventricular infusion in rats. *Neurochem. Res.* 25:1149–54.

Kim, E. H., Jang, M. H., Shin, M. C., Shin, M. S., and Kim, C. J. 2003. Protective effect of aqueous extract of *Ginseng radix* against 1-methyl-4-phenylpyridinium-induced apoptosis in PC12 cells. *Biol. Pharm. Bull.* 26:1668–73.

Kim, S., Kim, T., Ahn, K., Park, W. K., Nah, S. Y., and Rhim, H. 2004. Ginsenoside Rg_3 antagonizes NMDA receptors through a glycine modulatory site in rat cultured hippocampal neurons. *Biochem. Biophys. Res. Commun.* 323:416–24.

Kim, Y. C., Kim, S. R., Markelonis, G. J., and Oh, T. H. 1998. Ginsenosides Rb_1 and Rg_3 protect cultured rat cortical cells from glutamate-induced neurodegeneration. *J. Neurosci. Res.* 53:426–32.

Kim, J. H., Kim, S., Yoon, I. S., Lee, J. H., Jang, B. J., Jeong, S. M., Lee, J. H., et al. 2005. Protective effects of ginseng saponins on 3-nitropropionic acid-induced striatal degeneration in rats. *Neuropharmacology* 48:743–56.

Kim, S., Nah, S. Y., and Rhim, H. 2008. Neuroprotective effects of ginseng saponins against L-type Ca^{2+} channel-mediated cell death in rat cortical neurons. *Biochem. Biophys. Res. Commun.* 365:399–405.

Kim, Y. H., Park, K. H., and Rho, H. M. 1996. Transcriptional activation of the Cu, Zn-superoxide dismutase gene through the AP2 site by ginsenoside Rb_2 extracted from a medicinal plant, *Panax ginseng. J. Biol. Chem.* 271:24539–43.

Kim, K. T., Yoo, K. M., Lee, J. W., Eom, S. H., Hwang, I. K., and Lee, C. Y. 2007. Protective effect of steamed American ginseng (*Panax quinquefolius* L.) on V79-4 cells induced by oxidative stress. *J. Ethnopharmacol.* 111:443–50.

Kitts, D. and Hu, C. 2000. Efficacy and safety of ginseng. *Public Health Nutr.* 3:473–85.

Kitts, D. D., Wijewickreme, A. N., and Hu, C. 2000. Antioxidant properties of a North American ginseng extract. *Mol. Cell Biochem.* 203:1–10.

Kurimoto, H., Nishijo, H., Uwano, T., Yamaguchi, H., Zhong, Y. M., Kawanishi, K., and Ono, T. 2004. Effects of nonsaponin fraction of red ginseng on learning deficits in aged rats. *Physiol. Behav.* 82:345–55.

Lee, J. K., Choi, S. S., Lee, H. K., Han, K. J., Han, E. J., and Suh, H. W. 2003. Effects of ginsenoside Rd and decursinol on the neurotoxic responses induced by kainic acid in mice. *Planta Med.* 69:230–4.

Lee, J. H., Kim, S. R., Bae, C. S., Kim, D., Hong, H., and Nah, S. 2002. Protective effect of ginsenosides, active ingredients of *Panax ginseng*, on kainic acid-induced neurotoxicity in rat hippocampus. *Neurosci. Lett.* 325:129–33.

Lee, T. F., Shiao, Y. J., Chen, C. F., and Wang, L. C. 2001. Effect of ginseng saponins on beta-amyloid-suppressed acetylcholine release from rat hippocampal slices. *Planta Med.* 67:634–7.

Lei, K., Nimnual, A., Zong, W. X., Kennedy, N. J., Flavell, R. A., Thompson, C. B., Bar-Sagi, D. and Davis, R. J. 2002. The Bax subfamily of Bcl2-related proteins is essential for apoptotic signal transduction by c-Jun NH_2-terminal kinase. *Mol. Cell. Biol.* 22:4929–42.

Leung, K. W., Yung, K. K., Mak, N. K., Chan, Y. S., Fan, T. P., and Wong, R. N. 2007. Neuroprotective effects of ginsenoside-Rg_1 in primary nigral neurons against rotenone toxicity. *Neuropharmacology* 52:827–35.

Li, N., Liu, B., Dluzen, D. E., and Jin, Y. 2007. Protective effects of ginsenoside Rg_2 against glutamate-induced neurotoxicity in PC12 cells. *J. Ethnopharmacol.* 111:458–63.

Lian, X. Y., Zhang, Z., and Stringer, J. L. 2005. Protective effects of ginseng components in a rodent model of neurodegeneration. *Ann. Neurol.* 57:642–8.

Liao, B., Newmark, H. and Zhou, R. 2002. Neuroprotective effects of ginseng total saponin and ginsenosides Rb_1 and Rg_1 on spinal cord neurons *in vitro*. *Exp. Neurol.* 173:224–34.

Lim, J. H., Wen, T. C., Matsuda, S., Tanaka, J., Maeda, N., Peng, H., Aburaya, J., Ishihara, K., and Sakanaka, M. 1997. Protection of ischemic hippocampal neurons by ginsenoside Rb_1, a main ingredient of ginseng root. *Neurosci. Res.* 28:191–200.

Lipton, S. A. and Rosenberg, P. A. 1994. Excitatory amino acids as a final common pathway for neurologic disorders. *N. Engl. J. Med.* 330:613–22.

Liu, J. W., Tian, S. J., de Barry, J., and Luu, B. 2007. Panaxadiol glycosides that induce neuronal differentiation in neurosphere stem cells. *J. Nat. Prod.* 70:1329–34.

Lleo, A., Galea, E., and Sastre, M. 2007. Molecular targets of non-steroidal anti-inflammatory drugs in neurodegenerative diseases. *Cell. Mol. Life Sci.* 64:1403–18.

López, M. V., Cuadrado, M. P., Ruiz-Poveda, O. M., Del Fresno, A. M., and Accame, M. E. 2007. Neuroprotective effect of individual ginsenosides on astrocytes primary culture. *Biochim. Biophys. Acta.* 1770:1308–16.

Mehta, S. L., Manhas, N., and Raghubir, R. 2007. Molecular targets in cerebral ischemia for developing novel therapeutics. *Brain Res. Rev.* 54:34–66.

Miller, P. J. and Zaborszky, L. 1997. 3-Nitropropionic acid neurotoxicity: Visualization by silver staining and implications for use as an animal model of Huntington's disease. *Exp. Neurol.* 146:212–29.

Millikan, C. 1992. Animal stroke models. *Stroke* 23:795–7.

Miyazaki, I. and Asanuma, M. 2008. Dopaminergic neuron-specific oxidative stress caused by dopamine itself. *Acta Med. Okayama* 62:141–50.

Mook-Jung, I., Hong, H. S., Boo, J. H., Lee, K. H., Yun, S. H., Cheong, M. Y., Joo, I., Huh, K., and Jung, M. W. 2001. Ginsenoside Rb_1 and Rg_1 improve spatial learning and increase hippocampal synaptophysin level in mice. *J. Neurosci. Res.* 63:509–15.

Moore, D. J., West, A. B., Dawson, V. L., and Dawson, T. M. 2005. Molecular pathophysiology of Parkinson's disease. *Annu. Rev. Neurosci.* 28:57–87.

Musial, A., Bajda, M., and Malawska, B. 2007. Recent developments in cholinesterases inhibitors for Alzheimer's disease treatment. *Curr. Med. Chem.* 14:2654–79.

Nah, S. Y., Kim, D. H., and Rhim, H. 2007. Ginsenosides: Are any of them candidates for drugs acting on the central nervous system? *CNS Drug Rev.* 13:381–404.

Nishiyama, N., Cho, S. I., Kitagawa, I. and Saito, H. 1994. Malonylginsenoside Rb_1 potentiates nerve growth factor (NGF)-induced neurite outgrowth of cultured chick embryonic dorsal root ganglia. *Biol. Pharm. Bull.* 17:509–13.

Nitta, H., Matsumoto, K., Shimizu, M., Ni, X. H., and Watanabe, H. 1995a. *Panax ginseng* extract improves the scopolamine-induced disruption of 8-arm radial maze performance in rats. *Biol. Pharm. Bull.* 18:1439–42.

Nitta, H., Matsumoto, K., Shimizu, M., Ni, X. H., and Watanabe, H. 1995b. *Panax ginseng* extract improves the performance of aged Fischer 344 rats in radial maze task but not in operant brightness discrimination task. *Biol. Pharm. Bull.* 18:1286–8.

Nocerino, E., Amato, M., and Izzo, A. A. 2000. The aphrodisiac and adaptogenic properties of ginseng. *Fitoterapia* 71:S1–5.

Oh, G. S., Pae, H. O., Choi, B. M., Seo, E. A., Kim, D. H., Shin, M. K., Kim, J. D., Kim, J. B., and Chung, H. T. 2004. 20(*S*)-Protopanaxatriol, one of ginsenoside metabolites, inhibits inducible nitric oxide synthase and cyclooxygenase-2 expressions through inactivation of nuclear factor-kappaB in RAW264.7 macrophages stimulated with lipopolysaccharide. *Cancer Lett.* 205:23–9.

Park, E. K., Choo, M. K., Oh, J. K., Ryu, J. H., and Kim, D. H. 2004. Ginsenoside Rh_2 reduces ischemic brain injury in rats. *Biol. Pharm. Bull.* 27:433–6.

Park, I. H., Kim, N. Y., Han, S. B., Kim, J. M., Kwon, S. W., Kim, H. J., Park, M. K., and Park, J. H. 2002. Three new dammarane glycosides from heat processed ginseng. *Arch. Pharm. Res.* 25:428–32.

Park, E. K., Shin, Y. W., Lee, H. U., Kim, S. S., Lee, Y. C., Lee, B. Y., and Kim, D. H. 2005. Inhibitory effect of ginsenoside Rb_1 and compound K on NO and prostaglandin E_2 biosyntheses of RAW264.7 cells induced by lipopolysaccharide. *Biol. Pharm. Bull.* 28:652–6.

Qiao, C., Den, R., Kudo, K., Yamada, K., Takemoto, K., Wati, H., and Kanba, S. 2005. Ginseng enhances contextual fear conditioning and neurogenesis in rats. *Neurosci. Res.* 51:31–8.

Radad, K., Gille, G., Moldzio, R., Saito, H., Ishige, K., and Rausch, W. D. 2004a. Ginsenosides Rb_1 and Rg_1 effects on survival and neurite growth of MPP^+-affected mesencephalic dopaminergic cells. *J. Neural. Transm.* 111:37–45.

Radad, K., Gille, G., Moldzio, R., Saito, H., and Rausch, W. D. 2004b. Ginsenosides Rb_1 and Rg_1 effects on mesencephalic dopaminergic cells stressed with glutamate. *Brain Res.* 1021:41–53.

Radad, K., Gille, G., Liu, L., and Rausch, W. D. 2006. Use of ginseng in medicine with emphasis on neurodegenerative disorders. *J. Pharmacol. Sci.* 100:175–86.

Rausch, W. D., Liu, S., Gille, G., and Radad, K. 2006. Neuroprotective effects of ginsenosides. *Acta Neurobiol. Exp.* 66:369–75.

Rhule, A., Navarro, S., Smith, J. R., and Shepherd, D. M. 2006. *Panax notoginseng* attenuates LPS-induced pro-inflammatory mediators in RAW264.7 cells. *J. Ethnopharmacol.* 106:121–8.

Ribe, E. M., Serrano-Saiz, E., Akpan, N., and Troy, C. M. 2008. Mechanisms of neuronal death in disease: defining the models and the players. *Biochem. J.* 415:165–82.

Sakanaka, M., Zhu, P., Zhang, B., Wen, T. C., Cao, F., Ma, Y. J., Samukawa, K., et al. 2007. Intravenous infusion of dihydroginsenoside Rb_1 prevents compressive spinal cord injury and ischemic brain damage through upregulation of VEGF and Bcl-X_L. *J. Neurotrauma* 24:1037–54.

Salim, K. N., McEwen, B. S., and Chao, H. M. 1997. Ginsenoside Rb_1 regulates ChAT, NGF and trkA mRNA expression in the rat brain. *Brain Res. Mol. Brain Res.* 47:177–82.

Seo, J. Y., Lee, J. H., Kim, N. W., Her, E., Chang, S. H., Ko, N. Y., Yoo, Y. H., et al. 2005a. Effect of a fermented ginseng extract, BST204, on the expression of cyclooxygenase-2 in murine macrophages. *Int. Immunopharmacol.* 5:929–36.

Seo, J. Y., Lee, J. H., Kim, N. W., Kim, Y. J., Chang, S. H., Ko, N. Y., Her, E., et al. 2005b. Inhibitory effects of a fermented ginseng extract, BST204, on the expression of inducible nitric oxide synthase and nitric oxide production in lipopolysaccharide-activated murine macrophages. *J. Pharm. Pharmacol.* 57:911–8.

Shah, Z. A., Gilani, R. A., Sharma, P., and Vohora, S. B. 2005. Cerebroprotective effect of Korean ginseng tea against global and focal models of ischemia in rats. *J. Ethnopharmacol.* 101:299–307.

Shen, L. H. and Zhang, J. T. 2004. Ginsenoside Rg_1 promotes proliferation of hippocampal progenitor cells. *Neurol. Res.* 26:422–8.

Sloley, B. D., Pang, P. K., Huang, B. H., Ba, F., Li, F. L., Benishin, C.G., Greenshaw, A. J., and Shan, J. J. 1999. American ginseng extract reduces scopolamine-induced amnesia in a spatial learning task. *J. Psychiatry Neurosci.* 24:442–52.

Tohda, C., Matsumoto, N., Zou, K., Meselhy, M. R., and Komatsu, K. 2004. $A\beta_{25\text{-}35}$-induced memory impairment, axonal atrophy, and synaptic loss are ameliorated by M1, a metabolite of protopanaxadiol-type saponins. *Neuropsychopharmacology* 29:860–8.

Van Kampen, J., Robertson, H., Hagg, T., and Drobitch, R. 2003. Neuroprotective actions of the ginseng extract G115 in two rodent models of Parkinson's disease. *Exp. Neurol.* 184:521–9.

Wang, L. C., Wang, B., Ng, S. Y., and Lee, T. F. 2006a. Effects of ginseng saponins on β-amyloid-induced amnesia in rats. *J. Ethnopharmacol.* 103:103–8.

Wang, Z. J., Nie, B. M., Chen, H. Z., and Lu, Y. 2006b. Panaxynol induces neurite outgrowth in PC12D cells via cAMP- and MAP kinase-dependent mechanisms. *Chem. Biol. Interact.* 159:58–64.

Wang, X. Y. and Zhang, J. T. 2001. NO mediates ginsenoside Rg_1-induced long-term potentiation in anesthetized rats. *Acta Pharmacol. Sin.* 22:1099–102.

Wei, C., Jia, J., Liang, P., and Guan Y. 2008. Ginsenoside Rg_1 attenuates β-amyloid-induced apoptosis in mutant PS1 M146L cells. *Neurosci. Lett.* 443:145–9.

Wen, T. C., Yoshimura, H., Matsuda, S., Lim, J. H., and Sakanaka, M. 1996. Ginseng root prevents learning disability and neuronal loss in gerbils with 5-minute forebrain ischemia. *Acta Neuropathol.* 91:15–22.

Wong, W. W. and Puthalakath, H. 2008. Bcl-2 family proteins: The sentinels of the mitochondrial apoptosis pathway. *IUBMB Life* 60:390–7.

Xiang, Y. Z., Shang, H. C., Gao, X. M., and Zhang, B. L. 2008. A comparison of the ancient use of ginseng in traditional Chinese medicine with modern pharmacological experiments and clinical trials. *Phytother. Res.* 22:851–8.

Xue, J. F., Liu, Z. J., Hu, J. F., Chen, H., Zhang, J. T., and Chen, N. H. 2006. Ginsenoside Rb_1 promotes neurotransmitter release by modulating phosphorylation of synapsins through a cAMP-dependent protein kinase pathway. *Brain Res.* 1106:91–8.

Yamaguchi, Y., Higashi, M., and Kobayashi, H. 1997. Effects of ginsenosides on maze performance and brain choline acetyltransferase activity in scopolamine-treated young rats and aged rats. *Eur. J. Pharmacol.* 329:37–41.

Yamazaki, M., Hirakura, K., Miyaichi, Y., Imakura, K., Kita, M., Chiba, K., and Mohri, T. 2001. Effect of polyacetylenes on the neurite outgrowth of neuronal culture cells and scopolamine-induced memory impairment in mice. *Biol. Pharm. Bull.* 24:1434–6.

Yoshikawa, T., Akiyoshi, Y., Susumu, T., Tokado, H., Fukuzaki, K., Nagata, R., Samukawa, K., Iwao, H., and Kito, G. 2008. Ginsenoside Rb_1 reduces neurodegeneration in the peri-infarct area of a thromboembolic stroke model in non-human primates. *J. Pharmacol. Sci.* 107:32–40.

Yue, P. Y., Mak, N. K., Cheng, Y. K., Leung, K. W., Ng, T. B., Fan, D. T., Yeung, H. W., and Wong, R. N. 2007. Pharmacogenomics and the Yin/Yang actions of ginseng: anti-tumor, angiomodulating and steroid-like activities of ginsenosides. *Chin. Med.* 2:1–21.

Zhang, B., Hata, R., Zhu, P., Sato, K., Wen, T. C., Yang, L., Fujita, H., Mitsuda, N., Tanaka, J., Samukawa, K., Maeda, N., and Sakanaka, M. 2006. Prevention of ischemic neuronal death by intravenous infusion of a ginseng saponin, ginsenoside Rb_1, that upregulates Bcl-X_L expression. *J. Cereb. Blood Flow Metab.* 26:708–21.

Zhang, G., Liu, A., Zhou, Y., San, X., Jin, T., and Jin, Y. 2008. *Panax ginseng* ginsenoside-Rg_2 protects memory impairment via anti-apoptosis in a rat model with vascular dementia. *J. Ethnopharmacol.* 115:441–8.

Zheng, M., Qu, L., and Lou, Y. 2008. Effects of icariin combined with *Panax notoginseng* saponins on ischemia reperfusion-induced cognitive impairments related with oxidative stress and CA1 of hippocampal neurons in rat. *Phytother. Res.* 22:597–604.

Zhou, X. M., Cao, Y. L. and Dou, D. Q. 2006. Protective effect of ginsenoside-Re against cerebral ischemia/reperfusion damage in rats. *Biol. Pharm. Bull.* 29:2502–5.

Zipp, F. and Aktas, O. 2006. The brain as a target of inflammation: common pathways link inflammatory and neurodegenerative diseases. *Trends Neurosci.* 29:518–27.

8 Functional Foods from Green Tea

Amber Sharma, Rong Wang, and Weibiao Zhou

CONTENTS

8.1 INTRODUCTION

Tea (*Camellia sinensis*, Figure 8.1) has been discovered and utilized for its unique flavor and medicinal properties by man for a long time, which may be traced back to the Shen Nong era in ancient China, around 5000–6000 years ago (Chen, 2002). Although the tea plant grows in many countries, general consensus attributes the birth of the tea plant to the mountainous area of today's southwestern China (Hara, 2001). While Chinese legend tells how Sheng Nong found the antidote and healing power of tea, the first documented effects of tea were in the book Ben Cao Jing (本草经 *Materia Medica*), which was written during the era of the Han Dynasty (206

FIGURE 8.1 Tea plant (*Camellia sinensis*).

BC–AD 220). A comprehensive book on tea by Cha Jing (茶经 *The Classics of Tea*) was published in AD 780 by Yu Lu of Tang Dynasty, which described in detail the botany, cultivation, processing, quality, and preparation of tea, as well as proper methods and style of drinking tea including utensils and water quality (Chen, 2002; Hara, 2001).

Tea beverages are now the second most popular drinks and only next to water in terms of worldwide consumption. They are recognized as the best natural drink (Dew et al., 2005). Scientific research on the chemical components and functionalities of tea is relatively recent (Hara, 2001). Through research, tea has been linked with health benefits including protection of oxidative DNA damage, lowering the atherosclerotic index and improving blood flow, liver function, and oral health (Balentine et al., 1997; Dufresne and Farnworth, 2001; McKay and Blumberg, 2002). Meanwhile, tea antioxidants may improve food product quality without damaging its organoleptic properties or nutritional function. Nowadays, the utilization of tea has been extended not only to pharmaceutical products but also to toiletry, cosmetic, and food products (Wang et al., 2000a; Yamamoto et al., 1997).

Commercial tea leaf products can be classified into three major types: green tea, black tea, and oolong tea. Green tea is nonfermented processed tea, in which polyphenols in fresh tea leaves are less oxidized. Black tea and oolong tea are enzymatically fermented tea, with black tea being the most fermented and oolong tea partially fermented (or so called semifermented). Worldwide, about 80% of the consumers prefer black tea, while green tea is preferred in China, Japan, and Southeast Asian countries (Chen, 2002).

Besides tea beverages, the use of green tea or green tea extracts in foods such as cereal products, dairy products, and confectioneries gives a healthier appeal to the consumer; therefore, the marketing potential for these foods can be improved by the presence of green tea antioxidants. Indeed, there are a number of such products in the market. A good example might be mooncake, a traditional cake eaten during the Chinese Middle Autumn Festival, into which the incorporation of green tea extract is known to both increase its shelf life and improve its flavor.

In this chapter, first a description of major tea antioxidants, that is, tea catechins and their antioxidative activities is presented in Section 8.2. A comprehensive review of the literature results on the health benefits of green tea polyphenols, particularly catechins, via both *in vitro* and *in vivo* studies, is provided in Section 8.3. Section 8.4 discusses the stability and chemical changes of tea catechins under either storage conditions or processing conditions, including the latest results on their kinetics. Current status on the development of functional foods containing green tea or green tea antioxidants is discussed in Section 8.5, with an emphasis on the scientific issues.

8.2 GREEN TEA CATECHINS AND THEIR ANTIOXIDATIVE ACTIVITIES

The functional properties of tea are believed to be due to tea polyphenols. The principal tea polyphenols are tea catechins, which are mainly present in green tea, exhibiting the most effective antioxidative activity compared to other tea polyphenols.

8.2.1 Major Types of Green Tea Catechins

Four major tea catechins have been identified as: (–)-epigallocatechin (EGC), (–)-epigallocatechin gallate (EGCG), (–)-epicatechin (EC), and (–)-epicatechin gallate (ECG). Their corresponding epimers (–)are: (–)-gallocatechin (GC), (–)-gallocatechin gallate (GCG), (±)-catechin (C), and (–)-catechin gallate (CG), respectively (Chen and Chan, 1996; Pokorny et al., 2001; Wang et al., 2000a; Zhang et al., 1997). They share a common catechin backbone with variations in the substitutes at the C-3 and C-5 positions (Figure 8.2).

EGCG is the most abundant and active catechin compared to its homologues (Chen et al., 2001b; Dew et al., 2005; Higdon and Frei, 2003) and has been accepted as a quality indicator of green tea products (Pelillo et al., 2002; Wang et al., 2000b). Particularly, EGCG is the catechin only occurring in tea, while other catechins can also be found in foods such as dark chocolate, apples, grapes, and legumes (Arts et al., 2000a, 2000b).

There are some 8–30% of total catechins in dry green tea leaves (Chen et al., 2001b; Wang et al., 2000a), and 29–80% in green tea extract (GTE) (Miura et al., 2001; Pelillo et al., 2002; Wang et al., 2000b; Wang et al., 2003). The content of tea catechins is associated with species, climate, cultivation practices, production, preparation, and storage conditions (Wang et al., 2000a). The estimated daily intake of tea catechins based on 3 cups (600 mL) of green tea (1–4 g), which is brewed traditionally (1–5 min in boiling water), is in the range of 538–2594 mg of total catechins

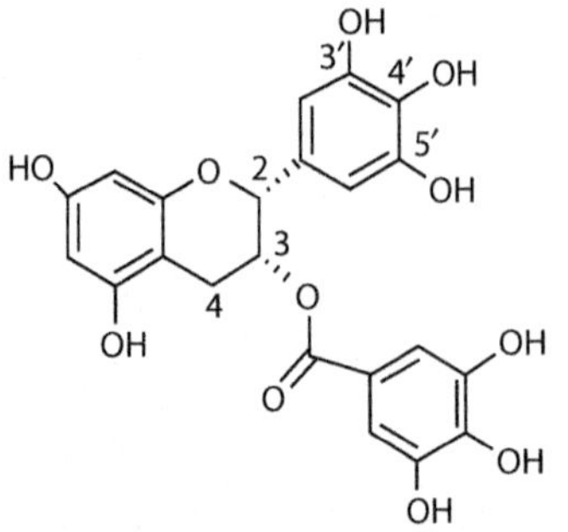

(–)-Epigallocatechin gallate (EGCG)

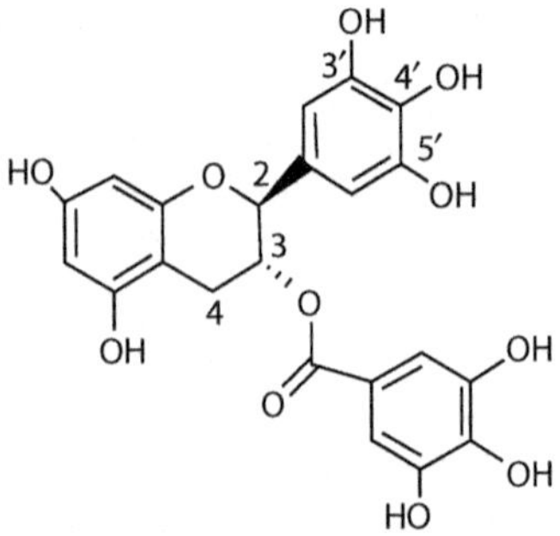

(–)-Gallocatechin gallate (GCG)

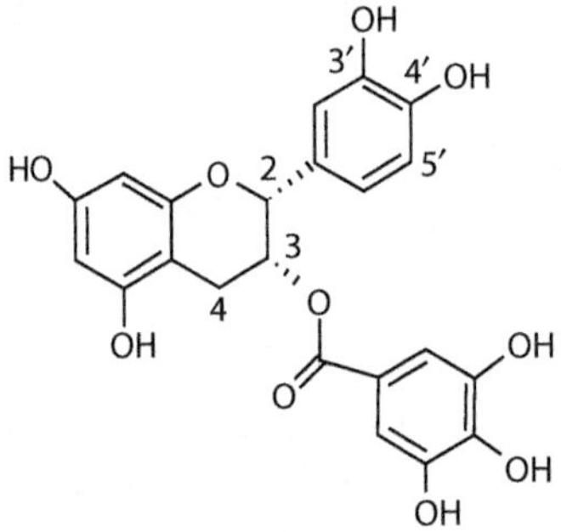

(–)-Epicatechin gallate (ECG)

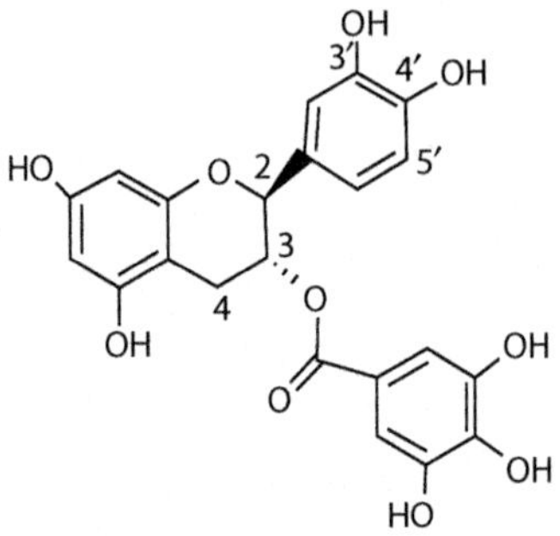

(–)-Catechin gallate (CG)

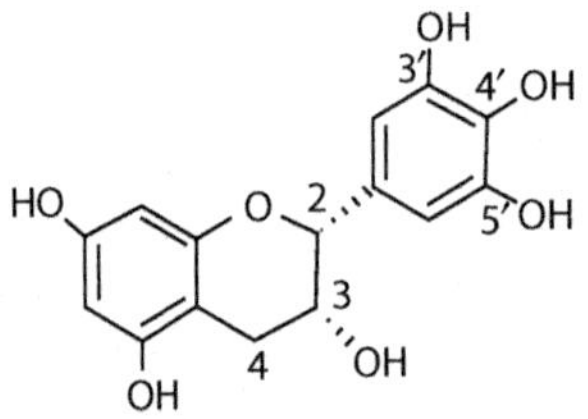

(–)-Epigallocatechin (EGC)

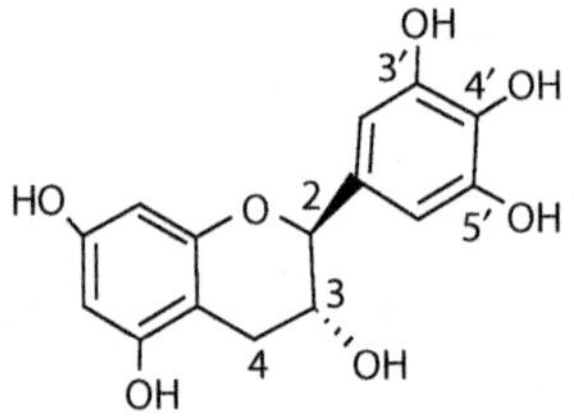

(–)-Gallocatechin (GC)

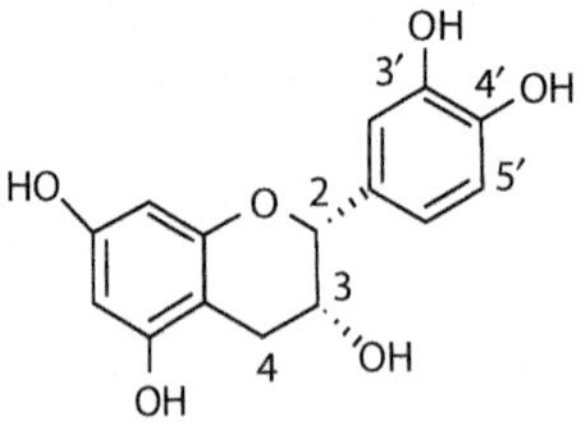

(–)-Epicatechin (EC)

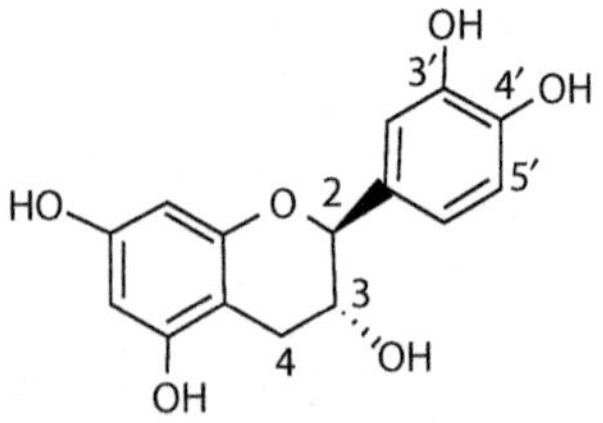

(±)-Catechin (C)

FIGURE 8.2 Chemical structures of tea catechins.

(Shishikura and Khokhar, 2005). Although a recommended daily intake (RDI) for the consumption of tea catechins is not available in the Food and Drug Administration (FDA) database, drinking green tea up to 10 cups per day was reported as "having no problem" (Fujiki et al., 1992) despite the size of the cup not having been specified. Additionally, the results of subchronic tests in rats showed that the dose of tea catechins at a level as high as 1000 mg/kg (bw) was safe (Madhavi et al., 1996).

8.2.2 Antioxidative Activity of Tea Catechins

The antioxidative activity of tea catechins is structure-dependent. The three adjacent hydroxyl (OH) groups at positions C-3′, 4′, and 5′ on the B ring of EGCG, GCG, EGC, and GC are more effective on scavenging free radicals than the two adjacent OH groups at C-3′ and 4′ in ECG, CG, EC, and C, respectively (Figure 8.2). Moreover, catechins with additional gallate moiety at C-3 generally hold stronger scavenging effects than nongallate catechins, that is, ECG > EC and EGCG > EGC (Chen and Chan, 1996; Guo et al., 1999; Mandel and Youdim, 2004; Mukai et al., 2005; Nanjo et al., 1996; Salah et al, 1995; Su et al., 2003; Yoshioka et al., 1991).

However, many studies in the literature demonstrate that the antioxidant activity of tea catechins is also radical-dependent and medium-dependent. The following is a summary of the antioxidative activity/free-radical scavenging ability of tea catechins in various systems.

1. Reduction potentials
   - EGC > EGCG > GCG > EC > ECG by saturated calomel electrode (SCE) at pH 6.15 (Balentine et al., 1997)
   - EGCG = EGC > ECG > EC by hydrogen electrode (NHE) at pH 7 (Higdon and Frei, 2003)
2. pH-associated system
   - EGCG > ECG > EGC at pH = 4–7 (Nanjo et al., 1996)
   - ECG > EGCG > EGC at pH = 10 (Nanjo et al., 1996)
   - EGCG > ECG = EGC > > EC at pH = 6–12 in the presence of linoleic acid (Kumamoto et al., 2001)
3. Free radicals/reactive oxygen species (ROS) induced system
   - 1,1-Diphenyl-2-picrylhydrazy (DPPH) or 2,2′-azobis(2-aminopropane) hydrochloride (AAPH):
     - EGCG = ECG > EGC > EC (Nanjo et al., 1996; Kondo et al., 1999)
     - EGCG > EGC > EC (Guo et al., 1999)
   - ABTS* radicals:
     - ECG > EGCG > EGC > EC = C (Salah et al., 1995)
     - ECG > EGCG > EGC > EC (Higdon and Frei, 2003)
   - Hydroxyl radicals (•OH):
     - ECG > EGCG > EC > GC > EGC > C (Wiseman et al., 1997)
     - ECG > EC > EGCG >> EGC (Guo et al., 1996)
   - Quenching singlet oxygen ($^1O_2$):
     - EGCG > ECG > EGC > EC > C (Mukai et al., 2005)

- Lipid peroxyl radicals:
  - ECG = EGCG = EC = C > EGC (Salah et al., 1995)

4. Lipid /lipophilic system
   - In canola oil: EGC > EGCG > EC > ECG (Chen and Chan, 1996)
   - In lard: EGCG > EGC > ECG > EC (Nanjo et al., 1996)
   - In cooked fish: EGCG ≈ ECG > EGC >> EC (He and Shahidi, 1997)
   - In bulk corn oil: ECG > EGCG > EGC (Huang and Frankel, 1997)
5. Metal ions induced lipid peroxidation
   - Cu^{2+} mediated oxidation of low-density lipoprotein (LDL):
     - EGCG = ECG > EC > EGC (Zhang et al., 1997)
     - EGCG > ECG > EC > C > EGC (Miura et al., 2001)
   - Fe^{2+}/Fe^{3+}-stimulated synaptosomal lipid peroxidation:
     - EGCG > ECG > EGC > EC (Guo et al., 1996)
   - Fe^{2+}/Fe^{3+}-induced lipid free radicals:
     - ECG > EGCG > EC > EGC (Guo et al., 1996)
6. Emulsion system
   - In soy lecithin liposomes:
     - EGCG > EC > C ≈ ECG > EGC (Huang and Frankel, 1997)
   - In lecithin with lipoxidase present:
     - ECG > EGCG > EGC > EC (Guo et al., 1996)
7. Micelle system
   - In sodium dodecyl sulfate (SDS) micelle initiated by di-*tert*-butyl hyponitrite (DBHN): EC > ECG > EGCG > EGC (Chen et al., 2001a)
   - In cetyltrimethylammonium bromide (CTAB) micelle:
     - ECG > EC > EGCG > EGC (Chen et al., 2001a).

The above summary clearly shows that the antioxidative activity/free radical scavenging ability of tea catechin varies with the type of radical species, ionization state, pH, polarity, and enzyme in the designated studies. Although many mechanisms have been proposed for the antioxidative activity of tea catechins, the precise oxidation pathway and free radical scavenging process of tea catechins are not well established (Hatano et al., 2005; Higdon and Frei, 2003; Kondo et al., 1999).

8.3 GREEN TEA ANTIOXIDANTS AND HEALTH BENEFITS

Tea antioxidants have drawn increased attention in recent years because of their potential health benefits, not only as an antioxidant agent but also as antiarteriosclerotic, anticarcinogenic, and antimicrobial agents (Wang and Zhou, 2004). This has encouraged a lot of research on the effect of green tea catechins on diseases associated with reactive oxygen species (ROS), such as cancer, cardiovascular, and neurodegenerative diseases. A number of epidemiological, animal model, and cell line studies have shown the preventive effect of green tea catechins on a number of diseases. Some of them include cancers such as those of the skin, breast, prostate, liver, and lung (Adhami et al., 2004; Yang et al., 2002), neurodegenerative diseases (Parkinson's disease, Alzheimer's disease, and ischemic damage) (Mandel and

Youdim, 2004; Sutherland et al., 2006; Weinreb et al., 2004; Zhao, 2005), and HIV (Nance and Shearer, 2003). Green tea is also known to be antiangiogenic (prevention of tumor blood vessel growth) (Cao and Cao, 1999; Pfeffer et al., 2003), antimutagenic (Kuroda and Hara, 1999), antidiabetic, and antiobesity (Kao et al., 2006), antibacterial (Stapleton et al., 2004), and anti-inflammatory (Dona et al., 2003).

8.3.1 Results from Studies Using Cell Lines

Tea catechins have been demonstrated to inhibit hydroperoxide-dependent toxicity, cell proliferation, cell cycle progression, early gene expression, and to display antimutagenic, antiallergenic, and apoptosis-inducing properties as observed in a variety of cell models. Cell culture studies have indicated that catechins are multifunctional molecules that may act through a wide variety of mechanisms in different cell types (Rijken et al., 2002).

8.3.1.1 Green Tea and Neurodegenerative Diseases

Green tea polyphenols, which are mainly catechins, formerly thought to be simple radical scavengers, are now considered to invoke a spectrum of cellular mechanisms of action related to their neuroprotective activity. These include pharmacological activities like iron chelation, scavenging of radicals, activation of survival genes and cell signaling pathways, and regulation of the mitochondrial function and possibly of the ubiquitin–proteasome system (Mandel and Youdim, 2004; Ramassamy, 2006).

8.3.1.1.1 Antioxidant Action: Radical Scavenging and Induction of Endogenous Antioxidants

Catechins chelate metal ions such as copper(II) and iron(III) to form inactive complexes and prevent the generation of potentially damaging free radicals. Another mechanism by which the catechins exert their antioxidant effects is through the ultra rapid electron transfer from catechins to ROS-induced radical sites on DNA. A third possible mechanism by which catechins scavenges free radicals is by forming stable semiquinone free radicals, thus preventing the deaminating ability of free radicals. In addition, after the oxidation of catechins, due to their reaction with free radicals, a dimerized product is formed, which has been shown to have increased superoxide scavenging and iron-chelating potential (Sutherland et al., 2006).

3-Hydroxykynurenine (3-HK) is an endogenous metabolite of tryptophan in the kynurenine pathway and is a potential neurotoxin in several neurodegenerative disorders. EGCG attenuated 3-HK-induced cell viability reduction and increase in the concentration of ROS and caspase-3 activity in neuronal culture, presumably via its antioxidant activity (Jeong et al., 2004). In the rat brain tissue, green tea and black tea extracts were shown to inhibit lipid peroxidation promoted by iron ascorbate in homogenates of brain mitochondrial membranes (IC_{50}: 2.44 and 1.40 μmol/L, respectively) (Levites et al., 2002a).

8.3.1.1.2 Modulation of Cell Survival/Death Genes

Studies based on customized cDNA array and confirmatory real-time PCR (Levites et al., 2002a; Weinreb et al., 2003) revealed that a low EGCG concentration decreased the expression of proapoptotic genes *bax*, *bad*, *caspase-1* and *-6*, cyclin-dependent kinase inhibitor *p21*, cell-cycle inhibitor *gadd45*, *fas-ligand*, and tumor necrosis factor-related apoptosis-inducing ligand *TRAIL*, in SH-SY5Y neuronal cells. However, antiapoptotic genes like the Bcl-2 family members *bcl-2* and *bcl-xL* were not affected by EGCG either 1.5 or 6 h postadministration (Weinreb et al., 2003), suggesting that the neuroprotective action of EGCG may implicate inactivation of cell-death-promoting genes, rather than induction of mitochondrial-acting antiapoptotic proteins (Mandel and Youdim, 2004).

In contrast to the neuroprotective effect observed with the low micromolar concentrations (<10 μM) of EGCG, a protoxic/proapoptotic pattern of gene expression was encountered with a high concentration (50 μM), elevating the levels of *bad*, *bax*, *gadd45*, *caspase* family members (3, 6, and 10), *p21/WAF1*, *fas*, and *fas-ligand*

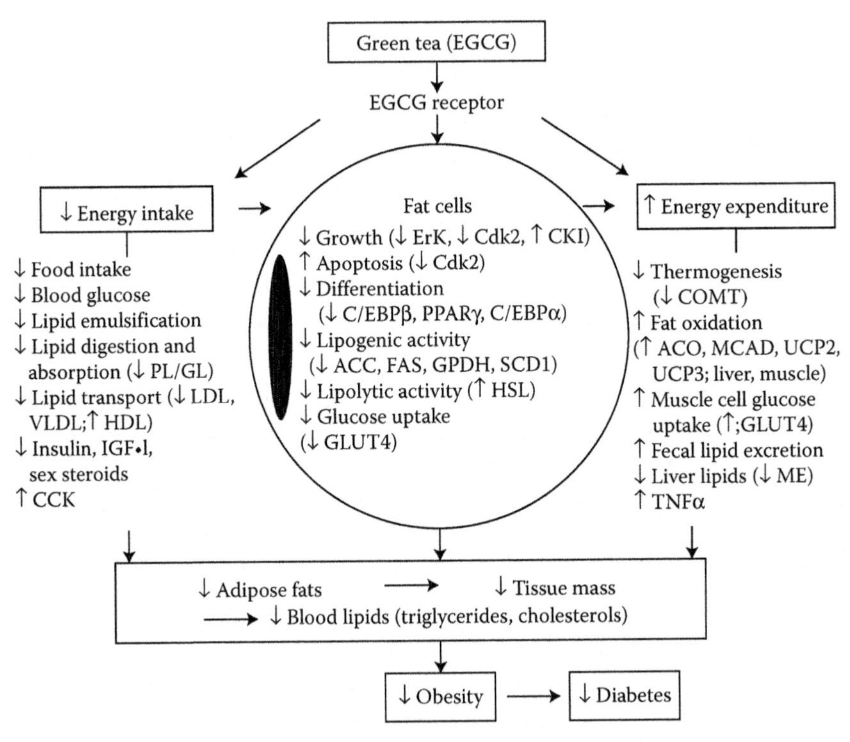

FIGURE 8.3 A proposed mechanism of the action of green tea EGCG on obesity. Signaling of EGCG in its modulation of body weight is mediated via a decrease in energy intake and stimulation of energy expenditure, both of which are dependent on the activity of fat cells as well as intestine, liver, and muscle cells. (From Moon, H.S. et al. 2007. *Chemicobiol. Interact.* 167:85–98. With permission.)

mRNAs while decreasing those of *bcl-2*, *bcl-xL*, and bcl-w (Levites et al., 2002a; Weinreb et al., 2003).

8.3.1.2 Green Tea and Obesity

The functions of tea polyphenols operate through many different mechanisms, and these mechanisms interact to alter the energy balance, the redox status, and the activities of obesity-related cells. As shown in Figure 8.3, an EGCG receptor has been identified and its widespread localization in the cells with many isoforms may explain the numerous biological effects of EGCG (Moon et al., 2007).

EGCG decreased the phosphorylated ERK1/2 in 3T3-L1 pre-adipocytes, but did not alter the total levels of MEK1, ERK-1, ERK-2, p38, phospho-p38, JNK, or phospho-JNK (Hung et al., 2005; Levites et al., 2002b) (where ERK, p38, and JNK refer to three major MAPK subfamilies), suggesting that EGCG acts on a specific type of MAPK, especially in the ERK MAPK family (Kao et al., 2006).

Hung et al. (2005) reported that EGCG downregulated adipocyte differentiation through the CDK2 signaling pathway. These results suggested that the antimitogenic effect of EGCG on 3T3-L1 pre-adipocytes is dependent on the ERK MAPK and CDK2 pathways and is likely mediated through decrease in their activities (Moon et al., 2007).

Hwang et al. (2005) suggested that several naturally occurring compounds have potential antiobesity effects, showing that either EGCG or capsaicin-activated AMPK and inhibited adipocyte differentiation in 3T3-L1 cells. Overall, the antiproliferatory and lipolytic effects of EGCG have been attributed to their ability to modulate various signaling pathways, specifically those that control cell proliferation and survival (Moon et al., 2007).

8.3.1.3 Green Tea and Cardiovascular Disease

Tea components may act at multiple molecular levels in cardiovascular relevant cells. The pathophysiology of cardiovascular diseases is multifactorial and comprises processes, which appear to be affected by tea ingredients: endothelial dysfunction, inflammation, migration and proliferation of smooth muscle cells, extra-cellular matrix formation, as well as thrombus formation. The effects of catechins and theaflavins may therefore be of potential therapeutic impact in protection and treatment of cardiovascular disease (Stangl et al., 2007).

8.3.1.4 Green Tea and Cancer

EGCG has various anticarcinogenic effects, including inhibition of oxidative stress, inhibition of carcinogen-induced mutagenesis, induction of apoptosis, and inhibition of angiogenesis (Shimizu and Weinstein, 2005). Green tea polyphenols have been shown to inhibit the proliferation of cultured mammalian cells, including colon carcinoma (Kautenburger et al., 2005), lung carcinoma, and breast carcinoma (Thangapazham et al., 2007). The anticancer effects of green tea antioxidants and their role in a number of cell signaling pathways is discussed in the rest of this section.

8.3.1.4.1 *Effect on NF-κB and AP-1 activities*

A plausible inhibitory mechanism underlying the suppression of anchorage-independent transformation by EGCG was examined in epidermal growth factor (EGF) and 12-*O*-tetradecanoylporbol-13-acetate (TPA)-stimulated mouse epidermal JB6 cells (Dong, 2000; Nomura et al., 2000). In these studies, EGCG was found to inhibit UVB-induced AP-1 activation as well as NF κB-dependent transcriptional activation. These results indicate that EGCG can inhibit both AP-1 and NF-κB activities, and provide evidence that these effects play a role in the inhibition of both cell growth and malignant cell transformation by this compound (Shimizu and Weinstein, 2005).

8.3.1.4.2 *Inhibitory Effect on Tumor Cell Invasion and Angiogenesis*

EGCG was found to inhibit tumor cell invasion and directly suppress the activity of matrix metalloproteases (MMP) 2 and MMP9, two proteases most frequently over expressed in cancer and angiogenesis that are essential in cutting through basement membrane barriers (Adhami et al., 2004).

8.3.1.4.3 *Effect on Cell Cycle Regulation*

Disruption of the cell cycle is the hallmark of a cancer cell (Adhami et al., 2004). The EGCG has been shown to upregulate Cdk inhibitors such as $p21^{WAF1/CIP1}$ and $p27^{KIP1}$ (negative cell cycle regulators), and inhibit the activities of Cdk 2 and Cdk 4 (positive cell cycle regulators) (Liang et al., 1999).

8.3.1.4.4 *Effect on Apoptotic Activity of Cells*

EGCG induces apoptosis and cell-cycle arrest in human epidermoid carcinoma (A431) cells (Ahmad et al., 1997). Importantly, this apoptotic response was specific for cancer cells; the EGCG treatment also resulted in the induction of apoptosis in human carcinoma cells DU145, and mouse lymphoma cells LY-R, but not in normal human epidermal keratinocytes.

8.3.2 Results from Studies through Animal Models and Clinical Trials

Administration of green tea to animal models of oxidative stress and oxidative stress-associated pathologies (e.g., cancer, inflammation, and atherosclerosis) elicits a range of responses that are consistent with the proposal that tea flavonoids or their metabolites are not only biologically available, despite the absorption being less than 1%, but are also active in affecting cellular processes *in vivo*, by mechanisms that may be related to their antioxidant functionalities (Rijken et al., 2002).

8.3.2.1 Green Tea and Cancer

Many experimental animal studies using biomarkers of cancer risk or cancer development have tested green tea extract (GTE) or EGCG. A number of these studies reported that GTE or EGCG protected against chemical carcinogens in various organs such as the intestine, lungs, liver, prostate, and breast, which were reviewed by Crespy and Williamson (2004), Khan and Mukhtar (2007), and Rao et al. (2003). Studies in animal models have demonstrated that green tea and EGCG can inhibit carcinogenesis at all stages, namely initiation, promotion, and progression. This multifaceted inhibition of the tumorigenic process is attributed to a combination of antioxidative, antiproliferative, and pro-apoptotic effects (Zaveri, 2005).

8.3.2.1.1 Skin Cancer

Skin is the largest body organ and serves as a protective barrier against environmental insults such as ultraviolet (UV) radiation-induced damage. Much of the deleterious effect of solar UV radiation is caused by ultraviolet B (UVB) (290–320 nm). UVB induces skin cells to produce reactive oxygen species (ROS), eicosanoids, proteinases, and cytokines; the inhibition of these mediators is considered to reduce skin damage (Adhami et al., 2004).

Studies have suggested that green tea polyphenols (GTPs) may afford protection against inflammatory responses and the risk of skin cancer. Skin tumorigenesis initiated by UVB and promoted by 12-*O*-tetradecanoylphorbol-13-acetate (TPA) was inhibited in SKH-1 mice by the administration of 1.25% green tea as the drinking fluid prior to and during the 10 days of the UVB treatment period (Yang et al., 1996).

8.3.2.1.2 Lung Cancer

The compound 4-(methylnitrosamino)-1-(3-pyridyl)-1-butanone (NNK) is generally used to induce lung tumorigenesis. Ingestion of green tea (2% of diet) decreased the number of lung tumors induced by NNK in mice, compared with a control group that was not treated with tea (Adhami et al., 2004; Kim and Masuda, 1997).

In a population-based case–control study in Shanghai, China (involving 649 women diagnosed with primary lung cancer from February 1992 to January 1994, and a control group of 675 women), consumption of green tea was associated with a reduced risk of lung cancer among nonsmoking women and the risks decreased with increasing consumption (Zhong et al., 2001).

8.3.2.1.3 Liver Cancer

A population-based case–control study was conducted in Taixing, Jiangsu province, China to explore the role of green tea in decreasing the risks of liver cancer among alcohol drinkers or cigarette smokers. Green tea drinking decreased the risk for the development of liver cancer by 78% among alcohol drinkers and 43% among cigarette smokers (Khan and Mukhtar, 2007).

Biological variables were measured after implantation of hepatoma cells in rats with and without ingestion of green tea. Green tea markedly suppressed hepatoma-induced hyperlipidemia (hypercholesterolemia and hypertriglyceridemia). Moreover, green tea increased biliary secretion into feces (Crespy and Williamson, 2004). Green tea in drinking water also inhibited AFB_1 -induced GST-P- positive hepatocytes in rats and enhanced the ability of the liver to metabolize aflatoxin to noncarcinogenic metabolites (Kuroda and Hara, 1999).

8.3.2.1.4 Breast Cancer

A case – control study conducted among Chinese-, Japanese-, and Filipino-American women demonstrated that green tea consumption was associated with a reduction in breast cancer risk, but only in women possessing a low-activity catechol-*O*-methyltransferase (COMT) allele (Wu et al., 2003). COMT is responsible for the rapid methylation of tea polyphenols and therefore differences in methylation capacity between individuals may alter the chemopreventative activity of green tea catechins. The findings of Wu et al. (2003) suggest that chemoprevention by green tea in women possessing low-activity COMT allele may result from increased bioavailability of catechins (Stuart et al., 2006). A recent study by Wu et al. (2005) conducted in 130 postmenopausal women observed significantly lower plasma levels of estrone in women regularly consuming green tea compared with non- or irregular tea drinkers (25.8 pg/mL vs 29.5 pg/mL). The findings of Wu et al. (2005) suggest that alteration of estrone levels may contribute to the chemopreventative activity of green tea polyphenols (Stuart et al., 2006).

8.3.2.1.5 Prostrate Cancer

Transgenic adenocarcinoma of the mouse prostate (TRAMP) is a model for prostate cancer that closely mimics progressive forms of the disease in humans. Green tea inhibits the growth and progression of prostate cancer in such mice, and furthermore inhibits metastasis of cancer to distant organ sites (lymph, lungs, liver, and bone) (Adhami et al., 2004; Caporali et al., 2004; Crespy and Williamson, 2004; Stuart et al., 2006).

A case–control study was conducted in southeast China assessing 130 patients with histologically confirmed incidental prostate cancer and 274 patients without cancer matched by age. It showed that the prostate cancer risk declined with increasing frequency, duration, and quantity of green tea consumed. This reduction was statistically significant, suggesting that green tea protects against prostate cancer (Jian et al., 2004).

8.3.2.2 Green Tea and Cardiovascular Diseases

The onset of cardiovascular disease (CVD) depends on numerous factors that can be modulated by components in the diet (Khan and Mukhtar, 2007). Some of the underlying mechanisms for the pharmacological effects of tea include vasculoprotective, antioxidative, antithrombogenic, anti-inflammatory, and lipid-lowering properties of tea flavonoids (Stangl et al., 2006). The development and progression of atherosclerosis has been inversely associated with the intake of green tea and it has been reported that dietary green tea intake preserves and improves arterial compliance and endothelial function (Murakami and Oshato, 2003).

Blood pressure increase was attenuated in spontaneously hypertensive rats by green tea extract. This effect was attributed to its antioxidant properties (Negishi et al., 2004). Green tea extract administered in drinking water in the apolipoprotein-E-deficient mouse model of artherosclerosis and prevented the development of artherosclerosis without affecting the plasma lipid or cholesterol levels (Miura et al., 2001).

8.3.2.3 Green Tea, Obesity, and Diabetes

Development of obesity and diabetes in an individual is characterized by increased the number and size of fat cells and by elevated blood glucose levels, respectively (Kao et al., 2006). Various animal models and treatments related to diabetes have been studied: Zucker rats (which are genetically obese), injection of streptozotocin or alloxan (which destroys pancreatic β-cells), and treatment with sucrose-rich diets (which induces obesity and insulin resistance). Tea catechins, especially EGCG, appear to have antiobesity and antidiabetic effects (Khan and Mukhtar, 2007).

Purified EGCG given to mice in their diet decreased diet-induced obesity by decreasing energy absorption and increasing fat oxidation (Klaus et al., 2005). Hepatotoxicity, fatty liver, and neurodegenerative disease were observed in oxidative stress-induced diabetes, and they were prevented by daily ingestion of green tea or EGCG. These observations support the reductive effects of green tea catechins on oxidative stress-induced diabetes (Kao et al., 2006).

8.4 STABILITY AND CHEMICAL CHANGES OF GREEN TEA CATECHINS DURING FOOD PROCESSING

The stability of tea catechins is dependent on pH and temperature. In acidic systems ($pH < 4$), tea catechins are fairly stable; in systems of $pH > 5$, that is, near neutral or alkaline systems, they degrade rapidly. On the contrary, tea catechins become less stable when processing temperature increases, where thermal degradation, oxidation, epimerization, and polymerization could occur (Komatsu et al., 1993; Seto et al., 1997; Wang et al., 2006a; Zhu et al., 1997). Ascorbic acid showed a significantly protective effect on the stability of tea catechins. The effective concentration ranged from 0.2 mg/mL (Zhu et al., 1997) to 11 mg/mL (Chen et al., 2001b). Some organic acids also demonstrated moderate protection by reducing the pH (Zhu et al., 1997).

The storage stability of tea catechins in aqueous systems including tea drinks prepared by conventional brewing and industrial canning processes has been examined by many researchers. It was reported that tea catechins remained stable at $pH < 4$, whereas they were relatively unstable at $pH > 6$. The stability of tea catechins decreased at higher storage temperatures (Chen et al., 2001b; Komatsu et al., 1993; Seto et al., 1997; Wang and Helliwell, 2000). Furthermore, tea catechins were sensitive to the concentration of oxygen present in aqueous systems (Zimeri and Tong, 1999).

The thermal stability of tea catechins in aqueous systems during processing has been reported in several publications (Chen et al., 2001b; Komatsu et al., 1993; Seto, et al., 1997; Su et al., 2003; Wang and Helliwell, 2000; Xu et al., 2003; Zimeri and Tong, 1999; Zhu et al., 1997). Komatsu et al. (1993) pointed out that the degradation and epimerization of tea catechins occurred simultaneously in thermal processes, and the thermal degradation followed first-order kinetics. A temperature of 82°C was reported as a turning point in thermal reactions of tea catechins. Above and below this point, tea catechins exhibited significantly different kinetics. However, "due to difficulties of qualification and complexity of the kinetics" (Komatsu et al., 1993), no mathematical model was established to predict tea catechins in an aqueous

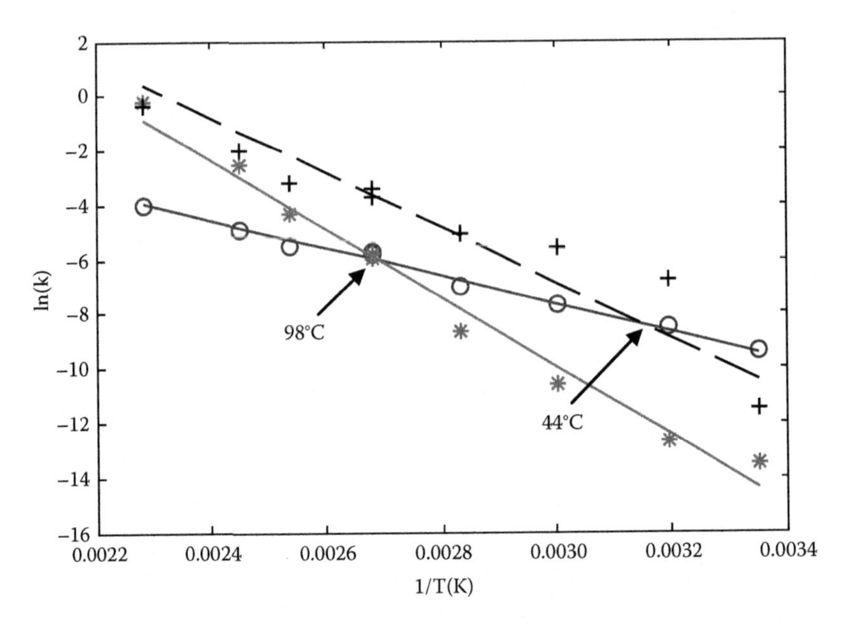

FIGURE 8.4 Arrhenius plot of the rate constants for the reactions of EGCG and GCG in aqueous solutions. ○: rate constant k_y for degradation of EGCG and GCG; ∗: rate constant k_1 for epimerization from EGCG to GCG; +: rate constant k_2 for epimerization from GCG to EGCG. (From Wang, R.; Zhou, W.; Jiang, X. 2008a. *J. Agric. Food Chem.* 56:2694–2701. With permission.)

system. Seto et al. (1997) reported that the epimerization was pH-dependent, and a maximum epimerization rate was at pH 5. Zimeri and Tong (1999) found that the stability of tea catechins was also oxygen-dependent. The oxidation followed a pseudo-first-order reaction. Activation energy (E_a) for the degradation was calculated as 18.7 kcal/mol (Zimeri and Tong, 1999), which is different from 4.7 kcal/mol (<82°C) and 37.9 kcal/mol (>82°C) reported by Komatsu et al. (1993).

It can be seen that activation energy of the degradation of tea catechins in an aqueous system differed among independent studies, which may be due to the lack of appropriate mathematical models to account for both degradation and epimerization of tea catechins during storage or thermal processing.

Recently, Wang et al. (2006a, 2008a) systematically examined the thermal stability of tea catechins, EGCG, and GCG in particular, in aqueous system including simultaneous degradation and epimerization reactions, using different concentrations of tea catechins and mathematical modeling techniques. The experimental temperatures ranged from 25°C to 165°C. The degradation and epimerization of EGCG and GCG complied with first-order reaction and their rate constants followed the Arrhenius equation. The activation energy (E_a) remained unchanged in catechin solutions with varied concentrations. The values of E_a were 43.09, 105.07, and 84.33 kJ/mol for the degradation, the epimerization from EGCG to GCG and the epimerization from GCG to EGCG, respectively. Two specific temperature points in the reaction kinetics were identified, at 44°C and 98°C, respectively (Figure 8.4). When the processing temperature was greater than 44°C, the epimerization from GCG to EGCG was the fastest reaction. The rate constants k can be ranked as: $k_2 > k_y > k_1$. When the temperature increased to 98°C and above, the rate of the epimerization from EGCG to GCG became faster than that of the degradation. The order of the rate constants became $k_2 > k_1 > k_y$ correspondingly. When the reaction temperature was below 44°C, the degradation, which was probably by oxidation, predominated the changes of EGCG and GCG in aqueous systems. The rate constants were in the order of $k_y > k_2 > k_1$. Based on these specific points, the reaction rates of the epimerization among tea catechins could be manipulated by adjusting the processing temperature so that a desired ratio between epi- and nonepi-catechins in the final product may be achieved. These results also indicated that the turning point of 82°C reported earlier in the literature for the reaction kinetics of catechins should be reexamined.

Mathematical models were further developed by Wang et al. (2008b) for the stability and chemical changes of EGCG and GCG during the baking process of bread containing the catechins. The models accounted not only for simultaneous thermal reactions but also varying moisture content and temperature profile in the bread crumb and crust. The corresponding rate constant (k) of the reaction kinetics followed Arrhenius equation. The activation energy (E_a) of the reactions previously obtained from aqueous systems remained unchanged in the bread baking system, while the frequency factor (A) changed significantly. The developed mathematical models enable prediction of the amount of tea catechins in the fortified bread under various baking conditions.

FIGURE 8.5 Mooncakes containing green tea.

8.5 FUNCTIONAL FOODS CONTAINING GREEN TEA OR GREEN TEA ANTIOXIDANTS

There are many green tea beverage products on the market, produced by major multi-international food companies as well as local food companies in various countries. Most of these products are either bottled or canned. Besides "pure" green tea beverage, there are some popular variants, for example, jasmine green tea and chrysanthemum green tea. It is worth to particularly mention that there are also catechin-enriched (or called catechin-plus) green tea beverages that contain additional tea catechins to those naturally brewed out of tea leaves.

There are dry green tea products (e.g., tea bags), which are mixtures of green tea and other herbs including ginseng, wolfberry, and *Ginkgo biloba*, among others. Such products are often marketed as herbal tea.

Besides tea beverages, green tea and green tea extracts can be incorporated in many other products (e.g., cereal, confectionary, dairy, edible oil) to either improve their shelf life, or provide new flavor, or give a healthier appeal to the consumer. As mentioned in the introduction section, green tea has successfully been used in moon-cake (Figure 8.5), which is a traditional cake eaten during the Chinese Middle Autumn Festival. There are ice cream and noodle products containing green tea or green tea extracts. Green tea extracts have also been incorporated in chocolates and chewing gums.

Although green tea and its extracts could be found in quite a number of food products in the market, there have been limited number of scientific reports in the literature examining the stability of tea polyphenols during processing and storage of such products (except those on green tea beverages and edible oils, many of which

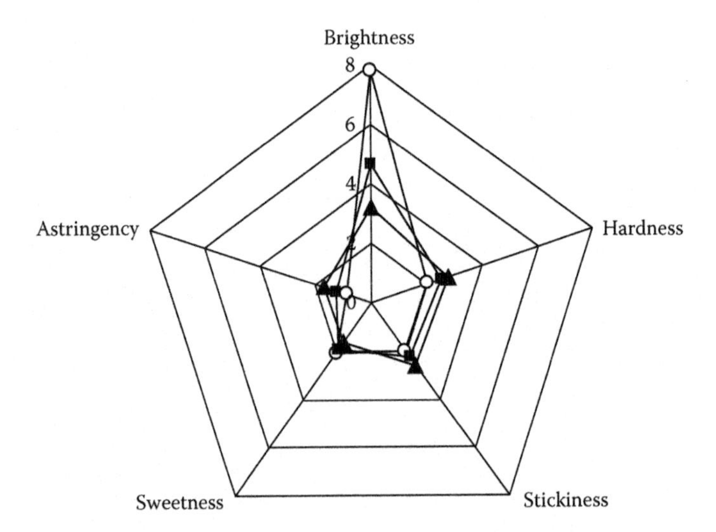

FIGURE 8.6 Radar plot of sensory results of the three bread variants by the trained panelists. ○: control bread containing no green tea extract (GTE); ■: bread with 150 mg GTE per 100 g flour; ▲: bread with 500 mg GTE per 100 g flour. (From Wang, R.; Zhou, W.; Isabelle, M. 2007. *Food Res. Int.* 40:470–479. With permission.)

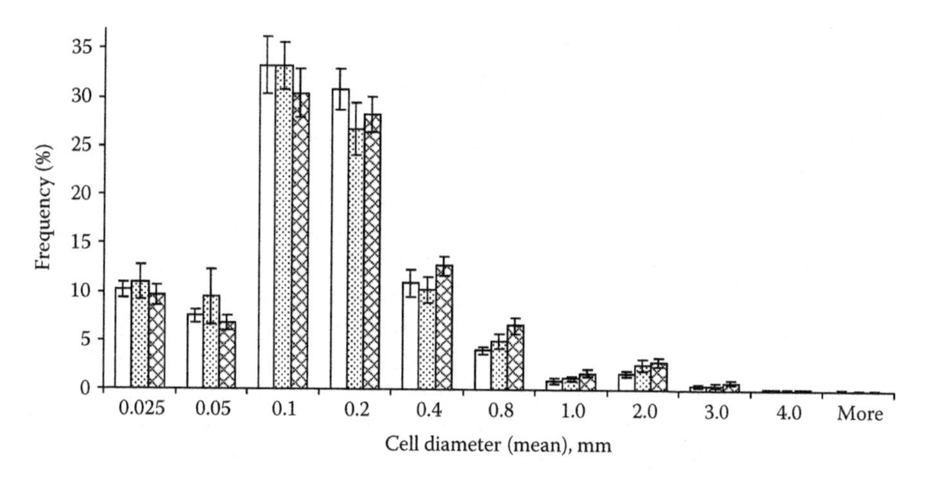

FIGURE 8.7 Histogram of crumb cell diameter (mean, $n = 10$). □: control bread containing no green tea extract (GTE); ▨: bread with 150 mg GTE per 100 g flour; ⊠: bread with 500 mg GTE per 100 g flour. (From Wang, R.; Zhou, W.; Isabelle, M. 2007. *Food Res. Int.* 40:470–479. With permission.)

were already cited in Sections 8.2 and 8.4), their interactions with other components in the products, and the consequent impact on the quality of the products.

A detailed study on the addition of green tea extract (GTE) to bread has been conducted (Bai and Zhou, 2006; Wang and Zhou, 2004; Wang et al., 2006b; 2007; 2008b). The stability of green tea catechins in frozen and unfrozen dough and in bread during their shelf life, as well as the effects of GTE on the quality of the corresponding dough and bread have been examined.

Results showed that green tea catechins were relatively stable in dough during freezing and frozen storage at –20°C for up to 9 weeks. There were no further detectable losses of tea catechins in bread during storage of four days at room temperature. It was also revealed that EGCG and EGC were more susceptible to degradation than ECG and EC. The retention levels of EGCG and ECG in freshly baked bread were ca. 83% and 91%, respectively (Wang and Zhou, 2004).

The addition of GTE yielded an adverse effect on bread quality. The actual change in quality depended on the amount of GTE added in dough. In general, low dose of GTE, for example, less than 100 mg per 100 g flour would not result in significant changes either in frozen or unfrozen dough processes. However, bread with the amount of GTE at 150 mg per 100 g flour exhibited significantly negative effects on bread volume and texture (Wang et al., 2006b). The precise mechanism behind these effects of tea catechins on bread/dough matrix, however, remains unclear.

The effects of GTE on the sensory qualities of the bread were examined. Both sensory evaluation techniques and instrumental analyses were used. Bread incorporating GTE at the levels of 150 mg and 500 mg per 100 g flour was analyzed concurrently with the control bread containing no GTE in a random order by panelists. Sensory analysis was carried out through a descriptive profiling test by both untrained panelists and trained panelists. A total of six sensory attributes including brightness, porosity, hardness, stickiness, sweetness, and astringency were evaluated. Instrumental analyses included image analysis for porosity, spectrophotometric measurement for brightness, and texture profile analysis for hardness and stickiness. Results showed that the sensory evaluation was generally correlated well with the instrumental analysis. With an increase in the level of GTE, the brightness and sweetness of the bread with GTE decreased, whereas the hardness, stickiness, and astringency increased (Figure 8.6). No significant difference in the histogram of cell diameter (porosity) was found (Figure 8.7). The threshold level of GTE was at 500 mg per 100 g flour for astringency and sweetness, and 150 mg per 100 g flour for brightness, hardness, and stickiness (Wang et al., 2007).

As mentioned in Section 8.4, based on the kinetic models for thermal degradation of EGCG and GCG and epimerization between them in aqueous solutions developed by Wang et al. (2006a, 2008a), mathematical models were developed in Wang et al. (2008b), which are capable of predicting the profile of EGCG and GCG in bread containing GTE during baking.

8.6 CONCLUSIONS

With an ever-increasing popularity of functional foods, foods containing natural antioxidants will be of high demand. Green tea offers a unique flavor and green tea antioxidants possess health benefits with an exceptionally long history Hence, there is no doubt that green tea would remain popular. It is predicted that there will be more food products appearing in the market that are fortified with green tea antioxidants. While incorporating green tea or green tea antioxidants into food products may help to convert many traditional products into modern functional food items, there are various scientific issues to be addressed. The most important ones include the stability of active compounds as well as their interactions with the other components of a food matrix that might be a novel environment for the compounds, which may significantly impact on the product quality. More scientific studies on these issues are necessary if green tea antioxidants are to be utilized as novel food ingredients in processed foods.

REFERENCES

Adhami, V.M.; Afaq, F.; Ahmad, N. et al. 2004. Tea polyphenols as cancer chemopreventive agents. In *Cancer Chemoprevention, Vol. 1: Promising Cancer Chemoprevention Agents*, eds. G.J. Kelloff, E.T. Hawk, and C.C. Sigman, pp. 437–449. Totowa, NJ: Humana Press Inc.

Ahmad, N.; Feyes, D.K.; Nieminen, A.L. et al. 1997. Green tea constituent epigallocatechin-3-gallate and induction of apoptosis and cell cycle arrest in human carcinoma cells. *J. Natl. Cancer Inst.* 89:1881–1886.

Arts, I.C.W.; Van de Putte, B.; Hollman, P.C.H. 2000a Catechin contents of foods commonly consumed in the Netherlands. 1. Fruits, vegetables, staples foods, and processed foods. *J. Agric. Food Chem.* 48:1746–1751.

Arts, I.C.W.; Van de Putte, B.; Hollman, P.C.H. 2000b. Catechin contents of foods commonly consumed in the Netherlands. 2. Tea, wine, fruit juices, and chocolate milk. *J. Agric. Food Chem.* 48:1752–1757.

Bai, X.; Zhou, W. 2006. Study of the bread oven rise by on-line image analysis. *Asia-Pacific J. Chem. Eng.* 1:104–109.

Balentine, D.A.; Wiseman, S.A.; Bouwens, C.M. 1997. The chemistry of tea flavonoids. *Crit. Rev. Food Sci. Nutri.* 37:693–704.

Cao, Y.; Cao, R. 1999. Angiogenesis inhibited by drinking tea. *Nature* 398:381.

Caporali, A.; Davalli, P.; Astancolle, S. et al. 2004. The chemopreventive action of catechins in the TRAMP mouse model of prostrate carcinogenesis is accompanied by clusterin over expression. *Carcinogenesis* 25:2217–2224.

Chen, M-L. 2002. Tea and health—an overview. In *Tea: Bioactivity and Therapeutic Potential*, ed. Y-S. Zhen. London: Taylor and Francis, pp. 1–13.

Chen, Z.Y.; Chan, P.T. 1996. Antioxidative activity of green tea catechins in canola oil. *Chemistry and Physics of Lipids* 82:163–172.

Chen, Z.H.; Zhou, B.; Yang, L.; et al. 2001a. Antioxidant activity of green tea polyphenols against lipid peroxidation initiated lipid-soluble radicals in micelles. *J. Chem. Soc. Perkin Trans.* 2:1835–1839.

Chen, Z.Y.; Zhu, Q.Y.; Tsang, D. et al. 2001b. Degradation of green tea catechins in tea drinks. *J. Agric. Food Chem.* 49:477–482.

Crespy, V.; Williamson, G. 2004. A review of the health effects of green tea catechins in *in vivo* animal models. *J. Nutr.* 134:3431S–3440S.

Dew, T.P; Day, A.J.; Morgan, M.R.A. 2005. Xanthine oxidase activity *in vitro*: effects of food extracts and components. *J. Agric. Food Chem.* 53: 6510–6515.

Dona, M.; Dell'Aica, I.; Calabrese, F. 2003. Neutrophil restraint by green tea: Inhibition of inflammation, associated angiogenesis, and pulmonary fibrosis. *J. Immunol.* 170: 4335–4341.

Dong, Z. 2000. Effects of food factors on signal transduction pathways. *Biofactors* 12: 17–28.

Dufresne, C.J.; Farnworth, E.R. 2001. A review of latest research findings on the health promotion properties of tea. *J. Nutr. Biol.* 12:404–421.

Fujiki, H.; Yoshiwaki, S.; Horiuchi, T. et al. 1992. Anticarcinogenic effects of (–)-epigallocatechin gallate. *Prevent. Med.* 21:503–509.

Guo, Q.; Zha, B.; Li, M. et al. 1996. Studies on protective mechanisms of four components of green tea polyphenols against lipid peroxidation in synaptosomes. *Biochim. Biophys. Acta* 1304:210–222.

Guo, Q.; Zhao, B.; Shen, S. et al. 1999. ESR study on the structure-antioxidant activity relationship of tea catechins and their epimers. *Biochim. Biophys. Acta* 1427: 13–23.

Hara, Y. 2001. *Green Tea: Health Benefits and Applications*. New York: Marcel Dekker.

Hatano, T.; Ohyaby, T.; Ito, H. et al. 2005. The structural variation in the incubation products of (–)-epigallocatechin gallate in neutral solution suggests its breakdown pathways. *Heterocycles* 65(2):303–310.

He, Y.; Shahidi, F. 1997. Antioxidant activity of green tea and its catechins in a model fish system. *J. Agric. Food Chem.* 45:4262–4266.

Higdon, J.V.; Frei, B. 2003. Tea catechins and polyphenols: Health effects, metabolism, and antioxidant functions. *Crit. Rev. Food Sci. Nutr.* 43(1): 89–143.

Huang, S.W.; Frankel, E.N. 1997. Antioxidant activity of tea catechins in different lipid system. *J. Agric. Food Chem.* 45(8):3033–3038.

Hung, P.F.; Wu, B.T.; Chen, H.C. et al. 2005. Antimitogenic effect of green tea (–)-epigallocatechin gallate on 3T3-L1 preadipocytes depends on the ERK and Cdk2 pathways. *Am. J. Physiol.—Cell Physiol.* 288:C1094–C1108.

Hwang, K.C.; Lee, K.H.; Jang, Y. et al. 2002. Epigallocatechin-3-gallate inhibits basic fibroblast growth factor induced intracellular signalling transduction pathway in rat aortic smooth muscle cells. *J. Cardiovasc. Pharmacol.* 39:271–277.

Jeong, J.H.; Kim, H.J.; Lee, T.J. et al. 2004. Epigallocatechin 3-gallate attenuates neuronal damage induced by 3-hydroxykynurenin. *Toxicology* 195:53–60.

Jian, L.; Xie, L.P.; Lee, A.H. et al. 2004. Protective effect of green tea against prostrate cancer: A case-control study in southeast China. *Int. J. Cancer* 108(1):130–135.

Kao, Y.H.; Chang, H.H; Lee, M.J. et al. 2006. Review: Tea, obesity and diabetes. *Mol. Nutr. Food Res.* 50:188–210.

Kautenburger, T.; Becker, T.W.; Pool-Zobel, B.L. 2005. Low concentrations of green tea catechins transiently induce and inhibit MAPK signal transduction and growth of human colon tumor cells. *Int. J. Cancer Prevent.* 2(2):117–127.

Khan, N.; Mukhtar, H. 2007. Tea polyphenols for health promotion. *Life Sci.* 81:519–533.

Kim, M.; Masuda, M. 1997. Cancer chemoprevention by green tea polyphenols. In *Chemistry and Applications of Green Tea*, eds. T. Yamamoto, L.R. Juneja, D-C. Chu, and M. Kim. Boca Raton, FL: CRC Press, pp. 61–73.

Klaus, S.; Pultz, S.; Thone-Reineke, C. et al. 2005. Epigallocatechin gallate attenuates diet induced obesity in mice by decreasing energy absorption and increasing fat oxidation. *Int. J. Obesity* 29(6):615–623.

Komatsu, Y.; Suematsu, S.; Hisanobu, Y. et al. 1993. Studies on preservation of constituents in canned drinks. Part II. Effects of pH and temperature on reaction kinetics of catechins in green tea infusion. *Bio. Biotech. Biochem.* 57:907–910.

Kondo, K.; Kurihara, M.; Miyata, N. et al. 1999. Scavenging mechanisms of (–)-epigallocatechin gallate and (–)-epicatechin gallate on peroxyl radicals and formation of superoxide during the inhibitory action. *Free Radic. Biol. Med.* 27(7/8):855–863.

Kumamoto, K.; Sonda, T.; Nagayama, K. et al. 2001. Effects of pH and metal ions on antioxidative activities of catechins. *Biosci. Biotechnol. Biochem.* 65(1):126–132.

Kuroda, Y.; Hara, Y. 1999. Antimutagenic activity of tea polyphenols. *Rev. Mutat. Res.* 436:69–97.

Levites, Y; Amit, T; Youdim, M.B.H. et al. 2002b. Involvement of protein kinase C activation and cell survival/cell cycle genes in green tea polyphenol (–)-epigallocatechin-3-gallate neuroprotective action. *J. Biol. Chem.* 277:30574–30580.

Levites, Y.; Youdim, M.B.H.; Maor, G. et al. 2002a. Attenuation of 6-hydroxydopamine (6-OHDA)-induced nuclear factor-kappaB (NF-kappaB) activation and cell death by tea extracts in neuronal cultures. *Biochem. Pharmacol.* 63:21–29.

Liang, Y.; Huang, Y.; Tsai, S. et al. 1999. Suppression of inducible cyclooxigenase and inducible nitric oxide synthase by apigenin and related flavonoids in mouse macrophages. *Carcinogenesis* 20:1945–1952.

Madhavi, D.L.; Singhal, R.S.; Kulkarni, P.R. 1996. Technological aspects of food antioxidants. In *Food Antioxidants*, eds. D.L. Madhavi, S.S. Deshpande, D.K. Salunke. New York: Marcel Dekker, Inc, pp. 159–265.

Mandel, S.; Youdim, M.B.H. 2004. Catechin polyphenols: Neurodegeneration and neuroprotection in neurodegenerative diseases. *Free Radic. Biol. Med.* 37:304–317.

McKay, D.L.; Blumberg, J.B. 2002. The role of tea in human health: An update. *J. Am. College Nutr.* 21(1):1–13.

Miura, Y.; Chiba, T.; Tomita, I. et al. 2001. Tea catechins prevent the development of atherosclerosis in apoprotein E-deficient mice. *J. Nutr.* 131(1):27–32.

Moon, H.S.; Lee, H.G.; Choi, Y.J. et al. 2007. Proposed mechanisms of –(-) epigallocatechin-3-gallate for antiobesity. *Chemicobiol. Interact.* 167:85–98.

Mukai, K.; Nagai, S.; Ohara, K. 2005. Kinetic study of the quenching reaction of singlet oxygen by tea catechins in ethanol solution. *Free Radic. Biol. Med.* 39:752–761.

Murakami, T.; Oshato, K. 2003. Dietary green tea intake preserves and improves arterial compliance and endothelial function. *J. Am. Coll. Cardiol.* 141:271–274.

Nance, C.L.; Shearer, W.T.S. 2003. Is green tea good for HIV-1 infection? *J. Allergy Clin. Immunol.* 112:851–853.

Nanjo, F.; Goto, K.; Seto, R. et al. 1996. Scavenging effects of tea catechins and their derivatives on 1,1-diphenyl-2-picrylhydrazyl radical. *Free Radic. Biol. Med.* 21:895–902.

Negishi, H.; Xu, J.W.; Ikeda, K. et al. 2004. Black and green tea polyphenols attenuate blood pressure I increases in stroke prone spontaneously hypertensive rats. *J. Nutr.* 134(1):38–42.

Nomura, M.; Ma, W.; Chen, N. et al. 2001. Inhibitory mechanisms of tea polyphenols on the ultraviolet B activated phosphatidylinositol 3-kinase—dependent pathway. *J. Biol. Chem.* 276:46624–46631.

Pelillo, M.; Biguzzi, B.; Bendini, A. et al. 2002. Preliminary investigation into development of HPLC with UV and MS-electrospray detection for the analysis of tea catechins. *Food Chem.* 78:369–374.

Pfeffer, U.; Ferrari, N.; Morini, M. et al. 2003. Antiangiogenic activity of chemopreventive drugs. *Int. J. Biol. Markers* 18:70–74.

Pokorny, J.; Yanishlieva, N.; Gordon, M. 2001. *Antioxidants in Food: Practical Applications.* Cambridge: Woodhead.

Ramassamy, C. 2006. Emerging role of polyphenolic compounds in the treatment of neurodegenerative diseases: A review of their intracellular targets. *Eur. J. Pharmacol.* 545:51–64.

Rao, T.P; Okubo, T; Chu, D-C. et al. 2003. Pharmacological functions of green tea polyphenols. In *Performance Functional Foods*, ed. D. Watson. Cambridge: Woodhead, pp. 140–167.

Rijken, P.J.; Wiseman S.A; Weisgerber U.M. et al. 2002. Antioxidant and other properties of green and black tea. In *Handbook of Antioxidants*, eds. E. Cadenas and L. Packer. New York: Marcel Dekker, Inc., pp. 371–399.

Salah, N.; Miller, N.J.; Paganga, G. et al. 1995. Polyphenolic flavonols as scavengers of aqueous phase radicals and as chain-breaking antioxidants. *Archiv. Biochem. Biophys.* 322:339–346.

Seto, R.; Nakamura, H.; Nanjo, F. et al. 1997. Preparation of epimers of tea catechins by heat treatment. *Biosci. Biotech. Biochem.* 61:1434–1439.

Shimizu, M.; Weinstein, I.B. 2005. Modulation of signal transduction by tea catechins and related phytochemicals. *Mutat. Res.* 591:147–160.

Shishikura, Y.; Khokhar, S. 2005. Factors affecting the levels of catechins and caffeine in tea beverage: Estimated daily intake and antioxidant activity. *J. Sci. Food Agric.* 85: 2125–2133.

Stangl, V.; Dreger, H.; Stangl, K. et al. 2007. Molecular targets of tea polyphenols in the cardiovascular system. *Cardiovasc. Res.* 73:348–358.

Stangl, V.; Lorenz, M.; Stangl, K. 2006. The role of tea and tea flavonoids in cardiovascular health. *Mol. Nut. Food Res.* 50(2):218–228.

Stapleton, P.D.; Shah, S.; Anderson, J.C. et al. 2004. Modulation of beta lactam resistance in *Staphylococcus aureus* by catechins and gallates. *Int. J. Antimicrobial Agents* 23(5): 462–467.

Stuart, E.C.; Scandlyn, M.J.; Rosengren, R.J. 2006. Role of epigallocatechin gallate (EGCG) in the treatment of breast and prostrate cancer. *Life Sci.* 79:2329–2336.

Su, Y.L.; Leung, L.K.; Huang, Y. et al. 2003. Stability of tea theaflavins and catechins. *Food Chem.* 83:189–195.

Sutherland, B.A.; Rahman, R.M.A.; Appleton, I. 2006. Mechanisms of action of green tea catechins. *J. Nutr. Biochem.* 17:291–306.

Thangapazham, R.L.; Singh, A.K.; Sharma, A. et al. 2007. Green tea polyphenols and its constituent epigallocatechin gallate inhibits proliferation of human breast cancer cells *in vitro* and *in vivo*. *Cancer Lett.* 245:232–241.

Wang, H.; Helliwell, K. 2000. Epimerisation of catechins in green tea infusions. *Food Chem.* 70:337–344.

Wang, H.; Helliwell, K.; You, X. 2000b. Isocratic elution system for the determination of catechins, caffeine and gallic acid in green tea using HPLC. *Food Chem.* 68:115–121.

Wang, H.; Provan, G.J.; Helliwell K. 2000a. Tea flavonoids: Their functions, utilisation and analysis. *Trends Food Sci. Technol.* 11:152–160.

Wang, H.; Provan, G.J.; Helliwell, K. 2003. HPLC determination of catechins in tea leaves and tea extracts using relative response factors. *Food Chem.* 81:307–312.

Wang, R.; Zhou, W. 2004. Stability of tea catechins in the breadmaking process. *J. Agric. Food Chem.* 52:8224–8229.

Wang, R.; Zhou, W.; Isabelle, M. 2007. Comparison study of the effect of green tea extract (GTE) on the quality of bread by instrumental analysis and sensory evaluation. *Food Res. Int.* 40:470–479.

Wang, R.; Zhou, W.; Jiang, X. 2008a. Reaction kinetics of degradation and epimerization of epigallocatechin gallate (EGCG) in aqueous system over a wide temperature range. *J. Agric. Food Chem.* 56:2694–2701.

Wang, R.; Zhou, W.; Jiang, X. 2008b. Mathematical modeling of the stability of green tea catechin epigallocatechin gallate EGCG during bread baking. *J. Food Eng.* 87: 505–513.

Wang, R.; Zhou, W.; Wen, R. H. 2006a. Kinetic study of the thermal stability of tea catechins in aqueous systems using a microwave reactor. *J. Agric. Food Chem.* 54:5924–5932.

Wang, R.; Zhou, W.; Yu, H. H. et al. 2006b. Effects of green tea extract on the quality of bread made from unfrozen and frozen dough processes. *J. Sci. Food Agric.* 86: 857–864.

Weinreb, O.; Mandel, S.; Amit, T. et al. 2004. Neurological mechanisms of green tea polyphenols in Alzheimer's and Parkinson's diseases. *J. Nutr. Biochem.* 15:506–516.

Weinreb, O.; Mandel, S.; Youdim, M. et al. 2003. cDNA gene expression profile homology of antioxidants and their antiapoptotic and pro-apoptotic activities in human neuroblastoma cells. *J. Fed. Am. Soc. Exp. Biol.* 17:935–937.

Wiseman, S.A.; Balentine, D.A.; Frei, B. 1997. Antioxidants in tea. *Crit. Rev. Food Sci. Nutr.* 37(8):705–718.

Wu, A.H.; Arakawa, K.; Stanczyk, F.Z. et al. 2005. Tea and circulating estrogen levels in postmenopausal Chinese women in Singapore. *Carcinogenesis* 26:976–980.

Wu, A.H.; Yu, M.C.; Tseng, C. et al. 2003. Green tea and risk of breast cancer in Asian Americans. *Int. J. Cancer* 106:574–579.

Xu, J.Z.; Leung L.K.; Huang, Y. et al. 2003. Epimerisation of tea polyphenols in tea drinks. *J. Sci. Food Agric.* 83:1617–1621.

Yamamoto, T.; Juneja, L.R.; Chu, D.C. et al. 1997. *Chemistry and Application of Green Tea.* Boca Raton, FL: CRC Press.

Yang, C.S.; Chen, L.; Lee, M.J. et al. 1996. Effects of tea on carcinogenesis in animal models and humans. In *Dietary Phytochemicals in Cancer Prevention and Treatment*, eds. N. Back, I.R. Cohen, D. Kritchevsky, A. Lajtha, and R. Paoletti. New York: Plenum Press, pp. 51–61.

Yang, C.S.; Maliakal, P.; Meng, X. 2002. Inhibition of carcinogenesis by tea. *Ann. Rev. Pharmacol. Toxicol.* 42:25–54.

Yoshioka, H.; Sugiura, K.; Kawahara, T. et al. 1991. Formation of radicals and chemiluminescence during the autoxidation of tea catechins. *Agric. Biol. Chem.* 55:2717–2723.

Zaveri, N.T. 2005. Green tea and its polyphenolic catechins: Medicinal uses in cancer and non cancer applications. *Life Sci.* 78:2078–2080.

Zhang, A.; Chan, P.T.; Luk, Y.S. et al. 1997. Inhibitory effect of jasmine green tea epicatechin isomers on LDL-oxidation. *J. Nutr. Biochem.* 8:334–340.

Zhao, B. 2005. Natural antioxidants for neurodegenerative diseases. *Mol. Neurobiol.* 31: 283–293.

Zhong, L.; Goldberg, M.S.; Gao, Y.T. et al. 2001. A population based case–control study of lung cancer and green tea consumption among women living in Shanghai, China. *Epidemiology* 12(6):695–700.

Zhu, Q.Y.; Zhang, A.; Tsang, D. et al. 1997. Stability of green tea catechins. *J. Agric. Food Chem.* 45:4624–4628.

Zimeri, J.; Tong, C.H. 1999. Degradation kinetics of (–)-epigallocatechin gallate as a function of pH and dissolved oxygen in a liquid model system. *J. Food Sci.* 64(5):753–758.

9 Polyphenols, Antioxidant Activities, and Beneficial Effects of Black, *Oolong*, and *Puer* Teas

Yueming Jiang, John Shi, and Sophia Jun Xue

CONTENTS

9.1 INTRODUCTION

Tea (*Camellia sinensis*) is a refreshing, popular, socially accepted, economical, and beneficial beverage (Cheng and Chen, 1994). Globally, the production and consumption of tea has increased over the last decade, and this increase is expected to remain sustained (Ho et al., 2008). Tea contains significant amount of phenolic compounds (Gramza and Korczak, 2005). During the last decade, epidemiological studies have shown that intake of tea catechins is associated with a lower risk of cardiovascular disease (Yang et al., 2001). Thus, tea polyphenols have exhibited health beneficial effects and potential uses. A growing body of evidence suggests that moderate consumption of tea may protect against several forms of cancer, cardiovascular diseases, bacterial infections, the formation of kidney stones, and dental cavities (Ho et al., 2008; Khan and Mukhtar, 2007). Furthermore, tea is considered as a functional food based on the reported beneficial effects on human health (An et al., 2004; Farhoosh et al., 2007; Halder et al., 2005; Han, 1997; Mello et al., 2004; Navas et al., 2005; Zhu et al., 2006), and, thus, it can be used as a readily accessible source of natural antioxidants and a possible supplement in the food or pharmaceutical industry.

Tea is usually classified into green, *Oolong*, black, and *Puer* in terms of the manufacturing process. This chapter focuses mainly on polyphenols and antioxidants of the black, *Oolong*, and *Puer* teas, with an emphasis on their beneficial effects on health. For more information on green tea, readers may refer to other chapters of this book.

9.2 CLASSIFICATION OF TEA

Tea is made from the tender leaves of two varieties of the plant *Camellia sinensis*: *Camellia assamica* and *Camellia sinensis*, and was first discovered in China where it has been consumed for its medicinal properties since 3000 BC (Balentine, 1997). Recently, many varieties of tea have been produced. This chapter includes *Oolong* tea (semifermented), black tea (fully fermented), and *Puer* tea (anaerobic bacterial fermented) (Cheng and Chen, 1994; Harbowy and Balentine, 1997). Black tea is made by a polyphenol oxidase that catalyzes oxidation of fresh leaf catechins through a process known as fermentation. This process results in the oxidation of simple polyphenols, that is, tea catechins converted into more complex condensed molecules that give black tea its typical color, and strong and astringent taste. *Oolong tea* is prepared by firing the leaves shortly after rolling, and then drying the leaves. The oxidation is completed by the firing process. Thus, *Oolong tea* is called semifermented tea and is the characteristics between black tea and green tea. *Puer* tea is a unique Chinese microbial fermented tea obtained through indigenous tea fermentation where microorganisms are involved in the manufacturing process.

9.3 TEA POLYPHENOLS

Tea polyphenols are known as catechins. Chemically, catechins are water-soluble, colorless compounds, which impart astringency to tea infusions (Lee and Ong, 2000; Peterson et al., 2005; Wheeler and Wheeler, 2004). The structures of the six major catechins, epigallocatechin gallate (EGCG), epigallocatechin (EGC), epicatechin gallate (ECG), epicatechin (EC), theaflavins (TFs), and thearubigens are shown in Figure 9.1 (Balentine et al., 1997; Harbowy and Balentine, 1997; Khan and Mukhtar, 2007). Table 9.1 lists the content of individual polyphenolic compounds in black, *Oolong* and *Puer* tea (Nishitani and Sagesaka, 2004), whereas Table 9.2 shows individual catechins, gallic acid, and caffeine present in three types of Chinese teas (Lin et al., 1998a). It is particularly noted that 3′-*O*-methyl-EGCG has been isolated from *Oolong* tea as a minor component.

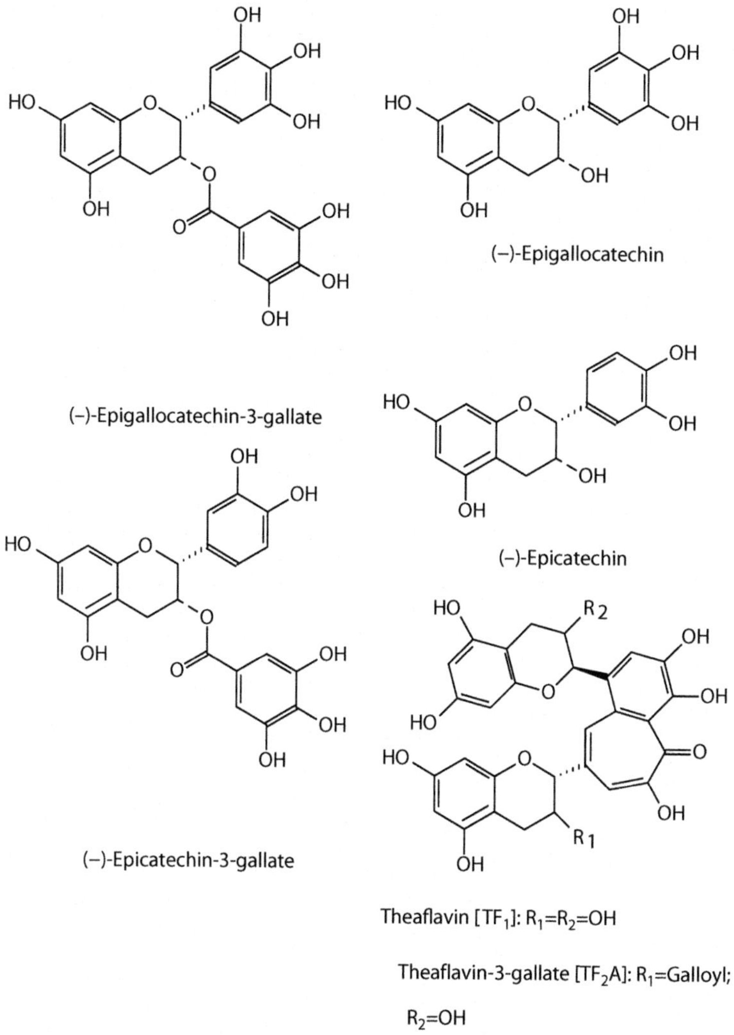

(–)-Epigallocatechin-3-gallate

(–)-Epigallocatechin

(–)-Epicatechin-3-gallate

(–)-Epicatechin

Theaflavin [TF_1]: $R_1=R_2=OH$

Theaflavin-3-gallate [TF_2A]: R_1=Galloyl;

$R_2=OH$

FIGURE 9.1 Structures of the major tea polyphenols. (Adapted from Balentine, D. 1997. *Critical Reviews in Food Science and Nutrition* 37:691–692; Khan, N. and Mukhtar, H. 2007. *Life Science* 81:519–533.)

9.3.1 Polyphenols in Black Tea

The major polyphenols in black tea are identified as theaflavins (TFs) (Table 9.1). The domination of TFs is simple theaflavin (TF), theaflavin-3-gallate (TF-3-g), theaflavin-3′-gallate (TF-3′-g), theaflavin-3,3′-digallate (TF-3,3′-dg), which comprise 0.3–2% of the dry matter of black tea (Łuczaj and Skrzydlewska, 2005; Steinhaus and Englehardt, 1989). Major factors affecting the content and the composition of polyphenols in black tea involve species, geographical location, and climate (Yao et al., 2006). For instance, Ding et al. (1992) reported that the content of polyphenols

TABLE 9.1
The Content of Individual Polyphenolic Compounds in Black, *Oolong*, and *Puer* Teas (mg/100 g Dry Tea; Water Extraction Data Only)

Polyphenolic Compound	Black Tea	*Oolong* Tea	*Puer* Tea	References
Flavan-3-ols				
Catechin	167	17	0	Collier and Mallows (1971); Kuhr and Engelhardt (1991); Lin et al. (1998a)
Epicatechin	316	259	81	Collier and Mallows (1971); Kuhr and Engelhardt (1991); Lee and Ong (2000); Lin et al. (1998a)
Epicatechin-3-gallate	923	707	40	Bronner and Beecher (1998); Collier and Mallows (1971); Kuhr and Engelhardt (1991); Lee and Ong (2000); Lin et al. (1998a)
Epigallocatechin	1257	510	157	Bronner and Beecher (1998); Collier and Mallows (1971); Kuhr and Engelhardt (1991); Lee and Ong (2000); Lin et al. (1998a)
Epigallocatechin-3-gallate	1393	3861	120	Bronner and Beecher (1998); Collier and Mallows (1971); Kuhr and Engelhardt (1991); Lee and Ong (2000); Lin et al. (1998a)
Gallocatechin	126			Arts et al. (2000)
Gallocatechin-3-gallate		93	13	Lin et al. (1998a)
Total catechins by sum of means	4182	5447	411	Lin et al. (1998a)
Total catechins	3937	6037	493	Kuhr and Engelhardt (1991); Lin et al. (1998a)
Theaflavin	162			Peterson et al. (2005); Steinhaus and Englehardt (1989); Takeo (1974)

Theaflavin 3-gallate	105			Steinhaus and Englehardt (1989); Takeo (1974)
Theaflavin 3′-gallate	178			Steinhaus and Englehardt (1989); Takeo (1974)
Theaflavin 3,3′-digallate	144			Steinhaus and Englehardt (1989); Takeo (1974)
Total theaflavins by sum of means	589			Peterson et al. (2005)
Total theaflavins	568			Peterson et al. (2005)
The arubigins	12,490			Owuor and Orbanda (1995); Steinhaus and Englehardt (1989); Takeo (1974)
Total flavan-3-ols	17,262	5447	411	Peterson et al. (2005)
Flavonols				
Kaempferol	132			Hertog et al. (1993); Pricee et al. (1998); Toyoda et al. (1997)
Myricetin	25			Hertog et al. (1993); Price et al. (1998); Toyoda et al. (1997)
Quercetin	210			Hertog et al. (1993); Price et al. (1998); Toyoda et al. (1997)
Total flavonols	367	269	115	Peterson et al. (2005)
Flavones				
Apigenin	54			Engelhardt et al. (1993); Toyoda et al. (1997)
Luteolin	5			Engelhardt et al. (1993); Toyoda et al. (1997)
Total flavones	59	119	5	Peterson et al. (2005)
Total flavonoids	17,687	5834	531	Peterson et al. (2005)

TABLE 9.2
The Contents of Individual Catechin, Gallic Acid, and Caffeine in the Black, *Oolong*, and *Puer* Teas (mg/g Tea)

Tea Name	GA	EGC	EGCG	EC	ECG	CG	CA
Puer tea	5.53	6.23	1.99	3.24	1.32	–	22.4
Fujian *Oolong* tea	1.42	10	22.2	2.63	6.06	0.27	7.44
Fujian black tea	2.06	5.71	3.79	1.36	4.45	–	21.6

Source: From Lin, J. K. et al., 1998a. *Journal of Agriculture and Food Chemistry* 46:3635–3642. With permission.

in black tea markedly varied with respect to different production areas such as Kenya, India, and China. Additionally, processing stages such as withering, rolling, and fermentation, play important roles in the structure of TFs in black tea.

9.3.2 Polyphenols in *Oolong* Tea

Oolong tea exhibits a similar phenolic profile to that of green tea or black tea but has higher levels of EGCG, EC, ECG, and EGC than those found in black tea (Table 9.2), as the fresh leaves of *Oolong* tea are subjected to a partial fermentation stage prior to drying (Zuo et al., 2002). *Oolong* tea contains gallic acid (GA), EGCG, EC, ECG, EGC, caffeine (CA), and catechin gallate (CG) (Lin et al., 1998a). Among these polyphenols, EGCG is the most abundant present in *Oolong* tea. However, a relatively low caffeine level in Guangdong *Oolong* tea was observed. The mechanisms involved in the reduction of the levels of polyphenolics by a series of biochemical activities during *Oolong* tea production are interesting, but require to be studied further.

9.3.3 Polyphenols in *Puer* Tea

Puer tea is a unique Chinese microbial fermented tea. It contains GA, EGC, EGCG, EC, ECG, and CA (Zuo et al., 2002). However, *Puer* tea has the lowest level of polyphenols among these three tea types, because the fermentation process during the tea manufacturing reduced tea polyphenols significantly (Table 9.3). Kuo et al. (2005) reported that *Puer* tea can lower the levels of triacylglycerol more significantly than green and black tea. Unfortunately, very little information on polyphenols in *Puer* tea is available compared with black tea. Furthermore, clones, geographical locations, climate, and manufacturing processes such as microorganisms and fermentation greatly affect the content and the composition of polyphenols in *Puer* tea and thus need to be investigated.

9.3.4 Bioavailability

It is important to look at the bioavailability of flavonoids from tea including absorption, distribution, metabolism, and elimination to get a comprehensive understanding of their possible impact on living organisms (Rawel and Kulling, 2007; Scholz and

TABLE 9.3
The Contents of Polyphenols in Black, *Oolong*, and *Puer* Teas (mg/100 g Dry Tea Water Extraction Only)

Polyphenols	*Oolong* Tea	Black Tea	*Puer* Tea
Catechin	17	167	0
Epicatechin	259	316	81
Epicatechin 3-gallate	707	923	40
Epigallocatechin	510	1257	157
Epigallocatechin 3-gallate	3861	1393	120
Gallocatechin 3-gallate	93	126	13
Total catechins by sum of means	5447	4183	411
Total catechins	6037	3937	493
Flavonols			
Kaempferol	90	132	23
Myricetin	49	25	40
Quercetin	130	210	52
Flavones			
Apigenin	110	54	0
Luteolin	9	5	5

Source: From Peterson, J. et al., 2005. *Journal of Food Composition and Analysis* 18:487–501. With permission.

Williamson, 2007). In the tea polyphenols, pure flavan-3 ols are poorly absorbed, but their glycosides show moderate to rapid absorption in humans, probably because an active glucose transport occurs in the small intestine. Catechins and catechin-condensed products from black tea can be absorbed in humans (Kyle et al., 2007; Manach et al., 2005). Catechins are metabolized extensively, but the absorption and metabolism mechanism of larger molecules present in black tea remains unclear (Lambert et al., 2007; Mennen et al., 2007). Catechins could pass through glucuronidation, sulfation, and *O*-methylation in the liver (Lambert et al., 2007). Polyphenols have a strong affinity for proteins via the various phenolic groups particularly when proteins have high proline content such as caseins, gelatin, and salivary proteins. However, the addition of milk to tea does not affect the polyphenol concentration in plasma (Kyle et al., 2007). Tea flavonoids have also a strong affinity for iron and form insoluble complexes which reduce the bioavailability of nonheme iron (Mennen et al., 2007; Nelson and Poulter, 2004). Absorption of ascorbic acid inhibits this complex formation. This finding has important implications mainly for people following a vegetarian diet.

9.4 ANTIOXIDANT ACTIVITY

Tea preparations have been shown to trap reactive oxygen species (ROS), such as superoxide radical, singlet oxygen, hydroxyl radical, peroxyl radical, nitric oxide, nitrogen dioxide, and peroxynitrite, and reduce their damage to lipid membranes,

TABLE 9.4
A Comparative Analysis of Total Antioxidant Activity and 1,1-Diphenyl-2-Picrylhydrazyl (DPPH) Radical Scavenging Activity of Black, *Oolong*, and *Puer* Teas (5 g/200 mL Distilled Water at 80° C)

Tea Type	Scavenging Activity (%) against DPPH Radicals
Black tea (Guangdong)	77.3
Oolong tea (Fujian)	81.9
Puer tea (Yunnan)	75.8

proteins, and nucleic acids in cell-free systems (Gramza and Korczak, 2005). Tea catechins can act as antioxidants by the donation of hydrogen atoms, acceptors of free radicals, interrupters of chain oxidation reactions, by chelating metals or annexation of hydroxide groups to catechin molecules (Gramza and Korczak, 2005; Hou et al., 2005; Wanasundara and Shahidi, 1996). Among these tea polyphenols, flavonoids, which are a class of antioxidants, constitute the relative majority of components of the leaves of *C. sinensis.* Tea is the richest source of flavonoids while EGCG is the most effective in reacting with most ROS. As shown in Table 9.4, differences in antioxidant activities of black, *Oolong*, and *Puer* teas are due to analytical methods, the kind of solvent used (Wang et al., 2000a, 2000b), and drying conditions as well as technological processes during tea production (Chu and Juneja, 1997; Fernandez et al., 2002; Lin et al., 1998b). The strongest antioxidant activity was detected in Oolong tea, followed by black tea and *Puer* tea.

9.4.1 Antioxidant Activity in Black Tea

Black tea contains a significant amount of theaflavin (TF) compounds that make an important contribution to antioxidant activity, but their effectiveness varies with the individual TF. For example, Miller et al. (1996) found that the radical scavenging activity of these compounds decreased as follows: TF-digallate > TF-3′-g=TF-3-g > TF-f. Similar results were also obtained by Leung et al. (2001), who demonstrated the antioxidant activity of individual TFs using human LDL oxidation as a model. Furthermore, Stewart et al. (2005) reported that flavan-3 ols exhibited similar levels in green tea and black tea, indicating that they are relatively stable during the fermentation process. Additionally, black tea processing stages, such as withering, rolling, and fermentation, play important roles in the structure of TFs and thus affect the antioxidant activity in black tea.

9.4.2 Antioxidant Activity in *Oolong* Tea

Oolong tea has characteristics of both black tea and green tea (Gadow et al., 1997). Catechins present in *Oolong* tea can act as antioxidants. Zhu et al. (2002) reported that the extract of *Oolong* tea using boiling water has a strong antioxidant ability and free radical scavenging activity. Furthermore, the *Oolong* tea infusion for 10 min in 100° C water produced the great ability to inhibit the peroxidation of peanut oil (Su et al., 2006). However, more investigations into the polyphenolic compounds and antioxidant ability of *Oolong* tea are required before a final conclusion can be drawn because of the various evaluation conditions and different sample sources employed.

9.4.3 Antioxidant Activity in *Puer* Tea

Puer tea can act as an inhibitor of lipid and nonlipid oxidative damage and exhibit metal-binding ability, reducing power, and scavenging effect for free radicals (Duh et al., 2004; Jie et al., 2006). Liang et al. (2005) reported that *Puer* tea suppressed the genotoxicity induced by nitroarenes, lowered the atherogenic index, and increased high-density lipoprotein. However, the antioxidant activity of *Puer* tea was lower than that of *Oolong* tea or black tea (Table 9.4). Although *Puer* tea has a history of production and consumption for hundreds of years, few studies can be found on its antioxidant ability related to microbiological properties, especially during the fermentation process. Thus, a standard evaluation for antioxidant capacities of black tea, *Oolong* tea, and *Puer* tea under the same condition is needed to make a final decision.

9.5 TEA AND HEALTH

In addition to the enjoyable, safe, and economical aspects of tea, this drink also provides a natural source of compounds preventing a wide array of diseases (Arts et al., 2000; Dufresne and Farnworth, 2001; Khan and Mukhtar, 2007; Toyoda et al., 1997). For centuries, people have been consuming tea in order to recover from flu and other related illnesses. Tea provides biologically active ingredients (e.g., flavonoids, vitamins, and fluoride) (Wheeler and Wheeler, 2004). Many biological activities of tea can be attributed to the antioxidant properties of the polyphenolic fraction through its metabolism (Mukhtar and Ahmad, 2000). Protection against cardiovascular diseases, atherosclerosis, cancer, gene mutations, bacterial growth, and diabetes are growing concerns. Among the tea catechins, EGCG is most effective in reacting with most reactive oxygen species. As reactions of EGCG and other catechins with peroxyl radicals lead to the formation of anthocyanin-like compounds as well as seven B ring anhydride dimers and ring-fission compounds, the B ring appears to be the principal site of antioxidant reactions (Kondo et al., 1999). The protective activities of tea polyphenols against the oxidation of lipoproteins have been suggested to contribute to the prevention of atherosclerosis and other cardiovascular diseases (Cheng and Chen, 1994; Ho et al., 2008; Tijburg et al., 1997). The beneficial effects of tea on human health can be grouped into major and minor categories. In the subsequent sections, the major effects, namely those affecting cardiovascular disease, cancer, antibacterial and antiviral activity, and diabetes are extensively discussed. The minor effects, namely those affecting gastrointestinal functions, the inflammation component, kidney stones, diarrhea, and the immune function, are briefly discussed.

9.5.1 Prevention of Cardiovascular Disease

Overall, cardiovascular health appears to be better among tea drinkers than nontea drinkers (Kuroda and Hara, 1999; Yang et al., 2001). Thus, tea intake exhibits a protective effect in cardiovascular disease (Kuroda and Hara, 1999; Yang, 1997; Yang and Koo, 1997). In a large cohort study conducted in seven countries over 25 years, the average flavonol and flavone intake appears to be inversely related with mortality rates from coronary heart disease (Hertog et al., 1993; Hollman et al., 1999). Tea flavonoids, mainly gallocatechins, protect low-density and very low-density lipoproteins (LDL and VLDL) against oxidation by aqueous and lipophilic radicals, copper ions, and macrophages (Vinson and Dabbagh, 1998; Vinson et al., 1995; Wiseman et al., 1997; Yokozawa et al., 1998a) against the proliferation of vascular smooth muscle cells, which leads to sclerosis of the artery (Yokozawa et al., 1995). In experiments conducted with rats, a reduction of triacylglycerol, total cholesterol, LDL-cholesterol (Yang and Koo, 1997), and the enhancement of superoxide dismutase (SOD) in serum and of gluthatione-*S*-transferase (GST) and catalase in the liver were observed (Lin et al., 1998a). Increases of SOD, GST, and catalase improve removal of the superoxide anion radical, the peroxides, and other free radicals responsible for LDL oxidation (Yokozawa et al., 1998b). Tea catechins effectively reduce cholesterol absorption from the intestine, lowering the solubility of the cholesterol and enhancing the excretion of cholesterol and total lipids. In addition, tea extract induces an anti-inflammatory and capillary strengthening effect (Tijburg et al., 1997). Tea components, mainly quercetin (Tijburg et al., 1997) and theanine (Yokogoshi et al., 1995), reduce blood pressure in animals and man, thus decreasing the risk for the development of cardiovascular diseases. Further human studies are needed to validate these effects with more data.

9.5.2 Tea and Cancer

In addition to the beneficial effects on the cardiovascular system, tea also appears to provide protective effects against several types of cancers. A large number of experimental studies suggest a beneficial effect of tea polyphenols on some cancers (Inoue et al., 1998; Yang et al., 2001). In one of the most carefully conducted studies, Yu et al. (1995) assessed the types of tea, age when tea drinking started, frequency of new batches of tea leaves per day, number of cups from each batch, duration per batch, and strength and temperature of the tea. Black tea was found to be inversely associated with stomach cancer among women in a population-based case-control study in Poland (Chow et al., 1999). However, absence of association between tea and cancer was found in some studies conducted among men and women (Chow et al., 1999; Goldbohm et al., 1996; Zheng et al., 1996). Thus, in contrast to the experimental evidence, which is relatively strong, the epidemiological evidence for a protective effect of tea against cancer is weak and inconsistent, with the exception of stomach cancer prevention. Additional studies are needed, preferentially using a prospective design and collecting detailed information on tea consumption patterns.

9.5.3 Antibacterial and Antiviral Activity

Antimicrobial activity against cariogenic and periodontal bacteria has been reported (Almajano et al., 2008). Tea extracts inhibit enteric pathogens such as *Staphylococcus aureus, S. epidermis, Plesiomonas shigelloides* (Toda et al., 1989), *Salmonella typhi, S. tiphimurium, S. enteritidis, Shigella flexneri, S. dysenteriae, Vibrio cholerae, V. parahaemolyticus* (Toda et al., 1989, 1991; Mitscher et al., 1997), *Campylobacter jejuni* and *C. coli* (Diker et al., 1991), but are not effective against *Escherichia coli, Pseudomonas aeruginosa* or *Aeromonas hydrophila* (Toda et al., 1989). Tea polyphenols also inhibit bacteria responsible for tooth decay (Kakuda et al., 1994). Tea polyphenols has a synergistic effect on the antibacterial activity and the anticariogenic properties (Kakuda et al., 1994). Furthermore, black and green tea extracts can kill *Helicobacter pylori* associated with gastric, peptic, and duodenal ulcer diseases (Diker and Hascelik, 1994). However, the tea concentration used in these studies exceeded normal human consumption levels.

Some results indicate that tea catechins are potentially antiviral and antiprotozoiac agents (Gutman and Ryu, 1996; Isaacs et al., 2008). EGCG inhibits influenza A and B viruses in animal cell culture (Mitscher et al., 1997). An antiviral activity has been found against HIV virus enzymes and against rotaviruses and anteroviruses in monkey cell culture when previously treated with EGCG (Mitscher et al., 1997). Furthermore, EGCG and ECG were found to be potent inhibitors of influenza virus replication in cell culture (Łuczaj and Skrzydlewska, 2005). Quantitative analysis revealed that, at high concentration, EGCG and ECG also suppressed viral RNA synthesis in cells, whereas EGC failed to show similar effect. Similarly, EGCG and ECG inhibited the neuraminidase activity more effectively than the EGC.

9.5.4 Diabetes

Tea consumption is associated with an increase in urine volume and electrolyte elimination, notably sodium, along with a blood pressure decrease in hypertensive adenine-induced rats (Yokozawa et al., 1994). Black tea extracts can decrease significantly the blood glucose level of aged rats by reducing the glucose absorption and uptake in different ways (Zeyuan et al., 1998). Tea polyphenols inhibit α-amylase activity in saliva, reduce the intestinal amylase activity, which in turn lowers the hydrolysis of starch to glucose and glucose assimilation (Hara et al., 1995) and decrease the glucose mucosal uptake because polysaccharides inhibit the glucose absorption and the diphenylamine of tea promotes its metabolism (Zeyuan et al., 1998). Tea polyphenols can reduce digestive enzyme activity and glucose absorption (Zeyuan et al., 1998), and decrease uremic toxin levels and the methylguanidine of hemodialysis patients (Sakanaka and Kim, 1997). They also protect against oxidative stress associated with late complications in diabetes pathology and are useful to maintain a balance between pro- and antioxidants in the organism (Zeyuan et al., 1998).

9.5.5 Protective Effects of Tea in Other Diseases

Tea can act on many cellular functions and can be helpful in other pathologies. It is reported that tea can improve gastrointestinal functions (Yamamoto et al., 1997), ethanol metabolism, kidneys, liver, pancreas, stomach injuries, protect skin and eyes, alleviate arthritis, allergies, dental caries, and can improve other diseases that have an inflammation component. It has a beneficial protective activity on several life-sustaining systems in the human body and drinking tea has positive effects to maintain a healthy condition and to delay action of the aging process (Zhu et al., 2006). However, the impact of tea consumption on longevity is not well covered by scientific literature.

9.6 MECHANISMS OF ACTION

Several recent review articles proposed mechanisms by which tea drinking confers protection against cardiovascular disease and cancer (Antony and Shankaranaryana, 1997; Dreostic et al., 1997; Dufresne and Farnworth, 2001; Khan and Mukhtar, 2007; Shankar et al., 2007; Yang et al., 2001, 2007). Black tea or *Oolong* tea is the richest source of flavonoids in the Northern European diet while tea contributes approximately 63% of dietary flavonoids in the diet. Flavonoids first enter the digestive tract, then the cardiovascular system, and finally diffuse into several tissues. The effect of flavonoids as antioxidants on the cardiovascular system appears to be linked to several modes of action. Antioxidants block free radicals, thus preventing such damage and avoiding the repair mechanism that causes smooth muscle cells to proliferate (Matés and Sánchez-Jiménez, 2000). Antioxidants have been observed to prevent LDL damage derived from hydroperoxides or other free radicals. Antioxidants have been shown to inhibit the cytotoxic activity of oxidized LDL. However, the mechanism of action of tea flavonoids on the digestive system is poorly understood and may be related to the absorption of tea flavonoids into the mucosal lining of the gastrointestinal tract. One such mode of action could account for the proven activity of tea flavonoids in blocking heterocyclic aromatic amines from promoting gastric and colorectal carcinogenesis (Weisburger, 1999). Other intracellular mechanisms are also possible. For instance, flavonoids were shown to protect mitochondria, control the expression of oncogenes, and prevent the loss of 5-methylcytosine (i.e., DNA demethylation or hypomethylation) (Nair and Salvi, 2008; Weisburger, 1999). Furthermore, EGCG suppresses gene expression of matrix metalloproteinases in HT-1080 cells and hepatic gluconeogenic enzymes in mice, which provides the basis of the protective effects of tea against cancer metastasis, diabetes, and hepatitis (Isemura et al., 2007). However, in contrast to antioxidant activity, black tea polyphenols exhibit pro-oxidative properties in certain conditions and a possible mechanism of oxidative isolated DNA damage by catechins has also been proposed by Oikawa et al. (2003). As antioxidant/pro-oxidant activity of tea polyphenols is dependent on many factors such as metal-reducing potential, chelating behavior, pH, solubility characteristics, the bioavailability, and stability in tissues and cells (Decker, 1997), the beneficial effects caused by tea could be due to the maintenance of a balance between pro- and antioxidants in the organism (Hou et al., 2005). Therefore, mechanisms of action by tea polyphenols concerning their bioactivity, antioxidant properties and beneficial effects require further studies, especially under various evaluation conditions.

9.7 CONCLUSION

Tea (*Camellia sinensis L.*) is an important source of dietary polyphenols. They are a large family of compounds with differing chemical structures and also possessing varying properties. Tea polyphenols known as catechins can act as antioxidants by the donation of hydrogen atoms, as an acceptor of free radicals, interrupter of chain oxidation reactions, by chelating metals or annexation of hydroxide groups to catechin molecules. Tea polyphenols exhibit health beneficial effects. A growing body of evidence suggests that moderate consumption of tea may protect against several forms of cancer, cardiovascular diseases, bacterial infections, and the formation of kidney stones and dental cavities. Although research provides many promising possibilities, detailed studies are still needed to understand the possible benefits of tea polyphenols to human health and food products. Future research needs to define the actual magnitude of health benefits, establish the safe range of tea consumption associated with these benefits, and elucidate potential mechanisms of action. Additionally, exploration at the cellular level will allow a better understanding of the underlying mechanisms regulating functions in normal and pathologic states by tea polyphenols and development of more specific and sensitive methods with more representative models will give a better understanding of how tea interacts with endogenous systems and other exogenous factors.

REFERENCES

An, B. J., Kwak, J. H., Son, J. H., Park, J. M., Lee, J. Y., Jo, C., and Byun, M. W. 2004. Biological and anti-microbial activity of irradiated green tea polyphenols. *Food Chemistry* 88:549–555.

Almajano, M. P., Carbo, R., Jimenez, J. A. L., and Gordon, M. H. 2008. Antioxidant and antimicrobial activities of tea infusions. *Food Chemistry* 108:55–63.

Antony, J. I. X. and Shankaranaryana, M. L. 1997. Polyphenols of green tea. *International Food Ingredients* 5:7–50.

Arts, I. C., van de Putte, B., and Hollman, P. C. 2000. Catechin contents of foods commonly consumed in the Netherlands. 2. Tea, wine, fruit juices, and chocolate milk. *Journal of Agriculture and Food Chemistry* 48:1752–1757.

Balentine, D. 1997. Tea and health. *Critical Reviews in Food Science and Nutrition* 37:691–692.

Balentine, D. A., Wiseman, S. A., and Bouwens, L. C. M. 1997. The chemistry of tea flavonoids. *Critical Reviews in Food Science and Nutrition* 37:693–704.

Bronner, W. E. and Beecher, G. R. 1998. Method for determining the content of catechins in tea infusions by high-performance liquid chromatography. *Journal of Chromatography A* 805:137–142.

Cheng, Q. K. and Chen, Z. M. 1994. *Tea and Health*. Beijing, China: Chinese Agricultural Science Press.

Chow, W. H., Swanson, C. A., Lissowska, J., Groves, F. D., Sobin, L. H., Nasierowska-Guttmejer, A., Radziszewski, J., Regula, J., Hsing, A. W., Jagannatha, S., Zatonski, W., and Blot, W. J. 1999. Risk of stomach cancer in relation to consumption of cigarettes, alcohol, tea and coffee in Warsaw, Poland. *International Journal of Cancer* 81:871–876.

Chu, D. C. and Juneja, L. R. 1997. General chemical composition of green tea and its infusion. In *Chemistry and Applications of Green Tea*, ed. T. Yamamoto, L. R. Juneja, D. C. Chu, and K. Mujo, New York: CRC Press, pp. 13–22.

Collier, P. D. and Mallows, R. 1971. The estimation of flavanols in tea by gas chromatography of their trimethylsilyl derivatives. *Journal of Chromatography* 57:29–45.

Decker, E. A. 1997. Phenolics: Prooxidants or antioxidants? *Nutrition Reviews* 55:396–398.

Diker, K. S. and Hascelik, G. 1994. The bactericidal activity of tea against *Helicobacter pylori. Letters in Applied Microbiology* 19:299–300.

Diker, K. S., Akan, M., Hascelik, G., and Yurdakök, M. 1991. The bactericidal activity of tea against *Campylobacter jejuni* and *Campylobacter coli. Letters in Applied Microbiology* 12:34–35.

Ding, Z., Kuhr, S., and Engelhardt, U. H. 1992. Influence of catechins and theaflavins on the astringent taste of black tea brews. *Zeitschrift für Lebensmittel Untersuchung und-Forschung* 195:108–111.

Dreostic, I. E., Wargovich, M. J., and Yang, C. S. 1997. Inhibition of carcinogenesis by tea: The evidence from experimental studies. *Critical Reviews in Food Science and Nutrition* 37:761–770.

Dufresne, C. J. and Farnworth, E. R. 2001. A review of latest research findings on the health promotion properties of tea. *The Journal of Nutritional Biochemistry* 12:404–421.

Duh, P. D., Yen, G. C., Yen, W. J., Wang, B. S., and Chang, L. W. 2004. Effects of pu-erh tea on oxidative damage and nitric oxide scavenging. *Journal of Agriculture and Food Chemistry* 52:8169–8176.

Engelhardt, U. H., Finger, A., and Kuhr, S. 1993. Determination of flavone *C*-glycosides in tea. *Zeitschrift fur Lebensmittel-Untersuchung und-Forschung* 197:239–244.

Farhoosh, R., Golmovahhed, G. A., and Khodaparast, M. H. H. 2007. Antioxidant activity of various extracts of old tea leaves and black tea wastes (*Camellia sinensis* L.). *Food Chemistry* 100:231–236.

Fernandez, P. L., Pablos, F., Martin, M. J., and Gonzales, A. G. 2002. Study of catechin and xantine tea profiles as geographical tracers. *Journal of Agriculture and Food Chemistry* 50:1833–1839.

Gadow, A. V., Joubert, E., and Hansmann, C. F. 1997. Comparison of the antioxidant activity of rooibos tea (*Aspalathus linearis*) with green, oolong and black tea. *Food Chemistry* 60:73–77.

Goldbohm, R. A., Hertog, M. G. L., Brants, H. A. M., van Poppel, G., and vanden Brandt, P. A. 1996. Consumption of black tea and cancer risk: A prospective cohort study. *Journal of the National Cancer Institute* 88:93–100.

Gramza, A. and Korczak, J. 2005. Tea constituents (*Camellia sinensis* L.) as antioxidant in lipid systems. *Trends in Food Science and Technology* 16:351–358.

Gutman, R. L. and Ryu, B. H. 1996. Rediscovering tea. An exploration of the scientific literature. *Herbal Gram* 37:33–48.

Halder, B., Pramanick, S., Mukhopadhyoy, S., and Giri, A. K. 2005. Inhibition of benzo pyrene induced mutagenicity and genotoxicity multiple test systems. *Food and Chemical Toxicology* 43:591–597.

Harbowy, M. E. and Balentine, D. A. 1997. Tea chemistry. *Critical Reviews in Food Science and Nutrition* 16:415–480.

Han, C. 1997. Screening of anticarcinogenic ingredients in tea polyphenols. *Cancer Letters* 114:153–158.

Hara, Y., Luo, S. J., Wickremashinghe, R. L., and Yamanishi, T. 1995. IX. Uses and benefits of tea. *Food Review International* 11:527–542.

Hertog, M. G. L., Hollman, P. C. H., and van de Putte, B. 1993. Content of potentially anticarcinogenic flavonoids of tea infusions, wines, and fruit juices. *Journal of Agriculture and Food Chemistry* 41:1242–1248.

Ho C. T, Jen-kun Lin J. K., and Fereidoon Shahidi F. 2008. *Tea and Tea Products: Chemistry and Health-Promoting Properties*. London: Taylor & Francis.

Hollman, P. C. H., Feskens, E. J., and Katan, M. B. 1999. Tea flavonols in cardiovascular disease and cancer epidemiology. *Proceedings of the Society for Experimental Biology and Medicine* 220:198–202.

Hou, Z., Sang, S. M., You, H., Lee, M. J., Hong, J., Chin, K. V., and Yang, C. S. 2005. Mechanism of action of (−)-epigallocatechin-3-gallate: Auto-oxidation-dependent inactivation of epidermal growth factor receptor and direct effects on growth inhibition in human esophageal cancer KYSE 150 cells. *Cancer Research* 65:8049–8056.

Inoue, M., Tajima, K., Hirose, K., Hamajima, N., Takezaki, T., Kuroishi, T., and Tominaga, S. 1998. Tea and coffee consumption and the risk of digestive tract cancers: Data from a comparative case-referent study in Japan. *Cancer Causes & Control* 9:209–216.

Isaacs, C. E., Wen, G. Y., Xu, W., Jia, J. H., Rohan, L., Corbo, C. D., Maggio, V., Jenkins, E. C., and Hillier, S. 2008. Epigallocatechin gallate inactivates clinical isolates of herpes simplex virus. *Antimicrobial Agents and Chemotherapy* 52:962–970.

Isemura, O., Abe, I. K., Kinae, N., Yamamoto-Maeda, M., and Koyama, Y. 2007. Modulation of gene expression by green tea in relation to its beneficial health effects. *Yakugaku Zasshi Journal of the Pharmaceutical Society of Japan* 127:1–3.

Jie, G. L., Lin, Z., Zhang, L. Z., Lv, H. P., He, P. M., and Zhao, B. L. 2006. Free radical scavenging effect of Pu-erh tea extracts and their protective effect on oxidative damage in human fibroblast cells. *Journal of Agriculture and Food Chemistry* 54:8058–8064.

Kakuda, T., Takihara, T., Sakane, I., and Mortelmans, K. 1994. Antimicrobial activity of tea extracts against peridontopathic bacteria. *Journal of the Agricultural Chemical Society of Japan* 68:241–243.

Khan, N. and Mukhtar, H. 2007. Minireview: Tea polyphenols for health promotion. *Life Science* 81:519–533.

Kondo, K., Kurihara, M., Miyata, N., Suzuki, T., and Toyoda, M. 1999. Scavenging mechanisms of (−)-epigallocatechin gallate and (−)-epicatechin gallate on peroxyl radicals and formation of superoxide during the inhibitory action. *Free Radical Biology and Medicine* 27:855–863.

Kuhr, S. and Engelhardt, U. H. 1991. Determination of flavanols, theogallin, gallic acid, and caffeine in tea using HPLC. *Zeitschrift fur Lebensmittel-Untersuchung und-Forschung* 192:526–529.

Kuo, K. L., Weng, M. S., Chiang, C. T., Tsai, Y. J., Lin-Shiau, S.Y., and Lin, J. K. 2005. Comparative studies on the hypolipidemic and growth suppressive effects of oolong, black, pu-erh, and green tea leaves in rats. *Journal of Agriculture and Food Chemistry* 53:480–489.

Kuroda, Y. and Hara, Y. 1999. Antimutagenic and anticarcinogenic activity of tea polyphenols. *Mutation Reserach* 43:69–97.

Kyle, J. A. M., Morrice, P. C., McNeill, G., and Duthie, G. G. 2007. Effects of infusion time and addition of milk on content and absorption of polyphenols from black tea. *Journal of Agriculture and Food Chemistry* 55:4889–4894.

Lambert, J. D., Sang, S. M., and Yang, C. S. 2007. Biotransformation of green tea polyphenols and the biological activities of those metabolites. *Molecular Pharmacology* 4: 819–825.

Lee, B. L. and Ong, C. N. 2000. Comparative analysis of tea catechins and theaflavins by high-performance liquid chromatography and capillary electrophoresis. *Journal of Chromatography A* 881:439–447.

Leung, L. K., Su, Y., Chen, R., Zhang, Z., Huang, Y., and Chen, Z. 2001. Theaflavins in black tea and catechins in green tea are equally effective antioxidants. *Journal of Nutrition* 131:2248–2251.

Liang, Y. R., Zhang, L. Y., and Lu, J. L. 2005. A study on chemical estimation of pu-erh tea quality. *Journal of Agriculture and Food Chemistry* 85:381–390.

Lin, J. K., Lin, C. L., Liang, Y. C., Lin, S. Y., and Juan, I. M. 1998a. Survey of catechins, gallic acid and methylxanthines in green, oolong, pu-erh and black teas. *Journal of Agriculture and Food Chemistry* 46:3635–3642.

Lin, Y. L., Cheng, C. Y., Lin, Y. P., Lau, Y. W., Juan, I. M., and Lin, J. K. 1998b. Hypolipidemic effect of green tea leaves through induction of antioxidant and phase II enzymes including superoxide dismutase, catalase, and glutathione-*S*-transferase in rats. *Journal of Agriculture and Food Chemistry* 46:1893–1899.

Łuczaj, W. and Skrzydlewska, E. 2005. Review: Antioxidative properties of black tea. *Preventive Medicine* 40:910–918.

Manach, C., Williamson, G., Morand, C., Scalbert, A., and Remesy, C. 2005. Bioavailability and bioefficacy of polyphenols in humans. I. Review of 97 bioavailability studies. *American Journal of Clinical Nutrition* 81:230S–242S.

Matés, J. M. and Sánchez-Jiménez, F. M. 2000. Role of reactive oxygen species in apoptosis: Implications for cancer therapy. *The International Journal of Biochemistry & Cell Biology* 32:157–170.

Mello, L. D., Alves, A. A., Macedo, D. V., and Kubota, L. T. 2004. Peroxidase-based biosensor as a tool for a fast evaluation of antioxidant capacity of tea. *Food Chemistry* 92:515–519.

Mennen, L., Hirvonen, T., Arnault, N., Bertrais, S., Galan, P., and Hercberg, S. 2007. Consumption of black, green and herbal tea and iron status in French adults. *European of Clinical Nutrition* 61:1174–1179.

Miller, N. J., Castelluccio, C., Tijburg, L., and Rice-Evans, C. 1996. The antioxidant properties of theaflavins and their gallate esters-radical scavengers or metal chelators? *FEBS Letters* 392:40–44.

Mitscher, L. A., Jung, M., Shankel, D., Dou, J. H., Steele, L., and Pillai, S. 1997. Chemoprotection: A review of the potential therapeutic antioxidant properties of green tea (*Camellia sinensis*) and certain of its constituents. *Medicinal Research Reviews* 17:327–365.

Mukhtar, H. and Ahmad, N. 2000. Tea polyphenols: Prevention of cancer and optimizing health. *American Journal of Clinical Nutrition* 71 (6 Suppl): 1698S–1702S.

Nair, C. K. K. and Salvi, V. P. 2008. Protection of DNA from γ-radiation induced strand breaks by epicatechin. *Mutation Research/Genetic Toxicology and Environmental Mutagenesis* 650:48–54.

Navas, P. B., Carrasquero-Durán, A., and Iraima, F. 2005. Effect of black tea, garlic and onion on corn oil stability and fatty acid composition under accelerated oxidation. *International Journal of Food Science and Technology* 40:1–5.

Nelson, M. and Poulter, J. 2004. Impact of tea drinking on iron status in the UK: A review. *Journal of Human Nutrition and Dietetics* 17:43–54.

Nishitani, E. and Sagesaka, Y. M. 2004. Simultaneous determination of catechins caffeine and other phenolic compounds in tea using new HPLC method. *Journal of Food Composition and Analysis* 17:675–685.

Oikawa, S., Furukawa, A., Asada, H., Hirakawa, K., and Kawanishi, S. 2003. Catechins induced oxidative damage to cellular and isolated DNA through the generation of reactive oxygen species. *Free Radical Research* 37:881–890.

Owuor, P. O. and Orbanda, M. 1995. Comparison of the spectrophotometric assay methods for theaflavins and the arubigins in black tea. *Tea* 16:41–47.

Peterson, J., Dwyer, J., Bhagwat, S., Haytowitz, D., Holden, J., Eldridge, A. L., Beecher, G., and Aladesanmi, J. 2005. Major flavonoids in dry tea. *Journal of Food Composition and Analysis* 18:487–501.

Price, K. R., Rhodes, M. J. C., and Barnes, K. A. 1998. Flavonol glycoside content and composition of tea infusions made from commercially available teas and tea products. *Journal of Agriculture and Food Chemistry* 46:2517–2522.

Rawel, H. M. and Kulling, S. E. 2007. Nutritional contribution of coffee, cacao and tea phenolics to human health. *Journal fur Verbraucherschutz und Lebensmittelsicherheit* 2:399–406.

Sakanaka, S. and Kim, M. 1997. Suppressive effect of uremic toxin formation by tea polyphenols. In *Chemistry and Applications of Green Tea*, eds. T. Yamamoto, L. R. Juneja, D. C. Chu, and M. Kim. Salem: CRC Press, pp. 75–86.

Scholz, S. and Williamson, G. 2007. Interactions affecting the bioavailability of dietary polyphenols *in vivo*. *International Journal for Vitamin and Nutrition Research* 77:224–235.

Shankar, S., Ganapathy, S., and Srivastava, R. K. 2007. Green tea polyphenols: Biology and therapeutic implications in cancer. *Frontiers in Bioscience* 12:4881–4899.

Steinhaus, B. and Englehardt, U. H. 1989. Theaflavins in black tea. *Zeitschrift fur Lebensmittel-Untersuchung und-Forschung* 188:509–511.

Stewart, A. J., Mullen, W., and Crozier, A. 2005. On-line high-performance liquid chromatography analysis of the antioxidant activity of phenolic compounds in green and black tea. *Molecular Nutrition & Food Research* 49:52–60.

Su, X. G., Duan, J., Jiang, Y. M., Shi, J., and Kakud, Y. 2006. Effects of soaking conditions on the antioxidant potentials of oolong tea. *Journal of Food Composition and Analysis* 19:348–353.

Takeo, T. 1974. Photometric evaluation and statistical analysis of tea infusion. *Japan Agricultural Research Quarterly* 8:159–164.

Tijburg, L. B. M., Mattern, T., Folts, J. D., Weisgerber, U. M., and Katan, M. B. 1997. Tea flavonoids and cardiovascular diseases: A review. *Critical Reviews in Food Science and Nutrition* 37:771–785.

Toda, M., Okubo, S., Hiyoshi, R., and Tadakatsu, S. 1989. The bactericidal activity of tea and coffee. *Letters in Applied Microbiology* 8:123–125.

Toda, M., Okubo, S., Ikigai, H., Suzuki, T., Suzuki, Y., and Shimamura, T. 1991. The protective activity of tea against infection by *Vibrio cholerae* 01. *The Journal of Bacteriology* 70:109–112.

Toyoda, M., Tanaka, K., Hoshino, K., Akiyama, H., Tanimura, A., and Saito, Y. 1997. Profiles of potentially antiallergic flavonoids in 27 kinds of health tea and green tea infusions. *Journal of Agriculture and Food Chemistry* 45:2561–2564.

Vinson, J. A. and Dabbagh, Y. A. 1998. Effect of green and black tea supplementation on lipids, lipid oxidation and fibrinogen in hamster: Mechanisms for the epidemiological benefits of tea drinking. *FEBS Letters* 433:44–46.

Vinson, J. A., Jang, J., Dabbagh, Y. A., Serry, M. M., and Cai, S. 1995. Plant polyphenols exhibit lipoprotein-bound antioxidant activity using an *in vitro* oxidation model for heart disease. *Journal of Agriculture and Food Chemistry* 43:2798–2799.

Wanasundara, U. N. and Shahidi, F. 1996. Stabilization of seal blubber and menhaden oils with green tea. *Journal of the American Oil Chemists' Society* 73:1183–1190.

Wang, H., Provan, G. J., and Helliwell, K. 2000a. Tea flavonoids: Their functions, utilisation and analysis. *Trends in Food Science and Technology* 11:152–160.

Wang, H., Provan, G. J., and You, X. 2000b. Isocratic elution system for the determination of catechins, caffeine and gallic acid in green tea using HPLC. *Food Chemistry* 68:115–121.

Wheeler, D. S. and Wheeler, W. J. 2004. The medicinal chemistry of tea. *Drug Development Research* 61:45–65.

Wiseman, S. A., Balentine, D. A., and Frei, B. 1997. Antioxidants in tea. *Critical Reviews in Food Science and Nutrition* 37:705–718.

Weisburger, J. H. 1999. Second international scientific symposium on tea and health: An introduction. *Proceedings of the Society for Experimental Biology and Medicine* 220:193–194.

Yamamoto, T., Juneja, L. R., Chu, D. C., and Kim, M. 1997. *Chemistry and Applications of Green Tea*. Boca Raton, FL: CRC Press.

Yang, C. S. 1997. Inhibition of carcinogenesis by tea. *Nature* 389:134–135.

Yang, T. T. C. and Koo, M. W. L. 1997. Hypocholesterolemic effects of Chinese tea. *Pharmaceutical Research* 35:505–512.

Yang, C. S., Lambert, J. D., Ju, J., Lu, G., and Sang, S. 2007. Tea and cancer prevention: Molecular mechanisms and human relevance. *Toxicology and Applied Pharmacology* 224:265–273.

Yang, C. S., Landau, J. M., Huang, T., and Newmark, H. L. 2001. Inhibition of carcinogenesis by dietary polyphenolic compounds. *Annual Review of Nutrition* 21:381–406.

Yao, L., Jiang, Y., Caffin, N., D'Arcy, B., Datta, N., Xu, L., Singanusong, R., and Xu, Y. 2006. Phenolic compounds in tea from Australian supermarkets. *Food Chemistry* 96:614–620.

Yokozawa, T., Oura, H., Sakanaka, S., and Kim, M. 1995. Effects of a component of green tea on the proliferation of vascular smooth muscle cells. *Bioscience, Biotechnology and Biochemistry* 59:2134–2136.

Yokozawa, T., Oura, H,, Sakanaka, S., Ishigaki, S., and Kim, M. 1994. Depressor effect of tannin in green tea on rats with renal hypertension. *Bioscience, Biotechnology and Biochemistry* 58:855–858.

Yokozawa, T., Dong, E., Nakagawa, T., Kim, D. W., Hattori, M., and Nakagawa, H. 1998a. Effects of Japanese black tea on artherosclerotic disorders. *Bioscience, Biotechnology and Biochemistry* 62:44–48.

Yokozawa, T., Dong, E., Nakagawa, T., Kashiwagi, H., Nakagawa, H., Takeuchi, S., and Chung, H. Y. 1998b. *In vitro* and *in vivo* studies on the radical-scavenging activity of tea. *Journal of Agriculture and Food Chemistry* 46:2143–2150.

Yu, G. P., Hsieh, C. C., Wang, L. Y., Yu, S. Z., Li, X. L., and Jin, T. H. 1995. Green-tea consumption and risk of stomach cancer: A population-based case–control study in Shanghai, China. *Cancer Causes Control* 6:532–538.

Zeyuan, D., Bingying, T., Xiaolin, L., Jinming, H., and Yifeng, C. 1998. Effect of green tea and black tea on the blood glucose, the blood triglycerides, and antioxidation in aged rats. *Journal of Agriculture and Food Chemistry* 46:875–878.

Zheng, W., Doyle, T. J., Kushi, L. H., Sellers, T. A., Hong, C. P., and Folsom, A. R. 1996. Tea consumption and cancer incidence in a prospective study of postmenopausal women. *American Journal of Epidemiology* 144:175–182.

Zhu, Y., Huang, H., and Tu, Y. 2006. A review of recent studies in China on the possible beneficial health effects of tea. *International Journal of Food Science and Technology* 41:333–340.

Zhu, Q. Y., Hackman, R. M., Ensunsa, J. L., Holt, R. R., and Keen, C. L. 2002. Antioxidative activities of oolong tea. *Journal of Agriculture and Food Chemistry* 50:6929–6934.

Zuo, Y., Chen, H., and Deng, Y. 2002. Simultaneous determination of catechins, caffeine and gallic acids in green, oolong, black and Pu-erh teas using HPLC with a photodiode array detector. *Talanta* 57:307–316.

10 Sesame for Functional Foods

Mitsuo Namiki

CONTENTS

10.1 INTRODUCTION: OPEN SESAME

Sesame seed and its oil have been utilized as important foodstuffs for about 6000 years. A cultivated sesame species, *Sesamum indicum* L., is believed to have originated in the Savanna of central Africa spreading to Egypt, India, the Middle East, China, and elsewhere. Sesame seed and oil have been evaluated as representative health foods and widely used for their good flavor and taste (Namiki and Kobayashi, 1989).

Historically, an old text, the Thebes Medicinal Papyrus (1552 BC), found in Egypt, describes the medicinal effect of sesame seed as a source of energy. Hippocrates in Greece noted its high nutritive value. A Chinese book (300 BC), which explains medicinal effects of various plants, describes sesame as a good food having various physiological effects, especially useful for providing energy, a tranquil frame of mind, and retarding the process of aging when eaten over a long period. Further, in traditional Indian medicine, Ayurveda, sesame oil has been used as the basal oil for human body massage since 700–1100 BC (Weiss, 1983; Joshi, 1961). The magic words, "Open, sesame!" from the *Arabian Night* stories are very familiar throughout the world and may have originated from the abrupt opening of the sesame capsule to scatter seeds and also to show sesame's magic power. In Japan people have traditionally believed that sesame is very good for health (Namiki, 1990, 1995).

Despite such high values placed on sesame seed and oil, there have been few scientific studies to elucidate their functions. Initiated from the studies on the protective effects of food components against the lethal effects of radiation (Sumiki et al., 1958), many antioxidants and antimutagens in various foods have extensively studied by author's group (Kada, 1978; Namiki, 1990; Osawa, 1997). Among them, studies on the traditionally believed health functions of sesame have been especially noted, and various interesting functions such as the antioxidation and antiaging effects with tocopherols, serum lipid-lowering effect, blood pressure-lowering effect, and other functions have been elucidated (Namiki, 1998, 2007).

10.2 PRODUCTION, PROCESSING, AND UTILIZATION OF SESAME SEED AND OIL

10.2.1 Cultivation and Production

Sesame is cultivated mainly in China, India, Myanmar, Sudan, Central America, and other tropical and subtropical countries. The total amount of seed produced was about 3500 kt and that of oil about 2000 kt in 2006 (Oil World Annual, 2006). The main countries for production are India (700 kt), China (625 kt), and Myanmar (570 kt), those for exporting are India (200 kt), Sudan (113 kt), and Nigeria (65 kt), and those for importing are China (264 kt), Japan (164 kt), Turkey (95 kt), and Korea (86 kt). In recent years, China has changed from the largest exporting country to the largest importing country, due to its rapid increase in domestic needs. Regarding the worldwide per capita consumption of sesame seed, South Korea is highest, at 6–7 g/day and Japan follows with about 2–3 g/day. The amount of sesame seed production is small compared with other oil sources such as soybean, rapeseed, and oil palm, but the value of sesame seed and its oil is considerably higher.

Despite the increasing demand for this highly valued food, the production of sesame seed has not increased, probably due to problems in cultivation. Sesame is an erect annual herb having simple or branching stems. The growth period for sesame usually ranges from 3 to 4 months but flowering begins as early as 30–40 days after sowing. Blooming continues until maturity, and the seeds scatter suddenly from the capsule, as illustrated by the magic words, "Open, sesame!" As a result, harvesting of sesame cannot be done mechanically but requires extensive manual labor (Namiki and Kobayashi, 1989). Recently, a farm in Texas (United States) has succeeded in large-scale production. After extensive breeding, they have developed excellent new varieties that are resistant to scattering, as well as to diseases, insects, and drought. By using these strains and improved cultivation techniques, production costs have been sharply lowered (Smith, 2000).

10.2.2 Sesame Seed as Food

There are many reasons why people use sesame seed, the main ones being that it contains much oil (about 50%), which is very stable against oxidative deterioration, and a superior nutritional value containing about 20% protein plus various minor nutrients. The good flavor generated by roasting sesame seeds is also a highly desirable characteristic.

Sesame seed varies considerably in color, size, and texture of the seed coat. The most commonly used are of a white-brown shade, but in Asian countries it is believed that the black seeds are best for health concerns. Gold and violet seeds are also highly valued. After cleaning by sieving, washing with water, and drying, the seeds are usually roasted at 120°C to 150°C for about 5 min. Roasting brings out their characteristic rich flavor. In processing, they are sometimes hulled mechanically, and the hulled seeds are said to be more easily digested, but they lose some useful nutrients such as calcium in the process (Namiki, 1998).

Sesame seed is used in various ways around the world. In East Asia, people usually like the roasted flavor, so in China, Korea, Japan, and other Asian countries, roasted sesame seed is generally used as a topping for many baked foods such as breads, biscuits, and crackers. In China, roasted seed may be ground into a paste-like product. In Japan, roasted seeds are mixed with common salt and used as a topping on cooked rice. Sesame mash and paste made by grinding seed in a conical ceramic mortar are widely utilized as seasoning for salads, cooked rice, boiled meat, and other foods. Recently, a large amount of various kinds of dressings containing mashed sesame seeds have flooded the Japanese food market. Sesame-tofu (*goma-dofu* in Japanese) prepared from mashed sesame seed and a starch such as arrowroot starch, is a popular and unique food in Japan (Sato et al., 1995).

The production of sesame seed in India is the largest in the world. Sesame oil is consumed as a hardened oil in margarine, and also used for a massage oil in Ayurveda, India's traditional medicine. Unlike in East Asia, the oil is not roasted. Other uses in India are for cakes, seasoning, and toppings. In the Near East and North Africa, sesame paste is generally used. For example, the paste is the base for much of the cooking (called *tahina*) and for cake (*halva*). In these uses, a weak roast flavor is preferred. In Africa, the leaves of sesame are eaten. In North America, sesame seed is generally hulled and used for toppings on bread and hamburger buns (Namiki, 1998).

10.2.3 Sesame Seed Oil

Two kinds of sesame seed oil are generally used: raw oil and roasted oil. The former can be prepared by expelling steamed seeds and refining through the usual refining process of common to vegetable oils (decoloration with acid clay and deodorization by steam distillation *in vacuo*). The product is a clear and mild flavored oil, mainly used for frying and as a salad dressing.

Roasted oil is produced by expelling the seeds previously roasted at 180–200°C for 10–20 min and refining simply by the filtration of precipitates. It has a characteristic roasted flavor and a deep yellow-brown color. It is widely used in China, Korea, and Japan as an important seasoning and cooking oil, for example, for tempura in Japan (Namiki et al., 2002).

Roasting is the most important form of processing of sesame foods and oils; it brings out a characteristic pleasant flavor. The roast flavor depends greatly on the roasting temperature and time exposed, ranging from a somewhat grassy, mild, sweet, and desirable aroma to an irritating, scorched smell. This change occurs in the temperature range of 140–230°C. Optimum production conditions are 150–175°C for 15–20 min for roasted seeds and 180–200°C for 10–20 min for the roasted seed oil.

10.3 COMPOSITION OF SESAME SEED

The major constituents of sesame seed are oil, protein, and carbohydrates, and the minor ones are various vitamins and minerals. The contents of the common brown seed are shown in Table 10.1 (Namiki, 1998).

10.3.1 Oil

The average content of oil was 55% in white-seed strains and that of black-seed strains was 47.8% (Tashiro et al., 1990), although the content can vary considerably, depending on the species and cultivation conditions. Fatty acids in the oil are mainly oleic (18:1 = 39.1%) and linoleic (18:2 = 40.0%) acids, with palmitic (16:0 = 9.4%) and stearic (18:0 = 4.76%) acids in smaller amounts, and linolenic (18:3 = 0.46%) acid in a trace amount. Linoleic, linolenic, and arachidonic acids are considered to be essential fatty acids for humans. According to recent studies on prostaglandins, *n*–3 (e.g., linolenic) and *n*–6 (e.g., linoleic) fatty acids have independent roles in the synthesis of prostaglandins, and the ratio of *n*–3 to *n*–6 fatty acids in the composition of fatty acids is important. From this standpoint, sesame oil in which *n*–3 fatty acid content is low will be inferior to soybean and corn oils (Numa, 1984). Variation in fatty acid composition of different sesame species including a cultivated type and three wild types has been studied (Kamal-Eldin and Appelqvist, 1994a).

TABLE 10.1
Composition of Sesame Seed (per 100 g)

Energy (calories)	578	Fe (mg)	9.6
Moisture (%)	4.7	Na (mg)	2
Fat (g)	51.9	K (mg)	400
Protein (g)	19.8	Vitamin A (IU)	0
Carbohydrate (g)	18.4	Carotene (μg)	17
Fiber (g)	10.8	Vitamin B_1 (mg)	0.95
Ash (g)	5.2	Vitamin B_2 (mg)	0.25
Ca (mg)	1200	Niacin (mg)	5.1
Mg (mg)	370	Vitamin C	0
P (mg)	540		

10.3.2 Protein and Carbohydrate

Sesame seed contains about 20% protein, with an average of 22.3%. In the defatted seed, it is about 50%. Protein can be extracted with 10% NaCl and separated into 13S (70%), 5S (10%), and 3S (20%) by centrifugation (Namiki and Kobayashi, 1989). The contents of amino acids vary somewhat among species, but no significant differences are found between the white and black species of sesame. Compared with the standard values recommended by FAO/WHO (1973), the amino acid composition of sesame seed protein is slightly lower in lysine (31 mg/g protein), but higher in other amino acids, especially methionine (36 mg), cystine (25 mg), arginine (140 mg), and leucine (75 mg). The protein in soybean is rich in lysine (68 mg) but low in methionine (16 mg), so that the simultaneous feeding of both proteins produces good growth in rats, as well as the feeding of casein (Namiki, 1995). A study on the nutritional quality of the protein fraction extracted with *iso*-propanol (DSS-Iso) showed that rats fed a DSS-Iso-based diet showed a significant decrease in total plasma cholesterol, triacylglycerols,and VLDL + LDL cholesterol in comparison with the rats fed a diet containing casein (Sen and Bhattacharyya, 2001). Unroasted seed contained considerable amounts of such tasty amino acids as glutamic acid, arginine, aspartic acid, and alanine, but the amounts were significantly reduced by roasting at 170°C for 20 min or 200°C for 5 min (Namiki, 1998). As shown later, it has been recently demonstrated that some sesame peptides have a marked inhibitory activity on the angiotensin 1-converting enzyme effective to suppress hypertension (Nakano et al., 2006).

The carbohydrate content in sesame seed is about 18–20 wt.%. The presence of small amounts of glucose and fructose, and also an oligo sugar planteose [*O*-α-D-galactopyranosyl-(1,6)-β-D-fructofuranosyl-α-D-glucopyranoside] has been reported (Wankhede and Tharanathan, 1976), but no starch is present. Most carbohydrates seem to be present as dietary fibers, and the content of the dietary fibers has been reported to be 10.8% (STA Japan, 2001). Further studies on the carbohydrates in relation to some functional activities of the water–alcohol extractable fraction of defatted sesame seed may be necessary.

10.3.3 Vitamins and Minerals

As shown in Table 10.1, sesame seed contains a significant amount of the vitamin B group. Since the vitamin B group is contained only in the coat or hull of the seed, it is necessary to use sesame flour or paste of whole sesame seed for utilization of the vitamin B group in sesame seed (Brito and Nunez, 1982). Among the vitamins in sesame seed, the presence of vitamin E is very interesting in relation to the effectiveness of sesame seed as a health food. As shown in Table 10.2, the content of tocopherols in sesame is less than that in soybean and corn oil (Speek et al., 1985). Moreover, the tocopherols in sesame seed are mostly present as the γ-homologue, and the content of α-tocopherol is very low. It is known that the vitamin E activity of γ-tocopherol is less than 10% that of α-tocopherol (Bieri and Evarts, 1974). Thus, sesame seed appears to contain very little vitamin E activity. Despite this, however, sesame shows clear antiaging effects in mice.

TABLE 10.2
Tocopherol Content of Various Vegetable Oils (mg/100 g)

Oil	α-Toco	β-Toco	γ-Toco	δ-Toco	Vitamin E Value
Sesame	0.4	Trace	43.7	0.7	4.3
Safflower	27.1	0.6	2.3	0.3	27.6
Soybean	10.4	2.1	80.9	20.8	19.9
Corn	17.1	0.3	70.3	3.4	24.3
Olive	7.4	0.2	1.2	0.1	7.5
Rapeseed	15.2	0.3	31.8	1.0	16.9
Sunflower	38.7	0.8	2.0	0.4	39.2

Vitamin E = $(\alpha \times 1) + (\beta \times 0.4) + (\gamma \times 0.1) + (\delta \times 0.01)$.

Sesame seed is rich in various mineral constituents, but few studies have been carried out on their nutritive values. Among them, calcium and iron, which are often deficient in modern diets, are found in high concentrations (1200 and 9.6 mg/100 g, respectively). Calcium is contained mainly in the hull as oxalate salt, so nutritionally available calcium may be reduced. A study showed that the bioavailability of calcium in sesame for rats is about 65% (Poneros-Schneier and Erdman, 1989), but another study estimated it to be about one-seventh of the total content, that is, about 168 mg/100 g due to a large amount of combined calcium (Ishii and Takiyama, 1994). It is also noted that most of the calcium contained in the hull is lost during hulling and only about 20% of the total calcium remains (Namiki, 1995). The presence of selenium is also unusual, that is, 36.1 mg/g in isolated sesame protein (Namiki, 1995). As is commonly known, selenium is a constituent of glutatione peroxidase, which is involved in the prevention of physiological peroxidation.

10.3.4 Roasted Sesame Seed Flavor

The characteristic flavor of roasted sesame seeds is a very important factor in the deliciousness of various sesame foods. As a result, many chemical studies have been done on the roasted sesame flavor and more than 400 components have been isolated and identified. They have been classified as pyrazines (number of identified compounds: 49), pyridines (17), pyrroles (16), furans (32), thiophens (16), thiazoles (23), carbonyl compounds, and others. Among them, alkylpyrazines are the main components and provide the representative deep-roasted flavor, whereas thiazoles and thiophenes are assumed to contribute to the characteristic roasted sesame flavor (Takei and Fukuda, 1995; Schieberle, 1995; Schieberle et al., 1996; Shimoda et al., 1996, 1997; Shahide et al., 1997; Namiki, 1998).

In addition to the pleasant aroma of the roasted sesame flavor, the antithrombosis effect of these flavor components has been investigated and is discussed further by Namiki et al. (2006).

10.3.5 Sesame Lignans

Lignans, a group of natural compounds and oxidative coupling product of β-hydroxyphenylpropane, are widely distributed as a minor component in the plant kingdom, especially in the bark of wood (Smeda, 2007). Some lignans are known to have antitumor, antimitotic, and antivirus activities. Interestingly, sesame seed contains significant amounts of characteristic lignans such as sesamin, sesamolin, sesaminol, and others, as shown in Figure 10.1.

As described hereinafter, sesame lignans are being noted as the most important and characteristic components of sesame seed in view of their various functional activities. Sesamin and sesamolin have been known as major lignans in sesame seed (Budowsky, 1964), and sesaminol was later identified as another major lignan (Osawa et al., 1985).

10.3.5.1 Sesamin and Sesamolin

Sesamin has a typical lignan structure of the β-β′(8-8′)-linked product of two coniferyl alcohol radicals, and it has been found in other plants such as beech in small amounts. Sesame seed contains sesamin in large amounts, usually about 0.4% in sesame oil (Tashiro et al., 1990). Sesamin is highly hydrophobic and obtained as a crystalline product from sesame oil, and commercially it can extracted from the scum obtained by the vacuum-deodorization process in the purification of unroasted sesame oil production. However, the product thus obtained is a 1:1 mixure of sesamin and its artifact product episesamin formed during the purification process of the oil.

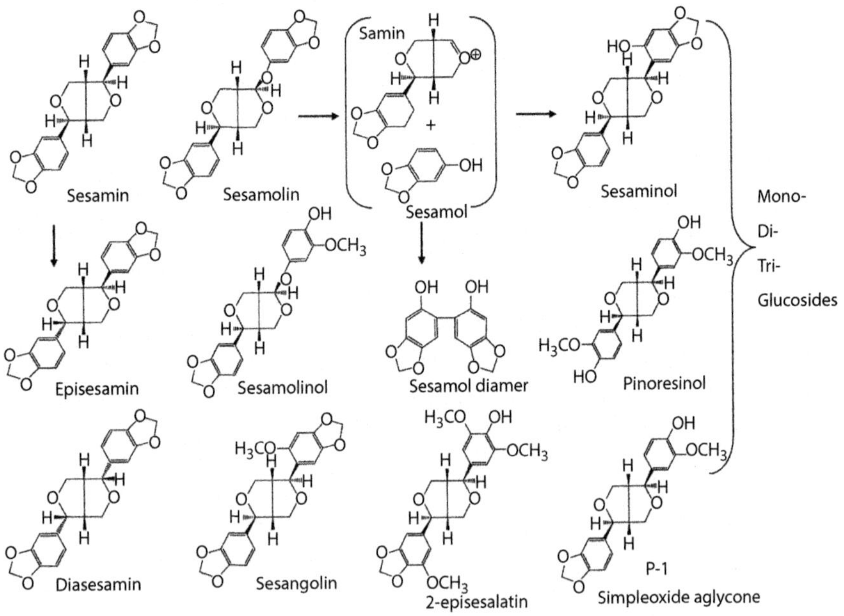

FIGURE 10.1 Sesame lignans and related compounds.

Except for some physiological effects, episesamin exhibited weaker activity than native sesamin (see further discussion later). So, to obtain a pure sesamin specimen, further purification using columun chromatography is required (Kushiro et al., 2002). Recently, details of the stereochemical structure and various thermodynamic properties of the sesamin molecule were elucidated using spectroscopic data and theoretical calculations (Hsieh et al., 2005).

Sesamolin has a unique structure involving one acetal oxygen bridge in a sesamin-type structure and seems to be a characteristic lignan of sesame seed. Its content is usually about 0.3% in the oil (Tashiro et al., 1990).

10.3.5.2 Sesaminol

Sesaminol was first found as one of the antioxidative components in sesame seed, along with active components including sesamolinol, pinoresinol, and others (Fukuda et al., 1985a; Osawa et al., 1985). It was isolated as the principal antioxidative factor of the refined unroasted sesame oil and verified as being produced as an artifact from sesamolin during the decolorization process of that oil (Fukuda et al., 1986b). This positive change in the antioxidative lignan molecules during decolorization was demonstrated to occur by a novel intermolecular rearrangement of sesamolin catalyzed by acid clay. The reaction was initiated by splitting the acetal oxygen bridge of sesamolin with an acidic catalyst in a nonaqueous system to produce the oxonium samin ion and sesamol, and this oxonium ion is immediately linked electrophilically to the *ortho*-carbon of sesamol to give a new product, sesaminol (see Figure 10.1). Here, the presence of water in even trace amounts leads to the hydrolysis of sesamolin to samin and sesamol (Fukuda et al., 1986c). Sesaminol thus obtained from the scum contained four isomers but it is present mainly in two of the isomers. The content of sesaminol in the refined raw sesame oil is about 120–140 mg/100 g, although that in the seed is very small, that is, about 1.0 mg/100 g as the free lignan form (Nagata et al., 1987). However, considerable amounts of di- or triglucosides of sesaminol were isolated and identified from water–alcohol extracts of the defatted meal of sesame seed (Katsuzaki et al., 1992). Determination of the sesaminol glucoside contents performed by solvent extraction, followed by HPLC analysis using MS and NMR, on 65 different sesame seeds indicates the content of triglucosides to range from 36 to 1560 mg/100 g, and that of diglucoside from 0 to 493 mg/100 g. No significant differences were observed between black and white seeds (Moazzami et al., 2006a).

It was reported on the determination of free sesaminol in sesame seeds that the deglucosidation of the sesame lignan glucosides is not easily done with the usual β-glucosidase alone, but it was necessary to use it in combination with cellulase (Kuriyama and Murui, 1993). Also, the yields of sesaminol glucosides from defatted sesame meal increased by pretreatment with some microbes (Ohtsuki et al., 2003), and also with some *Aspergillus* (Koizumi et al., 2007). Recently, a rapid and nondestructive method to determine the contents of lignans and liganan glucosides in sesame seeds by using near-infrared reflectance spectroscopy (NIRS) was developed, which is especially convenient in the quality determination of sesame seed, for example, the breeding programs of sesame (Kim et al., 2006).

10.3.5.3 Other Lignans

Two new lignans were isolated from the perisperm (coat) of *Sesamum indicum*, and identified as (+)-saminol and (+)-episesaminone-9-*O*-beta-D-sophoroside (Grougnet et al., 2006). The presence of sesamolinol diglucoside in sesame seed was newly isolated and identified by MS and NMR as [2-(3-methoxy)-4-(*O*-β–D-glucopyranosyl (1-→6)-*O*-β-D-glucopyranoside)phenoxy)-6-(3,4-methylenedioxyphenyl)-*cis*-3,7-dioxyabicyclo-(3.3.0)octane] (Moazzami et al., 2006b).

In studies on various functional activities of sesamin and sesaminol, much attention is being focused on the content of these lignans in sesame seeds. A significant positive correlation was observed between the oil content of sesame seed and sesamin content in the oil, whereas no correlation was found between the oil content and the sesamolin content (Tashiro et al., 1990). Large differences exist in lignan contents among the wild types of sesame. For example, an Indian type has a very low sesamolin content (sesamin: 256.1 mg/100 g; sesamolin: 35.6 mg/100 g) while another type from Borneo, Indonesia, has a remarkably high lignan content (sesamin: 1152.3 mg/100 g; sesamolin: 1360.7 mg/100 g) (Namiki, 1995). Interestingly, differences in lignan composition and content were observed among four *Sesamum* species, including three wild types. Sesamin was found in considerable amounts in *S. indicum* cultivars and *S. angustifolium*, in very large amounts (2.40% in oil) in *S. radiatum*, but only in very small amounts in *S. alantum*. Sesamolin was detected in considerable amounts in *S. indicum* and *S. angustifolium*, but in very small amounts in *S. alantum* and *S. radiatum*. The major lignan of *S. angustifolium* was sesangolin (3.1 5% in oil) and that of *S. alantum* was 2-episesalatin (1.37% in oil) (refer to Figure 10.1) (Jones et al., 1962; Kamal-Eldin and Appelqvist, 1994b).

To elucidate these differences in lignan content, the biosynthesis of these antioxidative lignans in sesame seeds was studied (Kato et al., 1998), but the biosynthetic route of sesamolin and sesaminol is not yet clear. As to the biosynthesis of sesamin, formation from (+)-pinoresinol via (+)-piperitol has been estimated, and it was recently demonstrated by isolation and identification of cytochrom SiP450, CYP81Q1 protein, and their roles in the formation of two methylenedioxy bridges (Ono et al., 2006).

Due to the high functional activities of sesame lignans, cultivation of sesame is focused on the development of new varieties having high lignan content and functional activities. Shirato-Yasumoto et al. (2000, 2001) have established several new sesame lines with high sesamin and sesamolin contents, about twice that of conventional varieties.

10.4 ANTIOXIDATIVE FUNCTIONS

10.4.1 Roasted Sesame Seed Oil

The characteristic of sesame oil, which has been proved empirically, is its high resistance to oxidative deterioration. A high antioxidative stability was well known in ancient Egypt, and in Japan it has been evaluated as the best oil for deep-frying tempura because of its superior stability against deterioration by heating. As noted earlier, there are two different kinds of sesame oil, roasted and unroasted. The antioxidative activities of these oils were demonstrated in experiments with other common vegetable oils that were stored in an open dish at 60°C, and autoxidation was determined by the increase in weight caused by peroxidation. Soybean oil, rapeseed oil, and others showed rapid increase after about 10 days, whereas both roasted and unroasted sesame oils were very stable. The unroasted oil remained unchanged for 30 days, while no oxidation was observed even after 50 days in the roasted oil (Fukuda and Namiki, 1988).

However, for a long time, no chemical investigations explaining the antioxidative factors contained in these oils were conducted, except for two reporting that sesamol was a strong antioxidative phenol and a degradation product of sesamolin in the roasted oil (Budowsky and Narkley, 1951; Budowsky, 1964).

Antioxidants in food have been found to play an important role in preventing damage due to active oxygens *in vivo*, which lead to various life style diseases such as circulatory disorders, carcinogenesis, and aging (Namiki, 1990, 1998). Namiki and Kobayashi, (1989) conducted investigations on the chemistry and function of antioxidative factors in sesame seed and oil. In their studies, several antioxidative components involving the phenolic lignans sesaminol and sesamolinol have been isolated and identified (Fukuda et al., 1985a; Namiki, 1995) (Figure 10.1).

Roasted sesame seed oil has a characteristic flavor and red-brown color probably caused by the Maillard-type reaction during roasting. The antioxidative activity increases mainly in proportion to the roasting temperature along with the brown color, indicating that some products of the roast reaction contribute to the antioxidative activity. Antioxidation tests on the ether and subsequent methanol extracts of the strongly roasted oil indicated that each fraction alone was not so effective but a combination of the extracts shows significant synergistic effect, and those of three or four fractions exhibit even stronger activity. Thus, the very strong antioxidative activity of the roasted oil might result from the synergistic effect of the combination of such effective factors as sesamol produced from sesamolin, γ-tocopherol, sesamin, and roasted products like melanoidin (Fukuda et al., 1986a; Koizumi et al., 1996; Fukuda et al., 1996).

10.4.2 Unroasted Sesame Seed Oil

As mentioned above, unroasted seed oil is obtained by expelling steamed raw seed and refining the product by decolorization with acid clay and deodorization by steam distillation *in vacuo*. It has much stronger antioxidative activity than common vegetable oils, but unlike oil from roasted sesame seeds, it contains no Maillard-type products or sesamol. Investigation of its antioxidative factors showed that sesamolin in the raw oil was almost completely lost during the decolorization process, whereas a significant amount of a new phenolic lignan named sesaminol, identified as the principal factor of the antioxidative activity of unroasted oil, could be produced by a novel intermolecular rearrangement of sesamolin, as noted above (Fukuda et al., 1986b, 1986c).

Sesame oil from the whole seed was somewhat more stable than that from the hulled seed, and the oil extracted with hexane–isopropanol was better than that extracted with hexane (Kamal-Eldin and Appelqvist, 1995).

It has also been shown that when foodstuff covered with wet material is fried in roasted seed oil, as in the case of Japanese tempura, sesamol is produced by the splitting of sesamolin, resulting in the formation of a strong antioxidative coating on the fried food (Fukuda et al., 1986a).

10.4.3 Supercritical Carbon Dioxide Fluid Extraction of Sesame Seed

Recently, a new extraction process using supercritical carbon dioxide fluid (SFE-CO_2) has been introduced. Interestingly, it was found that the lignans were extracted prior to the oil (triacylglycerols), so the lignans could be obtained easily in a concentrated form at an early stage of the extraction, along with the concentration of antioxidative factors and characteristic roast flavor. Such extraction of lignans had never been accomplished before (Namiki et al., 2002). More detailed investigation on the extraction conditions of sesame oil by SFE-CO_2 also indicated superiority of the SFE over *n*-hexane extraction, especially regarding the antioxidative activities of sesame oil (Hu et al., 2004).

Recently, SFE-CO_2 extraction of sesame oil is being spread in Korea and China due to its excellent quality in flavor, taste, and antioxidative stability, and so on (Kang, 2006).

10.4.4 Sesame Lignan

10.4.4.1 Sesaminol

This lignan, identified as antioxidative factor in unroasted sesame oil, includes sesamol as a moiety and has far stronger antioxidative activity than sesamol because sesamol is easily dimerized and its products have lower activity (Fukuda et al., 1986a). This markedly strong and stable antioxidative property may be provided by the presence of a bulky samin group at the *ortho* position of the phenol group in sesamol, similar to the BHT (butylated hydroxytoluene) molecule. As noted above, the content of sesaminol in the seed is very small in the free form, whereas there are considerable amounts of its di- or tri-glucosides in the seed as well as in the defatted meal or sesame flour. The glucosides are weak as active oxygen scavengers in *in vitro* tests but expected to act as potential antioxidants through deglucosidation by the action of β-glucosidase or intestinal microbes (Katsuzaki et al., 1994).

Sesaminol suppressed lipid peroxidation in model systems of *in vivo* peroxidation using the ghost membranes of rabbit erythrocytes and the liver microsome in rats (Osawa et al., 1990). Antioxidative activity of sesaminol was also observed in the lipid peroxidation of low-density lipoprotein induced by 2,2-azobis(2,4-dimethyl valeronitrile), where the effective radical scavenging activity of sesaminol was demonstrated by isolation and identification of the reaction products of sesaminol with the reagent radical (Kang et al., 1998b). In a similar study on the Cu^{2+}-induced lipid peroxidation of low-density lipoprotein, sesaminol inhibited the peroxidation dose-dependently and showed stronger inhibiting activity than α-tocopherol and probcol in the radical scavenger and in the production of lipid peroxidation products such as 4-hydroxy-nonenal and malonaldehyde adduct products (Kang et al., 2000).

10.4.4.2 Sesamin and Sesamolin

Sesamin is the main characteristic lignan of sesame seed, with a content of about 0.4% in seed oil. However, it has no free phenol group and showed very weak or no antioxidative effect in conventional *in vitro* tests, but sesamin exhibits significant *in vivo* physiological activities assumed to be due to antioxidative activity. To elucidate this discrepancy in the mode of action of sesamin, metabolic changes of sesamin in both *in vitro* and *in vivo* systems were investigated.

In the *in vitro* reaction of sesamin with rat liver homogenate, it was shown that sesamin changed into two kinds of metabolites involving the dihydroxyphenyl (catechol type) moiety, and these metabolites were also detected in rat bile after oral administration of sesamin. These products showed strong radical-scavenging activities. Thus sesamin that is ingested may be incorporated specifically into the liver and metabolized there into materials having strong antioxidative activities (Nakai et al., 2003).

Recently, the metabolism of sesamin was investigated *in vivo* using four humans and *in vitro* using human fecal microflora (Penalvo et al., 2005b; Moazzami et al., 2007). After ingestion of sesame seed, only small amounts of sesamin, pinoresinol, and other minor lignans contained in sesame seed were detected in blood plasma, while after 24 h, a rapid increase of large amounts of enterolacton and enterodiol was observed. On the contrary, the fermentation of sesamin with human intestinal microflora gave enterolacton and enterodiol as the main metabolites along with some intermediates named M1, M2, and M3, which were degraded lignans involving a catechol-type moiety, and M2 and M3 were identical to those observed in Nakai et al.'s (2003) study. These findings suggest that sesamin is not completely absorbed as such but is metabolized by the intestinal microflora into a series of demethylenated intermediates, M2 (3-demethylpiperitol) and M3 (3,3′-didemethylpinoresinol), which might be absorbed as such or transferred in situ to mammalian lignans such as enterolacton and enterodiol. The production of demethylenated catechol type metabolites was also observed on sesamin and sesaminol triglucoside by culturing with some *Aspergillus* fungi, and these products, especially sesaminol-6 catechol, showed strong antioxidative activity in the DPPH test (Miyake et al., 2005). Direct chemical preparation of the catechol-type active products from sesamin was performed by short time treatment of sesamin with supercritical water (Nakai et al., 2006).

The chemical structures of the main metabolic products of sesamin produced by liver enzymes or microorganisms are shown in Figure 10.2. These results suggest that sesamin is a kind of prodrug giving potent antioxidative polyphenol-type lignan derivatives *in vivo*, which in turn give rise to various antioxidative physiological activities.

Sesamolin is the second major lignan in sesame seed and it has a characteristic structure involving one oxygen bridge, but it shows no significant antioxidative activity in *in vitro* tests. However, some activities *in vivo*, such as the supression of lipid peroxidation and 8-hydropxy-2-deoxyguanosine (8-OhdG) excretion in urine, were observed (Kang et al., 1998a).

The above results indicate that sesamin and sesamolin have little or no antioxidative activity as such but act as proantioxidants, which become strong antioxidants via

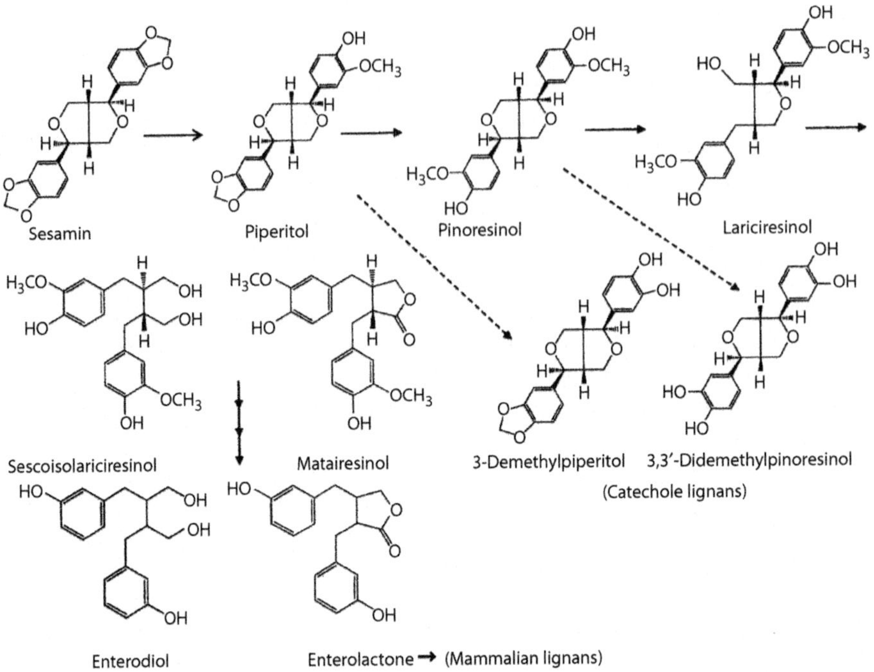

FIGURE 10.2 Main metabolic products of sesamin by liver enzymes and microorganisms.

in vivo metabolic changes in their structure. Another major lignan, sesaminol, acts as a strong antioxidant *in situ*. Moreover, it changes into stronger antioxidative catechol-type derivatives through metabolic action. There are thus considerable differences in the mode of antioxidative action *in vivo* between sesamin and sesaminol. This is an interesting problem to be elucidated in future functional studies on sesame.

10.4.5 Other Antioxidative Components

It is generally assumed that black sesame seed is superior to white or brown seed in antioxidative and antiaging effects. In order to ascertain this traditional belief, various chemical investigations were undertaken on black and white sesame seed and especially on their hulls (Fukuda et al., 1991). No significant differences in their whole seeds were observed in the composition and content of fatty acid, tocopherol, and sesame lignan, and also in the antioxidative activity of 80% ethanol–water extracts of crushed whole sesame seed. However, in the case of short time water extracts of crushed whole sesame seed, all the black and brown strains showed strong antioxidative activity but the white strain had little or no activity (Fukuda et al., 1991). It was also observed that in the various antioxidative factors, such as total phenols and radical scavenging activity, black sesame, especially its hulls, showed superior activity than those of white sesame (Shahide et al., 2006).

To isolate and identify the active components in the black sesame seed, the black pigment extracts obtained from the washed water of black sesame seed in the factory were chromatographed and four products were isolated, which were identified to be the following lignans: pinoresinol (compound 1), lariciresinol (compound 2) (Figure 10.2), hydroxymatairesinol (compound 3), and allohydroxymatairesinol (compound 4). Compounds 2, 3, and 4 were newly isolated and identified in the sesame seeds, and compounds 1 and 2 were far more abundant in black sesame than in white. These compounds had appreciable superoxide radical scavenging effect (Nagashima et al., 1999; Fukuda and Nagashima, 2001), though they were not a black compound. The chemical structure of the black pigment is assumed to be a kind of tannin but its structure is not yet clear. Most of the antioxidative lignans mentioned above are insoluble in water and transferred into the extracted oil. Therefore, it seems that there would remain no appreciable antioxidative activity in the defatted sesame meal. However, as described previously, a considerable amount of the water-soluble sesaminol glucosides found in the defatted sesame meal may act as radical scavengers (Kuriyama et al., 1995). Moreover, they are assumed to play an important role as potential antioxidants *in vivo* through deglucosidation by intestinal microorganisms.

On the contrary, a marked increase in the antioxidative activity was observed at an early stage of germination along with increase in the methanol-soluble phenolic compounds (Fukuda et al., 1985b). It was also observed that the amount of sesaminol glucosides markedly increased whereas those of sesamin and sesamolin rapidly decreased (Kuriyama et al., 1995). New lignan derivatives were also formed during the germination, and this may be related to the increase in antioxidative activity (Kuriyama and Murui, 1995). Investigation on changes in various components of sesame seed by germination showed that there are significant increases in the contents of Ca and other minerals, linoleic acid, sesamol, and α-tocopherol along with marked decrease in fat content (Hahm et al., 2008).

An increase in the antioxidative activity of methanol extracts of the defatted sesame meal was observed by far-infrared irradiation of sesame seed, along with an increase in the extractable phenol compounds (Lee et al., 2005).

10.5 ANTIAGING EFFECTS AND SYNERGISTIC FUNCTIONS WITH VITAMIN E

10.5.1 Senescence-Accelerated Mice

Habitual ingestion of food containing sesame seed in any form has long been believed to prevent aging, particularly in China and Japan. However, until fairly recently no scientific data existed to support this belief. Based on chemical knowledge of various antioxidative activities of sesame seed and oil, scientific studies on their antiaging effects have been conducted by Yamashita et al. (1990) using senescence-accelerated mice (SAM). SAM were developed by Takeda et al. (1981) at Kyoto University from an AKR strain, which has two series, a P series and an R series. The P series are a specific type with varieties from P-1 to P-11, each of which show different characteristic aging symptoms in behavior and appearance as well as some pathological changes such as inflamed tissues and amyloidosis. The R series are senescence-resistant mice that undergo the normal aging process.

In their studies, P-1 was used as the SAM and R-1 as the control, and the degree of aging was evaluated according to a grading system developed by Takeda et al. (1981). Changes in SAM aging scores during long-term feeding with either a standard diet or a diet supplemented with 20% black sesame seed powder were determined. With the standard diet, each index of aging increased after about 4–5 months, but in mice fed the diet containing sesame seed, the progress of aging was slow and suppressed. This was evident especially in periophthalmic lesions, hair glossiness, and skin coarseness. The overall appearance was quite different between the two groups, and that of the group fed the diet containing sesame seed was superior. In the mice sacrified after 7 months of feeding, levels of the lipid peroxide and aging pigment lipofustin in the livers were found to be slightly suppressed, and superoxide dismutase (SOD) activity in the group fed the diet containing the seed was increased.

These results suggest that there exists an ingredient in sesame seed which prevents the accumulation of factors causing aging such as lipid peroxidation (Yamashita et al., 1990; Namiki et al., 1993; Yamashita et al., 1994). Unlike soybean and other foods, the major components of sesame seed such as oil, protein, and carbohydrate do not have such antioxidation activities. Therefore, the characteristic lignans in sesame seed are considered to be the probable candidates responsible for the antiaging effect.

10.5.2 Synergistic Effect of Sesame Lignans with Tocopherols

10.5.2.1 Synergism with γ-Tocopherol

As shown in Table 10.2, tocopherol in sesame seed exists mostly as the γ-homologue, and sesame seed shows little vitamin E activity. However, like vitamin E, it prevents aging, probably via its antioxidative activity. To solve this discrepancy, the effect of sesame seed as vitamin E was compared with that of α- or γ-tocopherol. In this study, rats were fed a vitamin E-free diet (control), α- or γ-tocopherol, or sesame seed diets. After 8 weeks of feeding, the lipid peroxide levels in liver (expressed by TBARS value), oxidative hemolysis, and pyruvate kinase activity in plasma were determined as indices of vitamin E activity (Yamashita et al., 1992). As shown in Figure 10.3, the control group showed higher values of lipid peroxide in their livers, and the γ-tocopherol group had essentially the same values, while in the α-tocopherol and sesame seed groups, the values were suppressed to the same level. Similar results were observed for pyruvate kinase activity. The most remarkable difference was observed in red cell hemolysis, which was increased in the control group but was almost completely suppressed in the sesame seed group as well as in the α-tocopherol group, but only weakly suppressed in the γ-tocopherol group. The level of α-tocopherol in plasma and liver was high only in the α-tocopherol group with trace amounts in the other three groups. On the contrary, γ-tocopherol in plasma and liver was found only in the sesame group and not found even in the group fed the γ-tocopherol diet (Yamashita et al., 1992).

The elevation of plasma levels of γ-tocopherol from sesame seed was also demonstrated in humans (Cooney et al., 2001). The study was conducted with subjects ($n = 9$) fed muffins containing equivalent amounts of γ-tocopherol from sesame seeds, walnuts, or soybean oil. The results showed that consumption of about 5 mg of γ-tocopherol per day over a 3-day period from sesame seed significantly elevated

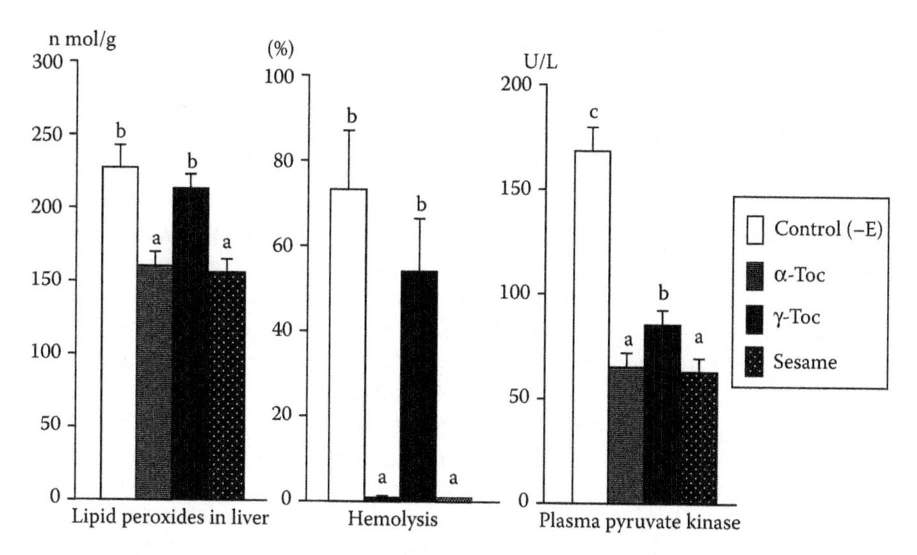

FIGURE 10.3 Effect of sesame seed on lipid peroxides in liver, red cell hemolysis, and plasma pyruvate kinase activity (rats fed for 6 weeks diets containing α- and γ-tocopherol equal to amounts of γ-tocopherol in the sesame diet) (50 mg/kg).

serum γ-tocopherol levels (19%) and depressed plasma β-tocopherol (34%), but this did not occur in subjects given walnut or soybean oil. No significant changes in baseline or postintervention plasma levels of cholesterol, triacylglycerol, or carotenoids were found for any of the intervention groups. Thus, the consumption of a moderate amount of sesame seed seems to cause significant increase in plasma γ-tocopherol and alter plasma tocopherol ratios, indicating further enhanced vitamin E bioactivity.

Moreover, it has been reported that γ-tocopherol concentrations in the serum of Swedish women increased with the intake of sesame oil (Lemcke-Norojarvi et al., 2001).

The above results suggest that there may exist a component in sesame seed that maintains the γ-tocopherol level in the body. The effect of the typical sesame lignans sesaminol and sesamin on lipid peroxidation was examined in rats. The results indicated that increase of thiobarbituric acid reactive substance (TBARS) value in liver in the control group was suppressed significantly with the addition of α-tocopherol or γ-tocopherol + lignans, and slightly suppressed with γ-tocopherol alone. Interestingly, considerably high levels of γ-tocopherol in plasma were observed only in the lignan-added groups, and sesaminol had a greater effect than sesamin (Yamashita et al., 1992). Thus, marked antioxidative and antiaging activities observed with sesame seed could not be obtained solely from γ-tocopherol contained in the seed, but developed synergistically in combination with γ-tocopherol and sesame lignans.

10.5.2.2 Synergism with α-Tocopherol

In normal diets, γ-tocopherol is usually ingested in greater amounts than α-tocopherol, but the tocopherol found in human plasma is chiefly α-tocopherol, with trace amounts of γ-tocopherol. To investigate the relationship of sesame seed with tocopherol in the body, the effect of sesame seed on α-tocopherol in plasma and vitamin E activity was examined using six groups of rats fed diets containing 10 (low), 50 (normal), and 250 (high) mg/kg α-tocopherol with or without 20% sesame seed. After 8 weeks of feeding, α-tocopherol in plasma increased linearly in proportion to the amount of α-tocopherol ingested, and the level increased remarkably upon the addition of sesame seed, even though it contained little α-tocopherol. On the contrary, γ-tocopherol was not present in plasma except in those cases where sesame seed alone or sesame seed containing a low amount of α-tocopherol was ingested (Yamashita et al., 1995).

These studies indicate that sesame seed increases the level of α-tocopherol in the body as well as that of γ-tocopherol. If excess α-tocopherol beyond a certain level is present, no γ-tocopherol remains, even though sesame is present in the body, indicating a clear preference of α-tocopherol over γ-tocopherol in relation to vitamin E *in vivo*.

To ascertain whether sesame lignans are related to the synergistic effect of sesame seed with tocopherol, a study has been conducted using sesaminol and sesamin. The addition of either of these sesame lignans reduced the level of lipid peroxide in liver and increased the levels of α-tocopherol in plasma and liver, and in these effects, sesaminol was more effective than sesamin (Yamashita et al., 1995).

The synergistic action of sesame lignan with α-tocopherol on lipid peroxidation was also observed in a study using docosahexaenoic acid (DHA) that is present in the central nervous system and is said to be related to learning and memory activities but has extremely high peroxidative susceptibility. The elevation of TBARS levels in liver and red cell hemolysis of rat, induced by the addition of DHA, was completely suppressed by the synergistic action of sesamin or sesaminol with a low level of α-tocopherol, along with marked elevation of tocopherol concentrations in liver, brain, and other organs. In these effects, sesaminol showed a higher activity than sesamin (Ikeda et al., 2003a). When a variety of sesame seeds containing higher sesamin and sesamolin contents was used, levels of γ-tocopherol increased greatly in the livers and brains of rats and also significantly in kidneys and serum (Ikeda et al., 2000b).

On the catalyst-induced lipid peroxidation in model biological systems using rat liver microsomes or mitochondoria, marked synergistic inhibiting effects of sesame lignans with tocopherol were also observed (Ghafoorunissa et al., 2004). Antioxidative studies were extended to iron-induced oxidative stress in rat by dietary sesame oil + sesamin (Hemalatha et al., 2004).

10.5.2.3 Synergism with Tocotrienols

In nature, tocopherols and tocotrienols are compounds having vitamin E activity, with the latter having lower activity. However, it has been reported that α-tocotrienol showed higher activity than α-tocopherol in the antioxidation of lipid peroxidation in rat microsome and mitochondria, and showed a higher antitumor activity as well (Theriault et al., 1999). Based on the fact that sesame seed and lignans markedly elevated the concentrations of γ-tocopherol in the tissues as well as vitamin E activity, the effects of sesame seed on tocotrienols were investigated (Ikeda et al., 2000a, 2001). In rats fed diets containing tocotrienol-rich fractions of palm oil (T-mix), accumulation of α- and γ-tocotrienols was observed only in the adipose tissues and skin, but not in plasma or other tissues. The addition of sesame seed also elevated tocotrienol concentration in adipose tissues and skin, but it was not affected in other tissues or in plasma. These data suggest that the transport and tissue uptake of vitamin E isoforms differ, and that sesame seeds elevate concentrations of tocotrienols.

Concerning the synergistic effect of tocotrienol with sesame lignan on antioxidation, a study using rats was undertaken with T-mix + sesaminol systems. In the antioxidative activities indicated by liver TBARS, red blood cell hemolysis, and plasma pyruvate kinase activity, a marked enhancement of antioxidation was observed with sesaminol. The concentrations of α-tocopherol in plasma, livers, and kidneys increased significantly by the addition of sesaminol, while α-tocotrienol increased less in kidneys than did tocopherol. These sesaminol-induced increases in the concentrations of α-tocopherol and α-tocotrienol in plasma and tissues were not the result of enhanced absorption due to sesaminol (Yamashita et al., 2002; Ikeda et al., 2003b).

As for damage caused by ultraviolet (UV-B) irradiation, sunburn induced on the back of hairless mice was very severe in the V-E-free control, decreased in the α-tocopherol group, and was less in the T-mix (rich in tocotrienols) group, while there was almost no damage in the T-mix + sesamin group. These results suggest that dietary tocotrienols are incorporated into skin where they prevent oxidative damage, and that the protection is enhanced by increasing the concentration of sesame lignans (Yamashita, 2004).

10.5.3 Effect on Metabolism of Tocopherols

As mentioned above, common vegetable oil is rich in γ-tocopherol, but its level in plasma and liver is much lower than that of the α-homologue. Differentiation between the isomers is thought to occur as a result of difference in binding capacity to the hepatic α-tocopherol transfer protein (α-TTP). In this respect, the mode of secretion of α- and γ-tocopherols in bile and the effect of sesame seed on the secretion were investigated using rats fed the following diets: tocopherol-free (control), α-tocopherol alone, γ-tocopherol alone, α- + γ-tocopherols, γ-tocopherol + sesamin, and sesame seed (Yamashita et al., 2000b; Ikeda et al., 2002b). The results indicated markedly higher concentrations of α-tocopherol in the liver as well as in plasma in the α-tocopherol group and α-+ γ-tocopherols group at essentially the same levels. Concentration of γ-tocopherol was only observed in the group fed γ-tocopherol alone at the lower level, and was significantly increased by the addition of sesamin or sesame seed. A similar tendency was observed in bile, although the concentrations of α- and γ-tocopherols were substantially lower than those in the liver. These results suggest that secretion into bile is not a major metabolic route for α- or γ-tocopherol.

In the metabolism of tocopherol, it is known that cytochrome *P-450* (CYP) enzyme mediates ω-hydroxylation of the tocopherol phytyl side chain, followed by stepwise removal of two or three carbon moieties and finally yielding carboxyethyl-hydroxychroman (CEHC) derivatives (Parker et al., 2000). In relation to the elevating effect of sesame on γ-tocopherol level in rats, the effect on urinary excretion of γ-CEHC was examined in rats fed diets with or without sesame seed or sesamin. Addition of sesame seed and sesamin markedly suppressed the urinary excretion of γ-CEHC, and elevated the levels of γ-tocopherol in liver, kidney, brain, and serum

(Uchida et al., 2007). A similar inhibiting effect on urinary excretion of γ-CEHC was also observed by ketoconazole, a potent and selective inhibitor of CYP3A. These results clarified that sesame seed and its lignans elevate γ-tocopherol concentrations by the inhibition of the CYP3A-dependent metabolism of γ-tocopherol (Ikeda et al., 2002b). These effects of sesame and its lignan on the metabolism of tocopherols were illustrated in Figure 10.4. It was also clarified that tocopherol ω-hydroxylase activity was associated only with CYP4F2, which also catalyzes ω-hydroxylation of LTB_4 and arachidonic acid. Sesamin strongly inhibited tocopherol ω-hydroxylation exhibited by CYP4F2 and microsomes in the livers of rats or humans, and resulted in elevated tocopherol levels *in vivo* (Sontag and Parker, 2002).

In addition to these elevating effects of sesame seed on tocopherol concentration in tissues, it was also demonstrated that sesame seed and its lignans elevated ascorbic acid concentration in the liver and kidney in the Wistar rats, stimulating ascorbic acid synthesis as a result of the induction of UGT IA- and 1B-mediated metabolism of sesame lignan in rats. The results suggest that intake of sesame seed significantly enhances antioxidative activity in the tissues by elevating the levels of two antioxidative vitamins, vitamin C and E (Ikeda et al., 2007).

10.5.4 Elevation of Tocopherol Concentration in the Brain

Based on the elevation effect of dietary sesame seed or its lignans on the concentration of tocopherols in serum, and related to the result that treatment with α-tocopherol slows the progression of Alzheimer's disease (Sano et al., 1997), the effect of sesame seed on tocopherol concentration in the brain was investigated. The rats were fed diets containing α-tocopherol of normal (50 mg/kg) or high levels (500 mg/kg) with or without sesame seeds for 8 weeks. The concentration of tocopherol in the liver was elevated significantly with the high tocopherol diet, while it was negligibly small in the brain. Interestingly, a significant elevation of tocopherol concentration in the brain was observed with the sesame seed-containing diet, even at normal concentrations of tocopherol, especially in the hippocampus, which is the region of the brain

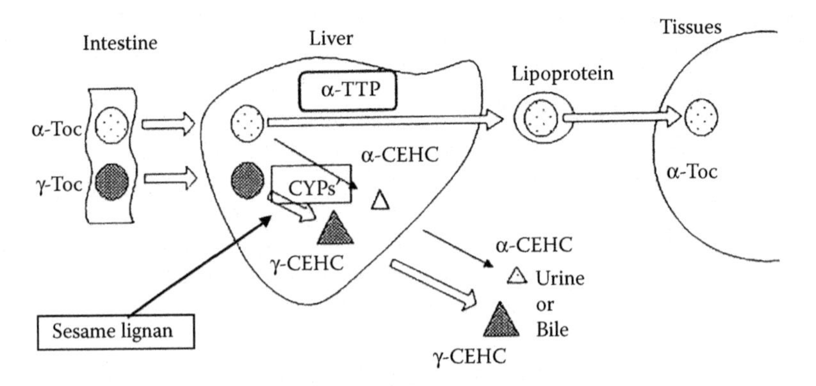

FIGURE 10.4 Metabolic pathways of α-tocopherol (α-Toc), γ-tocopherol (γ-Toc), and inhibition by sesame lignans. CYP = cytochrome P-450 enzymes, TTP = tocopherol transfer protein, CEHC = carboxyethyl hydroxychroman (degradation product of tocopherol).

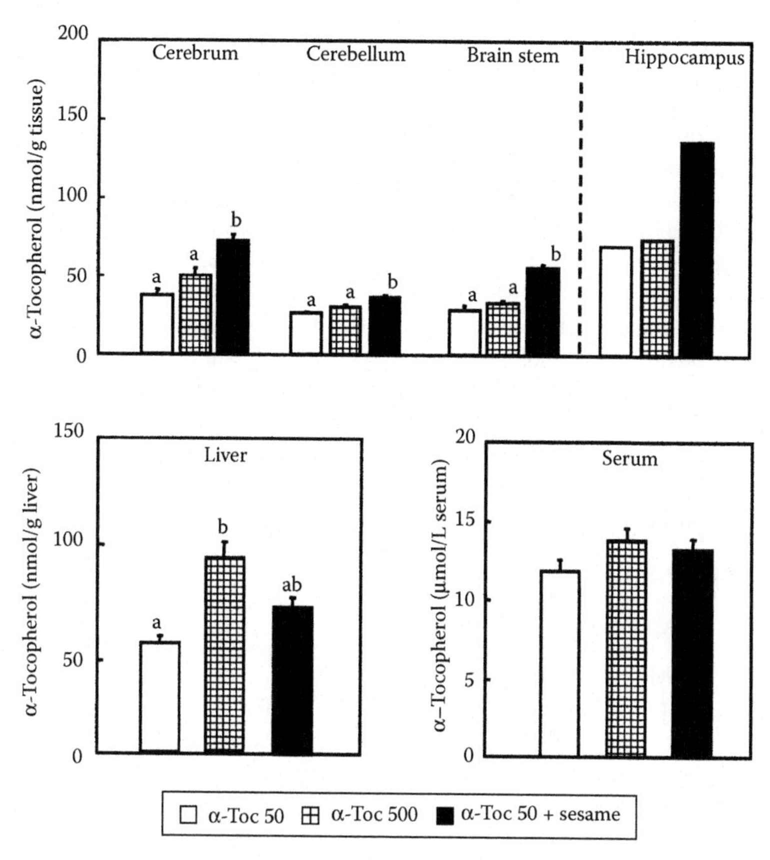

FIGURE 10.5 Effect of dietary α-tocopherol or sesame seeds on α-tocopherol concentrations in cerebrum, cerebellum, brain stem, hippocampus, liver, and serum of rats. Rats were fed for 8 weeks diets containing 50 mg α-tocopherol/kg (α-Toc 50), 500 mg α-tocopherol/kg (α-Toc 500), α-tocopherol 50 mg + sesame 200 g/kg (α-Toc+sesame).

related to memory activity (Abe et al., 2005) (Figure 10.5). It was noted that in the rats fed normal diets, significant concentrations of α-tocopherol but no γ-tocopherol were detected in the brain. However, when a sesame lignan was added to the diet, a significant amount of γ-tocopherol in the cerebral cortex was observed, and sesaminol showed superior activity than sesamin (Yamashita, 2004).

10.5.5 Comparison between Sesame Seed and Flaxseed on Elevation of Vitamin E Concentration

Flaxseed has received much attention as a health food that reduces chronic diseases such as cancer and coronary artery diseases. Flaxseeds resemble sesame seeds very closely in their consituents, for example, more than 40% oil, 20% protein, high in γ-tocopherol content (0.0151%) and almost no α-tocopherol, and also high in lignans. However, flaxseed contains much linolenic acid in the oil, and its primary lignans are secoisolariciresinol diglucosides (SDG) (about 2 mg/g) and secoisolariciresinol (about 1 mg/g), but no sesamin. The effects of flaxseed on tocopherol, TBARS, and cholesterol concentrations in the plasma and tissues of rats were compared with the effects of sesame seed. The results showed that no significant elevation of γ-tocopherol concentration or suppression of TBARS concentration in tissue was observed with flaxseed, indicating that the synergistic effect of tocopherols on vitamin E activity is characteristic only of sesame lignans (Yamashita et al., 2003). As for flaxseed lignans, SDG was shown to yield mammalian lignans such as enterodiol and enterolactone through metabolic changes (Figure 10.2), and these lignans and a related lignan, hydroxymatairesinol (HMR), were reported to be effective in preventing breast cancer and prostatic cancer. So the effects of these lignans and related compounds on the metabolism of tocopherol, especially γ-tocopherol, were investigated, and it was found that, unlike sesame lignans, these compounds have no elevating effect on vitamin E activity. Thus, the regulatory effect on tocopherol metabolism is indicated to be a property characteristic of sesame lignans (Yamashita et al., 2007).

10.6 EFFECT OF SESAME SEED AND LIGNANS ON LIPID METABOLISM

10.6.1 Effect on Fatty Acid Metabolism

10.6.1.1 In Microorganisms

As shown in Figure 10.6, the metabolism of essential fatty acids comprises two series, one series starting from linoleic acid (*n*–6, 18:2) and the other from α-linolenic acid (*n*–3, 18:3).

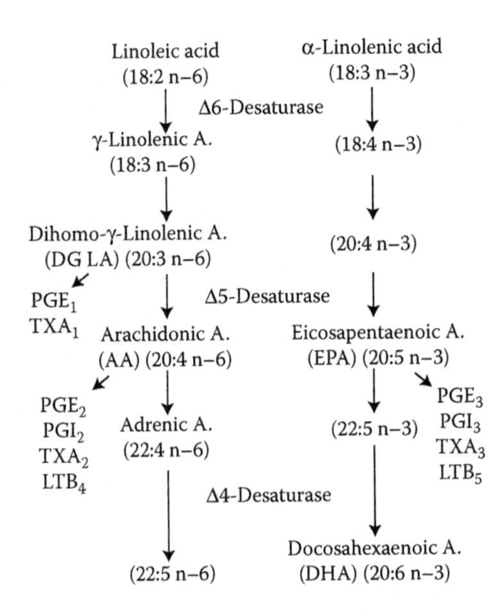

FIGURE 10.6 Desaturation and elongation of polyunsaturated fatty acids, and production of eicosanoids.

The fact that sesamin inhibits the course of fatty acid metabolism was first found in studies conducted by Shimizu et al. (1989a,b) focusing on production of polyunsaturated fatty acids such as DHA and EPA by fermentation from fatty acids. During investigation of the practical production of AA using *M. alpina* 1S-4, with vegetable oils as starting material, they found that incubation with sesame oil specifically increased the DGLA (precursor of AA) content, and decreased the yield of AA. This interesting effect of sesame oil was assumed to be because sesame oil contains a trace factor, which specifically inhibits the Δ5 desaturation enzyme. The inhibiting factor in sesame seed has been isolated and found to be sesamin and related compounds (Shimizu et al., 1989c). Sesamin also inhibits lysophosphatidylcholine acyltransferase in *M. alpina* (Chatrattanakunchai et al., 2000).

To study the effect of sesame lignans on the biosynthesis of DGLA and AA from stearic acid, the inhibiting activities on Δ5, Δ6, Δ9, and Δ12 desaturation enzymes and chain elongation enzymes were examined with noncell extract of *M. alpina* 1S-4. As shown in Table 10.3, every lignan compound strongly inhibited the Δ5 desaturation enzyme, even at lower concentrations (the effective concentration of sesamin to inhibit enzyme activities by 50% (IC_{50}) being 5.6 μM), while other Δ6, Δ9, and Δ12 unsaturation enzymes were not affected at all. The activities were stronger in the order sesamin > sesaminol > episesamin, whereas sesamol, a decomposed product of sesamolin, showed no activity (Shimizu et al., 1991). This study elucidated the unique action of sesame seed in inhibiting the activity of the Δ5 desaturation reaction, and spurred subsequent studies on fatty acid metabolism.

10.6.1.2 In Animal Cells, Rats, and Mice

Based on the above findings on microorganisms, a study was conducted to determine how sesame lignans affect fatty acid metabolism in animal cells (Fujiyama-Fujiwara et al., 1992). Primary cultured liver cells of rat were prepared, and DGLA and a fatty

TABLE 10.3
Specific Inhibition of Fungal and Rat Liver Δ5 Desaturases by Sesamin-Related Compounds

	Desaturase Activity (nmol/30 min/mg protein)						
	M. alpina				Rat Liver		
Compound Added	**Δ9**	**Δ21**	**Δ6**	**Δ5**	**Δ9**	**Δ6**	**Δ5**
None	0.30	0.25	0.55	0.64	1.12	0.45	3.48
Sesamin	0.32	0.28	0.57	0.08	1.05	0.49	2.15
Episesamin	0.31	0.25	0.55	0.16	1.08	0.53	2.76
Sesaminol	0.30	0.25	0.55	0.14	1.08	0.46	2.64
Sesamolin	0.30	0.26	0.53	0.12	1.13	0.56	2.60
Sesamol	0.30	0.24	0.52	0.50	1.03	0.44	3.44

M. alpina = *Moritierella alpina* 1S-4.

acid (20:4 in *n*–3 series) (FA) were added to media with sesamin. After incubation, in the *n*–6 series, the ratio of AA/DGLA clearly decreased in proportion to the concentration of sesamin added, whereas in the *n*–3 series, the ratio of EPA/FA slightly increased. These data suggest that sesamin inhibits the Δ5 desaturation enzyme reaction in the *n*–6 series but not in the *n*–3 series, and in fact may partially accelerate the reaction. Thus the effect of sesamin appears to be different between the *n*–6 and *n*–3 series in cultured animal cells. A similar study was conducted using liver microsome of rat. Only the Δ5 desaturation enzyme was specifically inhibited (IC_{50} 72 μM) (Table 10.3) (Shimizu et al., 1991). When rats were fed a diet containing 0.5% sesamin for 13 days, the levels of DGLA in liver, blood plasma, and blood cells were increased 2.1, 1.8, and 1.3 times, respectively, compared with a sesamin-free group. The composition of fatty acids was governed by the effect of inhibiting the Δ5 desaturation enzyme (Sugano et al., 1990; Akimoto et al., 1993).

Fatty acid composition in the liver membrane of mice was investigated using diets containing 0.25% sesamin and 15% safflower oil (providing 12% of the added fat as linoleic acid) or 15% of a mixture of linseed oil and safflower oil (2:1) (providing 6% α-linolenic acid and 6% linoleic acid) for 3 weeks (Chavali et al., 1998). Consumption of sesamin-supplemented safflower oil and a mixture of linseed oil and safflower oil diets resulted in a significant increase in the levels of DGLA, suggesting that sesamin inhibited Δ5 desaturation of the *n*–6 series fatty acids. In animals fed a diet of linseed and safflower oils, the levels of α-linolenic acid, EPA, and DHA were elevated with a concomitant decrease of AA in liver membrane phospholipids. Increase in the level of DGLA in the livers was also observed by the diet supplemented with sesamol (Chavali and Forse, 1999).

From these results, sesamin and sesamol may influence the production of each eicosanoid through inhibition of the Δ5 desaturation enzyme. Specifically, decrease in the concentration of prostaglandin PGE_2 in blood plasma was clearly seen compared with the control group (Hirose et al., 1992).

10.6.2 Controlling Action of Sesame Lignan on the *n*–6/*n*–3 Ratio of Polyunsaturated Fatty Acids

The *n*–3 series polyunsaturated fatty acids in the diets are assumed to be related to the onset of such diseases as arteriosclerosis, cancer and allergy diseases, and the ingestion ratio of fatty acids of the *n*–6 series to those of the *n*–3 series is considered to be significant. In many countries, this ratio is considered to be appropriate at 10:1 to 4:1. However, it may not be easy to control the kind of fat and oil ingested to maintain this ratio because the daily diet is very complicated.

To determine whether sesame lignan controls the balance of polyunsaturated fatty acids in the body, the effect of sesamin (sesamin and episesamin) on the ratio of *n*–6/*n*–3 was investigated (Umeda-Sawada et al., 1995). In the experiment, rats were fed diets containing α-linolenic acid as *n*–3 fatty acid control (ALA group) and EPA (EPA group), with or without sesamin for 4 weeks. As shown in Table 10.4, the intake of EPA increased the level of EPA in the liver compared with the control group, thus remarkably lowering the ratio of *n*–6/*n*–3. The addition of lignans suppressed a rise in the level of EPA so that the ratio of *n*–6/*n*–3 became close to the

TABLE 10.4
Effect of Sesame Lignan on Fatty Acid Metabolism in Liver

	EPA(n–3) Content and n–6/n–3 Ratio		
	Sesame Lignans	**EPA (µg/g Liver)**	**n–6/n–3**
Control (ALA)	–	2.40 ± 0.49^{a}	1.81 ± 0.16^{a}
	+	0.88 ± 0.08^{b}	2.88 ± 0.13^{b}
EPA Group	–	25.4 ± 9.15^{c}	0.50 ± 0.08^{c}
	+	11.5 ± 2.12^{d}	1.07 ± 0.11^{d}
	AA(n–6) Content and n–6/n–3 Ratio		
	Sesame Lignans	**AA(µg/g Liver)**	**n–6/n–3**
Control (LA)	–	29.8 ± 5.05^{a}	3.78 ± 0.61^{a}
	+	32.5 ± 3.16^{a}	4.27 ± 0.33^{a}
AA Group	–	71.7 ± 14.0^{b}	7.58 ± 1.09^{b}
	+	42.2 ± 1.64^{a}	5.95 ± 1.24^{c}

AA = Arachidonic acid; ALA = α-Linolenic acid; EPA = Eicosapentaenoic acid; LA = Linoleic acid.
a, b, c, and d are significantly different at $p < 0.05$.

value for the control group. Similar results were obtained using AA with n–6 series fatty acids in place of EPA (Table 10.4) (Umeda-Sawada et al., 1998).

From these studies, it was found that sesame lignans have the ability to adjust the ratio of n–6/n–3 in the organism to within the normal range, particularly in the case of excess intakes of EPA and AA. These results seem to be inconsistent with those observed *in vitro* in which Δ5 desaturase was inhibited in the n–6 series but not in the n–3 series, that is, if sesamin inhibits only Δ5 desaturase in the n–6 series, the level of AA, which is a product of Δ5 desaturase, could be reduced by the intake of sesame lignans but the level of EPA in the n–3 series would not be affected. Related to this, fatty acid analysis in the above two studies *in vivo* indicated that the level of DGLA (n–6) increased by the addition of sesame lignans, but a PUFA (20:4, n–3) was not detected independently of sesame lignans, indicating that the effect of sesame lignans on Δ5 desaturase in both series was the same as that obtained *in vitro*. Since the actual change in the level of DGLA by the intake of sesame lignans is very slight compared with that of EPA and AA, the effect of dietary sesame lignans on the level of fatty acids in the liver was considered to be stronger than on the level inhibiting Δ5 desaturase in the n–6 series.

To elucidate the effect of sesamin on the mRNA expressions of Δ6 and Δ5 desaturases, rat primary hematocytes were cultured in the presence of sesamin for 24 h. It was shown that sesamin had no effect on the mRNA expression of Δ6 and Δ5 desaturases, while it significantly reduced the index of Δ5 desaturation but not that of Δ6 desaturation. These results suggest that the effect of sesamin on the metabolism of polyunsaturated fatty acids exists only at the enzyme level and not at the gene expression level (Mizukuchi et al., 2003).

10.6.3 Effects of Sesamin on Fatty Acid Oxidation and Synthesis

In general, lipids from food are absorbed in the intestine, incorporated into the liver in the form of chylomicron remnants through the lymph duct, and carried to peripheral tissues via lipoproteins in the blood. Since the level of fatty acids in the liver was reduced by the intake of sesame lignans, the influence of sesame lignans on the absorption of fatty acids in the intestinal tract prior to incorporation into the liver was investigated (Umeda-Sawada et al., 1996). Two emulsions prepared with a mixture of an equal amount of ethyl esters of EPA and AA, with or without sesame lignans, were infused into the stomach. After 24 h, no differences in the amounts of EPA and AA absorbed in the intestinal tract were observed between the control and sesame lignan groups, indicating that the dietary sesame lignans did not have any significant effect on the absorption of fatty acids in the *n*–3 and *n*–6 series in the intestinal tract.

The effects of sesamin (sesamin:episesamin = 1:1) on secretion of triacylglycerol and the production of ketone bodies were studied using rats fed for four weeks with four diets; soybean and perilla oils (5:1, v/v) (control), AA-rich oil (AA), control + sesamin 0.5% (C + S) and AA + sesamin (AA + S).

In the AA and AA + S groups, the level of acetoacetate significantly increased, but the ratio of β-hydroxybutyrate/acetoacetate (an index of mitochondrial redox potential) decreased. On the contrary, in the AA group, the level of serum triacylglycerol was almost two times higher than in the other groups, and it was significantly reduced in the AA + S group. These results indicate that dietary sesamin mixtures promote ketogenesis and reduce the esterification of polyunsaturated fatty acid to triacylglycerol (Umeda-Sawada et al., 1998). A study using perfused livers isolated from rat also indicated that the effect of dietary sesamin is to exert its hypotriglyceridemic effect, at least in part, through the enhanced metabolism of exogenous-free fatty acids by oxidation at the expense of esterification in rat livers (Fukuda et al., 1998). Similar results were obtained using linolelaidic acid (di-*trans* isomer of linoleic acid) instead of oleic acid (Fukuda et al., 1999).

The effects of sesamin on hepatic fatty acid oxidation and synthesis were further investigated at the genetic level (Ashakumary et al., 1999). Rats were fed diets containing 0, 0.1, 0.2, and 0.5% sesamin (sesamin:episesamin = 1:1) for 15 days. Mitochondrial and peroxisomal palmitoyl Co-A oxidation rates increased dose-dependently in sesamin diets (Figure 10.7(I)). Mitochondrial activity almost doubled and peroxisomal activity increased more than 10-fold in rats fed 0.5% sesamin diets, compared with rats fed the sesamin-free diet. Dietary sesamin increased the hepatic activities of fatty acid oxidation enzymes as well as the enzymes involved in the auxiliary pathway of β-oxidation of unsaturated fatty acids dose-dependently.

Sesamin induced an increase in the gene expression of mitochondrial and peroxisomal fatty acid oxidation enzymes. In contrast, dietary sesamin lowered the hepatic activity and mRNA in fatty acid synthase and pyruvate kinase, the lipogenic enzymes. Dietary sesamin increased the activity and gene expression of malic enzyme, another lipogenic enzyme. These results suggest that dietary sesamin may change the metabolism of fatty acids in liver to lower serum lipids in rats.

Studies on the effect of dietary sesamin on mitochondrial and peroxisomal β-oxidation of AA and EPA indicated that sesamin and EPA significantly increased

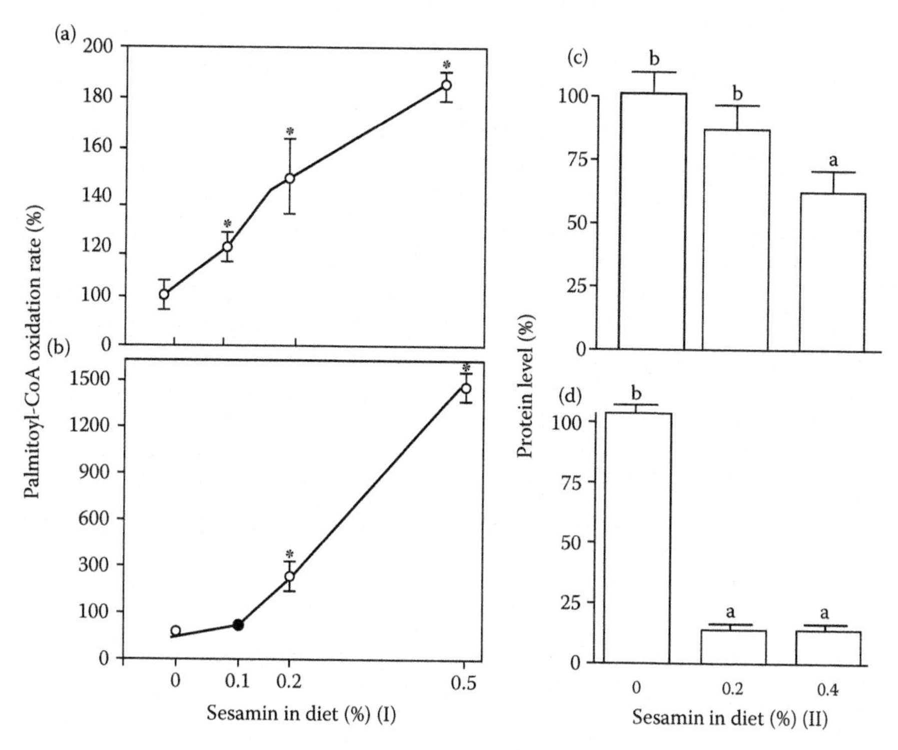

FIGURE 10.7 Effect of sesamin on hepatic fatty acid oxidation (I) and synthesis (II). (I) Effect of sessamin on mitochondria (a) and peroxisomal (b) palmitoyl-CoA oxidation rate in rate liver. Rats were fed experimental diets containing various amounts of sesamin (0-0.5%) for 15 days; (II) effect of sesamin on the sterol regulatory element binding protein-1 (SREBP-1) in the liver of rats fed diets containing 0.2 and 0.4% of sesamin; (c) microsomes, and (d) nuclear extracts.

the activities of fatty acid oxidation enzymes in hepatic mitochondria and peroxisome. In whole liver and triacylglycerol fractions, EPA and AA concentrations were significantly increased by dietary EPA and AA, respectively, and decreased by dietary sesamin. Thus, dietary sesamin increased β-oxidation enzyme activities and reduced hepatic EPA and AA concentrations, and its action was more significant in the EPA group than in the AA group (Umeda-Sawada et al., 2001).

DHA and EPA in fish oil are well known as effective components serving to increase hepatic fatty acid oxidation enzymes, but it has been demonstrated that sesamin has far stronger activity at the effective concentration. It has also been shown that there exists a significant synergistic effect between sesamin and fish oil for increasing hepatic fatty acid oxidation through upregulation of the gene expression of peroxisomal fatty acid oxidation enzymes (Ide et al., 2004).

Another study investigated the effect of sesamin on hepatic fatty acid synthesis in rats, in relation to the activity and gene expression of fatty acid synthesis enzymes such as acetyl-CoA carboxylase, fatty acid synthase, ATP-citrate lyase, and glucose-6-phosphate dehydrogenase (Ide et al., 2001). Rats were fed diets containing 0–0.4%

sesamin for 15 days. The 0.2% sesamin diet lowered these parameters to about one-half those in the sesamin-free control, but no further reduction was found in animals fed the 0.4% sesamin diet. Dietary sesamin lowered the sterol regulatory element binding protein-1 (SREBP-1) (related to the nutritional and physiological regulation of hepatic fatty acid synthesis) and mRNA level dose-dependently. The levels in rats fed a 0.4% sesamin diet were about one-half that in rats fed a sesamin-free diet.

As shown in Figure 10.7(II), the protein content of the membrane-bound precursor form of SREBP-1 decreased as dietary sesamin increased. Diets containing sesamin lowered the amount of the mature nuclear form of SREBP-1 protein to less than one-fifth that of the control. From these data, the decrease in lipogenic enzyme gene expression by sesamin may be caused by suppression of the gene expression of SREBP-1 and proteolysis of the precusor form to the mature form (Ide et al., 2001).

In relation to the effect of sesamin on decreasing the concentration of serum triacylglycerol, the potential body fat-reducing effect of sesamin may be expected by daily intake of sesame seed. It was demonstrated in an experiment using rats that sesamin in combination with conjugated linoleic acid, which has been evaluated as potential body fat-reducing substance, has a significant effect on the suppression of adiposity (Sugano et al., 2001; Akahoshi et al., 2002).

10.6.4 Differences in Physiological Effects between Stereo and Optical Isomers of Sesamin

Sesamin is naturally present in sesame seed, and episesamin is a geometrical isomer of sesamin produced in the refining process of unroasted sesame seed oil. The commercially available sesamin preparation is a 1:1 mixture of sesamin and episesamin by weight. Previous studies used this mixture, but it was not clear whether the two compounds had the same physiological properties. Several studies address this question.

Then, when a 1:1 mixture of sesamin and episesamin was administered, their concentrations in various tissues reached a plateau at about 3–6 h and then rapidly decreased. Although the rates of lymphatic transport of sesamin and episesamin were essentially the same, the concentrations of episesamin in all tissues and serum were more than twice that of sesamin 1–9 h after administration. These results suggest that sesamin and episesamin are absorbed in basically the same manner in the lymph, but sesamin is metabolized faster in the liver (Umeda-Sawada et al., 1999).

Differences in the two compounds in regarding cholesterol metabolism in serum and liver in rat were also observed (Ogawa et al., 1995). A study on the comparative effect of sesamin and episesamin on the activity and gene expression of ezymes in fatty acid oxidation and synthesis in rat liver was carried out with rats fed diets containing no sesamin, 0.2% sesamin, or 0.2% episesamin, for 15 days shown that as far as increasing the mitochondrial and peroxisomal palmitoyl-CoA oxidation rates, the effect of episesamin was about 1.5- to 3-fold greater than that of sesamin, and similar difference was also observed for the increase in the activity and gene expression of various fatty acid oxidation enzymes (Kushiro et al., 2002).

However, it was demonstrated that these differences in hepatic fatty acid metabolism observed between sesamin and episesamin in rats were not observed in the cases of mice and hamsters. In addition, there was considerable diversity in the specificity of the enzyme reaction toward sesamin and episesamin among animal species (Kushirto et al., 2004).

As noted above, sesame seed specifically inhibits Δ5 desaturation enzyme activity. Similar activities were also observed with peanut oil (Shimizu et al., 1989c), ethanol extract of turmeric (Shimizu et al., 1992b) and *Asiasari radix* (Shimizu et al., 1992a), although their activities were weaker than that of sesame seed. The activities of the lignans contained in sesame seeds were stronger in the order sesamin > sesaminol > episesamin, whereas sesamol showed no activity. Another study showed that (–)-asarinin and (–)-epiasarinin, which are enantiomers of (+)-episesamin and (+)-sesamin, respectively, and contained in *Asasari radix*, effectively inhibited the activity of the Δ5 desaturation enzyme of liver microsomes in rats (Shimizu et al., 1992a). The degree of inhibition was in the order (+)-sesamin > (–)-epiasarinin > (–)-asarinin > (+)-episesamin. Further, it was found that the inhibition mechanism of the Δ5 desaturation enzyme with sesamin and episesamin were noncompetitive, regardless of (+)-body or (–)-body. Sesamin is epimerized during the deodorization step in sesame oil refining, and a small amount of diasesamin, a third isomer of sesamin, is obtained. In the test using *M. alpina* 1S-4, diasesamin showed no inhibiting activity on Δ5 desaturation enzyme and no influence on the ratio of DGLA to AA. Similarly, it showed no activity in the PUFA synthesis by liver microsomes in rats.

The effect of the structure of lignan on the activity of inhibiting Δ5 desaturation enzyme was investigated for various sesamin-related compounds having a methylenedioxyphenol group or 3,7-dioxabicyclo[3.3.0]octane structure. No appreciable activity was found for any of them, suggesting that the methylenedioxyphenol group and dioxabicyclo [3.3.0] octane structure and its three-dimensional structure play an important role in the manifestation of the activity of sesamin. A similar specific inhibition of Δ5 desaturation enzyme was also observed with curcumin (Kawashima et al., 1996a) and propyl gallate (Kawashima et al., 1996b), although their activities were weak compared with sesamin. They also showed inhibition of Δ6 desaturase.

In comparative experiments on hepatic fatty acid metabolism between sesamin and sesamolin, it was shown that the increase in the activity of fatty acid oxidation enzymes was much greater with sesamolin than with sesamin; sesamolin accumulated far more in serum as well as in the liver than did sesamin. Thus, sesamolin can account for the potent physiological effects of sesame seeds in increasing hepatic fatty acid oxidation. However, sesamin, as compared with sesamolin, was more effective in reducing serum and liver lipid levels despite sesamolin increasing hepatic fatty acid oxidation more strongly (Lim et al., 2007).

These interesting results concerning changes in the physiological activities of various stereoisomers of sesame lignans will afford important information on the mechanism of the various activities of lignans in general. However, changes in the physiological activity of the artifacts formed during food processing should also be investigated from a food safety viewpoint.

10.7 HYPOCHOLESTEROLEMIC ACTIVITY OF SESAME LIGNANS

The effects of sesame seed oil in lowering serum cholesterol concentration and inhibiting the absorption of cholesterol in lymph have been noted (Koh, 1987; Satchithanandam et al., 1993). The results obtained could not be interpreted only from the composition of fatty acids in sesame oil, particularly the content of linoleic acid and P/S (polyunsaturate to saturate) ratio, suggesting the presence of an active ingredient other than the fatty acids. In fact, it was shown that sesamin has the effect of lowering serum cholesterol in rats independent of the addition of cholesterol into the diets (Sugano et al., 1990; Hirose et al., 1991). Similar results were obtained using normocholesterolemic and hypercholesterolemic stroke-prone spontaneously hypertensive rats (n-SHRSP and h-SHRSP, respectively) (Ogawa et al., 1995). In n-SHRSP, both sesamin and episesamin significantly increased the concentration of total cholesterol in serum by increasing apoE-HDL, and they also effectively decreased serum VLDL. In h-SHRSP rats fed high-fat and high-cholesterol diets, only episesamin improved serum lipoprotein metabolism with an increase in apoA-1 and a decrease in apoB. In the liver, both sesamin and episesamin significantly suppressed cholesterol accumulation. Only episesamin significantly increased the activity of microsomal cholesterol 7α-hydroxylase.

These results indicate that sesamin may be effective in preventing cholesterol accumulation in the liver, and episesamin may be more effective than sesamin in the regulation of cholesterol metabolism in serum and liver. This effect was also obtained in hamsters that are more suitable model animals for the study of cholesterol metabolism (Ogawa et al., 1994). In hamsters, the deposition of cholesterol in the liver for long periods of growth was prevented by sesamin. Sesaminol is also effective in protecting human LDL against oxidation *in vitro* (Kang et al., 2000). It inhibited also Cu^{2+}-induced lipid peroxidation in LDL dose dependently.

A synergistic effect of sesamin and α-tocopherol on lowering the concentration of serum cholesterol was observed (Nakabayashi et al., 1995). In the presence of α- and γ-tocopherols, sesamin lowered the concentrations of total cholesterol, VLDL + LDL and HDL, in plasma and liver of rats (Kamal-Eldin et al., 2000). Such synergistic action was also observed in humans (Hirata et al., 1996). Using patients with hypercholesterolemia as the control, sesamin and α-tocopherol were administered daily for 4 weeks. As a result of fasting, total plasma cholesterol and LDL-cholesterol concentrations decreased significantly. In addition, the concentration of ApoB, which is an essential apoprotein of LDL, also decreased. Sesamin lowers LDL cholesterol, which is a risk factor for arteriosclerosis, and inhibits the formation of foam cells by synergistic action with tocopherol in humans. These results indicate that sesamin plays a preventive role against arteriosclerosis.

The lowering effect of sesamin on cholesterol concentration may occur mainly through inhibition of the absorption of cholesterol from the intestine (Hirose et al., 1991), since it can selectively hinder the dissolution of cholesterol into micelles of cholic acid and increase the discharge of cholesterol into the feces. Here, sesamin does not influence the absorption of fat or the discharge of cholic acid through the feces. It lowers the concentrations of cholesterol and triacylglycerols in the liver. Interestingly, no acceleration of synthesis of cholesterol in the liver was observed when absorption of cholesterol was inhibited by sesamin, and in fact levels of synthesis were reduced (Hirose et al., 1991). Such an effect of sesamin is very specific, and it is possible that a combination of several mechanisms may be involved. The action of inhibiting cholesterol synthesis has also been observed in smooth muscle fibers in the artery (Umeda-Sawada et al., 1994).

Another mechanism, the inhibition of the activity of the Δ5 desaturation enzyme with sesamin, may influence the cholesterol mechanism because some eicosanoids (e.g., PGE_1) are known to have the function of controlling cholesterol metabolism. With an intake of sesamin, dihomo-γ-linolenic acid (DGLA) may accumulate in the liver and the production of prostaglandins in a series may be accelerated.

10.8 ENHANCEMENT OF LIVER FUNCTIONS BY SESAME

10.8.1 Acceleration of Alcohol Decomposition in the Liver

In Japan, it has traditionally been known that taking sesame foods such as sesame paste before drinking effectively prevent drunken sickness. It was elucidated by the following experiments in which a relatively large amount of alcohol was given to rats fed sesamin, the concentration of alcohol in the blood was little affected, while the rate of reducing the blood alcohol content was accelerated. This may have occurred because the enzyme activity in the alcohol metabolism system of the microsomes or peroxisomes increased with sesamin. It was also observed that the intake of 1% sesamin significantly suppressed ethanol-induced increases in the fat decomposition in the liver and also in the markers of liver function such as serum GOT and GPT (Akimoto et al., 1993).

On the basis of the findings of animal studies, in adult male humans deficient in alcohol dehydrogenase II, the effect of sesamin on the change in facial temperature as a result of drinking alcohol was studied. When 100 mg a day of sesamin was given for one week in advance, the decomposition of alcohol was more rapid compared with the control group. Thus, although sesamin had little effect on the rise in temperature on the face induced by alcohol intake, the rate of lowering temperature was clearly faster beginning about 30 min after intake. The other effects of alcohol intake such as respiratory effects and fluctuation of heart beat (suppression of lowering activity in the parasympathetic nerve) were reduced. In rats, sesamin lowered muscle relaxation caused by the intake of alcohol in a dose-dependent manner, and simultaneously accelerated recovery from relaxation (Yang et al., 1995). These results indicate that sesamin does not affect the absorption of alcohol, but increases decomposition of alcohol in the liver and reduces the acute toxicity of acetaldehyde that is an oxidation product of alcohol.

To elucidate the mechanism of these actions, the profiles of gene expression of the liver in rats given sesamin or vehicle were compared using DNA microarray analysis (Tsuruoka et al., 2005). It was shown that the ingestion of sesamin upregulated the genes coded for proteins involved in the metabolism of xenobiotic/endogeneous substances. Interestingly, sesamin increased the transcription of the gene of aldehyde dehydrogenase (ALDH1A7) about threefold, whereas it had no influence on the gene expression for other alcohol-metabolizing enzymes (Kiso, 2004). The toxic acetaldehyde is usually metabolized by ALDH of class 1 in cytoplasm or class 2 in mitochondria, with the former being the main metabolizer. The fact that sesamin specifically increased the class 1 ALDH is important in the detoxification of acetaldehyde. Sesamin thus appears to have a good influence on various functions induced by the ingestion of alcohol.

10.9 ANTIHYPERTENSIVE ACTIVITY OF SESAMIN

10.9.1 On Experimental Hypertensive Rat

Antihypertensive activity of sesamin was examined using DOCA-saline hypertensive model rats prepared by removing the right kidney of Sprague–Dawley (SD) rats followed by subcutaneous injection of a mineral corticosteroid, deoxycorticosterone acetate (DOCA) and 1% saline. Compared with the control group, the blood pressure of the DOCA-saline group rose while the rise in the blood pressure of the sesame group was effectively suppressed. Hypertrophy in heart and blood vessels observed in the DOCA-saline group was also significantly suppressed in the sesamin (1%) group, and the prevention of increase in blood pressure with sesamin was proved (Matsumura et al., 1995). The 2K, 1C type renal hypertensive models prepared by attaching silver clips on the left renal arteries of SD strain male rats also showed similar preventive action due to sesamin (Kita et al., 1995). In these models, the condition of hypertension occurs by promoting higher production of angiotensin II due to excessive secretion of renin from the left kidney having constricted renal arteries and increasing the contraction of smooth muscles.

In another study, altered vascular reactivity in aortic rings of DOCA-salt-induced rats improved with sesamin (Matsumura et al., 2000). The systolic blood pressure after 5 weeks of DOCA-salt treatment was much higher than that of sham-operated control animals. Acetylcholine (Ach)-induced endothelium-dependent relaxation of aortic rings was markedly decreased in the DOCA-salt hypertensive animals, compared with the control. These changes were significantly improved by sesamin feeding in the same manner.

In correlation to hypertension, the effect of sesamin on endothelial production of NO and endothelium-1 (ET-1) in the human umbilical vein endothelial cells (HUVECs) was investigated. Sesamin increased the NO concentration in the medium of HUVECs through induction of mRNA and protein expressions of NO synthase, while it lowered the ET-1 concentration in the medium through the inhibition of endothelium converting enzyme-1. These results suggest the potent ability of sesamin to improve hypertension (Lee, C.C. et al., 2004).

10.9.2 On Stroke-Prone Spontaneously Hypertensive Rats

More than 95% of stroke-prone spontaneously hypertensive rats (SHRSP) suffer from stroke lesion in the development of hypertension. Since it has been found that peroxidation of lipid plays an important role in the disorder of brains of the SHRSP, the effect of sesamin has been studied using these rats. Sesamin (0.5%) had no significant suppressive effect on blood pressure, but in several aspects it had some positive influence. Increase in brain weight and the crisis of stroke lesions were not observed in the sesamin group.

The rise of the level of uric acid in the cerebral cortex was suppressed, preventing decomposition of hypoxanthine and production of free radicals. The rise in Na/K-ATPase activity, which is considered to be a marker of disorder of cell membrane in the cerebral cortex, was such that the disorder of cell membranes would not be serious. From these results, it is believed that sesamin prevents ischemia by promptly scavenging free radicals (Murakami et al., 1995).

Among the rats in the experimental hypertensive model discussed above, sesamin effectively suppressed the rise of blood pressure in the DOCA–saline group. In an investigation on the effect of sesamin on SHRSP rats (Matsumura et al., 1998), significant suppression in the elevation of blood pressure and the formation of hypertrophy in heart and blood vessels was observed in the sesamin group. When the effect of sesamin on renovascular lesions was examined by histopathological investigation, both control and sesamin groups showed hypertrophy of the inner membranes of arterioles, but the number in the sesamin group was significantly smaller than in the control group.

A similar study was done using the same SHRSP (Noguchi et al., 2001). A closed cranial window was created and platelet-rich thrombi were induced *in vivo* using a helium–neon laser technique. In control rats, systolic blood pressure and the amount of urinary 8-hydroxy-2′-deoxyguanosine (8-OHdG) became significantly elevated with aging. However, these symptoms were significantly suppressed in rats administered vitamin E, sesamin, and especially vitamin E + sesamin. The number of laser pulses required to induce an occlusive thrombus in arterioles in the control group was significantly lower than in the other groups.

These results indicate that chronic ingestion of vitamin E and sesamin attenuated elevation in blood pressure, oxidative stress, and thrombotic tendency, suggesting that these treatments might be beneficial in the prevention of hypertension and stroke.

10.9.3 Angiotensin 1-Converting Enzyme Inhibitory Peptide in Sesame Protein

A noticeable fact is the finding of sesame peptide powder (SPP) which is effective in inhibiting the angiotensin 1-converting enzyme (ACE), and significantly and temporarily decreased the systolic blood pressure in spontaneously hypertensive rats (SHRs) by single administration (1 and 10 mg/kg), and six effective peptides were isolated and identified from SPP. The representative ones Leu-Val-Tyr, Leu-Gln-Pro, and Leu-Lys-Tyr, could competitively inhibit ACE activity at respective Ki values of 0.92, 0.50, and 0.48 μM. Repeated oral administration of SPP also lowered both SBP and the aortic ACE activity in SHR. These results demonstrate that sesame peptide powder would be a beneficial ingredient for preventing, and providing therapy against, hypertension and its related diseases (Nakano et al., 2006).

10.10 IMMUNOREGULATORY ACTIVITY

10.10.1 Effect of Sesamin on Chemical Mediator and Immunoglobulin Production

Sesamin affects the metabolism from linoleic acid to arachidonic acid and furthers the production of eicosanoids through inhibition of the Δ5 desaturation enzyme. The effects of sesamin have been studied for various factors, which may be related to food allergies in which their onset is linked with many chemical transmitters, including eicosanoids.

It is known that α-tocopherol acts on the arachidonic acid cascade to interfere with the production of eicosanoids. Therefore, by the synergistic action with sesamin, the balance of eicosanoids produced may be influenced and a possible immunofunctional effect may also be observed. In an experiment using rats sensitive to immunoresponse (Brown–Norway), simultaneous intake of sesamin and α-tocopherol (0.5% each) lowered the proportion of arachidonic acid to phospholipid in the tissue and inhibited the production of LTC_4 (a typical mediator in the lungs) (Gu et al., 1994). Even in the case of common rats (Sprague–Dawley), it induced not only a decrease in the production of LTC_4 in the lungs by reduction of arachidonic acid, but it also decreased the production of LTB_4 in the spleen. The concentration of histamine in the plasma may also be reduced (Gu et al., 1995). At the same time, a tendency to increase the concentrations of IgA, IgG, and IgM in the plasma and to decrease that of IgE was documented. The effect of the production of immunoglobulin may be related to the increase in the production of helper T-cells and to the decrease in the production of suppressor T-cells in lymphocytes in the spleen. Thus, the simultaneous intake of sesamin together with α-tocopherol is effective for the suppression of the chemical mediator as well as the inhibition of food allergies by controlling the production of immunoglobulins.

10.10.2 Effects of Sesame Lignans on Production of Eicosanoid and Interleukin

In response to an intraperitoneal injection of a lethal dose of lipopolysaccharide endotoxin in mice, all control animals died within 48 h, but 40% of the animals fed sesame oil survived, and the survival ratios were 27% and 50% in those fed Quil-A (a saponin potentiating the immune response) supplemented control or sesame oil diet, respectively. In these experiments, sesame oil and Quil-A reduced IL-1β, PGE_1, PGE_2 and thromboxane B_2, and elevated IL-6 and 10, which are associated with a marked increase in survival in mice (Chavali et al., 1997).

The effect of sesame oil (5% in the diet) on survival after cecal ligation and puncture in mice was studied (Chavali et al., 2001). Four days after the treatment, the survival rate was 20% in the controls, while it was about 65% in the sesame oil group. The levels of cytokine and dienoic eicosanoids were measured in response to an intraperitoneal injection of a nonlethal dose (50 μg/mouse) of endotoxin. IL-10 levels were markedly higher in the sesame oil group than in the control group. Plasma concentrations of PGE_1, PGE_2, TNF-α, IL-6, and -12 did not differ significantly between the two groups. From these data, sesamin, sesamol, and other lignans in sesame oil may be responsible for the increase in survival after the operation, and also for the increase in the IL-10 levels in response to a nonlethal dose of endotoxin in mice.

It was shown that chronic ethanol drinking significantly increased the plasma IgA and IgM levels, irrespective of the presence of 0.1% or 0.2% sesaminol, but the effects disappeared with 0.2% sesamin. The levels of IgG significantly increased, but the levels of plasma IgE were not affected by the dietary manipulation. Although ethanol drinking did not influence spleen leukotriene B_4 production, the addition of sesaminol tended to decrease its dose dependently, while the addition of sesamin increased the plasma PGE_2. These results suggest that sesaminol is favorable for ameliorating the disturbed immune circumstances created by ethanol, and sesaminol and sesamin seem to have different effects on the plasma levels of immunoglobulins and eicosanoids (Nonaka et al., 1997).

10.11 PREVENTION OF CANCER AND OXIDATIVE STRESS *IN VIVO* BY SESAME LIGNANS

Much attention is focused on food antioxidants, including sesame lignans, as having a potential anticancer function. In the growth of breast cancer induced by a chemical carcinogenic substance (dimethylbenzanthracene [DMBA]) in rats, a significant suppressive effect equivalent to that of rats administered α-tocopherol was observed when fed a sesamin-containing diet (0.2%) (Hirose et al., 1992).

In studies on the development of colon precancerous lesions induced by the azoxymethane (AOM), the effect of sesaminol glucoside (SG, 500 ppm) was investigated. The dietary SG significantly decreased the incidence of AOM-induced lesions, and also decreased the serum triacylglycerol level and mRNA expression of intestinal fatty acid-binding proteins in the colonic mucosa, as compared with the control, which indicated that the dietary SG inhibits AOM induced carcinogenesis and SG is useful as a possible chemopreventing agent (Sheng et al., 2007).

Although sesamin exhibited the same antioxidation value (as TBA value) as tocopherol, the activity of peripheral blood mononuclear cells increased significantly and the concentration of plasma PGE_2 decreased, in contrast to rats given tocopherol. Those effects are considered to be a mechanism of cancer prevention with sesamin and at the same time a key to the investigation of the stimulation of the immune mechanism. However, in the case of pancreatic cancer induced by a chemical carcinogen, the effect of sesamin was not clear (Ogawa et al., 1994). Test conditions such as the kind of chemical carcinogens and the level of sesamin intake should be studied.

In relation to the side effects induced by cisplatin, an effective drug for the tumor therapy, sesame oil attenuated potentially cisplatin-associated hepatic and renal injuries, and reduced the cisplatin-induced lipid peroxidation as well as the production of hydroxyl radicals, peroxynitrite, and nitrite in blood and tissue, and also it did not affect the antitumor capacity of cisplatin in mice with melanoma. Thus, sesame oil might provide a new approach for preventing cisplatin-induced multiple organ injury during the treatment of tumors (Hsu et al., 2007).

On the mechanism of the growth inhibitory effect of sesamin on human cancer cells, it was shown that sesamin induced downregulation of cyclin D1 protein expression through the activation of proteasome degradation (Yokota et al., 2007).

The effect of sesame lignans on the growth of human lymphoid leukemia Molt 4B cells was examined (Miyahara et al., 2001). Growth was inhibited with increasing concentration of lignan accompanied by the fragmentation of DNA into oligonucleosomal-size fragments. These are characteristic of apoptosis, which was observed to depend on lignan concentration and incubation time. Among the sesame lignans investigated, sesaminol was most effective, followed by sesamin, sesamolin, and episesamin. No induction of apoptosis was observed by sesame lignans in normal leukocytes prepared from healthy volunteers. These results suggest that sesame lignans may exert antitumor activity by triggering apoptosis.

Interestingly, the callus induced from the sprouts of sesame seeds grows fast even at temperatures as high as 35°C (Mimura et al., 1994), and methanol-water extracts of it yielded some antioxidative products involving esculentic acid and its coumaric acid derivative. The 100% methanol extracts (M-100) were effective in the anticarcinogenesis test using DMBA as the initiator and phorbol ester (TPA) as the promoter, and determined by the incidence of skin papilloma (initial stage of skin cancer). Application of M-100 at 30 min after phorbol ester decreased the incidence to 40% of the control, and was also effective in a similar experiment using ultraviolet light-B as the promoter.

10.12 OTHER FUNCTIONS OF SESAME SEED AND LIGNANS

10.12.1 Hypoglycemic Action of Sesame

The hypoglycemic action of sesame was found in an experiment using genetically diabetic (type II) KK-Ay mice fed a basal diet (BAS), a diet containing 4.0% hot-water extract from defatted sesame seed (HES), 1.4% of the water eluate fraction of HES (WFH), or 0.7% of the methanol eluate fraction of HES (MFH) from a HP-20 resin column (Takeuchi et al., 2001). After four weeks, the bovine serum albumin (BSA) group was divided into the MAL and MALH groups which were fed 1 mL per mouse of a 20% maltose solution with or without 4.0% HES, respectively. The plasma glucose concentration and amount of excreted urinary glucose were lower in the HES and MFH groups than in the BAS and WFH groups. The levels of plasma glucose and serum insulin were lower in the MALH group than in the MAL group. The hot-water extract group and the methanol eluate group with the defatted sesame showed decreased glucose concentration in plasma. This effect may be caused by delay in glucose absorption.

10.12.2 Prevention of Alzheimer's Diseases

Alzheimer's disease is said to be associated with the accumulation of β-amyloid (Aβ) in some part of the brain, generating reactive oxygen species and elevating intercellular calcium. In an experiment on the protective effect in Aβ-induced cell death in cultured rat pheochromocytoma (PC12) cells, sesaminol glucosides completely suppressed Aβ-induced generation of active oxygen species, formation of 8OdG from DNA, and elevation of calcium level concomitant with prevention of cell death and expression of apoptotic gene. These results suggest that sesaminol glucosides could be a useful therapeutic agent in the treatment of oxidative stress-induced neuronal degradation diseases (Lee S. Y. et al., 2005).

10.12.3 Antithrombosis Activity

Recently, the antithrombosis activity of food components is becoming the focus of attention as an important food factor in preventing myocardial and brain infarctions. Related to the traditionally believed health effect of the flavor of deep-roasted sesame seed oil, the effect on platelet aggregation of sesame flavor concentrates collected in water from a deep-roasted sesame seed oil factory was investigated. This activity was determined by the inhibitory effect on human blood platelet aggregation induced by collagen and measured by turbidimetry. The results indicated that the ether extracts of the concentrates, especially at alkaline pH, showed marked inhibitory activity. It was further demonstrated that alkylpyrazines, such as 2,3,5-trimethyl pyrazine, known as the main characteristic components of the deep-roasted sesame oil flavor, have very strong inhibitory activities comparable to that of the standard specimen, aspirin (Namiki et al., 2002).

In studies on the antithrombosis effect of various sesame seed and flour specimens by the fluidity test of human whole blood and the inhibition of platelet aggregation, it was shown that common black and white sesame seeds and defatted sesame flour had no significant activities in these tests, but the sesame flours cultured with a fungus (*Aspergillus niger*) or treated with hydrolysis enzymes exhibited strong activities in these tests (Koizumi et al., 2007). By HPLC analysis on these specimens, it was demonstrated that sesaminol produced from its glucosides in the flour by the fungus or the enzymes showed strong activity in both tests while sesamin and sesamol showed only weak activity in the platelet aggregation test (Namiki, 2006).

10.12.4 Mammalian Lignan

Relating to the fact that the incidence and mortality of many chronic diseases, such as breast, prostate, and colorectal cancer, as well as cardiovascular diseases are high in Western countries compared with those of Asia, the physiological effects of various phytochemical problems in the food styles were investigated, involving effects of intestinal microflora and fermentation activities.

Among various phytochemical problems in the food styles, it was noted that the so-called mammalian lignans, enterodiol, and enterolacton (Figure 10.2) detected in women's urine, are higher in the much more vegetable-eating Asian countries compared with the animal meat-eating Western countries, and may affect cancer incidence (Adlercreutz, 2007).

These mammalian lignans are confirmed to be produced from plant polyphenolic components, lignans, and isoflavones, through metabolic changes by internal bacteria. As to the lignans, sesamin in sesame seed and secoisolariciresinol in flaxseed are by far the highest in lignan content among various grains, vegetables, and fruits (Penalvo et al., 2005a).

At present, they are thought to have no hormonal effects, but to possess weak estrogenic and antiestrogenic activities that may protect against hormone-dependent diseases such as breast and prostate cancer and coronary heart diseases.

Concerning the effect of mammalian lignans, human experiments on the effects of various physiological conditions by sesame ingestion was undertaken by having 26 healthy postmenopausal women eat 50 g of sesame seed powder daily for five weeks. After the experiment, plasma total cholesterol (TC), LDL-C, LDL/HDL ratio, TBA reaction substance in oxidized LDL, and serum dehydroepiandrosterone sulfate decreased significantly by 5%, 10%, 6%, 23%, and 18%, respectively (Wu et al., 2006).

These results suggest that sesame ingestion benefits postmenopausal women by improvement of blood lipids, antioxidant status, and sex hormone status.

10.12.5 Sesame Allergy

Sesame food allergy (SFA) in children is being increasingly recognized, especially in the developing countries of Europe (Gangur et al., 2005), and also appears to be increasing in Asia in recent years (Chiag et al., 2007). Compared with the much higher incidence of allergies to animal foods such as egg and milk, plant food allergies are lower. The incidence rate of allergy to peanuts is highest, followed by sesame seed, and there seems to be some cross correlation between the two (Wallowitz et al., 2007). SFA is an emerging food allergy of a serious nature due to the high risk of systemic anaphylaxis. Recently, the 2S albumin allergen, Ses i 1 (Moreno et al., 2005), and the 11s globulins allergen, Ses i 3 (Navuluri et al., 2006) were purified from sesame seeds. The allergens were shown to be thermo-stable up to 90°C and highly resistant to digestion (Moreno et al., 2005).

However, as is well known, food allergies depend not only on a particular allergen but are also greatly influenced by various functions involving life-style customs, individual health conditions, and genetic factors (Bjoksten, 2005). A further systematic study of SFA and various related fields is required.

In another study, the products that inhibit allergen absorption through the intestinal tract were isolated from the enzyme hydrolysates of defatted black sesame. A tetradeca peptide and two kinds of sesaminol glucosides were identified as active components in an *in vitro* method using Caco-2 cells (Kobayashi et al., 2004).

10.12.6 Differences in Action of Sesame Seed by Animal Species

In some cases, the physiological action of sesame seed differs by animal species, for example, mouse and rat. The elevating effect on tocopherol concentration of sesame seed was relatively lower in mice than in rats (Ikeda et al., 2002a). Species differences were also observed in fatty acid metabolism, that is, significant increases in fatty acid oxidation and strong downregulations of lipogenic enzymes were observed only in rats and not so prominently in mice and hamsters. A difference was also observed in the specificity of enzyme activities toward sesamin and episesamin (Kushiro et al., 2004).

10.13 CONCLUDING REMARKS

As the magic words, "Open Sesame!" indicate, many excellent properties are hidden in the tiny sesame seed and these have been made clearer by the studies described above. Traditionally believed antiaging effect of sesame was elucidated to be due to the elevated vitamin E activity caused by a novel mechanism involving genetic inhibition of tocopherol decomposition enzymes. Sesame lignan accelerated fatty acid oxidation and suppression of fatty acid synthesis by gene expression, and also control cholesterol concentration in serum due to the inhibition of absorption from the intestine and suppression of synthesis in the liver.

Various useful functional activities of sesame were also clarified such as acceleration of alcohol decomposition, antihypertensive activity, immunoregulatory activities, prevention of cancer and oxidative stress, hypoglycemic action, prevention of Alzheimer's disease, and antithrombosis activity.

It was also elucidated that the main functional components, sesamin and sesaminol glucoside, were transformed into strong antioxidative products by metabolism with intestinal microorganisms or some enzymes in liver, that is, sesamin to free catechol-type compound and sesaminol glucosides change into antioxidative free sesaminol. Involving these metabolic changes, the mode of action of sesame lignans in various unique functional activities seems distinct from that of other phenolic functional products such as tea catechin and others (Hara, 2001).

All this important information presented herewith on the superior functionality of sesame will strongly promote the use of sesame seed in the daily diet worldwide. There are various forms and methods of using sesame seed and oil in Asian countries, and people in these countries enjoy many kinds of foods containing sesame seed and oil with superb taste and flavor. However, the use of sesame in Western countries is limited in variety and sesame is utilized mostly as topping on bread and biscuits, with low consumption. In this respect, it will be necessary to develop various sesame foods, which will suit many people's tastes throughout the world. For example, one recommended form of sesame may be used in salad dressing or seasoning containing ground sesame seed and oil, which can be used with various vegetables. This use of sesame is delicious in taste and has good digestibility with high nutritional value in combination with sesame lignans and various vegetable components.

In such ways, it is recommended that a person ingest at least three grams, preferably 10 g, per day of sesame seed and oil for promotion of health and prosperity of people throughout the entire world.

REFERENCES

Abe, C., Ikeda, S., and Yamashita, K. 2005. Dietarysesame seedselevate α-tocopherol concentration in rat brain. *J. Nutri. Sci. Vitaminol.*, **51**: 223–230.

Adlercreutz, H. 2007. Lignan and human health. *Crit. Rev. Clin. Lab. Sci.*, **44**: 483–525.

Akahoshi, A. et al. 2002. Conjugated linoleic acid reduces body fats and cytokine levels of mice. *Biosci. Biotech. Biochem.*, **66**: 916–920.

Akimoto, K. et al. 1993. Protective effects of sesamin against liver damage caused by alcohol or carbon tetrachloride in rodents. *Ann. Nutr. Metab.*, **37**: 218–224.

Ashakumary, L. et al. 1999. Sesamin, a sesame lignan, is a potent inducer of hepatic fatty acid oxidation in the rat. *Metabolism*, **48**: 1303–1313.

Bieri, J. G. and Evarts, R. P. 1974. Vitamin E activity of γ-tocopherol in the rat, chick, and hamster. *J. Nutr.*, **104**: 850–857.

Bjoksten, B. 2005. Genetic and environmental risk factors for the development of food allergy. *Curr. Opin. Allergy Clin. Immunol.*, **5**: 249–253.

Brito, O. J. and Nunez, N. 1982. Evaluation of sesame flour as a complementary protein source for combinations with soy and corn flours. *J. Food Sci.*, **47**: 457–460.

Budowsky, P. 1964. Recent research on sesamin, sesamolin, and related compounds. *J. Am. Oil Chem. Soc.*, **41**: 280–285.

Budowsky, P. and Narkley, K. S. 1951. The chemical and physiological properties of sesame oil. *Chem. Rev.*, **48**: 125–151.

Chatrattanakunchai, S., Fraser, T., and Stobart, K. 2000. Sesamin inhibits lysophosphatidylcholine acyltransferase in *Mortierella alpina. Biochem. Soc. Trans.*, **28**: 718–721.

Chavali, S. R. and Forse, R. A. 1999. Decreased production of interleukin-6 and prostaglandin E2 associated with inhibition of Δ-5 desaturation of ω6 fatty acids in mice fed safflower oil diets supplemented with sesamol. *Prostaglandins Leukot. Essent. Fatty Acids*, **61**: 347–352.

Chavali, S. R. et al. 1997. Decreased production of interleukin-1-beta, prostaglandin-E2 and thromboxane-B2, and elevated level of interleukin-6 and -10 are associated with increased survival during endotoxic shock in mice consuming diets enriched with sesame seed oil supplemented with Quil-A saponin. *Int. Arch. Allergy Immunol.*, **114**: 153–160.

Chavali, S. R., Utsunomiya, T., and Forse, R. A. 2001. Increased survival after cecal ligation and puncture in mice consuming diets enriched with sesame seed oil. *Crit. Care Med.*, **29**: 140–143.

Chavali, S. R., Zhong, W. W., and Forse, R. A. 1998. Dietary alpha-linolenic acid increases TNF-α, and decreases IL-6, IL-10 in response to LPS: Effects of sesamin on the Δ-5 desaturation of ω6 and ω3 fatty acids in mice. *Prostaglandins Leukot. Essent. Fatty Acids*, **58**: 185–191.

Chiag, W. C. et al. 2007. The changing face of food hypersensitive in an Asian country. *Clin. Exp. Allergy*, **37**: 1055–1061.

Cooney, R. V. et al. 2001. Effects of dietary sesame seeds on plasma tocopherol levels. *Nutr. Cancer*, **39**: 66–71.

Fujiyama-Fujiwara, Y., Umeda, R., and Igarashi, O. 1992. Effects of sesamin and curcumin on Δ5-desaturation and chain elongation of polyunsaturated fatty acid metabolism in primary cultured rat hepatocytes. *J. Nutr. Sci. Vitaminol.*, **38**: 353–363.

Fukuda, N. et al. 1998. Reciprocal effects of dietary sesamin on ketogenesis and triacylglycerol secretion by the rat liver. *J. Nutr. Sci. Vitaminol.*, **44**: 715–722.

Fukuda, N. et al. 1999. Effect of dietary sesamin on metabolic fate of an exogenous linolelaidic acid in perfused rat liver. *J. Nutr. Sci. Vitaminol.*, **45**: 437–448.

Fukuda, Y. et al. 1985a. Studies on antioxidative substances in sesame seed. *Agric. Biol. Chem.*, **49**: 301–306.

Fukuda, Y. et al. 1986a. Chemical aspects of the antioxidative activity of roasted sesame seed oil and the effect of using the oil for frying. *Agric. Biol. Chem.*, **50**: 857–862.

Fukuda, Y. et al. 1986b. Contribution of lignan analogues to antioxidative activity of refined unroasted sesame seed oil. *J. Am. Oil Chem. Soc.*, **63**: 1027–1031.

Fukuda, Y. et al. 1986c. Acidic transformation of sesamolin of sesame oil constituent into an antioxidant bisepoxylignan, sesaminol. *Heterocycles*, **24**: 923–926.

Fukuda, Y. et al. 1991. Antioxidative activities of fractions of components of black sesame seeds. *Nippon Shokuhin Kagaku Kogaku Kaishi*, **38**: 915–919.

Fukuda, Y. et al. 1996. Synergistic action of the antioxidative components in roasted sesame seed oil. *Nippon Shokuhin Kagaku Kogaku Kaishi*, **43**: 1272–1277.

Fukuda, Y. and Nagashima, M. 2001. Identification of a lignan antioxidant in dark brown wastewater from processing of black sesame seed. *Bull. Faculty of Education*, Shizuoka University, Natural Sci. Ser., **51**: 11–18.

Fukuda, Y. and Namiki, M. 1988. Recent studies on sesame seed and oil. *Nippon Shokuhin Kagaku Kogaku Kaishi*, **35**: 552–562.

Fukuda, Y., Osawa, T., and Namiki, M. 1985b. Studies on the enhancement in antioxidative activity of sesame seed induced by germination. *Nippon Shokuhin Kagaku Kogaku Kaishi*, **32**: 407–412.

Gangur V., Kelly C., and Navuluri, L. 2005. Sesame allergy: A growing food allergy of global proportions. *Ann Allergy Asthma Immunol*, **95**: 4–11.

Ghafoorunissa, S., Hemalatha, S., and Rao, M. V. 2004. Sesame lignans enhance antioxidant activity of vitamin E in lipid peroxidation systems. *Mol. Cell Biochem.*, **262**: 195–202.

Grougnet, R. et al. 2006. New lignans from the perisperm of *Sesamum indicum*. *J. Agric. Food Chem.*, **54**: 7570–7574.

Gu, J. Y. et al. 1994. Effect of sesamin and α-tocopherol on the production of chemical mediators and immunoglobulins in Brown-Norway rats. *Biosci. Biotech. Biochem.*, **58**: 1855–1858.

Gu, J. Y. et al. 1995. Effects of sesamin and alpha-tocopherol, individually or in combination, on the polyunsaturated fatty acid metabolism, chemical mediator production, and immunoglobulin levels in Sprague–Dawley rats. *Biosci. Biotech. Biochem.*, **59**: 2198–2202.

Hahm, T. S. et al. 2008. Effects of germination on chemical composition and functional properties of sesame (*Sesamum indicum* L.) seeds. *Bioresour Technol.* **100**(4): 1643–1647.

Hara, Y. 2001. *Green Tea: Health Benefits and Application*. Marcel Dekker, Inc., 255pp.

Hemalatha, S., Raghunath, M., and Ghafoorussa 2004. Dietary sesame oils inhibits iron-induced oxidative stress in rats. *Br. J. Nutr.*, **92**: 581–587.

Hirata, F. et al. 1996. Hypocholesterolemic effect of sesame lignan in humans. *Atherosclerosis*, **122**: 135–136.

Hirose, N. et al. 1991. Inhibition of cholesterol absorption and synthesis in rats by sesamin. *J. Lipid Res.*, **32**: 629–638.

Hirose, N. et al. 1992. Suppressive effect of sesamin against 7,12-dimethylbenz[α]anthracene-inducedrat mammary carcinogenesis. *Anticancer Res.*, **12**: 1259–1266.

Hsieh, T. J., Lu, L. H., and Su, C. C. 2005. NMR spectroscopic, mass spectroscopic, X-ray crystallographic and theoretical studies of molecular mechanics of natural products: Farformolide B and sesamin. *Biophys. Chem.*, **114**: 13–20.

Hsu, D. Z. et al. 2007. Sesame oil attenuates Cisplatin-induced hepatic and renal injuries by inhibiting nitric oxide-associated lipid peroxidation in mice. *Shock*, **27**: 199–204.

Hu, Q., Xu, J., Chen, S., and Yang, F. 2004. Antioxidant activity of extracts of black sesame seed (*Sesamum indicum* L.) by supercritical carbon dioxide extraction. *J. Agric. Food Chem.*, **52**: 943–947.

Ide, T. et al. 2001. Sesamin, a sesame lignan, decreases fatty acid synthesis in rat liver accompanying the down regulation of sterol regulatory element binding protein-1. *Biochim. Biophys. Acta.*, **1534**: 1–13.

Ide, T. et al. 2004. Interaction of dietary fat and sesamin on hepatic fatty acid oxidation in rats. *Biochim. Biophys. Acta*, **1682**: 80–91.

Ikeda, S. et al. 2002a. Increased tocopherol concentration by dietary sesame seeds: Comparison between rats and mice. *Nippon Kasei Gakkaishi*, **53**: 309–315.

Ikeda, S. et al. 2003a. Dietary sesame lignans decrease lipid peroxidation in rats fed docosahexanoic acid. *J. Nutr. Sci. Vitaminol.*, **49**: 270–276.

Ikeda, S. et al. 2003b. Dietary α-tocopherol decreases α-tocotrienol but not γ-tocotrienol concentration in rats. *J. Nutr.*, **133**: 428–434.

Ikeda, S. et al. 2007. Dietary sesame seed and its lignan increase both ascorbic acid concentration in some tissues and urinary excretion by stimulating biosynthesis in rats. *J. Nutr. Sci. Vitaminol.*, (Tokyo) **53**: 383–392.
Ikeda, S., Niwa, T., and Yamshita, K. 2000a. Selective uptake of dietary tocotrienols into rat skin. *J. Nutr. Sci. Vitaminol.,* **46**: 141–143.
Ikeda, S., Tohyama, T., and Yamashita, K. 2002b. Dietary sesame seed and its lignans inhibit 2, 7, 8-trimethyl-2(2′-carboxyethy-6-hydroxychroman excretion into urine of rats fed γ-tocopherol. *J. Nutr.*, **132**: 961–966.
Ikeda, S., Toyoshima, K., and Yamashita, K. 2001. Dietary sesame seeds elevate α- and γ-tocotrienol concentrations in skin and adipose tissue of rats fed the tocotrienol-rich fraction extracted from palm oil. *J. Nutr.*, **131**: 2892–2897.
Ikeda, S., Yasumoto-Shirato, S., and Yamashita, K. 2000. Increased vitamin E concentration in rats fed sesame seeds containing a high level of lignans. *Nippon Kasei Gakkaishi*, **51**: 1017–1025.
Ishii, Y. and Takiyama, K. 1994. Extraction of calcium, oxalate and calcium oxalate crystals from sesame seeds. *Bunseki Kagaku*, **43**: 151–155.
Jones, W. A., Beroza, M., and Pecker, E. D. 1962. Isolation and structure of sesangolin, a constituent of *Sesanum angolense. J. Org. Chem.*, **27**: 3232–3235.
Joshi, A. B. 1961. *Sesame*, 1st ed., Collier-Macmillan, London, 109pp.
Kada, K. 1978. Anti-mutagenic action of vegetable factor(s) on the mutagenic principle of tryptophan pyrolysate. *Mutation Res.*, **53**: 351–353.
Kamal-Eldin, A. et al. 2000. Effects of dietary phenolic compounds on tocopherol, cholesterol, and fatty acids in rats. *Lipids*, **35**: 427–435.
Kamal-Eldin, A. and Appelqvist, L. A. 1994a. Variation of fatty acid composition of the different acyl lipids in seed oils from four sesame species. *J. Am. Oil Chem. Soc.*, **71**: 135–139.
Kamal-Eldin, A. and Appelqvist, L. A. 1994b. Variation in the composition of sterols, and lignans in seed oils from four *Sesamum* species. *J. Am. Oil Chem. Soc.*, **71**: 149–156.
Kamal-Eldin, A. and Appelqvist, L. A. 1995. The effects of extraction methods on sesame oil stability. *J. Am. Oil Chem. Soc.*, **72**: 967–969.
Kang, M. H. 2006. Properties and commercialization of sesame oil extracted from supercritical fluids in Koria. *Sesame Newslett.*, **20**: 38–39.
Kang, M. H. et al. 1998a. Sesamolin inhibits lipid peroxidation in rat liver and kidney. *J. Nutr.*, **128**: 1018–1022.
Kang, M. H. et al. 1998b. Inhibition of 2,2-azobis(2,4-dimethylvaleronitril)-induced lipid peroxidation by sesaminols. *Lipids*, **33**: 1031–1036.
Kang, M. H. et al. 2000. Mode of action of sesame lignans in protecting low-density lipoprotein against oxidative damage *in vitro. Life Sci.*, **66**: 161–171.
Kato, M. J. et al. 1998. Biosynthesis of antioxidant lignans in *Sesamum indicum* seeds. *Phytochemistry*, **47**: 583–591.
Katsuzaki, H. et al. 1992. Structure of novel antioxidative lignan glucosides isolated from sesame seed. *Biosci. Biotech. Biochem.*, **56**: 2087–2088.
Katsuzaki, H., Osawa, T., and Kawakishi, S. 1994. Chemistry and antioxidative activity of lignan glucosides in sesame seed. In: *Food Phytochemicals for Cancer Prevention II.* Ho, C. T, Osawa, T., Hung, T., Eds.; American Chemical Society, Washington, DC, *Sym. Ser.* **547**: 275–280.
Kawashima, H. et al. 1996a. Inhibition of rat liver microsomal desaturases by curcumin and related compounds. *Biosci. Biotech. Biochem.*, **60**: 108–110.
Kawashima, H. et al. 1996b. Inhibitory effects of alkyl gallate and its derivatives on fatty acid desaturation. *Biochim. Biophys. Acta*, **1229**: 34–38.
Kim, K. S., Park, S. H., and Choug, M. G. (2006) Nondestructive determination of lignans and lignan glucosides in sesame seed by near infrared reflection spectroscopy, *J. Agric. Food Chem.*, **54**: 4544–4550.

Kiso, Y. 2004. Antioxidative roles of sesamin, a functional lignan in sesame seed, and its effect on lipid and alcohol metabolism in the liver: A DNA microarray study. *Biofactors*, **21**: 191–196.

Kita, S. et al. 1995. Antihypertensive effect of sesamin. II. Protection against two-kidney, one-clip renal hypertension and cardiovascular hypertrophy. *Biol. Pharm. Bull.*, **18**: 1283–1285.

Kobayashi, S. et al. 2004. A novel method for producing a foodstuff from defatted black sesame seed that inhibits allergen absorption. *Biosci. Biotechnol. Biochem.*, **68**: 300–305.

Koh, E. T. 1987. Comparison of hypolipemic effects of corn oil. *Nutr. Rep. Int.*, **36**: 903–917.

Koizumi, Y., Fukuda, Y., and Namiki, M. 1996. Marked antioxidative activity of sesame oils developed by roasting of sesame seeds. *Nippon Shokuhin Kagaku Kogaku Kaishi*, **43**: 689–694.

Koizumi, Y. et al. 2007. Antithrombosis effect of sesame seed and flour cultured with microorganisms. *Nippon Shokuhin Kagaku Kogaku Kaishi*, **54**: 9–11.

Kuriyama, S. and Murui, T. 1993. Effect of cellulase on hydrolysis of lignan glycosides in sesame seed by β-glucosidase. *Nippon Nougei Kagaku Kaishi*, **67**: 1701–1705.

Kuriyama, S. and Murui, T. 1995. Scavenging of hydroxyl radicals by lignan glucosides in germinated sesame seeds. *Nippon Nougei Kagaku Kaishi*, **69**: 703–705.

Kuriyama, S., Tsuchiya, K., and Murui, T. 1995. Generation of new lignan glucosides during germination of sesame seeds. *Nippon Nougei Kagaku Kaishi*, **69**: 685–693.

Kushiro, M. et al. 2002. Comparative effect of sesamin and episesamin on the activity and gene expression of enzymes in fatty acid oxidation and synthesis in rat liver. *J. Nutr. Biochem.*, **13**: 289–295.

Kushiro, M., Takahashi, Y., and Ide, T. 2004. Species differences in the physiological activity of dietary lignan (sesamin and episesamin) in affecting hepatic fatty acid metabolism. *Br. J. Nutr.*, **91**: 377–386.

Lee, C. C. et al. 2004. Sesamin induces nitric oxide and decreases endothelium-1 production in HUVECs: Possible implications for its antihypertensive effect. *J. Hypertens.* **22**: 2329–2338.

Lee, S. Y. et al. 2005. Effect of sesaminol glucosides on beta-amyloid-induced PC12 cell death through antioxidant mechanisms. *Neurosci. Res.*, **28**: 186–189.

Lee, S. C. et al. 2005. Effect of far-infrared irradiation on the antioxidant activity of defatted sesame meal extracts. *J. Agric. Food Chem.*, **53**: 1495–1498.

Lemcke-Norojarvi, M. et al. 2001. Corn and sesame oils increase serum γ-tocopherol concentrations in healthy Swedish women. *J. Nutr.*, **131**: 1195–1201.

Lim, J. S. et al. 2007. Comparative analysis of sesame lignans (sesamin and sesamolin) in affecting hepatic fatty acid metabolism in rats. *Br. J. Nutr.*, **97**: 85–95.

Matsumura, Y. et al. 1995. Antihypertensive effect of sesamin. I. Protection against deoxycorticosterone acetate-salt-induced hypertension and cardiovascular hypertrophy. *Biol. Pharm. Bull.*, **18**: 1016–1019.

Matsumura, Y. et al. 1998. Antihypertensive effect of sesamin. III. Protection against development and maintenance of hypertension in stroke-prone spontaneously hypertensive rats. *Biol. Pharm. Bull.*, **21**: 469–473.

Matsumura, Y. et al. 2000. Effect of sesamin on altered vascular reactivity in aortic rings of deoxycorticosterone acetate-salt-induced hypertensive rat. *Biol. Pharm. Bull.*, **23**: 1041–1045.

Mimura, A. et al. 1994. Antioxidative and anticancer components produced by cell culture of sesame. In: *Food Phytochemicals for Cancer Prevention II*. Ho, C. T., Osawa, T., Hung, T., Eds. *ACS Symposium Ser.*, **547**: 281–294.

Miyahara, Y. et al. 2001. Sesaminol from sesame seed induces apoptosis in human lymphoid leukemia Molt 4B cells. *Int. J. Mol. Med.*, **7**: 485–488.

Miyake, Y. et al. 2005. Antioxidative catechol lignans converted from sesamin and sesaminol triglucocide by culturing with *Aspergillus*. *J. Agric. Food Chem.*, **53**: 22–27.
Mizukuchi, A., Umeda-Sawada, R., and Igarashi, O. 2003. Effect of dietary fat level and sesamin on the polyunsaturated fatty acid metabolism in rats. *J. Nutr. Sci. Vitaminol.*, **49**: 320–326.
Moazzami, A. A., Andersson, R. E., and Kamal-Eldin, A. 2006a. HPLC analysis of sesaminol glucosides in sesame seeds. *J. Agric. Food Chem.*, **54**: 633–638.
Moazzami, A. A., Andersson, R. E., and Kamal-Eldin, A. 2006b. Characterization and analysis of sesamolinol diglucoside in sesame seed. *Biosci. Biotechnol. Biochem.*, **70**: 1478–1481.
Moazzami, A. A., Andersson, R. E., and Kamal-Eldin, A. 2007. Quantitative NMR analysis of a sesamin catechol metabolite in human urine. *J. Nutr.* **137**: 940–944.
Moreno F. J. et al. 2005. Thermostability and *in vitro* digestibility of a purified major allergen 2S albumin (Ses i 1) from white sesame seeds (*Sesamum indicum* L.) *Biochem. Biophys. Acta*, **25**: 142–153.
Murakami, T., Yoshizu, H., and Ito, H. 1995. Prophylactic effects of sesamin on ischemic cerebral injury in stroke-prone spontaneously hypertensive rats. *Nippon Eiyo-Shokuryo Gakkaishi*, **48**: 189–193.
Nagashima, M., Fukuda, Y., and Ito, R. 1999. Antioxidative lignans from industrial waste water in cleaning of black sesame seed. *Nippon Shokuhin Kagaku Kogaku Kaishi*, **46**: 382–388.
Nagata, M. et al. 1987. Stereochemical structures of antioxidative bisepoxylignans, sesaminol and its isomers, transformed from sesamolin. *Agric. Biol. Chem.*, **51**: 1285–1289.
Nakabayashi, A. et al. 1995. α-Tocopherol enhances the hypocholesterolemic action of sesamin in rats. *J. Vitam. Nutr. Res.*, **65**: 162–168.
Nakai, M. et al. 2003. Novel antioxidative metabolites in rat liver with ingested sesamin. *J. Agric. Food Chem.*, **51**: 1666–1670.
Nakai, M. et al. 2006. Decomposition reaction of sesamin in supercritical water. *Biosc. Biotechnol. Biochem.*, **70**: 1273–1276.
Nakano, D. et al. 2006. Antihypertensive effect of angiotensin I-converting enzyme inhibitory peptides from a sesame protein hydrolysate in spontaneously hypertensive rats. *Biosc. Biotechnol. Biochem.*, **70**: 1118–1126.
Namiki, K. et al. 2006. Antithrombosis effect of sesame lignans. *Hemorheol. Relat. Res.*, **9**: 23–30.
Namiki, M. 1990. Antioxidants/antimutagens in food. *Critical Reviews in Food Science and Nutrition*, **29**: 273–300.
Namiki, M. 1991. Antimutagen and anticarcinogen research in Japan. *Food Phytochemicals for Cancer Prevention I.* Eds. Hang, Osawa, Ho, and Rosen. *ACS Sym. Ser.*, **546**: 65–81.
Namiki, M. 1995. The chemistry and physiological functions of sesame. *Food Rev. Int.*, **11**: 281–329.
Namiki, M., Ed. 1998. *Goma, sono Kagaku to Kinousei (Advances in Sesame Science and Function)*. Maruzen Planet Co., Tokyo, 268pp.
Namiki, M. 2007. Nutraceutical functions of sesame: A review. *Crit. Rev. Food Sci. Nutr.*, **47**: 651–673.
Namiki, M. et al. 2002. Changes in functional factors of sesame seed and oil during various types of processing. In: *Bioactive Compounds in Foods*. Lee L. T., Ho C. T., Ed., American Chemical Society, Washington, DC, pp. 85–104.
Namiki, M., Kobayashi, T. Eds. 1989. *Goma no Kagaku (Sesame Science)*. Asakura Shoten Co., Tokyo, 246pp.
Namiki, M., Yamashita, K., and Osawa, T. 1993. Food-related antioxidants and their activities *in vivo*. In: *Active Oxygen, Lipid Peroxides, and Antioxidants*. Yagi, K. Ed. Sci. Soc. Press, Tokyo, Japan, pp. 319–332.

Navuluri, L. et al. 2006. Allergic and anaphylactic response to sesame seeds in mice: Identification of Ses I 3 and basic subunit of 11s globuline as allergens. *Int. Arch. Allergy Immunol.*, **140**: 270–276.

Noguchi, T. et al. 2001. Effects of vitamin E and sesamin on hypertension and cerebral thrombogenesis in stroke-prone spontaneously hypertensive rats. *Hypertens. Res.*, **24**: 735–724.

Nonaka, M. et al. 1997. Effects of dietary sesaminol and sesamin on eicosanoid production and immunoglobulin level in rats given ethanol. *Biosci. Biotech. Biochem.*, **61**: 836–839.

Numa, S. 1984. *Fatty Acid Metabolism and its Regulation*. Elsevier, Amsterdam.

Ogawa, H. et al. 1995. Sesame lignans modulate cholesterol metabolism in the stroke-prone spontaneously hypertensive rat. *Clin. Exp. Pharmacol. Physiol.*, **22**(Suppl. 1), S. 310–312.

Ogawa, T. et al. 1994. Lack of influence of low blood cholesterol levels on pancreatic carcinogenesis after initiation with *N*-nitroso-bis (2-oxopropyl) amine in Syrian golden hamsters. *Carcinogenesis*, **13**: 1663–1666.

Ohtsuki, T. et al. 2003. Increased production of antioxidative sesaminol glucosides from sesame oil cake through fermentation by *Bacillus circulans* strain YUS-2. *Biosci. Biotech. Biochem.*, **67**: 2304–2306.

Oil World Annual 2002/ISTA. Mielke GmbH, Germany.

Ono, E. et al. 2006. Formation of two methylenedioxy bridges by a *Sesamum* CYP81Q protein yielding a furofuran lignan, (+)-sesamin. *Proc. Natl. Acad. Sci. USA* **103**: 10116–10121.

Osawa, T. 1997. Recent progress of functional food research in Japan. In *Functional Foods for Disease Prevention II*. Eds. Shibamoto, Terao, and Osawa, *ACS Sym. Ser.*, **701**: 2–9.

Osawa, T. et al. 1985. Sesamolinol, a novel antioxidant isolated from sesame seeds. *Agric. Biol. Chem.*, **49**: 3351–3352.

Osawa, T., Namiki, M., and Kawakishi, S. 1990. Role of dietary antioxidants in protection against oxidative damage. In: *Antimutagenesis and Anticarcinogenesis Mechanisms II*. Kuroda, Y.; Shankel, D. M.; Waters, M. D. Eds., Plenum, New York, pp. 139–153.

Parker, R. S., Sontag, T. J., and Swanson, J. E. 2000. Cytochrome P450 3A-dependent metabolism of tocopherols and inhibition by sesamin. *Biochem. Biophys. Res. Commun.*, **277**: 531–534.

Penalvo, J. L. et al. 2005a. Quantification of lignan in food using isotope dilution gas chromatography—mass spectrometry. *J. Agric. Food Chem.*, **53**: 9342–9347.

Penalvo, J. L. et al. 2005b. Dietary sesamin is converted to enterolactone in humans. *J. Nutr.*, **135**: 1055–1062.

Poneros-Schneier, A. G. and Erdman, J. W. 1989. Bioavailability of calcium from sesame seeds, almond powder, whole wheat bread, spinach and nonfat dry milk in rats. *J. Food Sci.*, **54**: 150–153.

Sano, M. et al. 1997. A controlled trial of selegillin, alpha-tocopherol, or both as treatment for Alzheimer's disease. *N. Engl. J. Med.*, **336**: 1216–1222.

Satchithanandam, S. et al. 1993. Coconut oil and sesame oil affect lymphatic absorption of cholesterol and fatty acids in rats. *J. Nutr.*, **123**: 1852–1858.

Sato, E. et al. 1995. The effect of sesame contents on viscoelasticity and microstructure of goma-dofu (sesame tofu). *Nippon Shokuhin Kagaku Kougaku Kaishi*, **42**: 871–877.

Schieberle, P. 1995. Odor-active compounds in moderately roasted sesame. *Food Chem.*, **55**: 145–152.

Schieberle, P. et al. 1996. Structure determination of 4-methyl-3 thiazolin in roasted sesame flavour. *Food Chem.*, **56**: 369–372.

Sen, M. and Bhattacharyya, D. K. 2001. Nutritional quality of sesame seed protein fraction extracted with isopropanol. *J. Agric. Food Chem.*, **49**: 2641–2646.

Shahide, F. et al. 1997. Effect of processing on flavor precursor amino acids and volatiles of sesame paste (Tehina). *Am. Oil Chem. Soc.*, **74**: 667–678

Shahide, F. et al. 2006. Antioxidant activity of white and black sesame seeds and their hull fractions. *Food Chem.*, **99**: 475–483.

Sheng, H. et al. 2007. Modifying effect of dietary sesaminol glucosides on the formation of azoxymethane-induced premalignant lesions of rat colon. *Cancer Lett.*, **246**: 63–68.

Shimizu, S. et al. 1989a. Production of dihomo-γ-linolenic acid by *Mortierella alpina* 1S-4. J. *Am. Oil Chem. Soc.*, **66**: 237–241.

Shimizu, S. et al. 1989b. Microbial conversion of an oil containing γ-linolenic acid to an oil containing eicosapentaenoic acid. *J. Am. Oil Chem. Soc.*, **66**: 342–347.

Shimizu, S. et al. 1989c. Stimulatory effect of peanut oil on the production of dihomo-γ-linolenic acid by filamentous fungi. *Agric. Biol. Chem.*, **53**: 1437–1438.

Shimizu, S. et al. 1991. Sesamin is a potent and specific inhibitor of Δ5 desaturase in polyunsaturated fatty acid biosynthesis. *Lipids*, **26**: 512–515.

Shimizu, S. et al. 1992a. Inhibition of (5-desaturase on polyunsaturated fatty acid biosynthesis by (–)-asarinin and (–)-epiasarinin. *Phytochemistry*, **31**: 757–760.

Shimizu, S. et al. 1992b. Inhibitory effect of curcumin on fatty acid desaturation in *Mortierella alpina* 1S-4 and rat liver microsomes. *Lipids*, **27**: 509–512.

Shimoda, M. et al. 1996. Identification and sensory characterization of volatile flavor compounds in sesame seed oil. *J. Agric. Food Chem.*, **44**: 3909–3912.

Shimoda, M. et al. 1997. Quantative comparison of volatile flavor compounds in deep-roasted and light-roasted sesame seed oils. *J. Agric. Food Chem.*, **45**: 3193–3196.

Shirato-Yasumoto, S. et al. 2000. New sesame line having high lignan content in seed and its functional activity. *Breed. Res.*, **2**(Suppl. 2): 184.

Shirato-Yasumoto, S. et al. 2001. Effect of sesame seeds rich in sesamin and sesamolin on fatty acid oxidation in rat liver. *J. Agric. Food Chem.*, **49**: 2647–2651.

Smeds, A. I. et al. 2007. Quantification of a broad spectrum of lignans in cereals, oilseeds and nuts. *J. Agric. Food Chem.*, **55**: 1337–1134.

Smith, N. 2000. Production and utilization of sesame seeds in US. *Sesame News Lett.*, **14**: 14–15.

Sontag, T. J., and Parker, R. S. 2002. Cytochrome P450 omega-hydroxylase pathway of tocopherol catabolism; novel mechanism of regulation of vitamin E status. *J. Biol. Chem.*, **277**: 25290–25296.

Speek, A. J., Schrijiver, J., and Schreuers, W. H. 1985. Vitamin E composition of some seed oils as determined by high-performance liquid chromatography with fluorometric detection. *J. Food Sci.*, **50**: 121–124.

STA Japan 2001. Science and Technology Agency of Japan. *Standard Table of Food Composition in Japan 5th ed. 2001*. Science and Technology Agency of Japan, Tokyo.

Sugano, M. et al. 1990. Influence of sesame lignans on various lipid parameters in rats. *Agric. Biol. Chem.*, **54**: 2669–2673.

Sugano, M. et al. 2001. Dietary manipulations of body fat-reducing potential of conjugated linoleic acid in rats. *Biosci. Biotech. Biochem.*, **65**: 2535–2541.

Sumiki, Y. et al. 1958. Studies on protective substances against radiation lethal effect on mice. *Reports of the 2nd Radioisotope Congress of Japan* (in Japanese), pp. 218–220.

Takeda, T. et al. 1981. A new murine model of accelerated senescence. *Mech. Ageing Dev.*, **17**: 183–194.

Takei, Y. and Fukuda, Y. 1995. Effect of roasting temperature of sesame seed on quality of sesame oil. *Nihon Chourikagaku Kaishi*, **24**: 10–15.

Takeuchi, H. et al. 2001. Hypoglycemic effect of a hot-water extract from defatted sesame (*Sesamum indicum* L.) seed on the blood glucose level in genetically diabetic KK-Ay mice. *Biosci. Biotech. Biochem.*, **65**: 2318–2321.

Tashiro, T. et al. 1990. Oil and minor components of sesame (*Sesamum indicum* L.) strains. *J. Am. Oil Chem. Soc.*, **67**: 506–511.

Theriault, A. et al. 1999. Tocotrienol: A review of its therapeutic potential. *Clin. Biochem.*, **32**, 309–319.

Tsuruoka, N. et al. 2005. Modulating effect of sesamin, a functional lignan in sesame seeds, on the transcription levels of lipid- and alcohol-metabolizing enzymes in rat liver: A DNA microarray study. *Biosci. Biotech. Biochem.*, **69**: 179–188.

Uchida, T. et al. 2007. Dietary sesame seed decreases urinary excretion of alpha- and gamma-tocopherol metabolites in rats. *J. Nutr. Sci. Vitaminol. (Tokyo)*, **53**: 372–376.

Umeda-Sawada, R. et al. 1996. Effect of sesamin on the composition of eicosapentaenoic acid in liver and its lymphatic absorption in rats. *Biosci. Biotech. Biochem.*, **60**, 2071–2072.

Umeda-Sawada, R. et al. 2001. Effect of sesamin on mitochondrial and peroxisomal beta-oxidation of arachidonic and eicosapentaenoic acids in rat liver. *Lipids*, **36**: 483–489.

Umeda-Sawada, R., Fujiwara, Y., and Igarashi, O. 1994. Effect of sesamin on cholesterol synthesis and on the distribution of incorporated linoleic acid in lipid subfractions in cultured rat cells. *Biosci. Biotech. Biochem.*, **58**: 2114–2115.

Umeda-Sawada, R., Ogawa, M., and Igarashi, O. 1998. The metabolism and n-6/n-3 ratio of essential fatty acids in rats: Effect of dietary arachidonic acid and a mixture of sesame lignans (sesamin and episesamin). *Lipids*, **33**: 567–572.

Umeda-Sawada, R., Ogawa, M., and Igarashi, O. 1999. The metabolism and distribution of sesame lignans (sesamin and episesamin) in rats. *Lipids*, **34**: 633–637.

Umeda-Sawada, R., Takahashi, N., and Igarashi, O. 1995. Interaction of sesamin and eicosapentaenoic acid against Δ5 desaturation and n-6/n-3 ratio of essential fatty acids in rats. *Biosci. Biotech. Biochem.*, **59**: 2268–2273.

Wallowitz, M. L. et al. (2007) Sesi 6, the sesame 11S globulin, can activate basophils and shows cross-reactivity with walnut *in vitro. Clin. Exp. Allergy*, **37**: 929–938.

Wankhede, D. B. and Tharanathan, R. N. 1976. Sesame (*Sesamum indicum* L.) carbohydrates. *J. Agric. Food Chem.*, **24**: 655–659.

Weiss, E. A. 1983. *Oil Seed Crops*, pp. 282–340. Longman, London, UK.

Wu, W-H. et al. 2006. Sesame ingestion affects sex hormones, antioxidant status, and blood lipids in postmenopaused women. *J. Nutr.*, **136**: 1270–1275.

Yamashita, K. 2004. Enhancing effects on vitamin E activity of sesame lignans. *J. Clin. Biochem. Nutr.*, **35**: 17–27.

Yamashita, K. et al. 1990. Effects of sesame in the senescence-accelerated mouse. *Eiyo Shokuryou Gakkaishi*, **43**: 445–449.

Yamashita, K. et al. 1992. Sesame seed lignans and γ-tocopherol act synergistically to produce vitamin E activity in rats. *J. Nutr.*, **122**: 2440–2446.

Yamashita, K. et al. 1995. Sesame seed and its lignans produce marked enhancement of vitamin E activity in rats fed a low α-tocopherol diet. *Lipids*, **30**: 1019–1028.

Yamashita, K. et al. 2000a. Sesamin and α-tocopherol synergistically suppress lipid-peroxide in rats fed a high docosahexaenoic acid diet. *Bio Factors*, **11**: 11–13.

Yamashita, K. et al. 2002. Effect of sesaminol on plasma and tissue α-tocopherol and α-tocotrienol concentrations in rats fed a vitamin E concentrate rich in tocotrienols. *Lipids*, **37**: 351–358.

Yamashita, K. et al. 2007. Hydroxymatairesinol and sesaminol act differently on tocopherol concentration in rats. *J. Nutr. Sci. Vitaminol.*, (Tokyo), **53**: 393–399.

Yamashita, K. and Namiki, M. 1994. Suppressive effect of sesame seed and its lignans in senescence accelerated mouse (SAMP-1). In: *The SAM Model of Senescence*. Takeda, T., Ed. Elsevier Science B.V., Amsterdam, pp. 153–156.

Yamashita, K., Ikeda, S., and Ohbayashi, M. 2003. Comparative effect of flaxseed and sesame seed on vitamin E and cholesterol levels in rats. *Lipids*, **38**: 1249–1255.

Yamashita, K., Takeda, N., and Ikeda, S. (2000b.) Effects of various tocopherol-containing diets on tocopherol secretion into bile. *Lipids*, **35**: 163–170.

Yang, Z. et al. 1995. Effects of sesamin on ethanol-induced muscle relaxation. *Nippon Eiyo-Shokuryo Gakkaishi*, **48**: 103–108.

Yokota, T. et al. 2007. Sesamin, a lignan of sesame, down-regulates cyclin D1 protein expression in human tumor cells. *Cancer Sci.*, **98**: 1447–1453.

11 Fenugreek-Based Spice

A Traditional Functional Food Ingredient

Carani Venkatraman Anuradha

CONTENTS

11.1 INTRODUCTION

The use of dietary components as therapeutic agents for the modulation of disease states has become an emerging field of research. This has led to the identification of functional attributes of many food components, especially from plant sources such as fruits, vegetables, whole grains, and spices. Biologically active phytochemicals may impart health benefits and reduce the risk of disease. An important functional food derived from a plant source is the seed of fenugreek. It is used as a spice in Indian homes to impart flavor and color, enhance taste, and modify the texture of food. This spice, also a legume, has received attention as a useful antidiabetic agent and has undergone extensive research in clinical and animal models of diabetes which have clearly documented its blood glucose lowering properties. However, the seeds show many health benefits beyond that. A large body of scientific evidence confirms the health benefits of this spice. This chapter presents a review of the traditional use of fenugreek and recent data on its nutraceutical potential.

Fenugreek (*Trigonella foenum-graecum*) is an annual herb that belongs to the family *Leguminosae* and has a long history of traditional use as a condiment and a medicinal herb (Figure 11.1). The herb, a native of southern Europe and Western Asia, is now cultivated in Argentina, France, India, North Africa, the United States, and in the Mediterranean countries. For centuries, fenugreek has been used in folk medicine to heal ailments ranging from indigestion to baldness (Fazli and Hardman, 1968). The seeds are assumed to possess nutritive and restorative properties and to stimulate digestive processes (Moissides, 1939). It is supposed to be a tonic, carminative, aperient, and diuretic, useful in dropsy, chronic cough, external and internal swellings, and hair decay (Leela and Shafeekh, 2008). The seeds are described in the Greek and Latin *Pharmacopoeias* as possessing antidiabetic activity (Loeper and Lemaire, 1931;

FIGURE 11.1 Fenugreek (*Trigonella foenum-graecum*) plant and seeds.

Bever and Zahnd, 1979; Shani et al., 1974). In traditional Chinese medicine, fenugreek seeds are used as a tonic, as well as a treatment for weakness and edema of the legs (Yoshikawa et al., 1997). In India, fenugreek seeds are used as a stimulant for lactation (Mital and Gopaldas, 1986). They are known to be important constituents of the traditional food (*Methipak*) consumed during lactation. In Egypt, the seeds are used to supplement wheat and maize flour for bread-making (Morcos et al., 1981). In Arab countries, the seeds are used for the preparation of hot beverages after adding sugar (Elmadfa, 1975; Elmadfa and Kuhl, 1976). In Sudan, it is used in porridge and dessert and consumed mostly by lactating mothers (Abuzied, 1986).

11.2 FENUGREEK-BASED FOODS IN INDIA

The seeds are one of the healthiest spices used in the country. They are always kept handy in a spice box in the kitchen and regularly included in day-to-day cooking. It is a common practice for Indian women to use fenugreek seeds along with spices like cumin and mustard seeds while seasoning food items.

In many regions, the mother is given a special *laddoo* made of dried dates, coconut, nuts, and fenugreek seeds for 40 days after child birth to promote lactation. Besides, fenugreek seeds are also used for the preparation of pickles, curry powder, and a number of other dishes encountered in the daily cuisine of the Indian subcontinent (http://www.recipezaar.com/). The slightly bitter aroma of roasted seeds lends pickles its mouth-watering flavor that is very unique, and puts Indian pickles on the list of favorites across the globe.

The seeds are generally used after roasting, sprouting, or soaking. Sprouted seeds are used to prepare *salads* and *curries* to give a slightly pungent sweet flavor that also boosts the nutritional value.

Dry roasting increases the mellow flavor and reduces the bitterness of the spice. The dry roasted seeds are added as such or in a powdered form to pickles (a hot side-dish), gravy, or a dhal dish to leave a tangy milieu characteristic curry flavor.

The seeds are soaked in water overnight so they can be ground into a paste. This is one of the ingredients for preparing *idli* and *dosa*, which form a part of South Indian cuisine. Soaked and cooked seeds are used as an ingredient in the soup commonly known as *sambar* in south India.

The Indian *methi dosas* are traditionally prepared as a breakfast dish. For this, soaked rice and soaked fenugreek seeds (5:1) are ground separately and then mixed. The mixture is allowed to ferment for 8 h. Salt is added to taste, after which a small quantity of the fermented batter is spread in a circle on a hot nonstick griddle with a little oil poured over the *dosa*. The *dosa* is turned over till both sides become crispy and brownish. This dish is enjoyable with fried onion topping or when served with *sambar* or *chutney*.

Methi chutney is a delicious side-dish consumed with bread and *chapatti*. This is a coarse mixture prepared by grinding soaked methi seeds, coconut gratings, and red chilies into a paste. Lime juice, salt, and sugar are added to enhance the taste.

One of the Indian systems of medicine, *Ayurveda*, advocates the balance of the six tastes—sweet, salty, sour, bitter, astringent, and hot in the human diet for optimum health, good nutrition, and disease prevention.

11.3 COMPOSITION OF SEEDS

Fenugreek seeds are noncotyledenous but rather endospermic in nature, and are rich in protein, fiber, and gum. The typical nutrient composition of fenugreek seeds (g/100 g) is: moisture 2.9; protein 26.5; fat 7.9; saponins 4.9; total dietary fiber 57.8 (consisting of gum 19.0, hemicellulose 23.6, cellulose 8.9, lignin 2.4, and ash 3.9) (Shankaracharya and Natarajan, 1972).

The protein fraction contains essential amino acids and a unique nonprotein amino acid, 4-hydroxyisoleucine (4-OH Ile). The content of essential amino acids is comparable to that in soy protein.

Fenugreek seeds are reported to be rich in flavonoids (100 mg/g) (Gupta and Nair, 1999). Five different flavonoids namely vitexin, tricin, naringenin, quercetin, and tricin-7-*O*-β-D-glucopyranoside were reported to be present in fenugreek seeds (Shang et al., 1998a). Later, seven additional compounds, *N*,*N′*-dicarbazyl, glycerol monopalmitate, stearic acid, β-sitosteryl glucopyranoside, ethyl-α-D-glucopyranoside, D-3-*O*-methylchiroinsitol, and sucrose were identified by the same group (Shang et al., 1998b). HPLC analysis of an aqueous extract showed the presence of gallic acid, *O*-coumaric acid, *p*-coumaric acid, rutin, and caffeic acid (Dixit et al., 2005).

The presence of choline, essential fatty acids, lecithin, trimethylaminophytosterols, tannic acid, fixed and volatile essential oils and bitter extractive, diosgenin, alkaloids (trigonelline and gentianine), trigocoumarin, and trigomethyl coumarin in fenugreek seeds has also been reported (Jayaweera, 1981; Petit et al., 1995). The fenugreek leaf is rich in vitamins A, D, B_1, B_2, B_3, B_6, B_{12}, biotin, pantothenic acid, folic acid, para-aminobenzoic acid, and choline, and minerals such as iron, selenium, phosphorus, and potassium (Sharma, 1986a).

11.4 THERAPEUTIC APPLICATIONS OF FENUGREEK

Fenugreek is gaining popularity among a wider group of people. Researchers have convincingly shown several beneficial properties of fenugreek seeds in animals as well as in human trials. These include antidiabetic, hypolipidemic, anticancer, antioxidant, hepatoprotective, and gastroprotective effects as well as a host of others. These effects are mainly attributable to constituents like high fiber, protein, and flavonoid concentrations. In this chapter, most of the popular available reports published on fenugreek seeds will be summarized.

11.4.1 Hypoglycemic Action

The hypoglycemic properties of fenugreek seeds have been known for many years and are well documented in experimentally induced diabetic rats (Madar, 1984), dogs (Ribes et al., 1986), mice (Ajabnoor and Tilmisany, 1988) in healthy human volunteers (Sharma, 1986a), and in type 1 (Sharma et al., 1990) and type 2 diabetics (Madar et al., 1988). Both the seeds and leaves are found to possess antidiabetic properties (Sharma, 1986a).

In a clinical trial involving 60 patients, administration of fenugreek seeds improved clinical symptoms such as polydipsia and polyuria in a majority of the patients in spite of reducing the antidiabetic drug dose (Sharma and Raghuram, 1990). The improvement in clinical symptoms followed alterations in biochemical parameters such as a reduction in the blood glucose level and urinary excretion of glucose.

A double-blind placebo-controlled study by Gupta et al. (2001) showed that the seeds considerably improve insulin sensitivity as measured by the homeostatic model assessment (HOMA), glucose metabolism, and lipid profile. In this study, 25 patients with newly diagnosed type 2 diabetes daily received 1 g of hydro-alcoholic extract of fenugreek seeds. The mean fasting blood glucose levels were reduced. It was suggested that the use of fenugreek seed extract is an effective strategy for attaining glycemic control in type 2 diabetes. Raghuram et al. (1994) reported that the administration of 25 g powdered fenugreek seeds in a diet improved the glucose tolerance test scores and serum-clearance rates of glucose in type 2 diabetic patients. Neeraja and Rajalakshmi (1996) reported that fenugreek seeds lower postprandial hyperglycemia in human subjects with diabetes.

The effect of fenugreek on postprandial glucose and insulin levels following a meal tolerance test was studied in noninsulin-dependent diabetics (NIDDM) (Madar et al., 1988). The addition of powdered fenugreek seeds (15 g) soaked in water significantly reduced the subsequent postprandial glucose levels. Plasma insulin also tended to be lower, although the reduction was not statistically significant.

The seeds have been shown to lower the blood glucose level and to partially restore the activities of key enzymes for carbohydrate and lipid metabolism close to their normal values in various animal model systems. Ground seeds of fenugreek offered to diabetic rats decreased the postprandial glucose levels (Vats et al., 2002). Supplementation of fenugreek seeds in diets of alloxan-diabetic dogs also confirmed this property (Ribes et al., 1986).

Researchers have been interested in the identification of components and their hypoglycemic effects. In humans, fenugreek seeds exert hypoglycemic effects by stimulating glucose-dependent insulin secretion from pancreatic β-cells as well as by inhibiting the activities of α-amylase and sucrase, two intestinal enzymes involved in carbohydrate metabolism (Sharma et al., 1996b). Increases in the number of insulin receptors (Sharma et al., 1990) and the metabolic clearance rate of glucose, and a delay in gastric emptying with glucose absorption (Amin et al., 1987; Raghuram et al., 1994) are the mechanisms suggested. It was originally believed that the major alkaloid trigonelline, found in these seeds, was responsible for the hypoglycemic effect (Mishkinsky et al., 1967). However, investigations by Shani et al. (1974) on the effect of trigonelline in diabetic rats and humans led to the supposition that the active principle of *trigonella* might not be trigonelline. Nicotinic acid, coumarin, and many trace matters present in the seeds may exert a hypoglycemic action.

The high levels of fiber contribute to a secondary mechanism for the hypoglycemic effect of fenugreek seeds. The defatted powder of fenugreek seeds contains approximately 50% fiber (30% soluble fiber and 20% insoluble fiber). It is well known that gel-forming dietary fiber reduces the release of insulinotropic hormones and gastric inhibitory polypeptide (GIP), and slows down the rate of postprandial glucose absorption. Addition of fiber and gum to the diet was found to reduce the postprandial glycemia and urinary glucose excretion in diabetic patients (Jenkins et al., 1980; Anderson, 1985). The fenugreek seed fiber resembles guar gum in chemical structure and is very viscous when dissolved in water (Madar and Shomer, 1990). In addition to these properties, recent research suggests that fenugreek gum may also be surface active. Garti et al. (1997) found that stable emulsions with a relatively small droplet size (3 μm) could be formed using purified fenugreek gum. Thus, fiber contributes to a secondary mechanism for its hypoglycemic effect.

Much more work on 4-OH Ile, the unusual amino acid found in fenugreek seeds, and its hypoglycemic properties has been carried out. The 4-OH Ile extracted and purified from fenugreek seeds, displayed insulinotropic properties *in vitro* (Sauvaire et al., 1998), stimulated insulin secretion *in vivo* and improved glucose tolerance in normal rats and dogs and in a rat model of type 2 diabetes mellitus (Broca et al., 1999). The effect of 4-OH Ile was both dose and glucose-dependent, and was shown to stimulate insulin secretion as a result of direct β-cell stimulation (Broca et al., 2000). Besides 4-OH Ile, arginine and tryptophan are the other amino acids present in the seeds that have antidiabetic and hypoglycemic effects (Broca et al., 2000).

To elucidate the actions of the seed extract at the cellular and molecular levels, the seeds have been subjected to detailed scientific investigations by mechanism-based *in vitro* and *in vivo* assays (Vijayakumar et al., 2005). An aqueous extract of fenugreek seeds was dialyzed for 24 h to eliminate small molecules and the hypoglycemic potential of the fenugreek seed extract (FSE) was investigated *in vivo* in alloxan-induced diabetic mice. FSE significantly improved glucose homeostasis in diabetic mice and in normal glucose-loaded mice by effectively lowering blood glucose levels. This effect of FSE on glucose levels was found to be comparable to that of insulin.

FSE stimulated insulin-signaling pathways in adipocytes and liver cells and induces a rapid, dose-dependent stimulatory effect on cellular glucose uptake by activating cellular responses that led to GLUT4 translocation to the cell surface. It was suggested that fenugreek might act independent of insulin to enhance glucose transporter-mediated glucose uptake. FSE also activated the tyrosine phosphorylation of insulin receptor-β subunit (IR-β, subsequently enhancing tyrosine phosphorylation of the insulin receptor substrate (IRS-1) and the p85 subunit of phosphatidyl ionositol-3-kinase (PI3K). This suggests that adipocytes and liver cells could be target sites for FSE and exerts its effects by activating insulin signaling pathways. The authors also suggested that the seeds are capable of specifically activating the IR and its downstream signaling molecules in adipocytes and liver cells and do not act as general sensitizers of receptor tyrosine kinase domains (Vijayakumar et al., 2005).

Fenugreek seeds have an effect on the glyoxylase system and a link between antidiabetic action and the glyoxylase system has been suggested (Raju et al., 1999). Dietary administration of fenugreek seeds (at 1% and 2%) enhanced the glyoxylase I system and prevented the accumulation of methylglyoxal, a key player in advanced glycation end products (AGE) formation, oxidative stress, and diabetic complications (Choudhary et al., 2001).

11.5 LIPID-LOWERING EFFECTS

Fenugreek seeds also possess hypocholesterolemic effects. Several clinical and animal studies have shown that plasma lipid profiles and tissue lipid levels are markedly influenced both in human diabetics and experimental animals. For instance, Sharma (1986b) has shown a decrease in total cholesterol (TC), total low-density lipoprotein cholesterol (LDL-C) and very low-density lipoprotein cholesterol (VLDL-C), and in cholesterol and triacylglycerols (TG) in both type 1 and type 2 diabetic subjects by the administration of fenugreek seeds. Fenugreek seeds also lowered serum TG, total cholesterol (TC), and LDL-C (Sharma et al., 1996a). They investigated the hypolipidemic effect in 15 nonobese, asymptomatic, hyperlipidemic adults. Supplementation of 100 g defatted fenugreek powder per day for 3 weeks reduced TG and LDL-C levels from the baseline values. Slight decreases in high-density lipoprotein cholesterol (HDL-C) levels were also noted. In another study, Sharma et al. (1991) reported a decrease in TC levels in five diabetic patients treated with fenugreek seed powder (25 g per day oral) for 21 days. Bordia et al. (1997) administered 2.5 g twice daily for 3 months and observed significant decreases in TC and TG levels, with no change in the HDL-C level in a subgroup of 40 subjects affected by coronary artery disease and type 2 diabetes.

The active components responsible for the lipid-lowering effect of fenugreek seeds were found to be associated with the defatted part, rich in fiber, containing steroid saponins and protein (Gupta and Nair, 1999). The gum isolate constituting 19.2% of the defatted part had a strong hypocholesterolemic activity. The galactomannans present in the gum could increase the viscosity of the digesta, which may inhibit the absorption of cholesterol from the small intestine and also the reabsorption of bile acids from the terminal ileum, resulting in a decrease in serum cholesterol. The absorbed bile acids would be released by fecal excretion and this would be offset by the conversion of cholesterol into bile acids by the liver.

The saponins that are transformed in the gastrointestinal tract into sapogenins may contribute to the lipid-lowering effect. Saponins present in the defatted part (4.8%) reduced hypercholesterolemia and hypertriglyceridemia in alloxan-diabetic dogs (Sauvaire et al., 1991). It is suggested that saponins and cholesterol form insoluble complexes and retard cholesterol absorption in the intestine (Stark and Madar, 1993). The hypocholesterolemic effect was not observed with the saponin-free subfractions.

The lipid-lowering effect of fenugreek might also be attributed to its estrogenic constituent, indirectly increasing the thyroid hormone T4. Further, the flavonoids present in the seeds may also be responsible for these activities.

11.6 IMMUNOMODULATORY AND ANTI-INFLAMMATORY EFFECTS

An aqueous extract of fenugreek seeds showed immune-potentiating functions when administered to Swiss albino mice at different doses (Bin-Hafeez et al., 2003). The extract exhibited a stimulating effect on macrophages and showed a positive effect on specific and nonspecific immune functions of the lymphoid organs such as the thymus, bone marrow, spleen, and liver. It is suggested that high quantities of mucilage and iron in the organic form (Jonnalagadda and Seshadri, 2003) facilitate stimulation of macrophages and the hemopoietic system.

11.7 INOTROPIC AND CARDIOTONIC EFFECTS

Ion pump Na^+/K^+-ATPase is a ubiquitous membrane protein and is an active transporter of Na^+ and K^+ ions across the cell. Polyphenol fractions from fenugreek seeds inhibited dose-dependently the erythrocyte membrane Na^+/K^+-ATPase activity (Anuradha et al., 2003). While going through the mechanism of the action of polyphenols, it was suggested that the decreased activity of ATPase might be due to the conformational changes in the structure of the enzyme brought about by the extract. Flavones and flavonols containing hydroxyl groups inside the phenyl radical at ortho and vicinal positions exhibit high inhibitory effects on the membrane enzymes and positive inotropic effects (Umarova et al., 1998). The polyphenolic structures of flavonoids similar to cholesterol partition into the hydrophobic core of the membrane and cause a modulation in lipid fluidity (Arti et al., 2000). This could hinder the diffusion of ions and other transport processes.

Quercetin, a flavonol richly present in fenugreek seeds, produces inotropism in frog and rabbit hearts and in pig kidney medulla (Bhansali et al., 1989). Quercetin inhibits the Na^+/K^+-ATPase pump and stimulates β-adreno receptors involving the adenylate cyclase-cAMP system, ultimately increasing the availability of calcium from intracellular sites. These findings suggest that fenugreek has a positive inotropic effect due to the presence of quercetin and may be potentially useful as a cardiotonic agent.

11.8 ANTITUMOR EFFECTS

Most of the anti-inflammatory agents of plant origin show antitumor activity and it could be expected that fenugreek seeds exhibit antineoplastic effects. Sur et al. (2001) evaluated the antineoplastic activity of the seeds by studying the inhibition of tumor cell growth in Ehrlich ascites carcinoma model *in vivo* in mice by fenugreek seed extract. The alcoholic extract showed 70% inhibition of tumor cell growth, enhanced peritoneal exudation, and macrophage cell count, indicating the activation of macrophages and an anti-inflammatory action.

Epidemiological evidence suggests that dietary spices and fiber prevent colon carcinogenesis. Flavonoids and fiber can potentially act as anticarcinogenic agents by binding to free carcinogens and/or carcinogenic metabolites. This prevents their access to colonic mucosa and enhances fecal excretion (Fujiki et al., 1986; Bobek and Galbavy, 2001). The high contents of indigestible polysaccharides in the seeds can act as substrates to mucinase and this also helps to prevent the hydrolysis/degradation of mucin, thus contributing to the anticarcinogenic potential of the seeds (Devasena et al., 2003). Furthermore, the modulation of cholesterol and phospholipids metabolism in target organs by the seeds could in turn prevent 1,2-dimethylhydrazine-induced colon cancer (Devasena and Menon, 2003).

11.9 ANTIOXIDANT ACTIVITY

One desirable property of a dietary component is considered to be its antioxidant effect. Dietary antioxidants are micronutrients that have the ability to neutralize free radicals or their actions. Fenugreek seeds appear to have antiradical properties and an ability to prevent lipid peroxidation. The polyphenolic fraction of fenugreek seeds scavenged 2,2-diphenyl-1-picrylhydrazyl radical ($DPPH^{\bullet}$), 2,2′-azino-bis(3-ethylbenzthiazoline-6-sulfonic acid radical ($ABTS^{\bullet+}$) and hydroxyl radical ($OH^{\bullet}$) *in vitro* (Kaviarasan et al., 2007a). Supplementation studies have shown that dietary fenugreek seed (2%, w/w) can lower blood and tissue lipid peroxidation and augment the antioxidant potential in alloxan-diabetic rats (Anuradha and Ravikumar, 1999) and experimental colon cancer (Devasena and Menon, 2002).

An aqueous extract of the seeds inhibited rat liver lipid peroxidation stimulated by Fe^{2+}-ascorbate system and by glucose *in vitro* (Anuradha and Ravikumar, 1998), and the effect was comparable with that of α-tocopherol and reduced glutathione (GSH) (Thirunavukkarasu et al., 2003). A polyphenol-rich extract of fenugreek seeds protected erythrocytes from peroxide-induced oxidative hemolysis *in vitro* (Kaviarasan et al., 2003).

Dixit et al. (2005) have pointed out that germinated fenugreek seeds have several beneficial advantages over the ungerminated ones. Germination improved *in vitro* protein digestibility, as well as fat absorption capacity (Mansour and El-Adawy, 1994) and decreased the levels of total unsaturated fatty acids, total lipid, triacylglycerols, phospholipids, and unsaponifiable matter while saturated fatty acids are increased. They examined the antioxidant activities of different fractions from the powder of germinated seeds as well as that of two of its active chemical constituents, namely trigonelline and diosgenin. The aqueous extract of fenugreek had a high phenolic and flavonoid content with high antioxidant activity.

Dietary administration of fenugreek seeds resulted in an increased GSH, a major redox substance and glutathione-*S*-transferase (GST) activity in the liver of mice fed 1, 2, and 5% fenugreek seeds (Sharma, 1986b). The mode and magnitude of the effect seem to depend on the dose of fenugreek and the tissue studied.

11.10 ALCOHOL TOXICITY AND HEPATOPROTECTIVE ACTION

The effects of aqueous and polyphenolic extracts of fenugreek seeds on experimental ethanol toxicity in rats were tested. Administration of fenugreek seed extract minimized the effects of ethanol in tissues. Treatment with extract reduced fatty changes and portal inflammation in the livers of ethanol-treated rats and there was a reduction in spongiosis in their brains (Thirunavukkarasu et al., 2003). The polyphenol extract was effective in mitigating alcohol-induced collagen accumulation (Kaviarasan et al., 2007b) and protein and lipid damage (Kaviarasan et al., 2008) in rat liver. Treatment of Chang liver cells with a polyphenolic extract protected from ethanol-induced cytotoxicity. The extract significantly increased cell viability, prevented oxidative damage, redox changes, LDH enzyme leakage, and apoptosis (Kaviarasan et al., 2006). The presence of flavonoids in the seeds could be responsible for cytoprotection.

The seeds of fenugreek were found to modulate the detoxification process. Fenugreek seeds when fed at 2 dietary levels generally stimulated the hepatic mixed function oxygenase system (MFOS), the cytochrome P450-dependent aryl hydroxylase, cytochrome P_{450}, and b_5 (Sambaiah and Srinivasan, 1989). The stimulation of the hepatic xenobiotic metabolism under normal conditions by fenugreek may be implicated in its potential as a hepatoprotective/detoxifying agent.

11.11 GASTROINTESTINAL FUNCTION

In the Indian system of medicine, the seeds have been used to treat a number of gastrointestinal disorders and have been well recognized in stimulating digestion. They also possess carminative, tonic, and galactogogue properties. It has been used to check dysentery, diarrhea, and dyspepsia with loss of appetite. A study by Platel and Srinivasan (2000) revealed that diet containing 2% fenugreek seed improved the intestinal function by enhancing the activities of terminal enzymes of digestive process such as lipases, sucrase, and maltase. They are supposed to stimulate appetite and improve feeding behavior through an endocrine response (Petit et al., 1993).

The antiulcer potential of fenugreek seeds in gastric damage induced by aspirin and ethanol was also evaluated and compared with that of a proton pump inhibitor, omeprezole (Sujapandian et al., 2002; Thirunavukkarasu and Anuradha, 2006). The seeds showed antisecretory action and participated in the prevention of mucosal hyperemia and edema. The polysaccharide composition of the gel galactomannan, and or/the flavonoids are responsible for the gastroprotective and antisecretory activities of the seeds.

The functional properties of the protein from this legume, such as solubility, foaming, emulsifying properties, and nitrogen content, have been documented recently and were found to be more favorable when compared with other legumes (El Nasri and El Tinay, 2007). Being the richest source of both soluble and insoluble fiber, processed fenugreek fiber can be used in formulations for fortification intended to increase the medicinal and therapeutic efficacies of food. Furthermore, the antioxidant and free radical scavenging activities suggest that this health food can lower the risk for many human diseases in which free radicals are implicated.

11.12 SUMMARY

Functional foods are those dietary components that possess demonstrated health-promoting and disease-preventing properties, in addition to the simple nutritional function. An overwhelming body of evidence has been collected in recent years to show the health attributes and the immense medicinal value of fenugreek seeds. The wide range of benefits demonstrated in clinical and animal studies confirm the functional properties and substantiate the significance of this spice in maintaining good health and treating pathological conditions. This particular legume rarely appears in western diets and is often overlooked as a functional food. Concomitant research activities by both academic bodies and food industries are necessary to get wider acceptance and to exploit this oriental spice as a functional food ingredient.

REFERENCES

Abuzied, A.N. 1986. Al-hilba (fenugreek) In: *Al-nabatat Wa-alaashab al-tibeeva Das Albihar Beirut.* Lebanon (in Arabic), pp. 223–24.

Ajabnoor, M.A., Tilmisany, A.K. 1988. Effect of *Trigonella foenum graecum* on blood glucose levels in normal and alloxan-diabetic mice. *J Ethnopharmacol* 22: 45–9.

Amin, R., Abdul-Ghani, A.S., Suleiman, M.S. 1987. Effect of *Trigonella foenum graecum* on intestinal absorption. *Proc. of the 47th Annual Meeting of the American Diabetes Association* (Indianapolis, USA). *Diabetes* 36: 211.

Anderson, J.W. 1985. High-fiber diets for obese diabetic men on insulin therapy: Short-term and long term effects. In: Bjoerntorp, P., Vahouny, G.V., Kritchevsky, D., eds. *Current topics in nutrition and disease.* New York: Alan R Riss Inc, Vol. 14, pp. 49–68.

Anuradha, C.V., Kaviarasan, S., Vijayalakshmi, K. 2003. Fenugreek seed polyphenols inhibit RBC membrane Na^+/K^+-ATPase activity. *Orient Phar Exp Med* 3: 129–33.

Anuradha, C.V., Ravikumar, P. 1998. Anti-lipid peroxidative activity of seeds of fenugreek (*Trigonella foenum graecum*). *Med Sci Res* 26: 317–21.

Anuradha, C.V., Ravikumar, P. 1999. Effect of fenugreek seeds on blood lipid peroxidation and antioxidants in diabetic rats. *Phy Res* 13: 197–201.

Arti, A., Byren, T.M., Nair, M.G., Strasburg, G.M. 2000. Modulation of liposomal membrane fluidity by flavonoids and isoflavonoids. *Arch Biochem Biophys* 373: 102–9.

Bever, B.O., Zahnd, G.R. 1979. Plants with oral hypoglycemic action. *Quart. J Crude Drug Res* 17: 139–96.

Bhansali, B.B., Vyes, S., Goyal, R.K. 1989. Cardiac effects of quercetin on isolated rabbit and frog heart preparations. *Ind J Pharmacol* 19: 100–7.

Bin-Hafeez, B., Haque, R., Parvez, S., Pandey, S., Sayeed, I., Raisuddin, S. 2003. Immunomodulatory effects of fenugreek (*Trigonella foenum graecum* L.) extract in mice. *Int Immunopharmacol* 3: 257–65.

Bobek, P., Galbavy, S. 2001. Influence of insulin on dimethylhydrazine induced carcinogenesis and antioxidant enzymatic system in rat. *Biol Bratislava* 56: 287–91.

Bordia, A., Verma, S.K., Srivastava, K.C. 1997. Effect of ginger (*Zingiber officinale Rosc.*) and fenugreek (*Trigonella foenum graecum L.*) on blood lipids, blood sugar and platelet aggregation in patients with coronary artery disease. *Prosta Leukot Essen Fatty Acids* 56: 379–84.

Broca, C., Gross, R., Petit, P., Sauvaire, Y., Manteghetti, M., Tournier, M., Masiello, P., Gomis, R., Ribes, G. 1999. 4-Hydroxyisoleucine: Experimental evidence of its insulinotropic and antidiabetic properties. *Am J Physiol* 277: E617–23.

Broca, C., Manteghetti, M., Gross, R., Baissac, Y., Jacob, M., Petit, Y., Sauvaire, Y., Ribes, G. 2000. 4-Hydroxyisoleucine: Effects of synthetic and natural analogues on insulin secretion. *Eur J Pharm* 390: 339–45.

Choudhary, D., Chandra, D., Choudhary, S., Kale, R.K. 2001. Modulation of glyoxylase, glutathione-*S*-transferase and antioxidant enzymes in the liver, spleen and erythrocytes of mice by dietary administration of fenugreek seeds. *Food Chem Toxicol* 39: 989–97.

Devasena, T., Gunasekaran, G., Viswanathan, P., Menon, V.P. 2003. Chemoprevention of 1,2-dimethylhydrazine-induced colon carcinogenesis by seeds of *Trigonella foenum graecum* L. *Biol Bratislava* 58: 357–64.

Devasena, T., Menon, V.P. 2002. Enhancement of circulatory antioxidants by fenugreek during 1,2-dimethylhydrazine-induced rat colon carcinogenesis. *J Biochem Mol Biol Biophys* 6: 289–92.

Devasena, T., Menon, V.P. 2003. Fenugreek affects the activity of β-glucuronidase and mucinase in the colon. *Phy Res* 17: 1088–91.

Dixit, P., Ghaskadbi, S., Mohan, H., Devasagayam, T.P.A. 2005. Antioxidant properties of germinated fenugreek seeds. *Phy Res* 19: 977–83.

Elmadfa, I. 1975. Fenugreek (*Trigonella foenum graecum*) protein. *Die Nahrung* 19: 683–86.

Elmadfa, I., Kuhl, B.E. 1976. The quality of fenugreek seeds protein tested alone and in a mixture with corn flour. *Nutr Rep Int* 14: 165–72.

El Nasri, N.A., El Tinay, A.H. 2007. Functional properties of fenugreek (*Trigonella foenum graecum*) protein concentrate. *Food Chem* 103: 582–89.

Fazli, F.R.Y., Hardman, R. 1968. The spice, fenugreek (*Trigonella foenum graecum*. L.): Its commercial varieties of seed as a source of diosgenin. *Trop Sci* 10: 66–78.

Fujiki, H., Horiuchi, T., Yamashita, K. 1986. Inhibition of tumor promotion by flavonoids. *Prog Clin Biol Res* 213: 429–40.

Garti, N., Madar, Z., Aserin, A., Sternheim, B. 1997. Fenugreek galactomannans as food emulsifiers. *LWT-Food Sci Technol* 30: 305–11.

Gupta, R., Nair, S. 1999. Antioxidant flavonoids in common Indian diet. *South Asian J Prev Cardiol* 3: 83–94.

Gupta, A., Gupta, R., Lal, B. 2001. Effect of *Trigonella foenum-graecum* (fenugreek) seeds on glycaemic control and insulin resistance in type 2 diabetes mellitus: A double blind placebo controlled study. *J Assoc Phys India* 49:1057–61.

http://www.recipezaar.com/recipes.php?s_type=%2Frecipes.php&q=fenugreek+based+food & Search, Fenugreek based food. Recipezaar Website. Accessed September 18, 2008.

Jayaweera, D.M.A. 1981. *Medicinal Plant: Part III. Peradeniya*. Srilanka: Royal Botanic Garden, 225pp.

Jenkins, D.J.A., Wolever, T.M.S., Taylor, R.H., Reynolds, D., Nineham, R., Hockaday, T.D.R. 1980. Diabetic glucose control, lipids and trace elements on long term guar. *Br Med J* 280: 1353–54.

Jonnalagadda, S.S., Seshadri, S. 2003. *In vitro* availability of iron from cereal meal with the addition of protein isolates and fenugreek leaves. *Plant Foods Hum Nutr* 45: 119–25.

Kaviarasan, S., Naik, G.H., Gangabhagirathi, R., Anuradha, C.V., Priyadarsini, K.I. 2007a. *In vitro* studies on antiradical and antioxidant activities of fenugreek (*Trigonella foenum graecum*) seeds. *Food Chem* 103: 31–7.

Kaviarasan, S., Nalini, R., Gunasekaran, P., Varalakshmi, E., Anuradha, C.V. 2006. Fenugreek (*Trigonella foenum graecum*) seed extract prevents Chang liver cells against ethanol-induced toxicity and apoptosis. *Alc Alcohol* 41: 267–73.

Kaviarasan, S., Soundarapandiyan, R., Anuradha, C.V. 2008. Protective action of fenugreek (*Trigonella foenum graecum*) seed polyphenols against alcohol-induced protein and lipid damage in rat liver. *Cell Biol Toxicol* 24: 391–400.

Kaviarasan, S., Vijayalakshmi, K., Anuradha, C.V. 2003. A polyphenol-rich extract of fenugreek seeds protect erythrocytes from oxidative damage. *Plant Foods Hum Nutr* 59: 143–47.

Kaviarasan, S., Viswanathan, P., Anuradha, C.V. 2007b. Fenugreek seed (*Trigonella foenum graecum*) polyphenol inhibit ethanol-induced collagen and lipid accumulation in rat liver. *Cell Biol Toxicol* 23: 373–83.

Leela, N.K., Shafeekh, K.M. 2008. Fenugreek. In: Parthasarathy, V.A., Chempakam, B., Zachariah, T.J., eds. *Chemistry of Spices*. Biddles Ltd, King's Lynn, UK, CAB International, pp. 242–59.

Loeper, M., Lemaire, A. 1931. Sun quelques points delaction generale des amers. *Presse. Med.* 24: 433–35.

Madar, Z. 1984. Fenugreek (*Trigonella foenum graecum*) as a means of reducing postprandial glucose level in diabetic rats. *Nutr Rep Int* 29: 1267–73.

Madar, Z., Abel, R., Samish, S., Arad, J. 1988. Glucose-lowering effect of fenugreek in non-insulin dependent diabetics. *Eur J Clin Nutr* 42: 51–4.

Madar, Z., Shomer, I. 1990. Polysaccharide composition of a gel fraction derived from fenugreek and its effect on starch digestion and bile acid absorption in rats. *J Agr Food Chem* 38: 1535–39.

Mansour, E.H., El-Adawy, T.A. 1994. Nutritional potential and functional properties of heat-treated and germinated fenugreek seeds. *LWT-Food Sci Technol.* 27: 568–72.

Mishkinsky, J., Joseph, B., Sulman, F.G. 1967. Hypoglycaemic effect of trigonelline. *Lancet* 2: 1311–12.

Mital, N., Gopaldas, T. 1986. Effect of fenugreek (*Trigonella foenum graecum*) seed based diets on the birth outcome in albino rats. *Nutr Rep Int* 33: 363–69.

Moissides, M. 1939. Le fenugrec autrefois et aujourd'hui. *Janus*. 43: 123–30.

Morcos, S.R., Elhawary, Z., Gabrial, G.N. 1981. Protein-rich food mixtures for feeding the young in Egypt Formulation. *Z. Ernahrungswiss.* 20: 275–82.

Neeraja, A., Rajalakshmi, P. 1996. Hypoglycemic effect of processed fenugreek seeds in humans. *J Food Sci Techol* 33: 427–30.

Petit, P.R., Sauvaire, Y.D., Hillaire-Buys, D.M., Leconte, O.M., Baissac, Y.G., Ponsin, G.R., Ribes, G. 1995. Steroid saponins from fenugreek seeds: Extraction, purification and pharmacological investigation on feeding behavior and plasma cholesterol. *Steroids* 10: 674–80.

Petit, P.R., Sauvaire, Y., Ponsin, G., Manteghetti, M., Fave, A., Ribes, G. 1993. Effects of fenugreek seed extract on feeding behaviour in the rat: Metabolic-endocrine correlates. *Pharm Biochem Behav* 45: 369–74.

Platel, K., Srinivasan, K. 2000. Influence of dietary spices and their active principles on pancreatic digestive enzymes in albino rats. *Nahrung* 44: 42–6.

Raghuram, T.C., Sharma, R.D., Sivakumar, B., Sahay, B.K. 1994. Effect of fenugreek seeds on intravenous glucose disposition in non-insulin dependent diabetic patients. *Phy Res* 8: 83–6.

Raju, J., Gupta, D., Rao, A.R., Baquer, N.Z. 1999. Effect of antidiabetic compounds on glyoxylase I activity in experimental diabetic rat liver. *Ind J Exp Biol* 37: 193–95.

Ribes, G., Sauvaire, Y., Costa, C.D., Baccou, J.C., Mariani, L.M.M. 1986. Antidiabetic effects of sub fractions of fenugreek seeds in diabetic dogs. *Proc Soc Exp Biol Med* 182: 159–66.

Sambaiah, K., Srinivasan, K. 1989. Influence of spices and spice principles on hepatic mixed function oxygenase system in rats. *Ind J Biochem Biophys* 26: 254–58.

Sauvaire, Y., Petit, P., Broca, C., Manteghetti, M., Baissac, Y., Fernandez-Alvarez, J., Gross, R., Roye, M., Leconte, A., Gomis, R., Ribes, G. 1998. 4-Hydroxyisoleucine: A novel amino acid potentiator of insulin secretion. *Diabetes* 47: 206–10.

Sauvaire, Y., Ribes, G., Baccou, J., Loubafieres, M. 1991. Implication of steroid-saponins and sapogenins in the hypocholesterolemic effect of fenugreek. *Lipids* 26: 191–97.

Shang, M., Cai, S., Han, J., Li, J., Zhao, Y., Zheng, J., Namba, T., Kadota, S., Tezuka, Y., Fan, W. 1998a. Studies on flavonoids from fenugreek (*Trigonella foenum graecum* L.). *Zhongguo Zhong Yao Za Zhi* 23: 614–16.

Shang, M., Cai, S., Wang, X. 1998b. Analysis of amino acids in *Trigonella foenum graecum* seeds. *Zhong Yao Cai* 21: 188–90.

Shani, J., Goldshmied, A., Joseph, B., Ahronson, Z., Sulman, F.G. 1974. Hypoglycemic effect of *Trigonella foenum graecum* and *Luipnus termis* (*Leguminosae*) seeds and their major alkaloid in alloxan-diabetic and normal rats. *Arch Int Pharmacodyn Ther* 210: 27–36.

Shankaracharya, N.B., Natarajan, C.P. 1972. Fenugreek—Chemical composition and use. *Ind Spices* 9: 2–12.

Sharma, R.D. 1986a. Effect of fenugreek seeds and leaves on blood glucose and serum insulin responses in human subjects. *Nutr Res* 6: 1353–64.

Sharma, R.D. 1986b. An evaluation of hypocholesterolemic factor of fenugreek seeds (*Trigonella foenum graecum*). *Nutr Rep Int* 33: 669–77.

Sharma, R.D., Raghuram, T.C. 1990. Hypoglycemic effect of fenugreek seeds in non-insulin dependent diabetic subjects. *Nutr Res* 10: 731–39.

Sharma, R.D., Raghuram, T.C., Rao, N.S. 1990. Effect of fenugreek seeds on blood glucose and serum lipids in type 1 diabetes. *Eur J Clin Nutr* 44: 301–06.

Sharma, R.D., Raghuram, T.C., Rao, D.V. 1991. Hypolipidaemic effect of fenugreek seeds. A clinical study. *Phy Res* 3: 145–47.

Sharma, R.D., Sarkar, A., Hazra, D.K. 1996a. Hypolipidaemic effects of fenugreek seeds: A chronic study in non-insulin dependent diabetic patients. *Phy Res* 10: 332–34.

Sharma, R.D., Sarkar, A., Hazra, D.K., Mishra, B., Singh, J.B., Sharma, S.K., Maheshwari, B.B., Maheshwari, P.K. 1996b. Use of fenugreek seed powder in the management of non-insulin dependent diabetes mellitus. *Nutr Res* 16: 1331–39.

Stark, A., Madar, Z. 1993. The effect of an ethanol extract derived from fenugreek (*Trigonella foenum graecum*) on bile acid absorption and cholesterol levels in rats. *Br J Nutr* 69: 277–87.

Sujapandian, R., Anuradha, C.V., Viswanathan, P. 2002. Gastroprotective effect of fenugreek seeds (*Trigonella foenum graecum*) on experimental gastric ulcer in rats. *J Ethnopharmacol* 81: 393–97.

Sur, P., Das, M., Gomes, A., Vedasiromoni, J.R., Sahu, N.P., Banerjee, S., Sharma, R.N., Ganguly, D.K. 2001. *Trigonella foenum graecum* (fenugreek) seed extract as an antineoplastic agent. *Phy Res* 15: 257–59.

Thirunavukkarasu, V., Anuradha, C.V. 2006. Gastroprotective effect of fenugreek seeds (*Trigonella foenum graecum*) on experimental gastric ulcer in rats. *J Herbs Spices and Medicinal Plants* 12: 13–35.

Thirunavukkarasu, V., Viswanathan, P., Anuradha, C.V. 2003. Protective effect of fenugreek (*Trigonella foenum graecum*) seeds in experimental ethanol toxicity. *Phy Res* 17: 737–43.

Umarova, F.T., Khushbactova, Z.A., Batirov, E.H., Mekler, V.M. 1998. Inhibition of Na^+/K^+-ATPase by flavonoids and their inotropic effect. Investigation of the structure–activity relationship. *Cell Biol* 12: 27–40.

Vats, V., Grover, J.K., Rathi, S.S. 2002. Evaluation of anti-hyperglycemic and hypoglycemic effect of *Trigonella foenum graecum* Linn, *Ocimum sanctum* Linn and *Pterocarpus marsupium* Linn in normal and alloxan-diabetic rats. *J Ethnopharmacol* 79: 95–100.

Vijayakumar, M.V., Sandeep, S., Chhipa, R.R., Manoj, K.B. 2005. The hypoglycemic activity of fenugreek seed extract is mediated through the stimulation of an insulin signaling pathway. *Br J Phar* 146: 41–8.

Yoshikawa, M., Murakami, T., Komatsu, H., Murakami, N., Yamahara, J., Matsuda, H. 1997. Medicinal foodstuffs. IV. Fenugreek seed. (I): Structure of trigoneosides Ia, Ib, IIa, IIb, IIIa, and IIIb, new furostanol saponins from the seeds of Indian *Trigonella foenum graecum* L. *Chem Phar Bull* 45: 81–7.

12 Soybean as a Special Functional Food Formula for Improving Women's Health

Erin Shea Mackinnon and Leticia G. Rao

CONTENTS

The soybean is a source of protein, fats, oligosaccharides, and dietary fiber. It is considered to be a complete food because in addition to these macronutrients, it also contains minerals (Liu, 1999; Mateos-Aparicio et al., 2008), essential amino acids and beneficial secondary metabolites; phytochemicals such as isoflavones and other phenolic compounds (Sakthivelu et al., 2008). Soy intake is the highest in Asia, where isoflavone intake is estimated to be 20–50 mg/day (Adlercreutz et al., 1991; Cassidy, 2003). Amongst the Western population, soy consumption is much lower and infrequent; isoflavone intake is negligible at less than 1 mg/day on an average (Cassidy, 2003; Setchell et al., 1999). Table 12.1 shows commonly consumed soy products.

TABLE 12.1
Isoflavone Content in Foods in Commonly Consumed Foods, Listed from Lowest to Highest (mg/100 g) Showing Daidzein, Genistein, Glycitein, and Total Isoflavone Content, as Listed on USDA Database

	Amount Isoflavone (mg/100 g)			
Type of Food	**Daidzein**	**Genistein**	**Glycitein**	**Total Isoflavones**
Peanut butter (reduced fat)	1.30	0.69	0.08	2.09
Pistachio nuts (raw)	1.88	1.75	0.0	3.63
Donut, plain	2.58	2.44	0.29	5.34
Vegetarian hamburgers	2.36	5.01	0.55	6.39
Soy noodles	0.90	3.70	3.90	8.50
Oncom (raw)	6.60	3.10	0.0	9.70
Black bean sauce	5.96	4.04	0.53	10.26
Soymilk	4.84	6.07	0.93	10.73
Vegetarian hot dogs	5.78	6.43	0.06	12.27
Soybean seed sprouts (steamed)	5.00	6.70	0.80	12.50
Sufu	7.50	5.46	0.78	13.75
Bread (soy and linseed)	4.87	9.13	0.67	14.67
Soy lecithin	5.40	10.30	0.0	15.70
Mayonnaise (tofu based)	5.50	11.30	0.0	16.80
Silken Tofu	9.15	8.42	0.92	18.04
Red clover	11.00	10.00	0.0	21.00
Firm tofu (cooked)	10.26	10.83	1.35	22.05
Soy cheese	5.79	11.14	0.0	25.72
Soy yogurt	13.77	16.59	2.80	33.17
Soybean seed sprouts (raw)	12.86	18.77	2.88	34.39
Soy paste	19.71	17.79	6.05	38.24
Miso	10.43	23.24	3.00	41.45
Soy fiber	18.80	21.68	7.90	44.43
Soybean chips	26.71	27.45	0.0	54.16
Tempeh	22.66	36.15	3.82	60.61
Soybeans (boiled)	30.76	31.26	3.75	65.11
Miso soup	29.84	40.00	0.0	69.84
Soy protein drink	27.98	42.91	10.76	81.65
Natto	33.22	37.66	10.55	82.29
Soy protein isolate	30.81	57.28	8.54	91.05
Instant soy beverage (powdered)	40.07	62.18	10.90	109.51
Vegetarian bacon bits	64.37	45.77	8.33	118.50
Soy nuts	62.14	75.78	13.33	148.50
Soybeans (raw)	62.07	80.99	14.99	154.53
Soy flour (textured)	67.69	89.42	20.02	172.55

Source: Adapted from U.S. Department of Agriculture ARS. USDA Database for the Isoflavone Content of Selected Foods, Release 2.0. In Agriculture UDo, ed. Nutrient Data Laboratory Home Page, 2008.

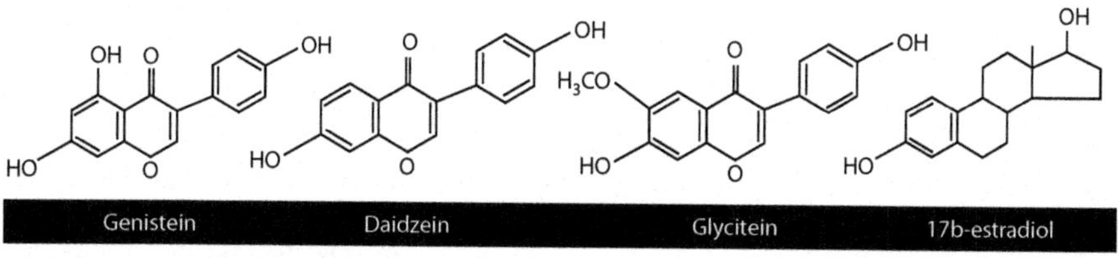

FIGURE 12.1 Chemical structures of the isoflavones genistein, daidzein and glycitein compared to the classical estrogen, 17 β-estradiol.

Among legumes, the soybean is one of the highest sources of protein, about 40%, while others contain only 20–25% protein (Liu, 1999; Xiao, 2008). Approximately 90% of the proteins in soybeans are composed of two storage globulins, 11S glycinin and 75 β-conglycinin (Torres et al., 2006), which in themselves contain the essential amino acids, making them a suitable replacement for protein from animal sources (Xiao, 2008). However, the quality of protein in soybeans may be diminished by other minor components including phytic acid, phenolics, and trypsin inhibitors (Liu, 1999).

Soybeans also contain 35% carbohydrates (Mateos-Aparicio et al., 2008) (Figure 12.1), including the soluble di- and oligosaccharides, sucrose (2.5–8.2%), raffinose (0.1–0.9%), stachyose (1.4–4.1%), and starch (1%) (Mateos-Aparicio et al., 2008). Raffinose and stachyose contribute greatly to soybean food functionality because they are prebiotics which enhance the growth of *Bifidobacterium* (Espinosa-Martos and Ruperez, 2006).

Almost half of the energy in soybeans is derived from fat (Messina, 1999). Containing 18–22% oil, the fat fraction is composed of 99% triglycerols, while the remaining components include phospholipids, free fatty acids, tocopherols, and phytosterols (Liu, 1999). However, soybeans are low in saturated fat and free of cholesterol. They contain high concentrations of the polyunsaturated fatty acids, linoleic (C18:2) (53% of soybean fat) and linolenic (C18:3) acids (7–8% soybean fat), and unsaturated fatty acids, including oleic acid (C18:1). They also contain the saturated fatty acids, palmitic (C16:0) and stearic (C18:0) (Mateos-Aparicio et al., 2008).

The main components of soybeans which contribute to their classification as a functional food are the polyphenols, including isoflavones, glycosides and malonate conjugates (Lee et al., 2008). Isoflavones are phytochemicals which are part of the flavonoid family—the nonnutritive substances with protective health benefits (Mateos-Aparicio et al., 2008), including antioxidant activity. (Lozovaya et al., 2005) (Patel et al., 2001; Yen et al., 2003). Their basic structure is composed of the flavone nucleus, two benzene rings linked with heterocyclic pyrane (Mateos-Aparicio et al., 2008). They are structurally similar to naturally occurring estrogens (Sakthivelu et al., 2008). They are highly polar, water soluble compounds (Mateos-Aparicio et al., 2008) with low molecular weight, hydrophobic peptides, or fatty acid components (Sakthivelu et al., 2008).

Soybean contains 0.2–1.6 mg of isoflavones per gram of dry weight (Setchell et al., 1999). In the soybean seed, more than 80% of the total isoflavone content is found in cotyledons (Sakthivelu et al., 2008), which contain more than 20 mg/g of isoflavones (Mateos-Aparicio et al., 2008), primarily in their storage form, malonyl glucoside (Sakthivelu et al., 2008). The three main isoflavones found in soybeans are genistein, daidzein, and glycitein (Lee et al., 2008; Mateos-Aparicio et al., 2008; Xiao, 2008). They may be derived from their precursors, biochanin A and formononetin (Sirtori, 2001). In most Western countries, these major isoflavones in soy-based foods are conjugated to sugars in glycosidic form, but in Asian soy foods, they tend to be more active and bioavailable because they contain higher levels of aglycone (Xiao, 2008). These three main isoflavones are present in the four following glycosidic forms in soy: aglycones, β-glucosides, acetylglucosides, and malonyl glucosides (Mateos-Aparicio et al., 2008; Sakthivelu et al., 2008).

The content of isoflavones in soybeans is affected by many variables including cultivar, tissue type, as well as certain growth conditions like planting location, crop year, soil nutrition, temperature, and storage time. Low temperature and high precipitation during seed development have been associated with higher isoflavone contents (Sakthivelu et al., 2008).

12.1 HISTORY OF THE SOYBEAN AND ITS DEVELOPMENT AND APPLICATION AS FUNCTIONAL FOOD

Soybean is classified as part of the order of *Rosaceae*, in the family of *Leguminosae* or *Papillonaceae*, subfamily of *Papilionoidae*, genus of *Glycine*, and the cultivar Glycine max. (Mateos-Aparicio et al., 2008). Soybean originated from central and northern China 4000 to 5000 years ago (Mateos-Aparicio et al., 2008) and has been an important crop for centuries (Malencic et al., 2008). In the East, where a rice-based diet is prevalent, frequent consumption of soybeans is an effective method of decreasing protein deficiency (Lee et al., 2008). However, the soybean was not introduced in Europe until 1712 by a German botanist, Engelbert Kaempfer. Later, it was classified as Glycine max by Carl von Linne (Liu, 1999; Mateos-Aparicio et al., 2008). At the time, however, inadequate soil and climatic conditions (Liu, 1999; Mateos-Aparicio et al., 2008) limited its European production to a minor crop of stored feed. It was not until the last century that the worldwide production of the soybean increased to provide a major source of oil and protein (Malencic et al., 2008). In 1931, high concentrations of isoflavones in soybeans (Walz, 1931) were discovered, and 10 years later genistein glycoside was isolated from soybeans (Walter, 1941).

For many years, the common belief was that the isoflavones in soy-based foods were affected by fermentation. It was thought that fermented foods like miso and tempeh contained unconjugated isoflavone aglucones, while nonfermented foods, such as soymilk, tofu, soy flour, soy protein conjugate, isolated soy protein, contained the β-glucoside conjugates. At this time, isoflavone analysis was performed by heated extraction in aqueous solvents including acetonitrile, ethanol, and methanol (Coward et al., 1998). It was not ascertained until 1991 that isoflavone extraction without heat resulted in a different type of isoflavone glucoside, the malonyl β-glucoside conjugates (Coward et al., 1998).

Recently, proposed health benefits of soybeans have increased their popularity and their use as a "functional food" in both Asia and Europe (Lee et al., 2008).

Official health claims have been made with respect to soy as a functional food in Japan (1996) (Xiao, 2008), the United States (1999) (Administration, 1999), the United Kingdom (2002), South Africa (2002), the Philippines (2004), and Korea, Indonesia, and Brazil (2005) (Xiao, 2008). Approved claims state that the soybean protein, in conjunction with a diet low in saturated fat and cholesterol, is effective for risk reduction of coronary heart disease (Administration, 1999). Similar health claims are currently being reviewed in Canada and France (Xiao, 2008).

12.2 DISTRIBUTION, PROCESSING, STORAGE CONDITIONS, AND STABILITY OF SOY

In Asia, soybeans are classified according to 100-seed weight into small, medium, or large. Typically, the small soybeans are most often used for sprouts, while the medium and large soybeans are used for either cooking with rice and vegetables, or to produce soybean curd, fermented soybean pastes, soy milk, or sauce (Lee et al., 2008). Further, they are considered a complete food in terms of nutritional benefit. Cultivars and hybrids, are both inexpensive and readily available as a raw material for large-scale pharmaceutical and food industries (Malencic et al., 2008).

Soybean export is an expanding industry. Most soybean imports and exports are done using ship or land-carriage, which involves transport time and climatic variations could adversely affect the quality of the bean (Kim et al., 2005). Once harvested, soybeans are stored on either the farm or in transport vehicles until food processing begins. The main factors influencing storage are seed moisture content, temperature, duration of storage, and humidity. Other factors include the amount of foreign materials present and the general conditions of the product prior to storage. Typically, storage conditions vary widely, from low temperature and low humidity to high temperature and high humidity. Studies suggest that storage results in decreased water absorption rate, solid extractability and tofu yield, accompanied by an increase in soymilk acidity, hardness, and resilience (Kong et al., 2008). Experimental evidence demonstrated that storage for up to four years yielded a range of total polyphenols of 730–1812 μmol g^{-1}, mainly enclosed within the embryo, cotyledon, and seed coat (Lee et al., 2008). Other studies on storage have determined that soybean deterioration was augmented by increasing heat and humidity, which is probably a result of decreased solid and protein content in soymilk, soymilk pH, tofu yield, water-holding ability of proteins and increased tofu hardness, resilience, and elasticity. Exposure to extremely high temperature and humidity was found to weaken the hydrophobic bond interactions and decreased disulfide cross-links among the soy protein molecules. Optimal storage conditions are a low temperature (<22°C) and humidity of 55–60%, which corresponds to 8–10% of the original moisture content of the beans (Kong et al., 2008). However, experimental evidence shows that the malonyl glucosides are only stable for 24 h at 4°C and further storage can lead to decomposition of this isoflavone (Coward et al., 1998).

In Asia, the most traditional foods made from soybeans are milk and tofu (Kong et al., 2008). Production of these foods involves soymilk pH, solid extractability, tofu texture, color, and yield (Kong et al., 2008). Many popular foods worldwide contain soy products, but the products are primarily soy oils and lecithin, which contain minimal isoflavones (Setchell et al., 1999). During processing, isoflavones migrate into the protein fraction of the soybean (Setchell et al., 1999). Glucoside conjugation is not altered by either grinding to make soy flour, or hexane extraction to remove fats. However, certain processing conditions for soy-based foods alter the chemistry of isoflavone conjugates; the glucosides in particular, are very easily altered. Malonyl glucosides are unstable when exposed to heat, and aqueous heating, as used in methanol extraction, results in their conversion into β-glucoside conjugates; extractions performed at room temperature reduce the rate of isoflavone loss. Glucoside conjugation is also altered by (1) dry heat to roast flour, (2) extrusion to produce textured vegetable protein (TVP) from loss of carbon dioxide, (3) hot aqueous extraction for tofu or soymilk, and (4) fermentation to make miso and tempeh. Ethanol extraction of soy flour to make soy protein concentrates completely diminishes isoflavones (Coward et al., 1998) (Table 12.2).

Cooking also changes glucoside conjugates. Malonyl glucosides are the most easily and commonly affected by cooking; they are converted into β-glucoside and acetyl-β-glucoside conjugates. The type of conjugate to which they are converted depends on the type of cooking performed; baking yields β-glucoside conjugates while frying yields acetyl-β-glucoside conjugates (Coward et al., 1998). Under normal cooking conditions, the total isoflavone content is not diminished; however if burning occurs, there is an increase in aglucones and a decrease in total isoflavones. This effect is observed in soy oil (Cassidy, 2003; Cassidy et al., 2006; Coward et al., 1998) (Table 12.2).

In addition to being inexpensive (Malencic et al., 2008) to market on a large scale, soybean have nutritional qualities, health benefits, and consumer popularity that make them good candidates for research and development in the field of biotechnology. Glyphosate-tolerant soybeans (GTSs, Roundup Ready soybeans, Montsanto)

TABLE 12.2
Comparison of Different Extraction Methods and the Affects on Total Isoflavone Content in Foods, Compared to Raw, Unprocessed Soybeans

Type of Food	Extraction Method Used	Total Isoflavones (mg/100 g)
Soybeans (raw)	None	154.53
Soy flour (full fat)	Grinding	178.10
Soy flour (defatted)	Grinding and hexane extraction	150.94
Silken Tofu	Hot aqueous extraction	18.04
Firm tofu (cooked)	Hot aqueous extraction	22.05
Tempeh	Fermentation	60.61
Miso	Fermentation	41.45
Soy protein concentrate	Aqueous extraction	94.65
Soy protein concentrate	Alcohol extraction	11.49
Soybean oil	Extreme heat	0.0

were the first biotechnologically improved soybeans to be marketed as early as 1996. Research indicates that it will be possible to develop and market soybeans with higher nutrient content, such as amino acids and energy. Another attractive possibility is its superior oil content, with greater oxidative stability and modified saturated fat content. In addition, from a farming perspective, crops of soybeans tolerant to herbicides would be beneficial. Each of these possibilities is being explored and it is thought that they would yield a soybean comparable to those which are currently being grown and marketed (Rogers, 1998).

12.3 BIOAVAILABILITY AND SAFETY OF SOY-BASED FOODS

Soy bioavailability is affected by the dose, chemical form of the compound, and the type of food matrix in which it is consumed (Cassidy, 2003; Cassidy et al., 2006) (Table 12.1). The proportion of each chemical isoflavone which is absorbed into the blood is still poorly understood (Coward et al., 1998).

Absorption and utilization of isoflavones comprise multiple deconjugation and reconjugation steps (Sirtori, 2001). After consumption, it is hypothesized that isoflavone glycosides are hydrolyzed by both gastric acid in the stomach and enzymes secreted by microflora in the intestine (Cassidy, 2003; Cassidy et al., 2006; Slavin et al., 1998), resulting in deconjugation (Xu et al., 1995). However, these glycosides cannot be metabolized unless they are hydrolyzed back into their bioactive forms—genistein, daidzein or aglycones—during digestion (Xiao, 2008). Glucosidases are intestinal enzymes (Cassidy, 2003; Cassidy et al., 2006), which release the aglycones into daidzein and genistein (Mateos-Aparicio et al., 2008). Different types of isoflavones are thought to be absorbed from different portions of the intestine: daidzein and genistein are readily absorbed from the upper portion of the small intestine, β-glucoside conjugates from the distal small intestine after hydrolysis to aglucone, and malonyl glucoside conjugates are absorbed from the large intestine after hydrolysis (Barnes et al., 1996). The hydrolyzed glycosides which are not absorbed by the intestinal lumen are further metabolized and degraded by intestinal microflora (Cassidy, 2003; Cassidy et al., 2006; Slavin et al., 1998; Xu et al., 1995) to further produce equol and *O*-desmethyl-anglolensin. These isoflavones metabolites are in the form of heterocyclic phenols, which are similar in structure to estrogen (Sirtori, 2001). After absorption, isoflavones are reconjugated mainly as glucuronic acid, with minor amounts of sulfuric acid. In blood, only a diminutive amount of free agylcone is detectable, demonstrating that this rate of reconjugation is quite high (Rowland et al., 2003). Conjugated isoflavone are excreted in the urine and bile (Sirtori, 2001).

Studies on excretion of isoflavones show that genistein and daidzein are present in the urine after consumption of soy-based foods, together with the following metabolites: daidzein metabolites; equol and *O*-desmethylangolensin (O-DMA) (Sirtori, 2001), and genistein metabolites; 6′-hydroxy-*O*-desmethylangolensin (Joannou et al., 1995; Slavin et al., 1998). Research also shows that urinary measurements of isoflavones are a useful tool to measure dietary isoflavone consumption (Slavin et al., 1998) because excretion of isoflavones is strongly correlated with their intake (Karr et al., 1997), and excretion is linear at low to moderate levels of soy consumption (Slavin et al., 1998). Foods which are fermented result in an increased recovery of urinary isoflavones, suggesting an increased bioavailability of isoflavones (Cassidy et al., 2006; Slavin et al., 1998).

The half-life of isoflavones in circulation is usually 8–10 h (Cassidy, 2003; Cassidy et al., 2006). However, differences in excretion rates of soy isoflavones have been found, which indicates that there are differences in the half-life of isoflavones in circulation, resulting in differing physiological effects (Slavin et al., 1998). Bioavailability of isoflavones is affected by dietary fiber (Cassidy et al., 2006) and use of antibiotics (Cassidy, 2003); in women excretion of isoflavone metabolites has been shown to be associated with higher intake of fiber and carbohydrates (Slavin et al., 1998).

Different sources of soy isoflavones have different bioavailability; isoflavone chemistry is altered by cooking, absorption, and metabolism (Cassidy et al., 2006). In addition, the metabolism of isoflavones has been shown to vary among individuals (Slavin et al., 1998), suggesting that there are multiple metabolic pathways which are differentially active (Sirtori, 2001), due to which it has been recommended that instead of a single daily dose, doses should be given in smaller amounts throughout the day for higher efficacy (Cassidy, 2003; Cassidy et al., 2006). All these variables need to be taken into account during the interpretation of clinical trial data (Coward et al., 1998), and may account for the high level of variability among results of studies associating consumption of soy with decreased risk of disease. Indeed, an extensive range of food sources is used in clinical studies, including pure compound (tablets or capsules), soybean germ or red clover (*Trifolium pratense* L.) extracts, soybean flour, soy beverages, isoflavone-enriched soybean protein, or soybean protein drinks (Cassidy et al., 2006). These different sources of soy-based foods result in different levels of isoflavone consumption during clinical trials. For uniformity in clinical studies, the amount of isoflavone in the provided food or supplement needs to be measured at the time of consumption (Cassidy et al., 2006), rather than at the time of production. This prevents misrepresentation of isoflavone consumption which may occur if measurements are done before consumption; storage and processing will affect isoflavone concentration in the final product, and stability of isoflavones is different depending on the matrix in which it is stored (Lee et al., 2003).

The safety of isoflavone consumption has been a focus of research mainly due to the estrogenic properties of isoflavones, particularly related to tissues which are regulated by estrogen such as endometrial, uterine, and breast (Cassidy, 2003). Currently, there is no data to support this concern that isoflavones exert the same risks as elevated estrogen (Cassidy, 2003; Sirtori, 2001), and past precedent has been established in Asian countries which have safely consumed these products for many generations (Cassidy, 2003). Genotoxic concentrations of genistein have been reported to be 25 mol/L (Sirtori, 2001), however, even high consumption of soy results in nontoxic plasma levels of only 0.2–1 mol/L (Setchell et al., 1997).

Isoflavones have other effects and mechanisms of action however, which may pose a safety risk, not limited to their estrogenic properties (Cassidy, 2003). Some potential adverse effects of isoflavones have been observed and these include antithyroid actions, endocrine disruption and carcinogenesis augmentation (Xiao, 2008). *In vitro*, genistein has been shown to inhibit tyrosine kinase, the phosphorylation of which is essential to cell cycle check points during DNA damage. Tyrosine kinase monitors the progression from G2 to the M-phase and if inhibited may allow mitosis of damaged cells (Sirtori, 2001). Genistein has also been shown to induce cell death in cell cultures (Kumi-Diaka et al., 1998, 1999), an effect which is enhanced in the presence of dexamethasone, which is a matter of concern (Sirtori, 2001). In oviduct cells, isoflavones are shown to induce synthesis of the leukemia inhibitory factor, which is an essential glycoprotein for embryo implantation, and may lead to infertility (Reinhart et al., 1999). These results have yet to be verified in animal or human studies (Sirtori, 2001), and although they are still poorly understood, they may pose a concern for consumers, health professionals, and policy makers (Xiao, 2008).

12.4 MECHANISMS OF ISOFLAVONE ACTION

Mechanisms of isoflavone action are multifactorial (Cassidy, 2003; Cassidy et al., 2006) and include different levels of ERβ receptor effect action, different types of response elements which receptors interact with in ER-regulated genes, and the varying presence of co-activators and co-repressors in different cell types. Table 12.3 summarizes the proposed mechanism of isoflavone action, which explains in part the high variability seen in different isoflavone or soy experimental studies (Cassidy et al., 2006).

12.4.1 Estrogenic Mechanisms of Action

Isoflavones are part of the phytoestrogen family; they have a similar chemical structure to estrogen and estradiol, but are nonsteroidal (Cassidy, 2003). Chemically, isoflavones and estrogen are almost identical; the distance between the hydroxyl groups at each end of both molecules is the same (Setchell et al., 1999) (Figure 12.1). Isoflavones also contain the phenolic ring which is the crucial component to estrogen receptor binding (Setchell et al., 1999). Although isoflavones bind to both types of estrogen receptors (ER), they bind preferentially to ERβ. In fact, they bind with 20 times more affinity to ERβ than to ERα, and are 500–850 times more effective in activating the binding of ERβ to estrogen response elements within target genes (Cassidy, 2003; Xiao, 2008). Despite this, overall, the estrogenic activity of isoflavones is 100–1000 times weaker than estradiol itself. Interestingly, it has been

TABLE 12.3
Summary of the Proposed Mechanisms of Isoflavone Action

Nonestrogenic	Estrogenic
• Antioxidant activities • Inhibition of tyrosine autophosphorylation • Modulation of TGF-β signaling pathways • Inhibition of DNA topoisomerase II activity • Regulation of cell cycle checkpoints • Antiangiogenic activity • Antiproliferative effects on transcriptional processes	• ER agonists • Activate binding of ERβ to estrogen response elements in target genes • ER antagonists • Inhibit metastases and angiogenesis in cancer development • Estrogenic-mimicking activities—affect growth of tumors

reported that high amounts of isoflavones in food cause highly elevated plasma estrogen levels relative to those normally seen from endogenous estrogen levels (Cassidy, 2003). Among isoflavones and their metabolites, the daidzein metabolite equol exhibits the highest estrogenic qualities (Slavin et al., 1998). Unlike estrogen, isoflavones do not accumulate in fat tissues (Cassidy, 2003; Cassidy et al., 2006).

The similarity of isoflavones to estrogen provides them with similar tissue-specific effects (Cassidy, 2003), such as those exerted in breast, endometrial and ovarian tissues. An ER-β-dependent mechanism of genistein action has been shown experimentally to inhibit metastases and angiogenesis in the development of cancer (Fotsis et al., 1993; Li et al., 1999a, 1999b; Schleicher et al., 1999; Zhou et al., 1999). Another *in vitro* study suggests that isoflavones influence tumor formation through their estrogenic-mimicking activities, by affecting the growth of benign and malignant tumors (Malencic et al., 2007). Human breast cancer cells convert isoflavones into phase I and phase II metabolites (Peterson et al., 1998), and under cell culture conditions which stimulate oxidative bursts, isoflavones are converted into halogenated and nitrated derivatives (Patel et al., 2001). These modifications of isoflavones might influence their binding to ER or other proteins (Sirtori, 2001).

Isoflavones may act similarly to selective estrogen receptor modulators (SERMs) (Sirtori, 2001), in that they act as partial agonists and antagonists to ER. It is possible that at particular concentrations they compete with estrogen binding both to antagonize and inhibit the effect of estrogen; a mechanism of action which is also credited to Raloxifene (Setchell et al., 1999).

Owing to the weak estrogenic properties of isoflavones and their estrogenic mechanisms of action, the consumption of soy has been recommended to manage osteoporosis and menopausal symptoms, such as hot flush (Lee et al., 2008; Blair et al., 1996).

12.4.2 Nonestrogenic Mechanisms of Action

The isoflavone genistein has been shown to modulate the growth factor- and cytokine-stimulated proliferation of both normal and cancerous cells (Kim et al., 1998). Experimental evidence shows two main mechanisms of genistein action: (1) in the mammalian cell, genistein is a strong and specific inhibitor of tyrosine autophosphorylation of epidermal growth factor (EGF) receptor, and (2) in cell culture and human studies, genistein may inhibit cell growth by modulating the transforming growth factor (TGF)-β signaling pathways. Other possible mechanisms of action include: inhibition of DNA topoisomerase II activity, regulation of cell cycle checkpoints, and antiangiogenic activity (Kim et al., 1998). Genistein has been shown to dose-dependently increase the expression and production of TGF-β (Kim et al., 1998; Sathyamoorthy et al., 1998), which is an inhibitor of epithelial cell growth in normal and cancerous breast epithelial cells. Another possibility is that genistein is antiproliferative through its effects on transcriptional processes. In other words, the varying effects of genistein and other isoflavones in estrogen-sensitive tissues may depend on the production of both paracrine and autocrine growth factors that cause proliferation of cells which do not express estrogen receptors (Sirtori, 2001).

The nonestrogenic mechanism of action which has been most extensively studied is the antioxidant mechanism. Isoflavones are antioxidants and the main focus of soybean research. However, it should be noted that soybeans contain other types of polyphenols, particularly flavonoids, predominantly within the seed extract, which possess a high antioxidant capacity (Malencic et al., 2007). In fact, different cultivars of soybeans have not only varying levels of polyphenols but varying radical scavenging activities. The total antioxidant activity correlates positively with total polyphenol content, as well as the size of the soybean (Lee et al., 2008). Polyphenols with the highest variability among cultivars include tannins and proanthocyanidins; experimental studies show that Chinese and Serbian genotypes of soybeans are richer in total phenolics, while American genotypes are rich in proanthocyanidins (Malencic et al., 2007). Isoflavones and other flavonoids in soybeans directly scavenge reactive oxygen species and peroxynitrite, a reactive oxidant that is formed when superoxide reacts with nitric oxide (Malencic et al., 2007).

Research shows that isoflavones act as antioxidants to enhance intrinsic antioxidant enzymes, either directly or indirectly, including catalase, superoxide dismutase, glutathione peroxidase, and glutathione reductase (Kurzer et al., 1997), which enhances the internal response to combat oxidative stress. Furthermore, *in vitro* and *in vivo* studies show that isoflavones are capable of decreasing lipid peroxidation (Kapiotis et al., 1997; Kerry and Abbey, 1998; Tikkanen et al., 1998; Wagner et al., 1997).

The antioxidant capacity of soybean polyphenols, including isoflavones, results in anticarcinogenic and antiatherogenic mechanisms. Cellular and molecular research shows that isoflavones modulate both the transcription factors of lipid metabolism and their downstream expression at transcriptional and posttranslational levels (Xiao, 2008). Isoflavones also have other antioxidant protective mechanisms related to cancer and cardiovascular disease including plasma lipid modifications, vascular reactivity changes, and hormonal actions (Sirtori, 2001). Indeed, animal and human studies show that consumption of isoflavones is associated with decreased risk of CVD, including lower liver and blood triglycerol, increased HDL-cholesterol and HDL:LDL ratio (Xiao, 2008), and molecular events in the initiation, promotion, and progression stages of cancer (Malencic et al., 2007).

12.5 HEALTH BENEFITS OF SOY IN WOMEN: NUTRITIONAL AND PHYSIOLOGICAL EFFECTS AND EFFICACIES IN HEALTH AND DIET

Till date, large epidemiological studies on the health benefits of soy have proven difficult, because there is an acute lack of easily applicable measurements to indicate isoflavone exposure (Sirtori, 2001). As previously mentioned, these issues which interfere with results from clinical studies stem from the fact that isoflavone excretion in urine varies with the quality of diet (Adlercreutz et al., 1986, 1991; Horn-Ross et al., 1997) and individuals metabolize isoflavones very differently (Sirtori, 2001; Slavin et al., 1998). In addition, concentrations of isoflavones in foods are variable due to differences in species, industrial processing and storage, and environmental and geographical conditions (Cassidy, 2003; Cassidy et al., 2006; Coward et al., 1998; Kong et al., 2008). Further, it is difficult to reconcile effects seen in animal models or *in vitro* conditions compared with human clinical studies, because the former typically reports effects seen after consumption of aglucones, while the latter tends to report effects seen from glucoside conjugates (Coward et al., 1998). However, here we outline a concise summary of clinical trial data, including meta-analyses, on the effect of soy consumption on women's health.

12.5.1 Endocrine Effects in Premenopausal Women

Recent studies related to healthy premenopausal women and a provision of 45 mg of isoflavones per day in food resulted in significant changes in hormone levels of the menstrual cycle. Lower dosages of 20 mg/day produced no effect, but higher doses over a period of three months resulted in significantly decreased estrogen concentrations (Cassidy et al., 1994, 1995; Cassidy and Faughnan, 2000). Therefore, a "critical" dose is required in order to produce biological effects on premenopausal women (Cassidy, 2003). These studies reported that ingestion of soy resulted in hormonal modification (Cassidy and Faughnan, 2000). It is proposed that this endocrine effect occurs through the hypothalamic-pituitary-gonadal axis, resulting in a marked decrease in the normal mid-cycle surge of follicle-stimulating hormone (FSH) and luteinizing hormone (LH) and an increased follicular phase of the menstrual cycle which increases the cycle length (Cassidy, 2003). The nature of the effect on the reproductive function in these women, is yet to be ascertained (Xiao, 2008).

12.5.2 Endocrine Effects in Menopausal and Postmenopausal Women

Currently, in postmenopausal women, soy consumption is used to provide a source of estrogen (Cassidy, 2003) and to manage menopausal symptoms. Indeed, epidemiological data suggest that in areas where consumption of soy is high, such as Asia, menopausal symptoms are reduced (Tang, 1994). This has been reported in China, where phases of hot flush typically occur in only 18% of women, compared to women in Europe, where the occurrence is as high as 70–80% (Tang, 1994). However, a meta-analysis on 25 trials conducted between 1966 and 2004 showed that overall, soy did not improve menopausal symptoms (Krebs et al., 2004). Overall, the general consensus is that while soy foods are not as effective as the hormone replacement therapy (HRT) in managing the symptoms, relative to the consumption of placebo there is a significant beneficial effect (Cassidy, 2003), and in combination with pharmaceuticals they may enhance the improvements typically seen.

12.5.3 Soy Consumption and the Risk of Breast Cancer

Epidemiological studies suggest that the lower incidence of breast cancer found in Asian women is associated with higher soy intake (Finkel, 1998). Despite this and the fact that some studies suggest supplementation with soy increases breast cell proliferation (Hargreaves et al., 1999; McMichael-Phillips et al., 1998), concrete evidence that isoflavones protect against breast cancer (Sirtori, 2001) is limited. Overall, animal studies currently prove to be inconclusive, (Sirtori, 2001) and reports on the ability of soy consumption to increase or decrease plasma estrogen levels in a manner effective in reducing cancer are conflicting (Lu et al., 1996; Nagata et al., 1998; Petrakis et al., 1996). Clinical studies show that consumption of soy did not affect endometrial biopsies or sex hormone-binding globulin levels (Cassidy et al., 1995; Nagata et al., 1998; Petrakis et al., 1996), but the nipple aspirate volume was affected (Petrakis et al., 1996). These findings, amalgamated with epidemiological evidence, indicate that if there is any effect of soy consumption on breast cancer, it is only minimally beneficial (Messina, 1999).

12.5.4 Soy Consumption and Bone Health

It is currently popular for women to use soy isoflavones as an alternative to HRT, believing that it may decrease the risk of osteoporosis without the reported negative side effects associated with the use of HRT (Brink et al., 2008). Epidemiological observational studies show a positive effect of soy on overall bone health, but long term studies are limited and have conflicting results (Brink et al., 2008). While some studies report positive effects on bone mineral density (BMD) and bone mineral content with soy consumption ranging from 35 to 56 mg of aglycone equivalents per day (Chen et al., 2003; Clifton-Bligh et al., 2001; Morabito et al., 2002; Potter et al., 1998), others report no effect with consumption as high as 103 mg of aglycone equivalents (Gallagher et al., 2004; Kreijkamp-Kaspers et al., 2004). A more recent study (Brink et al., 2008) explored the effects of isoflavone consumption not only on BMD, but also on bone metabolism and hormonal status in postmenopausal women. This randomized, double-blind, placebo-controlled, parallel, multicenter trial provided early postmenopausal women with 110 mg isoflavone aglycones or control products for 1 year and reported that although the plasma and urine levels of isoflavones increased there was no effect on BMD, bone turnover markers, or hormone concentrations. One suggestion for the conflicting results reported is that the effect of soy isoflavone on bone is dose dependent, and could be due to a balance between interactions with estrogen and peroxisome proliferatoractivated receptors (Brink et al., 2008).

A recent meta-analysis of randomized clinical trials investigated reports on soy isoflavone intake and spinal bone loss in 10 valid studies with over 600 menopausal subjects (Ma et al., 2008). Results showed that spine BMD in participants consuming soy isoflavone increased significantly by 20.6% mg/cm^2 versus those who consumed placebo. The intake was also shown to increase the mineral content in the spine by 0.93 g, an effect which was only marginally significant. Consumption for 6 months with more than 90 mg/day resulted in increases in spine BMD of >27 mg/cm^2. The authors concluded that in menopausal women isoflavone intervention significantly decreased bone loss in the spine (Ma et al., 2008).

12.5.5 Effect of Soy Consumption on Thyroid Function

Consumption of soy has been linked to an effect on thyroid function. Animal and human studies suggest that a high intake of soy-based foods, especially those not fortified with iodine, is associated with the development of thyroid enlargement and goitre. Further, in babies possessing hypothyroidism, those who were fed soy-based formulas required supplementation with 25% more synthetic thyroid hormone than those who were not. Research not only suggests that the intake of soy may decrease the thyroid hormone function, but also soybeans contain agents which interfere with iodine utilization or functioning of the thyroid gland, the consumption of which may lead to thyroid problems or goitre development (Xiao, 2008). It is speculated that these negative effects on the thyroid function are only seen in conjunction with iodine deficiency, and they can in fact be reversed with proper iodine supplementation (Schone et al., 1990). However, it has recently been shown that genistein and daidzein in combination with iodide ion can block thyroid peroxidase activity, causing catalysis of tyrosine iodination, and resulting in the formation of mono-, di- and triiodoisoflavones, which cause negative thyroid effects (Divi et al., 1997). However, further research into this area is required.

12.5.6 Soy Consumption and Risk of Cardiovascular Disease

The ability of soy to reduce the risk of cardiovascular disease has been extensively studied. Rat dietary studies suggest that isoflavones have a lipolytic and cholesteryl ester hydrolytic effect (Peluso et al., 2000). Most human studies report that consumption of soy isoflavones results in lower LDL cholesterol in hyperlipidemic subjects (Xiao, 2008), and a cholesterol-lowering effect has been described in patients with hypercholesterolemia (Descovich et al., 1980; Hermansen et al., 2001; Sirtori et al., 1977). However, there are some clinical studies which describe no effect of isoflavones in perimenopausal and postmenopausal women with both normal and high basal cholesterol levels (Sirtori, 2001). Thus, the magnitude of the effect which is possible and the exact dosage required for a beneficial effect on plasma lipid levels remains unclear (Xiao, 2008).

A previous meta-analysis concluded that diets rich in soy protein caused significantly reduced cholesterol levels, particularly in patients with elevated baseline levels; 60–70% of this effect is estimated to be credited to isoflavones (Anderson et al., 1995). However, more recent *in vitro* and *in vivo* studies indicate that this reported cholesterol-lowering mechanism is more likely a result of the LDL receptor activation induced, not by isoflavones, but rather by components within soy protein (Lovati et al., 1992, 1996, 2000; Sirtori et al., 1984). This idea has been verified in postmenopausal women with moderate heightened cholesterol (Baum et al., 1998) and in cells isolated from patients with high cholesterol (Lovati et al., 1987), but is contradicted in a primate dietary study (Anthony et al., 1997). A potential, and more probable, explanation is that there is a shared interaction between soy protein and isoflavones, together with a diet low in saturated fat and cholesterol (Sfakianos et al., 1997).

The effect of soy on vascular reactivity and cellular proliferation has also been studied. In primates, soy protein isolate has proved to increase vascular function (Sirtori, 2001) and isoflavones may enhance the dilator response to acetylcholine of atherosclerotic arteries (Honore et al., 1997). *In vitro*, isoflavones affect smooth muscle cells involved in atherosclerosis progression, through an ERβ mechanism (Sirtori, 2001). *In vivo*, intravenous administration of genistein significantly increased arterial flow in the forearm of women (Walker et al., 2001). A significant improvement was seen in postmenopausal women with abnormal endothelium-dependent, flow-mediated dilation in the brachial artery, who consumed soy protein, containing 80 mg of isoflavones, daily for 1 month (Anthony, 2000). Similarly a 26%

improvement in systemic arterial compliance has been found in a study with comparable parameters (Nestel et al., 1997), but a lack of association was found in another, more recent study (Simons et al., 2000).

12.5.7 Conclusions Regarding the Health Benefits of Soy in Women

As mentioned at the beginning of this section, large human studies on the health benefits of soy are challenging due to a lack of uniformity in the type and amount of soy product provided to participants. Recently, a critical review (Cassidy et al., 2006) was conducted in which the authors compiled all relevant studies and converted the soy foods consumed to glycone equivalents. This was done to minimize some of the issues related to the type and amount of food consumed. They reported that while whole soybean foods and soy protein isolates improve the lipid markers of cardiovascular risk and endothelial function, there is no effect on the blood lipid levels or blood pressure. In addition, current results indicate an improved bone health associated with soy consumption, but need to be verified by future research. Furthermore, isoflavones are effective in reducing the hot flush symptom, but evidence that isoflavones improve overall menopausal symptoms is incomplete. Finally, the authors concluded that in order to confirm the effects of isoflavones in diabetes, cognitive functions, and breast or colon cancer in postmenopausal women, more randomized clinical trials need to be conducted within groups considered to be at high risk for development of these diseases (Cassidy et al., 2006).

Research on the beneficial effects of soy on women's health is constantly and continually expanding along with the body of knowledge available to health care professionals and consumers who are both healthy and at the same time face the risk for development of diseases. Although there are no definitive answers as to whether soy would be an alternative to pharmaceutical intervention for disease prevention, it remains clear that consumption of soy will do no harm and may actually enhance the effects of currently recommended practices.

REFERENCES

Adlercreutz H, Fotsis T, Bannwart C, Wahala K, Makela T, Brunow G, Hase T. Determination of urinary lignans and phytoestrogen metabolites, potential antiestrogens and anticarcinogens, in urine of women on various habitual diets. *J Steroid Biochem* 25(5B): 791–7, 1986.

Adlercreutz H, Honjo H, Higashi A, Fotsis T, Hamalainen E, Hasegawa T, Okada H. Urinary excretion of lignans and isoflavonoid phytoestrogens in Japanese men and women consuming a traditional Japanese diet. *Am J Clin Nutr* 54(6): 1093–100, 1991.

Administration USFaD. Food labeling health claims: Soy protein and coronary heart disease. In *Food and Drug Administration* H, ed. Fed. Regist. 57700–33, 1999.

Anderson JW, Johnstone BM, Cook-Newell ME. Meta-analysis of the effects of soy protein intake on serum lipids. *N Engl J Med* 333(5): 276–82, 1995.

Anthony MS. Soy and cardiovascular disease. Cholesterol lowering and beyond. *J Nutr* 130(3): 662S–3S, 2000.

Anthony MS, Clarkson TB, Bullock BC, Wagner JD. Soy protein versus soy phytoestrogens in the prevention of diet-induced coronary artery atherosclerosis of male cynomolgus monkeys. *Arterioscler Thromb Vasc Biol* 17(11): 2524–31, 1997.

Barnes S, Sfakianos J, Coward L, Kirk M. Soy isoflavonoids and cancer prevention. Underlying biochemical and pharmacological issues. *Adv Exp Med Biol* 401(87–100), 1996.

Baum JA, Teng H, Erdman JW, Jr., Weigel RM, Klein BP, Persky VW, Freels S. et al. Long-term intake of soy protein improves blood lipid profiles and increases mononuclear cell low-density-lipoprotein receptor messenger RNA in hypercholesterolemic, postmenopausal women. *Am J Clin Nutr* 68(3): 545–51, 1998.

Blair HC, Jordan SE, Peterson TG, Barnes S. Variable effects of tyrosine kinase inhibitors on avian osteoclastic activity and reduction of bone loss in ovariectomized rats. *J Cell Biochem* 61(4): 629–37, 1996.

Brink E, Coxam V, Robins S, Wahala K, Cassidy A, Branca F. Long-term consumption of isoflavone-enriched foods does not affect bone mineral density, bone metabolism, or hormonal status in early postmenopausal women: A randomized, double-blind, placebo controlled study. *Am J Clin Nutr* 87(3): 761–70, 2008.

Cassidy A. Potential risks and benefits of phytoestrogen-rich diets. *Int J Vitam Nutr Res* 73(2): 120–6, 2003.

Cassidy A, Albertazzi P, Lise Nielsen I, Hall W, Williamson G, Tetens I, Atkins S. et al. Critical review of health effects of soyabean phyto-oestrogens in post-menopausal women. *Proc Nutr Soc* 65(1): 76–92, 2006.

Cassidy A, Bingham S, Setchell K. Biological effects of isoflavones in young women: Importance of the chemical composition of soyabean products. *Br J Nutr* 74(4): 587–601, 1995.

Cassidy A, Bingham S, Setchell KD. Biological effects of a diet of soy protein rich in isoflavones on the menstrual cycle of premenopausal women. *Am J Clin Nutr* 60(3): 333–40, 1994.

Cassidy A, Faughnan M. Phyto-oestrogens through the life cycle. *Proc Nutr Soc* 59(3): 489–96, 2000.

Chen YM, Ho SC, Lam SS, Ho SS, Woo JL. Soy isoflavones have a favorable effect on bone loss in Chinese postmenopausal women with lower bone mass: A double-blind, randomized, controlled trial. *J Clin Endocrinol Metab* 88(10): 4740–7, 2003.

Clifton-Bligh PB, Baber RJ, Fulcher GR, Nery ML, Moreton T. The effect of isoflavones extracted from red clover (Rimostil) on lipid and bone metabolism. *Menopause* 8(4): 259–65, 2001.

Coward L, Smith M, Kirk M, Barnes S. Chemical modification of isoflavones in soyfoods during cooking and processing. *Am J Clin Nutr* 68(6 Suppl): 1486S–1491S, 1998.

Descovich GC, Ceredi C, Gaddi A, Benassi MS, Mannino G, Colombo L, Cattin L. et al. Multicentre study of soybean protein diet for outpatient hyper-cholesterolaemic patients. *Lancet* 2(8197): 709–12, 1980.

Divi RL, Chang HC, Doerge DR. Anti-thyroid isoflavones from soybean: Isolation, characterization, and mechanisms of action. *Biochem Pharmacol* 54(10): 1087–96, 1997.

Espinosa-Martos I, Ruperez P. Soybean oligosaccharides. Potential as new ingredients in functional food. *Nutr Hosp* 21(1): 92–6, 2006.

Finkel E. Phyto-oestrogens: The way to postmenopausal health? *Lancet* 352(9142): 1762, 1998.

Fotsis T, Pepper M, Adlercreutz H, Fleischmann G, Hase T, Montesano R, Schweigerer L. Genistein, a dietary-derived inhibitor of *in vitro* angiogenesis. *Proc Natl Acad Sci USA* 90(7): 2690–4, 1993.

Gallagher JC, Satpathy R, Rafferty K, Haynatzka V. The effect of soy protein isolate on bone metabolism. *Menopause* 11(3): 290–8, 2004.

Hargreaves DF, Potten CS, Harding C, Shaw LE, Morton MS, Roberts SA, Howell A, Bundred NJ. Two-week dietary soy supplementation has an estrogenic effect on normal premenopausal breast. *J Clin Endocrinol Metab* 84(11): 4017–24, 1999.

Hermansen K, Sondergaard M, Hoie L, Carstensen M, Brock B. Beneficial effects of a soy-based dietary supplement on lipid levels and cardiovascular risk markers in type 2 diabetic subjects. *Diabetes Care* 24(2): 228–33, 2001.

Honore EK, Williams JK, Anthony MS, Clarkson TB. Soy isoflavones enhance coronary vascular reactivity in atherosclerotic female macaques. *Fertil Steril* 67(1): 148–54, 1997.

Horn-Ross PL, Barnes S, Kirk M, Coward L, Parsonnet J, Hiatt RA. Urinary phytoestrogen levels in young women from a multiethnic population. *Cancer Epidemiol Biomarkers Prev* 6(5): 339–45, 1997.

Joannou GE, Kelly GE, Reeder AY, Waring M, Nelson C. A urinary profile study of dietary phytoestrogens. The identification and mode of metabolism of new isoflavonoids. *J Steroid Biochem Mol Biol* 54(3–4): 167–84, 1995.

Kapiotis S, Hermann M, Held I, Seelos C, Ehringer H, Gmeiner BM. Genistein, the dietary-derived angiogenesis inhibitor, prevents LDL oxidation and protects endothelial cells from damage by atherogenic LDL. *Arterioscler Thromb Vasc Biol* 17(11): 2868–74, 1997.

Karr SC, Lampe JW, Hutchins AM, Slavin JL. Urinary isoflavonoid excretion in humans is dose dependent at low to moderate levels of soy-protein consumption. *Am J Clin Nutr* 66(1): 46–51, 1997.

Kerry N, Abbey M. The isoflavone genistein inhibits copper and peroxyl radical mediated low density lipoprotein oxidation *in vitro*. *Atherosclerosis* 140(2): 341–7, 1998.

Kim JJ, Kim S.H., Hahn S.J, Chung, I.M. Changing soybean isoflavone composition and concentrations under two different storage conditions over three years. *Food Res Int* 38(4): 435–444, 2005.

Kim H, Peterson TG, Barnes S. Mechanisms of action of the soy isoflavone genistein: Emerging role for its effects via transforming growth factor beta signaling pathways. *Am J Clin Nutr* 68(6 Suppl): 1418S–1425S, 1998.

Kong F, Chang SK, Liu Z, Wilson LA. Changes of soybean quality during storage as related to soymilk and tofu making. *J Food Sci* 73(3): S134–44, 2008.

Krebs EE, Ensrud KE, MacDonald R, Wilt TJ. Phytoestrogens for treatment of menopausal symptoms: A systematic review. *Obstet Gynecol* 104(4): 824–36, 2004.

Kreijkamp-Kaspers S, Kok L, Grobbee DE, de Haan EH, Aleman A, Lampe JW, van der Schouw YT. Effect of soy protein containing isoflavones on cognitive function, bone mineral density, and plasma lipids in postmenopausal women: A randomized controlled trial. *Jama* 292(1): 65–74, 2004.

Kumi-Diaka J, Nguyen V, Butler A. Cytotoxic potential of the phytochemical genistein isoflavone (4′,5′,7-trihydroxyisoflavone) and certain environmental chemical compounds on testicular cells. *Biol Cell* 91(7): 515–23, 1999.

Kumi-Diaka J, Rodriguez R, Goudaze G. Influence of genistein (4′,5,7-trihydroxyisoflavone) on the growth and proliferation of testicular cell lines. *Biol Cell* 90(4): 349–54, 1998.

Kurzer MS, Xu X. Dietary phytoestrogens. *Annu Rev Nutr* 17(353–81, 1997.

Li Y, Bhuiyan M, Sarkar FH. Induction of apoptosis and inhibition of c-erbB-2 in MDA-MB-435 cells by genistein. *Int J Oncol* 15(3): 525–33, 1999a.

Li D, Yee JA, McGuire MH, Murphy PA, Yan L. Soybean isoflavones reduce experimental metastasis in mice. *J Nutr* 129(5): 1075–8, 1999b.

Liu KS. Chemistry and nutritional value of soybean components. In: *Soybeans: Chemistry, Technology and Utilization.* Gaithersburg, MD: Aspen Publication, Inc., 25–113, 1999.

Lee SJ, Ahn JK, Kim SH, Kim JT, Han SJ, Jung MY, Chung IM. Variation in isoflavone of soybean cultivars with location and storage duration. *J Agric Food Chem* 51(11): 3382–9, 2003.

Lee SJ, Kim JJ, Moon HI, Ahn JK, Chun SC, Jung WS, Lee OK, Chung IM. Analysis of isoflavones and phenolic compounds in Korean soybean [*Glycine max* (L.) Merrill] seeds of different seed weights. *J Agric Food Chem* 56(8): 2751–8, 2008.

Lovati MR, Manzoni C, Canavesi A, Sirtori M, Vaccarino V, Marchi M, Gaddi G, Sirtori CR. Soybean protein diet increases low density lipoprotein receptor activity in mononuclear cells from hypercholesterolemic patients. *J Clin Invest* 80(5): 1498–502, 1987.

Lovati MR, Manzoni C, Corsini A, Granata A, Frattini R, Fumagalli R, Sirtori CR. Low density lipoprotein receptor activity is modulated by soybean globulins in cell culture. *J Nutr* 122(10): 1971–8, 1992.

Lovati MR, Manzoni C, Corsini A, Granata A, Fumagalli R, Sirtori CR. 7S globulin from soybean is metabolized in human cell cultures by a specific uptake and degradation system. *J Nutr* 126(11): 2831–42, 1996.

Lovati MR, Manzoni C, Gianazza E, Arnoldi A, Kurowska E, Carroll KK, Sirtori CR. Soy protein peptides regulate cholesterol homeostasis in Hep G2 cells. *J Nutr* 130(10): 2543–9, 2000.

Lozovaya VV, Lygin, A.V., Ulanov, A.V., Nelson, R.L., Dayde, J., Widhohn, J.M. Effect of temperature and soil moisture states during seed development on soybean seed isoflavones concentration and composition. *Crop Sci* 45(674–1677), 2005.

Lu LJ, Anderson KE, Grady JJ, Nagamani M. Effects of soya consumption for one month on steroid hormones in premenopausal women: Implications for breast cancer risk reduction. *Cancer Epidemiol Biomarkers Prev* 5(1): 63–70, 1996.

Ma DF, Qin LQ, Wang PY, Katoh R. Soy isoflavone intake increases bone mineral density in the spine of menopausal women: Meta-analysis of randomized controlled trials. *Clin Nutr* 27(1): 57–64, 2008.

Malencic D, Maksimovic Z, Popovic M, Miladinovic J. Polyphenol contents and antioxidant activity of soybean seed extracts. *Bioresour Technol* 99(14): 6688–91, 2008.

Malencic D, Popovic M, Miladinovic J. Phenolic content and antioxidant properties of soybean (*Glycine max* (L.) Merr.) seeds. *Molecules* 12(3): 576–81, 2007.

Mateos-Aparicio I, Redondo Cuenca A, Villanueva-Suarez MJ, Zapata-Revilla MA. Soybean, a promising health source. *Nutr Hosp* 23(4): 305–312, 2008.

McMichael-Phillips DF, Harding C, Morton M, Roberts SA, Howell A, Potten CS, Bundred NJ. Effects of soy-protein supplementation on epithelial proliferation in the histologically normal human breast. *Am J Clin Nutr* 68(6 Suppl): 1431S – 1435S, 1998.

Messina MJ. Legumes and soybeans: Overview of their nutritional profiles and health effects. *Am J Clin Nutr* 70(3 Suppl): 439S–450S, 1999.

Morabito N, Crisafulli A, Vergara C, Gaudio A, Lasco A, Frisina N, D'Anna R. et al. Effects of genistein and hormone-replacement therapy on bone loss in early postmenopausal women: A randomized double-blind placebo-controlled study. *J Bone Miner Res* 17(10): 1904–12, 2002.

Nagata C, Takatsuka N, Inaba S, Kawakami N, Shimizu H. Effect of soymilk consumption on serum estrogen concentrations in premenopausal Japanese women. *J Natl Cancer Inst* 90(23): 1830–5, 1998.

Nestel PJ, Yamashita T, Sasahara T, Pomeroy S, Dart A, Komesaroff P, Owen A, Abbey M. Soy isoflavones improve systemic arterial compliance but not plasma lipids in menopausal and perimenopausal women. *Arterioscler Thromb Vasc Biol* 17(12): 3392–8, 1997.

Patel RP, Boersma BJ, Crawford JH, Hogg N, Kirk M, Kalyanaraman B, Parks DA, Barnes S, Darley-Usmar V. Antioxidant mechanisms of isoflavones in lipid systems: Paradoxical effects of peroxyl radical scavenging. *Free Radic Biol Med* 31(12): 1570–81, 2001.

Peluso MR, Winters TA, Shanahan MF, Banz WJ. A cooperative interaction between soy protein and its isoflavone-enriched fraction lowers hepatic lipids in male obese Zucker rats and reduces blood platelet sensitivity in male Sprague–Dawley rats. *J Nutr* 130(9): 2333–42, 2000.

Peterson TG, Ji GP, Kirk M, Coward L, Falany CN, Barnes S. Metabolism of the isoflavones genistein and biochanin A in human breast cancer cell lines. *Am J Clin Nutr* 68(6 Suppl): 1505S–1511S, 1998.

Petrakis NL, Barnes S, King EB, Lowenstein J, Wiencke J, Lee MM, Miike R, Kirk M, Coward L. Stimulatory influence of soy protein isolate on breast secretion in pre- and postmenopausal women. *Cancer Epidemiol Biomarkers Prev* 5(10): 785–94, 1996.

Potter SM, Baum JA, Teng H, Stillman RJ, Shay NF, Erdman JW, Jr. Soy protein and isoflavones: Their effects on blood lipids and bone density in postmenopausal women. *Am J Clin Nutr* 68(6 Suppl): 1375S–1379S, 1998.

Reinhart KC, Dubey RK, Keller PJ, Lauper U, Rosselli M. Xeno-oestrogens and phyto-oestrogens induce the synthesis of leukaemia inhibitory factor by human and bovine oviduct cells. *Mol Hum Reprod* 5(10): 899–907, 1999.

Rogers SG. Biotechnology and the soybean. *Am J Clin Nutr* 68(6 Suppl): 1330S–1332S, 1998.

Rowland I, Faughnan M, Hoey L, Wahala K, Williamson G, Cassidy A. Bioavailability of phyto-oestrogens. *Br J Nutr* 89(Suppl 1): (S45–58, 2003.

Sakthivelu G, Akitha Devi MK, Giridhar P, Rajasekaran T, Ravishankar GA, Nikolova MT, Angelov GB, Todorova RM, Kosturkova GP. Isoflavone composition, phenol content, and antioxidant activity of soybean seeds from India and Bulgaria. *J Agric Food Chem* 56(6): 2090–5, 2008.

Sathyamoorthy N, Gilsdorf JS, Wang TT. Differential effect of genistein on transforming growth factor beta 1 expression in normal and malignant mammary epithelial cells. *Anticancer Res* 18(4A): 2449–53, 1998.

Schleicher RL, Lamartiniere CA, Zheng M, Zhang M. The inhibitory effect of genistein on the growth and metastasis of a transplantable rat accessory sex gland carcinoma. *Cancer Lett* 136(2): 195–201, 1999.

Schone F, Jahreis G, Lange R, Seffner W, Groppel B, Hennig A, Ludke H. Effect of varying glucosinolate and iodine intake via rapeseed meal diets on serum thyroid hormone level and total iodine in the thyroid in growing pigs. *Endocrinol Exp* 24(4): 415–27, 1990.

Setchell KD, Cassidy A. Dietary isoflavones: biological effects and relevance to human health. *J Nutr* 129(3): 758S-767S, 1999.

Setchell KD, Zimmer-Nechemias L, Cai J, Heubi JE. Exposure of infants to phyto-oestrogens from soy-based infant formula. *Lancet* 350(9070): 23–7, 1997.

Sfakianos J, Coward L, Kirk M, Barnes S. Intestinal uptake and biliary excretion of the isoflavone genistein in rats. *J Nutr* 127(7): 1260–8, 1997.

Simons LA, von Konigsmark M, Simons J, Celermajer DS. Phytoestrogens do not influence lipoprotein levels or endothelial function in healthy, postmenopausal women. *Am J Cardiol* 85(11): 1297–301, 2000.

Sirtori CR. Risks and benefits of soy phytoestrogens in cardiovascular diseases, cancer, climacteric symptoms and osteoporosis. *Drug Saf* 24(9): 665–82, 2001.

Sirtori CR, Agradi E, Conti F, Mantero O, Gatti E. Soybean-protein diet in the treatment of type-II hyperlipoproteinaemia. *Lancet* 1(8006): 275–7, 1977.

Sirtori CR, Galli G, Lovati MR, Carrara P, Bosisio E, Kienle MG. Effects of dietary proteins on the regulation of liver lipoprotein receptors in rats. *J Nutr* 114(8): 1493–500, 1984.

Slavin JL, Karr SC, Hutchins AM, Lampe JW. Influence of soybean processing, habitual diet, and soy dose on urinary isoflavonoid excretion. *Am J Clin Nutr* 68(6 Suppl): 1492S–1495S, 1998.

Tang GW. The climacteric of Chinese factory workers. *Maturitas* 19(3): 177–82, 1994.

Tikkanen MJ, Wahala K, Ojala S, Vihma V, Adlercreutz H. Effect of soybean phytoestrogen intake on low density lipoprotein oxidation resistance. *Proc Natl Acad Sci USA* 95(6): 3106–10, 1998.

Torres N, Torre-Villalvazo I, Tovar AR. Regulation of lipid metabolism by soy protein and its implication in diseases mediated by lipid disorders. *J Nutr Biochem* 17(6): 365–73, 2006.

U.S. Department of Agriculture ARS. USDA Database for the Isoflavone Content of Selected Foods, Release 2.0. In Agriculture UDo, ed. Nutrient Data Laboratory Home Page, 2008.

Wagner JD, Cefalu WT, Anthony MS, Litwak KN, Zhang L, Clarkson TB. Dietary soy protein and estrogen replacement therapy improve cardiovascular risk factors and decrease aortic cholesteryl ester content in ovariectomized cynomolgus monkeys. *Metabolism* 46(6): 698–705, 1997.

Walker HA, Dean TS, Sanders TA, Jackson G, Ritter JM, Chowienczyk PJ. The phytoestrogen genistein produces acute nitric oxide-dependent dilation of human forearm vasculature with similar potency to 17beta-estradiol. *Circulation* 103(2): 258–62, 2001.

Walz E. Isoflavon-und saponin-glucoside in Soja-Hispida. *Justus Liebigs Ann Chem* 489: 118–55, 1931.

Walter ED. Genistein (an isoflavone glucoside) and its aglucone, genistein, from soybeans. *J Am Oil Chem Soc* 63: 3273–6, 1941.

Xiao CW. Health effects of soy protein and isoflavones in humans. *J Nutr* 138(6): 1244S–9S, 2008.

Xu X, Harris KS, Wang HJ, Murphy PA, Hendrich S. Bioavailability of soybean isoflavones depends upon gut microflora in women. *J Nutr* 125(9): 2307–15, 1995.

Yen GC, Lai HH. Inhibition of reactive nitrogen species effects *in vitro* and *in vivo* by isoflavones and soy-based food extracts. *J Agric Food Chem* 51(27): 7892–900, 2003.

Zhou JR, Gugger ET, Tanaka T, Guo Y, Blackburn GL, Clinton SK. Soybean phytochemicals inhibit the growth of transplantable human prostate carcinoma and tumor angiogenesis in mice. *J Nutr* 129(9): 1628–35, 1999.

13 Southeast Asian Fruits and Their Functionalities

Lai Peng Leong and Guanghou Shui

CONTENTS

13.1 INTRODUCTION

Southeast Asian countries are particularly rich in their variety of fruits, some of which are available all year round while others are seasonal (Tables 13.1 and 13.2). However, due to commercialization, some of the seasonal fruits are now available around the year.

It is generally believed that the high consumption of fruits and vegetables is highly related to a lower risk of degenerative diseases; thus, considerable work has been done to understand the functional properties of this group of foods. One of the most studied functional characteristics of these fruits is their ability to scavenge free radicals and their antioxidant activities (Kim and Jo, 2006; Leong and Shui, 2002; Thaipong et al., 2006). Other functional characteristics include chemoprevention, hypoglycaemic activity, hypolipidemic activity and hypocholesterolemic activity.

TABLE 13.1
Nonseasonal Fruits Commonly Found in Southeast Asia

Nonseasonal Fruits
Salak (*Salacca edulis* Reinw)
Star fruit (*Averrhoea carambola*)
Banana (*Musa × paradisiaca*)
Jack fruit (*Artocarpus heterophyllus*)
Papaya (*Carica papaya*)
Ciku (*Manilkara achras*)
Watermelon (*Citrullus lanatus*)
Guava (*Psidium guajava* L.)
Pineapple (*Ananas comosus*)
Soursop (*Annona muricata)*
Dragon Fruit (*Hylocereus undatus*)
Custard Apple (*Annona reticulata)*

A host of different mechanisms including antioxidant activity may be responsible for the benefits of consuming fruits and vegetables. However, this chapter focuses only on the antioxidant properties of fruits in Southeast Asia.

13.1.1 Antioxidant Activities of Fruit Extracts

Among the most common methods used to assess the antioxidant properties of fruits, is one based on their abilities to scavenge free radicals. Such assays can be done easily when colored radicals such as 2,2′-azino-bis-3-ethylbenzothiazoline-6-sulfonic acid ($ABTS^{\bullet +}$) or 1,1-diphenyl-2-picrylhydrazyl (DPPH•) are used and when scavenged, the resulting products do not absorb at the visible range. Thus, the loss of color is used as an indication of the scavenging activities of the item in question.

TABLE 13.2
Seasonal Fruits Commonly Found in Southeast Asia

Seasonal Fruits
Duku, langsat, dokong (*Lansium domesticum*)—December and June
Rambutan (*Nephelium lappaceum*)—June and November
Durian (*Durio zibethinus*)—March and September
Pulasan (*Nephelium mutabile* Blume)—August and December
Mangosteen (*Garcinia mangostana*)—May
Cempedak (*Artocarpus integer*)
Mango (*Mangifera indica*)—June
Water apple (*Syzygium aqueum* Alst.)—March and October
Longan (*Dimocarpus longan* Lour.)—August

When such methods are used to assess the antioxidant properties of an extract, the antioxidant capacity and the antioxidant efficiency can be obtained.

The antioxidant capacity of a fruit extract refers to the quantity of radicals it is able to scavenge regardless of its rate of reaction. Therefore, the reaction is allowed to continue until it comes to a plateau or no change in absorbance is observed. This is taken as the maximum amount of radicals the extract is able to scavenge and thus termed as the antioxidant capacity. Table 13.3 shows the antioxidant capacity of fruits grown in Southeast Asia using ABTS•+ and DPPH• radical scavenging methods.

TABLE 13.3
Antioxidant Capacities of Some Fruits Found in Southeast Asia in Ascorbic Acid Antioxidant Capacity

Fruits	AEAC (mg/100 g)	Country of Origin
Ciku[a]	3396.4 ± 387.9	Malaysia
Star fruit[a]	277.5 ± 22.3	Malaysia
Guava[a]	270.2 ± 18.8	Thailand
Salak[a]	260.2 ± 32.5	Malaysia
Mangosteen[a]	149.5 ± 23.3	Malaysia
Avocado[a]	143.3 ± 16.5	Thailand
Papaya (solo)[a]	140.9 ± 26.7	Malaysia
Mango[a]	139.1 ± 21.5	Philippines
Cempedak[a]	126.2 ± 19.1	Malaysia
Pomelo[a]	103.6 ± 34.7	Malaysia
Pineapple[a]	85.6 ± 21.3	Malaysia
Papaya (foot long)[a]	72.5 ± 2.6	Malaysia
Rambutan[a]	71.5 ± 7.6	Malaysia
Pulasan[a]	70.6 ± 8.2	Malaysia
Banana[a]	48.3 ± 1.2	Philippines
Coconut pulp[a]	45.8 ± 6.5	Malaysia
Tomato[a]	38.0 ± 1.7	Malaysia
Honeydew[a]	19.6 ± 0.8	Malaysia
Watermelon[a]	11.9 ± 0.1	Malaysia
Coconut water[a]	11.5 ± 2.2	Malaysia
Jackfruit[b]	500 ± 7	Malaysia
Longan[b]	65 ± 28	Malaysia
Dragon fruit[c]	13.5 ± 2.1	Malaysia
Water apple[c]	31 ± 10	Malaysia
Langsat[c]	14.6 ± 4.3	Malaysia

[a] Leong et al. (2002)
[b] Soong and Barlow (2004) based on ABTS•+ radical scavenging assay.
[c] Lim et al. (2007) based on DPPH radical scavenging assay.
AEAC = Ascorbic acid equivalent antioxidant capacity.

While interpreting antioxidant capacity data using radical scavenging methods, one must ensure that the reaction has reached a plateau. For antioxidant values obtained from a radical scavenging assay after a fixed period of time, and not normally sufficient for all reactions to come to a plateau, a comparison of antioxidant activity may not be meaningful. An example is given in Figure 13.1 where the antioxidant activity of guava is higher than that of plum before 55 min, but the opposite is true after 55 min. A similar observation can be obtained for the kiwi fruit and mango. However, in the latter case, the change in the order of antioxidant activity occurred at 20 min. Although such examples are not very common, it is known that the antioxidant activity scale is not linear across different reaction times (Haruenkit et al., 2007).

Antioxidant capacity serves as an easy comparison between the fruit extracts for their antioxidant activities, and information on the antioxidant efficiency of fruit extracts is likely to be more relevant in terms of being able to prevent aging or degenerative diseases. This is because an antioxidant is only able to render protection to a biological molecule if it is able to scavenge radicals before it has the opportunity to cause any damage. The value of antioxidant efficiency can be obtained indirectly from assays such as oxygen radical absorbance capacity (ORAC). In this type of assay, a biological marker with the capability of producing a signal is used. Radicals are generated throughout the experimental period to cause damage to the biological marker. When damaged, the total signal recorded will reduce (a damaged biological marker will not produce signal). When antioxidants are present, it may scavenge the free radicals generated before it is able to cause any damage to the biological marker. This will delay the onset of signal loss until the antioxidants are either used up or are not efficient enough to prevent the damage due to the free radicals generated. One may record the lag time, that is, the time taken before the marker is damaged by the radicals or the area under curve showing the cumulative effect of all the antioxidants in an extract. Not much information is available on the antioxidant efficiency of

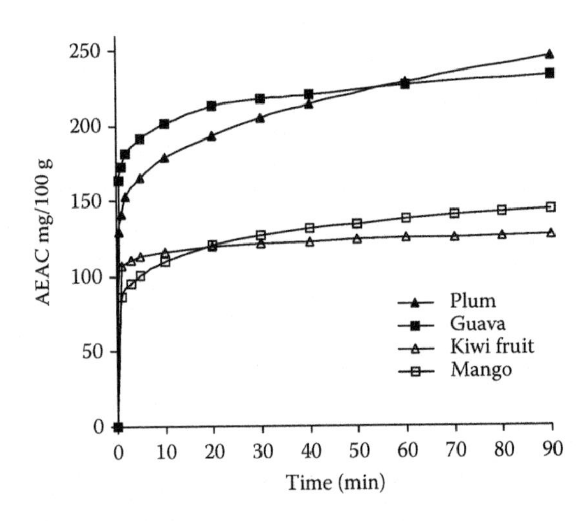

FIGURE 13.1 AEAC values vs. reaction time for crude extract of plum, guava, kiwi fruit, and mango.

TABLE 13.4
Antioxidant Efficiency of Some Southeast Asian Fruit Extracts

Name	Antioxidant Activity (μmol TE/g)
Mango[a]	7.4 to 21.0 ± 7.7
Guava[a]	18.4 ± 0.4
Papaya[a]	3.0 ± 0.5
Mangosteen[a]	25.1 ± 8.0
Banana[a]	2.6 ± 0.5
Guava[b]	21.3 ± 3.1

[a] Patthamakanokporn et al. (2008).
[b] Thaipong et al. (2006).

Southeast Asian fruit extracts. Table 13.4 provides the results for some fruits in trolox equivalent antioxidant activity. More accurately, the antioxidant efficiency data can be obtained from kinetic studies which take into account the mechanism and the rate of reaction. In such studies, the antioxidant compound needs to be isolated and purified before it can be associated with a particular rate of reaction (Madsen et al., 2000).

13.1.2 Components in Fruits Contributing to Antioxidant Activities

Many studies have been conducted to correlate antioxidant activities to the total phenolic content of fruit and plant extracts (Leong and Shui, 2002; Maisuthisakul et al., 2008; Patthamakanokporn et al., 2008; Shui et al., 2004; Thaipong et al., 2006). These assays are most commonly conducted using the Folin–Ciocalteu reagent which contains phosphotungstic and phosphomolybdic acids. A complex redox reaction occurs between these compounds found in the reagent and phenolic compounds as well as some other compounds found in the fruit extract. Therefore, this assay is not the most specific or ideal but nevertheless gives a good estimate of the total phenolic content in the extracts. All studies reporting the correlation between antioxidant activities and total phenolic content show a very high correlation between the two. Principal component analysis also shows that the total phenolic compound contributes the most toward antioxidant activities compared to other components studied (Maisuthisakul et al., 2008; Wong et al., 2006). Some active compounds identified in fruits and their activities are described in the following sections.

13.1.2.1 *Artocarpus* sp.

Several compounds such as cycloheterophyllin and artonins A and B (Figure 13.2) have been isolated from *Artocarpus heterophyllus* and are found to exhibit strong antioxidant activities (Ko et al., 1998). Cycloheterophyllin and artonin B are found to exhibit anti-inflammatory properties by inhibiting the release of chemical mediators from inflammatory cells (Wei et al., 2005).

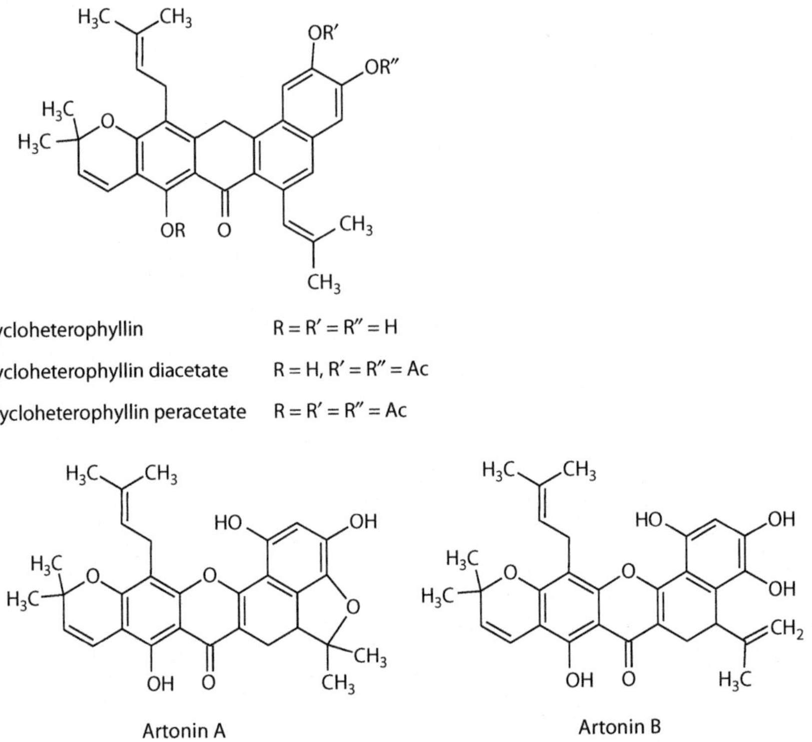

Cycloheterophyllin R = R′ = R″ = H

Cycloheterophyllin diacetate R = H, R′ = R″ = Ac

Clycloheterophyllin peracetate R = R′ = R″ = Ac

Artonin A

Artonin B

FIGURE 13.2 Chemical structures of some prenylflavones found in *Artocarpus heterophyllus*. (Data from Ko et al. 1998. *Free Radical Biology and Medicine*, 25(2), 160–168.)

13.1.2.2 *Carica papaya*

The main contribution to antioxidant activities in papaya is due to ascorbic acid (Kondo et al., 2005). However, the contribution by polyphenolics remains high.

Fermented papaya prepared by yeast fermentation has also been found to contain antioxidant activities (Imao et al., 1998). It has been found to have the potential to prevent contact hypersensitivity immune response (Hiramoto et al., 2008). In a trial involving cirrhosis patients, fermented papaya preparations have been found to improve the redox status of patients (Marotta et al., 2007). Fermented papaya preparations have been found to be more effective than vitamin E in the improvement of redox status.

13.1.2.3 *Durio zibethinus*

In a study on the health properties of durian, researchers identified that the most important phenolic acids and flavonoids in durian are caffeic acid, *p*-coumaric acid, cinnamic acid, vanillic acid, campherol, quercetin, apigenin, morin, and myricitin (Haruenkit et al., 2007). The structures of these compounds are given in Figures 13.3 and 13.4. It was also concluded from their rat studies that the consumption of durian is able to prevent the rise in plasma lipid levels. The reduction of plasma antioxidant activity is also hindered.

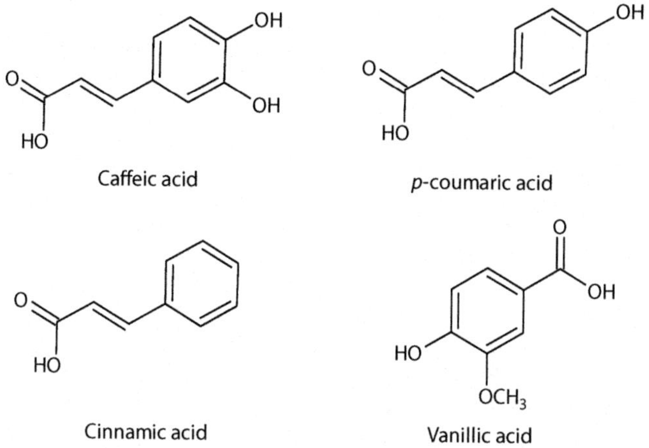

Caffeic acid

p-coumaric acid

Cinnamic acid

Vanillic acid

FIGURE 13.3 Some phenolic acids found in durian.

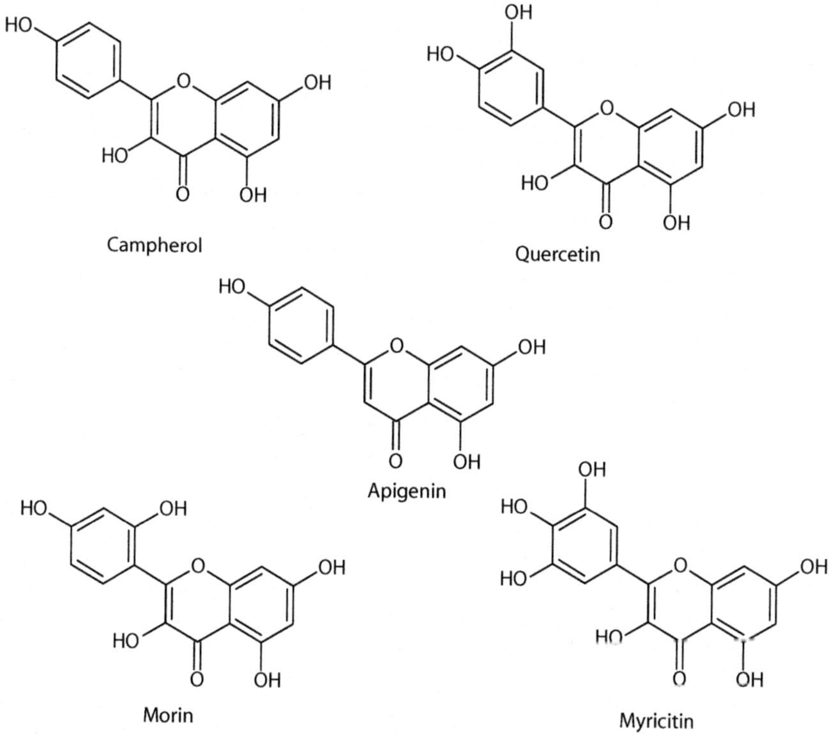

Campherol

Quercetin

Apigenin

Morin

Myricitin

FIGURE 13.4 Some flavonoids found in durian.

13.1.2.4 *Manilkara achras*

Some antioxidants in ciku are identified to be polyphenolics with basic blocks of gallocatechin or catechin (Shui et al., 2004) and 4-*O*-galloylchlorogenic acid (Ma et al., 2003). There are no reports on the benefits of consuming ciku as a fruit although the benefits of gallocatechin and catechin from other sources have been reported. The content of antioxidants in ciku was found to vary as the fruit ripens. This will be discussed in Section 13.1.3.

13.1.2.5 *Citrullus lanatus*

Watermelon is found to be rich in carotenoids. Lycopene is found to be higher in seedless varieties compared to those with seeds (Perkins-Veazie et al., 2006). Watermelon extract has been found to improve the antioxidant status of the body by increasing levels of antioxidant enzymes (Micol et al., 2007). It has also shown improvement in the glycemic and lipid metabolism. In addition, watermelon rind has been found to contain citrulline (Figure 13.5), a nonessential amino acid and precursor to arginine which is important in the nitric oxide system to help maintain endothelial cells (Rimando and Perkins-Veazie, 2005).

13.1.2.6 *Salacca edulis* Reinw

Chlorogenic acid, epicatechin, proanthocyanidins existing as dimers through hexamers have been isolated in snake fruit or Salak (Figure 13.6) (Shui and Leong, 2003). When fed to mice, Salak is found to reduce blood lipid levels and prevent the reduction of antioxidant activity in the blood (Leontowicz et al., 2006).

13.1.2.7 *Psidium guajava* L.

Antioxidant activity in guava was found to be highly correlated with total phenolic contents and ascorbic acid (Thaipong et al., 2006). Active antioxidants in guava include ferulic acid, gallic acid, and quercetin. The extract of guava is found to inhibit the glycation of LDL, which may be useful to prevent a variety of diseases associated with glycation.

13.1.2.8 *Averrhoea carambola*

Star fruit contains a very high level of antioxidants in its residue compared to its juice. The main antioxidants in star fruit are contributed by singly linked proanthocyanidins that exist as dimers, trimers, tetramers, and pentamers of catechin or epicatechin (Shui and Leong, 2004).

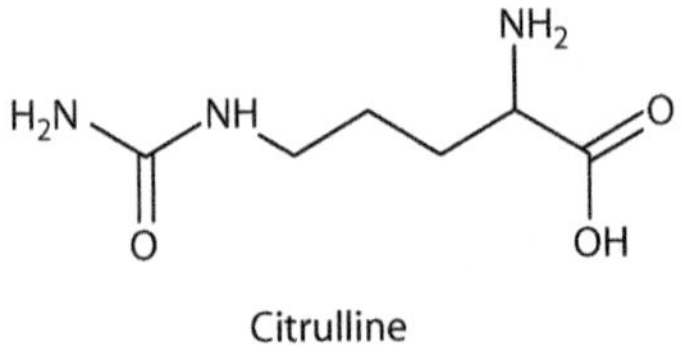

NH_2 H_2N NH O O OH

Citrulline

FIGURE 13.5 Citrulline.

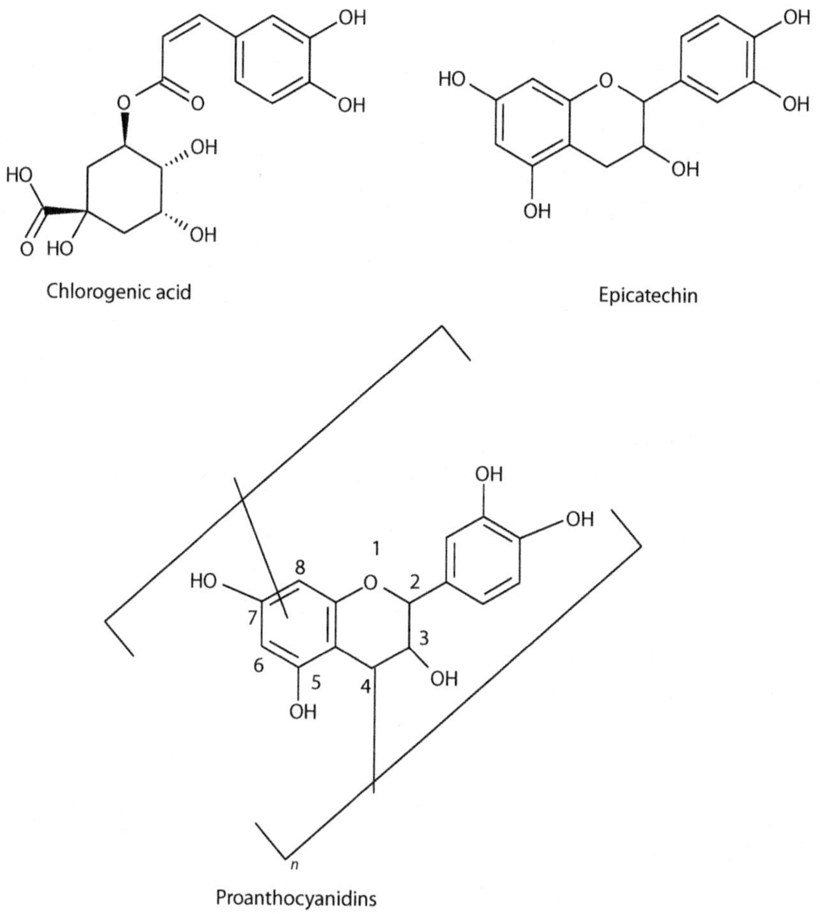

Chlorogenic acid

Epicatechin

Proanthocyanidins

FIGURE 13.6 Some antioxidants found in Salak (snake fruit) ($n = 1 - 6$).

13.1.2.9 *Musa* sp.

The content of gallocatechin (Figure 13.7) is found to be correlated with the antioxidant activity in bananas (Someya et al., 2002). Like in most fruits, the content of antioxidants in the peel is found to be higher than that in the pulp. Dopamine, an antioxidant known to be as good as gallocatechin gallate and ascorbic acid has also been isolated from banana (Kanazawa and Sakakibara, 2000).

13.1.2.10 *Mangifera indica* L.

Mangiferin, isomangiferin (Figure 13.8), mangiferin gallate, isomangiferin gallate, quercetin and its glycosides, and kaempferol glycoside have been identified as antioxidants in mango (Ribeiro et al., 2008). Mangiferin, together with epigallocatechin gallate has been found to be able to have a protective effect on lipid peroxidation in red blood cells, possibly due to its antioxidant activity (Rodriguez et al., 2006). Mangoes are also found to contain high levels of β-carotene and vitamin C (ascorbic

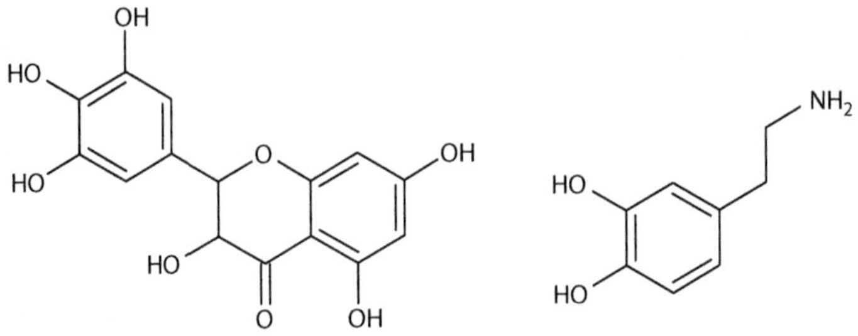

Gallocatechin

Dopamine

FIGURE 13.7 Antioxidants found in bananas.

acid + dehydroascorbic acid) (Ribeiro et al., 2007). Extracts from mango have been found to prevent oxidative stress on the liver and improve the overall antioxidant condition in rats (Pardo-Andreu et al., 2008).

In another experiment where rats are fed with cholesterol, the oral administration of flavonoids from mango is able to increase radical scavenging enzymes and reduce lipid peroxide in the serum (Anila and Vijayalakshmi, 2003).

13.1.2.11 *Garcinia mangostana*

Extracts of mangosteen were found to possess high antioxidant activities and to be able to suppress the production of proinflammatory cytokines (Chomnawang et al., 2007). Oligomeric proanthocyanidins have been extracted from their pericarps which contribute to their antioxidant activity (Fu et al., 2007). Three other antioxidants were also isolated from the skin of mangosteen, that is, 1,3,6,7-quadhydroxy-3-methoxy-2,8-(3-methyl-2-butenyl) xanthone, 1,3,6-trihydroxy-3-methoxy-2,8-(3-methyl-2-butenyl) xanthone (Figure 13.9) and epicatechin. Most methods of assessing antioxidant activities revealed that 1, 3, 6, 7-quadhydroxy-3-methoxy-2,8-(3-methyl-2-butenyl) xanthone is a better radical scavenger than the other two. However, due to the method used, it is hard to justify the comparison of the effectiveness of the three isolated antioxidants.

Other compounds isolated from mangosteen also showed antioxidant activities including 8-hydroxycudraxanthone G, α-mangostin, γ-mangostin, and smeathxanthone A (Figure 13.9).

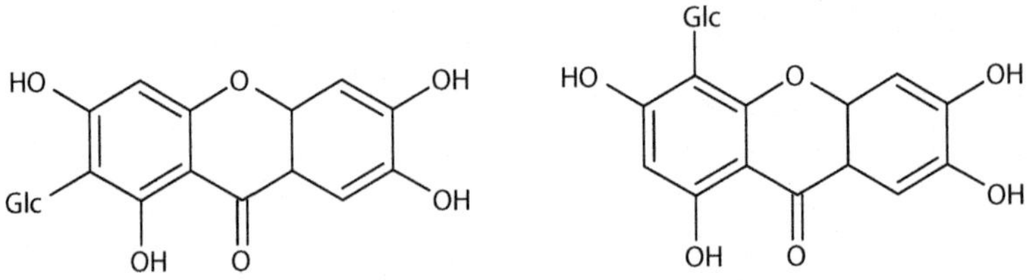

Mangiferin

Isomangiferin

FIGURE 13.8 Antioxidants found in mangoes.

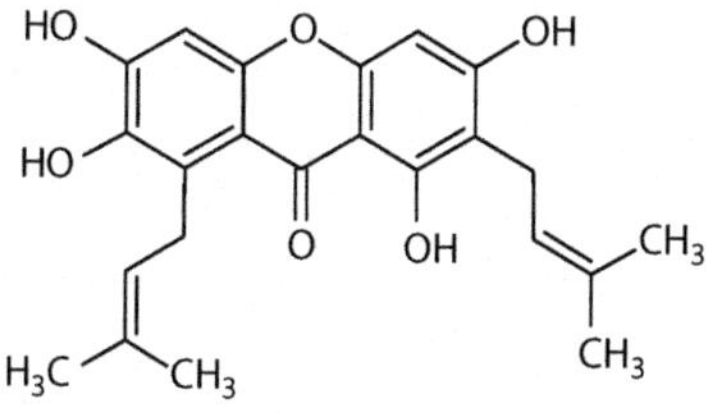

1,3,6,7-quadhydroxy-3-methoxy-2, 8-(3-methyl-2-butenyl) xanthone

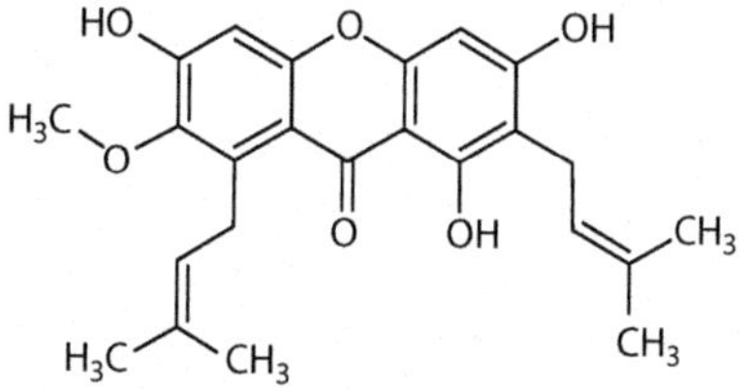

1,3,6-trihydroxy-3-methoxy-2,8-(3-methyl-2-butenyl)xanthone

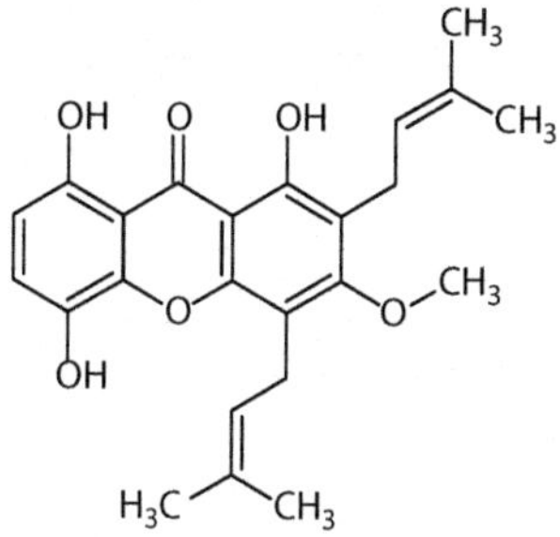

8-hydroxycudraxanthone G

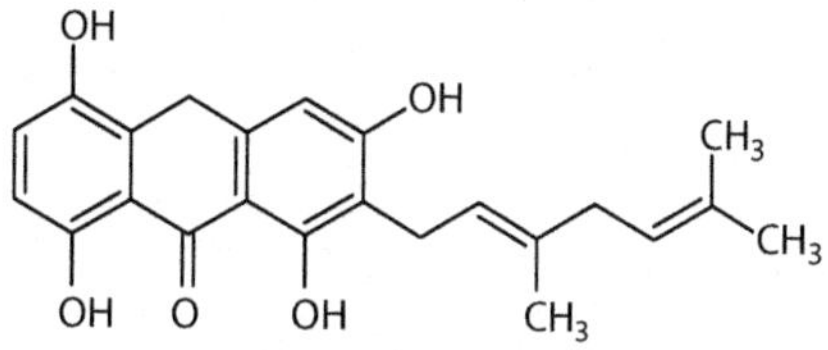

Smeathxanthone A

FIGURE 13.9 Antioxidant compounds isolated from mangosteen.

13.1.2.12 *Dimocarpus longan*

Two compounds, 4-*O*-methylgallic acid and (–)-epicatechin have been identified in longan fruits (Sun et al., 2007). The authors also showed that the reducing power and antioxidant activity of 4-*O*-methylgallic acid was higher than that of (–)-epicatechin, based on its ability to scavenge DPPH, hydroxyl, and superoxide radicals.

13.1.3 Changes of Antioxidant Levels in Fruits during Storage

As a fruit ripens, normally the total phenolic content is reduced, concurrent with a loss of astringency and antioxidant activity (Patthamakanokporn et al., 2008; Shui et al., 2004). Although unripe fruits have very high level of antioxidants, their astringency is so high that it renders the fruit inedible. Besides, the fruit is too hard to be consumed. In order to benefit the most from the antioxidants in such fruits such as ciku, it should be consumed as soon as it ripens or else the total phenolic content and hence the antioxidant activity is reduced drastically (Shui et al., 2004).

Owing to the fact that the level of antioxidants changes during the ripening process of fruits, the study on the effect of processing techniques of fruits will have to take into consideration the ripening stage of the fruits. In a study where the effect of hot water immersion and controlled atmosphere storage on the antioxidant levels in mango was examined, it was found that the overall rate of degradation of phenolic antioxidants was decelerated by the delay of the ripening process (Kim et al., 2007). In an earlier study by the same research group, the concentration of gallic acid and four gallotannins were hardly affected by postharvest treatment with hot water submersion at 50 or 60°C, as well as storage at 5 or 20°C. An increase of 34% of these compounds was observed as the fruits ripened. Also as the fruits ripened, the concentration of carotenoids and hydrolysable tannins increased (Talcott et al., 2005).

Another study reported that there was no change in total phenolic compounds using the Folin method as well as the antioxidant activity using ORAC when the mango was treated with an electron-beam ionizing radiation (Reyes and Cisneros-Zevallos, 2007). However, a reduction in ascorbic acid during storage was observed. The authors also concluded that the carotenoid profile remained unchanged during the storage period due to the delay in the ripening process.

When UV-C is used to treat fresh cut mangoes for 10 min, phenols and flavonoids were found to accumulate, possibly due to a defense response generated (Gonzalez-Aguilar et al., 2007). As a result, antioxidant assays like ORAC and DPPH showed an increase of antioxidant activities. On the contrary, it was noted that as the irradiation time was increased, the β-carotene and ascorbic acid contents of the freshly cut mangoes were decreased during storage at 5°C.

In the development of functional food products using fruits with high antioxidant activities as an ingredient, the study of the change of antioxidant activities upon homogenization and cold storage is useful. One such study on *Psidium guajava* L. shows that homogenization of the fruit allows enzymes such as polyphenol oxidase to be released and as a result, drastic reduction of antioxidant activities was observed even when the mixture was stored at 20°C over a period of 3 months (Patthamakanokporn et al., 2008). In the same study, the researchers found that a whole guava when stored at 5°C over a period of 10 days showed an increase of antioxidant activity for the first few days, which reduced upon further storage. It is likely that over a period of 10 days, the fruit undergoes a ripening process and the polyphenolic compounds get reduced. Again this shows that fruits should be consumed when they are just ripe and not stored for longer period of time to get the most benefit. In another study, the level of ascorbic acid, a strong antioxidant in guava, increased immediately after harvesting but remained constant upon longer storage time until 10 days (Gomez and Lajolo, 2008). On the other hand, the same researchers reported that the level of ascorbic acid in mango was reduced after more than about five days of storage after harvesting.

13.1.4 Food Products from Asian Fruits

Southeast Asian fruits are normally eaten fresh as discussed above. These fruits are also canned for local and export markets. Juices of Asian fruits are also very common though juices cannot be extracted from all fruits. Fruits are also used in dairy products such as yoghurt, yoghurt drinks, and ice cream. Some fruits are consumed unripe. Unripe fruits can be pickled or used in salads. Pomelo and mango are often used in refreshing salad dishes in Thailand. Fruits like papaya, durian, and banana may also be fermented.

In Asia, a number of fruits are cooked before eating. One of the most common methods of cooking is deep-frying the battered fruit. For example, banana and cempedak fritters are most popular. Some fruits are also sliced thinly and deep fried to make chips. Lately, vacuum-fried sliced fruits are gaining popularity not just among the locals but tourists to the region as it can retain its flavors along with its functional components (unpublished).

Some fruits are dried so that they can be kept for a long period of time. Dry fruits may be eaten as they are or used for making other products. It is very common to use dry fruits in preparing soup, both for the enhancement of the taste as well as for their functional properties. Dried fruits can also be found in desserts . Fruits can also be preserved in sugar or acid solution.

Some fruits are used for making confectionary products. For example, durian is used to make a product called "dodol" which contains glutinous rice flour, coconut milk, and palm sugar. Fruits are used in the fillings of baked food items such as tarts, cakes, bread, and biscuits.

13.2 SUMMARY

Southeast Asian fruits contain many chemical compounds that are potentially beneficial to health. A number of compounds have already been identified and are known to have high antioxidant activities. Since most of the antioxidant compounds are polyphenolics, their composition undergoes change based on the level of ripeness of the fruit. The number of functional food products developed using fruits is increasing in the marketplace. Some examples of these products have been cited. Other functionalities of most of these compounds are yet to be discovered.

REFERENCES

Anila, L. and Vijayalakshmi, N. R. 2003. Antioxidant action of flavonoids from *Mangifera indica* and *Emblica officinalis* in hypercholesterolemic rats. *Food Chemistry*, *83*(4), 569–574.

Chomnawang, M. T., Surassmo, S., Nukoolkarn, V. S., and Gritsanapan, W. 2007. Effect of *Garcinia mangostana* on inflammation caused by *Propionibacterium acnes*. *Fitoterapia*, *78*(6), 401–408.

Fu, C., Loo, A. E. K., Chia, F. P. P., and Huang, D. 2007. Oligomeric proanthocyanidins from mangosteen pericarps. *Journal of Agricultural and Food Chemistry*, *55*(19), 7689–7694.

Gomez, M., and Lajolo, F. M. 2008. Ascorbic acid metabolism in fruits: Activity of enzymes involved in synthesis and degradation during ripening in mango and guava. *Journal of the Science of Food and Agriculture*, *88*(5), 756–762.

Gonzalez-Aguilar, G. A., Villegas-Ochoa, M. A., Martinez-Tellez, M. A., Gardea, A. A., and Ayala-Zavala, J. F. 2007. Improving antioxidant capacity of fresh-cut mangoes treated with UV-C. *Journal of Food Science*, *72*(3), S197–S202.

Haruenkit, R., Poovarodom, S., Leontowicz, H., Leontowicz, M., Sajewicz, M., Kowalska, T., Delgado-Licon, E. 2007. Comparative study of health properties and nutritional value of durian, mangosteen, and snake fruit: Experiments *in vitro* and *in vivo*. *Journal of Agricultural and Food Chemistry*, *55*(14), 5842–5849.

Hiramoto, K., Imao, M., Sato, E. F., Inoue, M., and Mori, A. 2008. Effect of fermented papaya preparation on dermal and intestinal mucosal immunity and allergic inflammations. *Journal of the Science of Food and Agriculture*, *88*(7), 1151–1157.

Imao, K., Wang, H., Komatsu, M., and Hiramatsu, M. 1998. Free radical scavenging activity of fermented papaya preparation and its effect on lipid peroxide level and superoxide dismutase activity in iron-induced epileptic foci of rats. *Biochemistry and Molecular Biology International*, *45*(1), 11–23.

Kanazawa, K., and Sakakibara, H. 2000. High content of dopamine, a strong antioxidant, in Cavendish banana. *Journal of Agricultural and Food Chemistry*, *48*(3), 844–848.

Kim, W. S. and Jo, J. A. 2006. Fruit characteristics, phenolic compound, and antioxidant activity of Asian pear fruit from an organically cultivated orchard. *Hortscience, 41*(4), 1029.

Kim, Y., Brecht, J. K., and Talcott, S. T. 2007. Antioxidant phytochemical and fruit quality changes in mango (*Mangifera indica* L.) following hot water immersion and controlled atmosphere storage. *Food Chemistry, 105*(4), 1327–1334.

Ko, F. N., Cheng, Z. J., Lin, C. N., and Teng, C. M. 1998. Scavenger and antioxidant properties of prenylflavones isolated from *Artocarpus heterophyllus. Free Radical Biology and Medicine, 25*(2), 160–168.

Kondo, S., Kittikorn, M., and Kanlayanarat, S. 2005. Preharvest antioxidant activities of tropical fruit and the effect of low temperature storage on antioxidants and jasmonates. *Postharvest Biology and Technology, 36*(3), 309–318.

Leong, L. P., and Shui, G. 2002. An investigation of antioxidant capacity of fruits in Singapore markets. *Food Chemistry, 76*(1), 69–75.

Leontowicz, H., Leontowicz, M., Drzewiecki, J., Haruenkit, R., Poovarodom, S., Park, Y. S., Jung, S. T., Kang, S. G., Trakhtenberg, S., and Gorinstein, S. 2006. Bioactive properties of snake fruit (*Salacca edulis* Reinw) and mangosteen (*Garcinia mangostana*) and their influence on plasma lipid profile and antioxidant activity in rats fed cholesterol. *European Food Research and Technology, 223*(5), 697–703.

Lim, Y. Y., Lim, T. T., and Tee, J. J. 2007. Antioxidant properties of several tropical fruits: A comparative study. *Food Chemistry, 103*(3), 1003–1008.

Ma, J., Luo, X. D., Protiva, P., Yang, H., Ma, C. Y., Basile, M. J., Weinstein, I. B., and Kennelly, E. J. 2003. Bioactive novel polyphenols from the fruit of *Manilkara zapota* (Sapodilla). *Journal of Natural Products, 66*(7), 983–986.

Madsen, H. L., Andersen, C. M., Jorgensen, L. V., and Skibsted, L. H. 2000. Radical scavenging by dietary flavonoids. A kinetic study of antioxidant efficiencies. *European Food Research and Technology, 211*(4), 240–246.

Maisuthisakul, P., Pasuk, S., and Ritthiruangdej, P. 2008. Relationship between antioxidant properties and chemical composition of some Thai plants. *Journal of Food Composition and Analysis, 21*(3), 229–240.

Marotta, F., Yoshida, C., Barreto, R., Naito, Y., and Packer, L. 2007. Oxidative-inflammatory damage in cirrhosis: Effect of vitamin E and a fermented papaya preparation. *Journal of Gastroenterology and Hepatology, 22*(5), 697–703.

Micol, V., Larson, H., Edeas, B., and Ikeda, T. 2007. Watermelon extract stimulates antioxidant enzymes and improves glycemic and lipid metabolism. *Agro Food Industry Hi-Tech, 18*(1), 22–26.

Pardo-Andreu, G. L., Barrios, M. F., Curti, C., Hernandez, I., Merino, N., Lemus, Y., Martinez, L., Riano, A., and Delgado, R. 2008. Protective effects of *Mangifera indica* L. extract (Vimang), and its major component mangiferin, on iron-induced oxidative damage to rat serum and liver. *Pharmacological Research, 57*(1), 79–86.

Patthamakanokporn, O., Puwastien, P., Nitithamyong, A., and Sirichakwal, P. P. 2008. Changes of antioxidant activity and total phenolic compounds during storage of selected fruits. *Journal of Food Composition and Analysis, 21*(3), 241–248.

Perkins-Veazie, P., Collins, J. K., Davis, A. R., and Roberts, W. 2006. Carotenoid content of 50 watermelon cultivars. *Journal of Agricultural and Food Chemistry, 54*(7), 2593–2597.

Reyes, L. F., and Cisneros-Zevallos, L. 2007. Electron-beam ionizing radiation stress effects on mango fruit (*Mangifera indica* L.) antioxidant constituents before and during post-harvest storage. *Journal of Agricultural and Food Chemistry, 55*(15), 6132–6139.

Ribeiro, S. M. R., Barbosa, L. C. A., Queiroz, J. H., Knodler, M., and Schieber, A. 2008. Phenolic compounds and antioxidant capacity of Brazilian mango (*Mangifera indica* L.) varieties. *Food Chemistry, 110*(3), 620–626.

Ribeiro, S. M. R., de Queiroz, J. H., Lopes, M. E., de Queiroz, R., Campos, F. M., and Sant'ana, H. M. P. 2007. Antioxidant in mango (*Mangifera indica* L.) pulp. *Plant Foods for Human Nutrition*, *62*(1), 13–17.

Rimando, A. M., and Perkins-Veazie, P. M. 2005. Determination of citrulline in watermelon rind. *Journal of Chromatography A*, *1078*(1–2), 196–200.

Rodriguez, J., Di Pierro, D., Gioia, M., Monaco, S., Delgado, R., Coletta, M., and Marini, S. 2006. Effects of a natural extract from *Mangifera indica* L., and its active compound, mangiferin, on energy state and lipid peroxidation of red blood cells. *Biochimica et Biophysica Acta-General Subjects*, *1760*(9), 1333–1342.

Shui, G., Wong, S. P., and Leong, L. P. 2004. Characterization of antioxidants and change of antioxidant levels during storage of *Manilkara zapota* L. *Journal of Agricultural and Food Chemistry*, *52*(26), 7834–7841.

Shui, G. H., and Leong, L. P. 2003. Rapid screening and identification of antioxidants of salak (*Salacca edulis* Reinw) using high performance liquid chromatograph coupled with mass spectrometry. *Free Radical Biology and Medicine*, *35*, S47–S48.

Shui, G. H., and Leong, L. P. 2004. Analysis of polyphenolic antioxidants in star fruit using liquid chromatography and mass spectrometry. *Journal of Chromatography A*, *1022*(1–2), 67–75.

Someya, S., Yoshiki, Y., and Okubo, K. 2002. Antioxidant compounds from bananas (*Musa cavendish*). *Food Chemistry*, *79*(3), 351–354.

Soong, Y. Y., and Barlow, P. J. 2004. Antioxidant activity and phenolic content of selected fruit seeds. *Food Chemistry*, *88*(3), 411–417.

Sun, J., Shi, J., Jiang, Y. M., Xue, S. J., and Wei, X. Y. 2007. Identification of two polyphenolic compounds with antioxidant activities in longan pericarp tissues. *Journal of Agricultural and Food Chemistry*, *55*(14), 5864–5868.

Talcott, S. T., Moore, J. P., Lounds-Singleton, A. J., and Percival, S. S. 2005. Ripening associated phytochemical changes in mangoes (*Mangifera indica*) following thermal quarantine and low-temperature storage. *Journal of Food Science*, *70*(5), C337–C341.

Thaipong, K., Boonprakob, U., Crosby, K., Cisneros-Zevallos, L., and Byrne, D. H. 2006. Comparison of ABTS, DPPH, FRAP, and ORAC assays for estimating antioxidant activity from guava fruit extracts. *Journal of Food Composition and Analysis*, *19*(6–7), 669–675.

Wei, B. L., Weng, J. R., Chiu, P. H., Hung, C. F., Wang, J. P., and Lin, C. N. 2005. Antiinflammatory flavonoids from *Artocarpus heterophyllus* and *Artocarpus communis*. *Journal of Agricultural and Food Chemistry*, *53*(10), 3867–3871.

Wong, S. P., Leong, L. P., and Koh, J. H. W. 2006. Antioxidant activities of aqueous extracts of selected plants. *Food Chemistry*, *99*(4), 775–783.

14 Health Benefits of *Kochujang* (Korean Red Pepper Paste)

Kun-Young Park and In-Sook Ahn

CONTENTS

14.1 INTRODUCTION

Korea has a long history of eating fermented foods and fermentation skills have been developed for more than 1500 years. Major fermented foods consumed in Korea include kimchi, jeotgal (salted fish and shellfish), vinegar, soy-based products such as doenjang (fermented soybean paste), chungkukjang (quick fermented soybean paste), kanjang (soy sauce), and *kochujang* (red pepper paste). These foods are prepared through various fermenting methods using a wide spectrum of raw materials.

Red pepper powder (RPP) is one of the most popular condiments in Korea in the same way as Indian curry powder and European black pepper. Its strong pungent taste differentiates Korean food from Japanese and Chinese foods, *Kochujang* is a Korean traditional fermented food that has been eaten with doenjang as a seasoning spice for a long time. *Kochujang* is fermented from a mixture of RPP with fermented soybeans (meju) and a starch source including rice, glutinous rice, and/or wheat. It has played an important role in providing specific taste, color, and flavor in foods. *Kochujang* has a sweet taste from the starch hydrolyzates, a hot taste of RPP, a salty taste of salt, and savory tastes from the soybean protein hydrolyzates and nucleic acids (Moon and Kim, 1988; Woo and Kim, 1990; Lee et al., 1993). Its special flavors are due to various alcohols and organic acids that are generated during fermentation by microorganisms, molds, bacteria, or yeasts. It is used for enhancing the taste of soups and seasonings in Korean foods.

The functional characteristics of *kochujang* have been reported in the scientific literature only in the past decade, although it has a long history of use. These reports revealed that capsaicin, which is the active compound of RPP, and meju endow *kochujang* with various functionalities including antimutagenic, anticancer, and antiobesity effects. However, recent studies have opened the possibility that some compounds generated during the *kochujang* fermentation process might exhibit functional properties as well.

14.2 *KOCHUJANG* HISTORY

Red pepper (*Capsicum annuum*) is the most consumed spice in the world. There is evidence that Mexican Indians were consuming capsicum as early as 7000 BC. After Europeans made explorative expeditions to the American continents in 1492–1493, cultivation of red pepper spread throughout the Mediterranean and Central European regions. Within a comparatively short time, Spain and Portugal established capsicum farms in many tropical and subtropical regions. In Korea, even though there are opinions that red pepper was introduced from the continent or southward, it is believed that it was distributed from Japan during the Japanese invasion of the Chosun Dynasty (1592–1599). Red pepper had the greatest impact on the dietary lifestyle of the Chosun Dynasty. Therefore, it is believed that *kochujang* was first consumed in Korea between the end of the sixteenth century and the early seventeenth century when red pepper came to Korea.

Domundaejak (1611), the book written by Hurkyun critically describing Korean cuisine, used the word Choshi, which is a Chinese character meaning pungent and fermented soybean, to describe an early type of *kochujang*. Sanlimkungje (1715), written by Hongmansun, mentioned a *kochujang*-making method for the first time. Jeungbosanlimkungje (1766), written by Yujoonglim, described a different processing method, which added dried fish and sea tangle into *kochujang* to improve its taste. Sumunsasul, written by Leepyo in 1740, described a preparation method for Sunchang *kochujang* in the book of Sikchibang, which deals with traditional remedies. This book also indicated that *kochujang* could additionally contain abalone, shrimp, mussel, and ginger, which increased its nutritional value. Sunchang *kochujang* is famous for being one of the royal presents given to the king in the past, and it tastes as excellent today as it has since old times. Matching the wise ancestral saying, "Relish of paste is dependent on taste of water," the Sunchang area is blessed with pure water and a clean environment, high-quality of raw materials, and a particular climate. These factors all positively affect the production or fermentation of *kochujang*. Therefore, people in different areas cannot produce the same taste and flavor of Sunchang *kochujang*, even though the *kochujang* is prepared with the same ingredients which come from the Sunchang area.

Later, many preparation methods for *kochujang* became available and various styles of high-quality *kochujang* were developed. In the middle 1800s, Yeukjubangmoon described *kochujang* made with barley, and thin soy sauce was used for a salty taste. In Gyhapchongseo (1815), which is considered the first home encyclopedia of the Chosun Dynasty, more *kochujang* developments were introduced. For example, *kochujang* meju and salt were used for *kochujang* preparation, which is the same as the modern method. Also, the addition of honey, dried meat, and jujube improved the flavor of *kochujang*. Nonggawollyeongga (1861), Songs of Monthly Events of Farm Families, described the proper season for *kochujang* preparation, which was March.

14.3 MANUFACTURING METHODS AND CHARACTERISTICS OF *KOCHUJANG*

According to the Korean Food Sanitation Act, *kochujang* (Korean government recently changed the English name of *kochujang* to gochujang) is supposed to be a fermented food in which steamed rice, hot RPP, and salt are mixed with meju made with grains or soybeans, and then fermented and aged. RPP and salt may be mixed either before or after the fermentation process. RPP and steamed rice content should be more than 6% and 15%, respectively. *Kochujang* should be homogenized and have its unique flavor and color (red or dark red color derived from red pepper). Crude protein content should be more than 4.0% (w/w), and no tar colorant should be detectable. Preservatives including sorbic acid, should not be present in higher concentrations than 1 g/kg of *kochujang*.

Kochujang can be classified into two groups, traditional *kochujang* using meju (by the conventional methods) and a commercial one (by the convenient factory method) using koji or bacterial enzymes. The use of meju is the difference between a traditional *kochujang* and the commercial one.

14.3.1 Traditional *Kochujang*

Generally, traditional *kochujang* is made of glutinous rice, meju, RPP, and salt, which is fermented by the enzymatic reactions of bacteria or yeast. Malt is an optional ingredient that may be used to saccharify glutinous rice.

Sunchang-kun, Jeollabuk-do (province) in Korea is a famous region that has preserved the traditional methods for preparing *kochujang*. They make a few kinds of traditional *kochujang* and Sikhae *kochujang*, a representative kind of *kochujang* using malt for saccharifying the starch source such as glutinous rice. Figure 14.1 shows the processing method for making Sikhae *kochujang*, which includes two steps; meju making and *kochujang* making. To make meju, rice and soybeans are soaked in water for a day. After being drained, they are mixed, steamed until they become soft, and then pounded in a wooden mortar. Doughnut-shaped cakes (diameter 10–15 cm) are prepared, dried, and then fermented for 40 days wrapped with rice straw (meju). Dried meju is ground into powder, which is called meju powder. To make *kochujang*, the glutinous rice is soaked for a day and then ground. On the other hand, the powdered malt is soaked for a day to get the enzyme extracted with water. The ground glutinous rice and malt water are mixed and kept warm at 45°C to digest, and then boiled down. Finally, it is mixed with all the other ingredients such as the

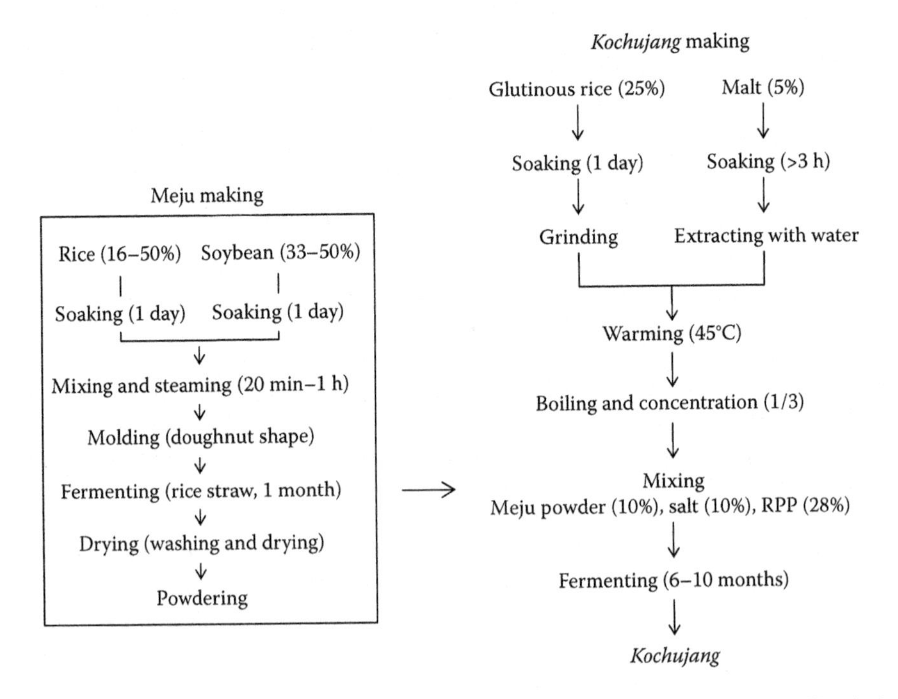

FIGURE 14.1 Traditional method for preparing *kochujang* in Sunchang-kun in Chunbuk Province.

meju powder, salt, and RPP, and then placed into an earthen pot and fermented for 6–10 months. The standardized ingredients for traditional *kochujang* making are 25% glutinous rice, 28% RPP, 10% meju powder, and 10% salt. For making Sikhae *kochujang*, 5% malt is additionally added.

14.3.2 Commercial *Kochujang*

The big difference between commercial *kochujang* and traditional is the addition of saccharogenic amylase or koji. Figures 14.2 and 14.3 show the processing methods for making commercial *kochujang*, which includes koji making, mixing ingredients, and aging.

The commercial preparation of this *kochujang* includes several steps. Steamed wheat flour is first inoculated with *Aspergillus oryzae* and incubated at 35°C for 3 days. It is then mixed with steamed wheat grains and salt (this mixture is called the first-stage fermented product). The mixture is matured in the presence of *Zygosaccharomyces rouxii* at 30°C for a week (second-stage fermented product), and aged for 30–40 days (final-stage fermented product). *Kochujang* preparation is completed after the final-stage fermented product is mixed with RPP (ratio is 4:1), starch syrup, MSG, and so on. It is sterilized at 70°C for 8 min and packaged (Figure 14.2). In other cases, koji can be made with starch materials such as wheat flour, rice, and soybeans with *Aspergillus oryzae* and *Bacillus subtilis* at 30–35°C for 2 days. Koji is mixed with RPP, salt, starch syrup, MSG, and so on, which is further ripened at 30–35°C for 7–15days. It is sterilized 3 times at 90°C for 10 min and then packaged (Figure 14.3).

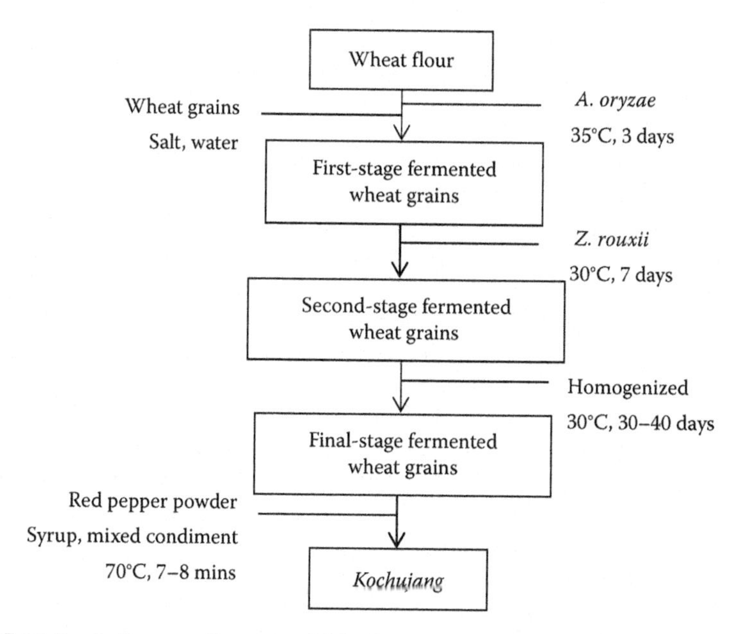

FIGURE 14.2 A diagram of commercial *kochujang* preparation method I. (From Ahn, I.S. 2007. *In vitro* antiobesity effect of *kochujang* and isolation and identification of active compounds. PhD dissertation, Pusan National University, Busan, Korea. With permission.)

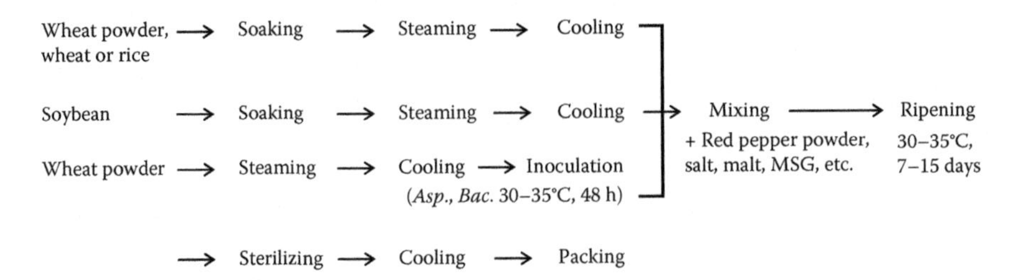

FIGURE 14.3 A diagram of commercial *kochujang* preparation method II.

In Korea, people have typically prepared *kochujang* at home by the traditional methods. Nowadays, to meet the increased demand and for the sake of convenience, a large proportion of *kochujang* is produced by mass production systems in factories. The main differences between them are in the microorganisms, carbon source, and fermentation period, which may induce some differences in taste, flavor, functionality, and so on.

Commercial *kochujang* is prepared under controlled conditions; using selected pure culture (*Aspergillus oryzae*) for koji making and 15–30 days of aging period. The main carbon source is wheat flour. However, traditional *kochujang* includes natural microbes for fermentation, at least a 6-month-long fermentation period, and rice as the main carbon source.

14.4 NUTRITIONAL AND FUNCTIONAL PROPERTIES OF *KOCHUJANG*

14.4.1 Components and Nutrition of *Kochujang* and Its Ingredients

14.4.1.1 *Kochujang*

According to a study of traditional *kochujang* samples collected from 55 families in different regions, they comprise 46.9 ± 8.8% total sugar, 27.5 ± 7.3% reducing sugar, 11.8 ± 3.9% crude protein, 0.3 ± 0.1% amino nitrogen, 2.7 ± 2.4% ethanol, and 15.0 ± 6.5% salt on a dry basis. They contained 46.7 ± 6.0% moisture. The pH and water activity were 4.60 and 0.79, respectively (Shin et al., 1996). Succinic acid was found to be the major organic acid, followed in decreasing order by citric acid, lactic acid, oxalic acid, and formic acid. The major free sugars were glucose and maltose, and the minors were fructose and sucrose.

The traditional *kochujang* samples contained large amounts of proline, glutamic acid, aspartic acid, serine, and lysine. Among the nucleotides and their related compounds in traditional *kochujang*, CMP, hypoxanthine, IMP, inosine, and GMP were detected (Kim et al., 1996). Linoleic acid and linolenic acid comprise 61.3–85.0% of the free fatty acids in the *kochujang*.

The quality of *kochujang* is dependent on the combination of ingredients and processing methods. Meju preparation is an important process for traditional *kochujang*. Meju is naturally fermented, and includes microorganisms such as *Bacillus subtilis*, *Mucor*, *Rhizopus*, and *Aspergillus oryzae*. They secrete amylase, protease, lipase, and other enzymes, and help meju/*kochujang* to get its unique taste and flavor. Lee et al. (1976) isolated *Aspergillus oryzae* and *Bacillus subtilis* which have high enzyme activities from traditional *kochujang*. Lee et al. (1970) also reported that they selected useful yeasts including *Saccharomyces cerevisiae, S. oviformis, S. steineri, S. rouxii,* and *S. melis* from koji *kochujang* and rice koji.

14.4.1.2 Glutinous Rice

Glutinous rice typically contains 13.2% moisture, 8% protein, 1.2% fat, 75.7% carbohydrate, 1.2% fibers, 1.0% ash, and other minerals and vitamins. Glutinous rice is softer and whiter than nonglutinous rice. Starch is composed of amylopectin, which makes it more easily digested than nonglutinous rice.

14.4.1.3 Wheat Grains

Wheat is one of the most important cereal crops, along with rice, in Korean meals. Wheat contains mostly carbohydrates and protein (Kim et al., 1997), Vitamin B_1 can be found in the embryo and vitamin B_2 is located throughout the whole grain. Wheat contains many phenolics. Ferulic acid is the most common phenolic acid found in wheat bran, and it contributes 57–78% of the phenolic acids in wheat (Smith and Harttley, 1983; Onyeneho and Hettiarachchy, 1992; Abdel-Aal et al., 2001; Zhou et al., 2004; Mpofu et al., 2006). Fiber intake from cereal grains, which is more effective than that from fruits and vegetables decreases the risk of heart diseases (Wolk et al., 1999).

Phytochemicals contained in wheat grains are reported to have various functional properties. Adom et al. (2003) examined the phytochemical profiles of 11 varieties of wheat. From this and another study (Adom et al., 2005), it is known that wheat bran contains more phenolics than any other part of the wheat grain, and it showed high antioxidant activity. Wheat bran reduces incidences of breast cancer and colorectal cancer, and it also reduces blood glucose after meals and improves type 2 diabetes (Vaaler et al., 1986). Wheat grains also show antioxidant activity, which inhibits LDL oxidation and improves lipid peroxidation in the liver of high-cholesterol-fed subjects (Bu et al., 2002; Choe and Kim, 2002). A wheat diet fed to obese women for 12 weeks reduced weight gain, which was more significant than a low caloric diet (Fordyce-Baum et al., 1989). Adam et al. (2002) reported that the intake of whole wheat powder reduced fat accumulation in the liver as well as blood lipids.

14.4.1.4 Soybean and Meju

Soybean (*Glycine max*) is a good source of protein. Therefore, soybean is called "beef produced from the field" in Korea. Soybeans contain many functional compounds that improve physiological activities: protease inhibitors, dietary fiber, lecithin, oligosaccharides (raffinose, stachiose), and phytochemicals such as phytic acid, triterpenes (saponin, squalene, sterols), phenolics (isoflavones, phenolic acids), lignan, carotenoids, courmarin, and so forth (Messina and Messina, 1991).

Meju (Figure 14.1) is dried fermented soybean–rice cake, which is one of the major ingredients of traditional *kochujang*. Meju is made by steaming and mashing soybeans, shaping the mashed soybeans into rectangular or donut cakes, and then drying and fermenting them. The meju is used for kanjang (fermented soy sauce), doenjang, or *kochujang* in Korea. During fermentation of steamed soybean, the moisture decreases from 51.40% to 11.4% and it loses 48% of its weight. Ahn et al. (1987) isolated 36 microorganisms from meju including *Bacillus, Aspergillus*, and *Saccharomyces* species. Changes in microflora and enzyme activities during fermentation influence the taste, flavor, and chemical characteristics of *kochujang*. Park, J.M. et al. (1995) reported that 40 days of fermentation of *kochujang* meju produced the highest quality since it resulted in the greatest increase in the number of microflora resulting in high protease and amylase enzyme activities.

14.4.1.5 Red Pepper

Red pepper (*C. annuum* L.) is one of the most abundantly consumed spices in the world. It is used for its hot pungent flavor and as a natural food color while some other species are used as food preservatives. The principal pungent ingredient of red pepper is the phenolic compound, called capsaicin (*trans*-8-methyl-*N*-vanyll-6-nonenamide), and others are dihydrocapsaicin and nor-dihydrocapsaicin. Capsaicin and dihydrocapsaicin comprise 50–69% and 20–30% of the total capsaicinoids, respectively. Distribution of chemical constituents varies among different parts (whole, placenta, pericarp, and seed) of the red pepper, and capsaicin is located mostly in the pericarp. The red color of red pepper comes from carotenoid pigments in the pericarp, notably xanthophylls, which are oxygenated derivatives of carotenes, and this is made up of capsanthin, β-carotene, violaxanthin, cryptoxanthin, capsorubin, and cryptocapsin (Nagle and Burns, 1979). The pungent taste and color of red pepper are very important factors in the production of *kochujang*.

14.4.2 Antimutagenic and Anticancer Effects of *Kochujang*

Kochujang posseses cancer chemopreventive properties since it uses meju, glutinous rice, and RPP as major ingredients, all of which contain various functional compounds. We employed samples of representative traditional *kochujang* (prepared as Figure 14.1 at M. Corp., Jeonbuk, Korea) from a soy products folk village in Sunchang-kun and commercial *kochujang*s (prepared as Figures 14.2 and 14.3 at H. Corp. Chungnam, Korea and C. Corp., Jeonbuk, Korea, respectively) from large soybean products factories, in order to determine their antimutagenic and antitumor effects.

14.4.2.1 Antimutagenic Effects of *Kochujang*

The active components in traditional *kochujang* (prepared as Figure 14.1), commercial *kochujang* (prepared as Figure 14.3), and RPP were extracted with methanol. We employed the Ames test and SOS chromotest to evaluate the antimutagenicities of the samples. As shown in Table 14.1, traditional *kochujang* (TK I) which is not fermented, but has all ingredients before the fermentation, did not show antimutagenicity against aflatoxin B_1. However, when the fresh *kochujang* was fermented for 6 months (TK II), the *kochujang* showed the highest antimutagenicity (42–47% inhibition). Commercial *kochujang* also exhibited antimutagenicity, but the inhibition rate was somewhat lower (27–31%) than TKII (Kong, 2001). This is because of the short period of fermentation during mass production in the factory. Jung et al. (2006) indicated that a prolonged fermentation period increased chemopreventive effects in doenjang. Two-year fermentation of doenjang significantly increased

TABLE 14.1
Effect of Methanol Extracts from Various Kinds of *Kochujang* and Red Pepper Powder on the Mutagenicity Induced by Aflatoxin B_1 (AFB_1, 0.3 µg/plate) in *Salmonella typhimurium* TA100

	Revertant/Plate (mg/Plate)	
Treatment	**2.5**	**5**
AFB_1 (Control)	$1367 \pm 24^{b,A}$	
AFB_1 + TKI[1]	1534 ± 62^{a} (−13)	1181 ± 144^{B} (15)
+ TKII[2]	833 ± 35^{d} (42)	775 ± 11^{D} (47)
+ CK[3]	1027 ± 117^{c} (27)	976 ± 66^{C} (31)
+ RPP[4]	962 ± 56^{cd} (32)	783 ± 16^{D} (46)

Source: From Kong, G.R. 2001. Standardization of *kochujang* preparation and its effects of cancer prevention and lipid metabolism in rat. MS thesis, Pusan National University, Busan, Korea. With permission.

Note: Spontaneous revertant number was 98 ± 3.

[1] Traditional *kochujang* I: 0-day fermented *kochujang*, Moonokrye Co.

[2] Traditional *kochujang* II: 6-month fermented *kochujang*, Moonokrye Co.

[3] Commercial *kochujang*: Chungjungwon Co.

[4] Red pepper powder.

[a–d,A–D] Means with different letters in the same column are significantly different ($p < 0.05$) by Duncan's multiple range test.

antimutagenicity and anticancer effects as compared with 6-month fermented doenjang. The RPP also showed higher antimutagenicity against aflatoxin B1. Capsaicin, polyunsaturated fatty acids, and carotenoids in the RPP could be the major compounds that decreased the mutagenicity. In the SOS chromotest system, the traditional *kochujang* fermented for 6 months showed 94% inhibition of the MNNG mutagenicity, and commercial *kochujang* showed 65% inhibition. However, fresh traditional *kochujang* exhibited lower antimutagenicity of 58% (Kong, 2001). Thus, the fermentation process is an important factor in increasing antimutagenicity.

We attempted to identify the active compounds from the traditional *kochujang*. The dichloromethane fraction showed the highest antimutagenic activity. Hexanedioic acid dioctyl ester, 2-imidazolidinone, 2,5-dihydroxy-α-methyl-phenethyl alcohol, and 10-hydroxy-hexadecanoic acid methyl ester were identified as the active compounds by GC–MS analysis (Kim, 1999). Other active compounds seemed to be involved that might reveal antimutagenic effects in that experiment. More studies are needed to identify the active compounds, where the compounds originate, and their chemopreventive mechanisms.

14.4.2.2 Antimutagenicity of Major Ingredients of Kochujang

The antimutagenicities of methanol extracts from traditional and commercial *kochujang*s and their ingredients were evaluated in *Salmonella*/mammalian microsome assay system (Table 14.2). The traditional *kochujang* exhibited higher antimutagenic

TABLE 14.2
Effect of Methanol Extracts from Two Kinds of *Kochujang* and their Ingredients on the Mutagenicity Induced by Aflatoxin B_1 (AFB_1, 0.5 μg/Plate) in *Salmonella typhimurium* TA100

		Revertants/Plate (Level of Sample, mg/Plate)	
Treatment	**Ingredient**	**1.25**	**2.50**
	Control	$1174 \pm 26^{a,3}$	1174 ± 26^{A}
Traditional[1]	*Kochujang*	566 ± 23^{f} (56)[4]	485 ± 16^{G} (63)
Kochujang	RPP	682 ± 20^{d} (45)	888 ± 17^{C} (26)
	Meju	374 ± 17^{g} (74)	253 ± 26^{H} (85)
	Glutinous rice powder	617 ± 18^{e} (51)	516 ± 31^{FG} (60)
Commercial[2]	*Kochujang*	764 ± 25^{c} (38)	666 ± 23^{D} (47)
Kochujang	RPP	765 ± 30^{c} (38)	940 ± 13^{B} (22)
	Koji	452 ± 13^{g} (66)	513 ± 23^{FG} (61)
	Glutinous rice powder	691 ± 10^{d} (44)	586 ± 25^{E} (54)
	Wheat flour	915 ± 23^{b} (24)	846 ± 16^{C} (30)
	Wheat grain	672 ± 26^{d} (46)	543 ± 15^{F} (58)

Source: From Jung, K.O. et al., 2000. *J. Korean Assoc. Cancer Prev.* 5:209–216. With permission.
Note: Spontaneous revertant numbers were 86 ± 3.
[1] Sunchang traditional *kochujang* prepared with glutenous rice powder and malt (Moonokrye Co.).
[2] Chungjungwon Co.
[3] The values are means of three replicates ±SD.
[4] The values within parentheses are the inhibition rates (%).
[a–f,A–H] Means with different letters in the same column are significantly different at the $p < 0.05$ level of significance as determined by Duncan's multiple range test.

effects than the commercial one against aflatoxin B_1 (AFB_1) on *Salmonella typhimurium* TA100. Among the ingredients of the traditional and commercial *kochujang*s, meju (74–85%), koji (61–66%), and glutinous rice powder (44–60%) and wheat grains (46–58%) effectively reduced the mutagenicity induced by AFB_1. Antimutagenic effects of meju for traditional *kochujang* were higher than those of koji for commercial *kochujang*. Glutinous rice powder had strong inhibitory effects on the mutagenicity induced by AFB_1. However, RPP showed a lower inhibition rate than did the *kochujang*s. Meju, koji, and glutinous rice powder also strongly inhibited effects on the mutagenicity induced by MNNG (Jung et al., 2000). These results indicate that meju, koji, and glutinous rice powder seem to be the major antimutagenic components in *kochujang*.

14.4.2.3 Antitumor Effect of *Kochujang* in Sarcoma-180 Transplanted Mice

In an experiment conducted to examine the effect of *kochujang* (Park et al., 2001) on tumor formation and growth, Sarcoma-180 tumor cells were transplanted to the left groin of Balb/c mice. The mice were injected i.p. with *kochujang* once a day for 20 days from 24 h after the transplantation. After 32 days, the solid tumors formed from the left groin of the mice were removed and weighed. The average weight of sarcoma-180 cells transplanted into the control group was 6.0 ± 0.1 g. Traditional *kochujang* I (nonfermented) and commercial *kochujang* inhibited the tumor formation by 17% and 23%, respectively. However, the traditional *kochujang* II (6-month fermented) showed the highest tumor inhibition rate among the groups, with tumor weights of 3.3 g (45% inhibition). Therefore, these results revealed that fermentation of *kochujang* seemed to increase its antitumor activity and that the traditional *kochujang* has more antitumor activity than the commercial *kochujang* (Table 14.3).

GST activity: GST (glutathione *S*-transferase) is a phase II enzyme found in the liver. It binds toxic compounds and converts them to water-soluble compounds that can be removed from the body. GST uses reduced glutathione and removes toxic and peroxide compounds. In our study GST activities were reduced by the transplantation of S-180 cells, but a 6-month fermented traditional *kochujang*-treated sample significantly increased the GST activity ($p < 0.05$, Park et al., 2001).

NK cell activity: NK (natural killer) cells are cytotoxic lymphocytes that kill microorganisms, viruses, and cancer cells (Parveen et al., 1994) and thus any lower levels and activities of the NK cells result in cancer and increased rates in the growth of cancer cells (Terry et al., 1990). NK cell activity is significantly reduced by tumor cell transplantation. However, *kochujang* samples significantly increased NK cell

TABLE 14.3
Antitumor Activities of Methanol Extracts from Various Kinds of *Kochujang* and Red Pepper Powder (RPP) in Tumor Bearing Balb/c Mice with Sarcoma-180 Cells[1]

Sample	Tumor wt. (g)	Inhibiton Rate (%)
S-180 + PBS	6.0 ± 0.1^{a}	–
+ CK[2]	4.5 ± 0.1^{b}	23
+ TK I[3]	5.0 ± 0.7^{ab}	17
+ TK II[4]	3.3 ± 0.3^{c}	45
+ RPP[5]	4.7 ± 0.3^{b}	22

Source: From Park, K.Y. et al., 2001. *J. Food Sci. Nutr.* 6:187–191. With permission.

[1] 7-days sarcoma-180 ascites cells were s.c. transplanted into the left groin of inbred strain. 1.0 mg/kg of methanol extract from various kinds of *kochujang*, RPP or the equal volume of phosphate-buffered saline (control) was i.p. injected once a day for 20 days from 24 h following transplantation. All mice were sacrificed at 5 weeks following the transplantation, and tumor, spleen, and liver weights were measured.

[2] Commercial *kochujang*: Chungjungwon Co.

[3] Traditional *kochujang* I: 0-day fermented *kochujang*, Moonokrye Co.

[4] Traditional *kochujang* II: 6-month fermented *kochujang*, Moonokrye Co.

[5] RPP: the same sample that added in TK I and II.

[a–c] Means with different letters are significantly different ($p < 0.05$) by Duncan's multiple range test.

activities, and TK II showed the highest activity ($p < 0.05$). The unfermented *kochujang* (TK I) and RPP did not stimulate the NK cell activity (Park et al., 2001).

14.4.2.4 Inhibitory Effect of *Kochujang* on Lung Metastasis

Most deaths caused by cancer are not the result of primary tumor growth but, rather, are due to the dissemination of tumor cells to secondary sites by a process known as the metastatic cascade (Fidler, 1991). Inhibition of tumor metastasis by *kochujang* extracts was studied using an experimental metastasis model in mice. Table 14.4 shows the inhibitory effect of *kochujang* on tumor metastasis. To do this, the number of tumors disseminated into the lungs was measured after injecting colon 26-M3.1 cells into Balb/c mice tails. The methanol extract from traditional *kochujang* I, II, and commercial *kochujang* products significantly reduced tumor metastasis by 48%, 84%, and 58% respectively when they were injected in mice at 1.25 mg/day. Traditional *kochujang* II (6-month fermented) was the most effective in inhibiting lung metastasis of colon 26-M3.1 cells. Therefore, it is considered that the well-ripened traditional *kochujang* has more suppressive effects on lung metastasis than RPP, commercial *kochujang* (CK), and traditional *kochujang* (TK) without fermentation (Park et al., 2001). Several studies indicated that fermented foods increased antimutagenic and anticarcinogenic activities, depending on the increase in the

TABLE 14.4
Inhibitory Effect of Methanol Extracts from Various Kinds of *Kochujang* and Red Pepper Powder Sample on Tumor Metastasis Produced by Colon 26-M3.1 Cells

		No. of Lung Metastasis (Inhibition%)	
Treatment	**Dose (mg/mouse)**	**Mean ± SD**	**Range**
Control		318 ± 11[a]	310–330
Commercial *Kochujang*[1]	0.25	208 ± 56[ab](34)	163–271
	1.25	133 ± 15[c](58)	124–150
Traditional *Kochujang* I[2]	0.25	121 ± 105[bc](62)	90–260
	1.25	165 ± 87[bc](48)	60–242
Traditional *Kochujang* II[3]	0.25	52 ± 63[c](79)	4–123
	1.25	66 ± 88[c](84)	3–166
RPP[4]	0.25	308 ± 12[a](3)	297–321
	1.25	258 ± 34[ab](19)	233–296

Source: From Park, K.Y. et al., 2001. *J. Food Sci. Nutr.* 6:187–191. With permission.

[1] Commercial *Kochujang*: (Chungjungwon. Co.).

[2] Traditional *Kochujang* I: 0-day fermented (Moonokrye. Co.).

[3] Traditional *Kochujang* II: 6-month fermented (Moonokrye. Co.).

[4] Red pepper powder.

[5] Eu = (1000 × A420)/time (min).

[a–c] Means with different letters are significantly different ($p < 0.05$) by Duncan's multiple range test.

fermentation time (Park J.M. et al., 1995; Park K.Y. et al., 1995; Choi et al., 1997; Ko et al., 1999). It has also been found that chungkookjang or doenjang, which are Korean fermented soybean products, show higher antimutagenic effects than nonfermented cooked soybeans (Ko et al., 1999; Hwang, 2004). It seems that the higher inhibition rate in fermented *kochujang* probably results from some end products produced by the action of microorganisms during *kochujang* fermentation.

14.4.2.5 Antimutagenic Effects of Fermented Wheat Grains in Commercial *Kochujang*

Whole wheat grains, especially, are the major starch source for *kochujang* preparation in factories which have mass production systems operating under controlled conditions as shown in Figure 14.2. One of the major Korean *kochujang* factories (H. Corp.) uses whole wheat grains as the starch source. They first make wheat koji (first-stage fermented wheat grains; FFWG) using *Aspergillus oryzae*, and then add more steamed wheat grains for further fermentation by using *Zygosaccharomyces rouxii* for a week (second-stage fermented wheat grains; SFWG). It is then ripened for about a month (final-stage fermented wheat grains; FiFWG). Finally, FiFWG is mixed with RPP to make *kochujang*. They provide very good materials to examine whether fermentation affects the antimutagenic activity. The whole wheat grains and fermented wheat grains at different stages of fermentation were tested to compare their antimutagenic activities using the Ames test (Table 14.5). The methanol extract of whole wheat grains inhibited mutagenesis against MNNG in *Salmonella typhimurium* TA100 by 38% compared with the control group when it was treated at the concentration of

TABLE 14.5
Antimutagenic Effects of Wheat Grains, Fermented Wheat Grains, and *Kochujang* in *Salmonella typhimurium* TA100

	Revertants/Plate (mg/Plate)	
Treatment	**1.25**	**2.5**
Control (MNNG)	933 ± 76[a]	
Wheat grain	738 ± 51[b] (24)[4]	655 ± 74[b] (34)
FFWG[1]	581 ± 14[cd] (43)	539 ± 34[c] (49)
SFWG[2]	514 ± 8[d] (52)	501 ± 18[c] (53)
FiFWG[3]	550 ± 40[cd] (47)	511 ± 27[c] (52)
Kochujang	561 ± 28[cd] (46)	490 ± 5[c] (55)

Source: From Kim, J.Y. 2004. Studies on the promotion of anti-obesity and anti-cancer effects of *kochujang*. MS thesis, Pusan National University, Busan, Korea. With permission.

Note: Spontaneous revertant numbers were 122 ± 24.

[1] First-stage fermented wheat grains (Aspergillus oryzae).

[2] Second-stage fermented wheat grains (Zygosaccharomyces).

[3] Final-stage fermented wheat grains (30–40 days fermentation).

[4] Inhibition rate (%).

[a–d] Means with different letters are significantly different ($p < 0.05$) by Duncan's multiple range test.

2.5 mg/plate. However, FiFWG inhibited mutagenesis by 59%, which means the inhibition rate of FiFWG was higher than that of whole wheat grains ($p < 0.05$).

The anticancer effects of wheat grains at different fermentation stages were tested by 3-(4,5-dimethylthiazol-2yl)-2,5-diphenyl-2H-tetrazolium bromide (MTT) assay in HT-29 human colon carcinoma cells and AGS human gastric cancer cells (Kim, 2004; Kim, S.O. et al., 2005). Whole wheat grains inhibited the growth of both cell lines (45–62%). However, the inhibition rate gradually increased according to the degree of fermentation from whole wheat grains to FiFWG. The growth of AGS human gastric cancer cells was especially inhibited more than the HT-29 human colon carcinoma cells (Figure 14.4). In this experiment, commercial *kochujang*, the mixture of FiFWG and RPP, showed almost the same inhibitory effect as FiFWG. Taken together, the fermentation of whole wheat grains is an important factor for developing anticancer activity, rather than just the wheat grains and RPP, in commercial *kochujang*.

14.4.3 Antiobesity Effect of *Kochujang*

The obese population is increasing rapidly and obesity is becoming a big problem in most developing countries. It also causes us special concern since it is related to many other diseases including type II diabetes, cancer, cardiovascular disease, and osteoarthritis. (Kopelman, 1994). Many studies on capsaicin have verified its antiobesity effects, which reduces weight gain and the lipid levels in adipose tissue and serum in rats. Capsaicin triggers catecholamine secretion into the blood, which stimulates β-adrenergic receptors, resulting in stimulation of energy expenditure and lipolysis. Capsaicin decreases body weight through inhibition of hepatic lipogenic enzyme activities such as acetyl CoA carboxylase (ACC). (Kawada et al., 1986, 1996; Watanabe et al., 1987; Choo and Shin, 1999). Owing to the antiobesity activity of capsaicin contained in red pepper, it is reasonable to hypothesize that *kochujang* also has antiobesity activity.

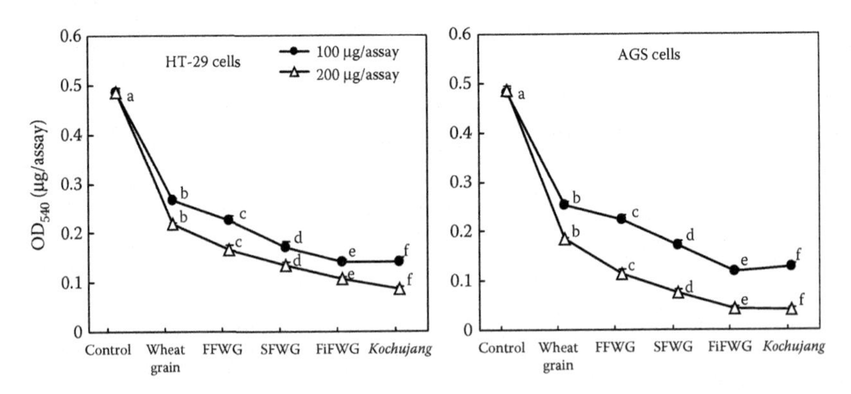

FIGURE 14.4 Inhibitory effect of methanol extracts from fermented wheat grains and *kochujang* on the growths of HT-29 human colon carcinoma cells and AGS human gastric adenocarcinoma cells in MTT assay. Data are expressed as means. Means with different letters are significant different (*p*, 0.05) by Duncan's multiple range test. (From Kim, J.Y. 2004. Studies on the promotion of anti-obesity and anti-cancer effects of *kochujang*. MS thesis, Pusan National University, Busan, Korea. With permission.)

14.4.3.1 Antiobesity Effect of Traditional *Kochujang* and Red Pepper

Choo (2000) studied the antiobesity effects of traditional *kochujang* and red pepper. Sprague–Dawley rats were fed high-fat diets to which *kochujang* and red pepper were added at 95 g/kg and 22 g/kg (same amount contained in *kochujang*) in the diet, respectively. After 3 weeks, the *kochujang* diet decreased body fat gain of rats by 30% compared with high-fat diet (HFD) as well as increased protein and DNA contents of interscapular brown adipose tissue, which suggests that *kochujang* decreased body fat gains by stimulating energy expenditure through activation of brown adipose tissue. However, the red pepper-containing diet decreased body fat gain by only 15% and did not affect the energy expenditure of rats.

Another study (Kong, 2001) also showed similar results as the former study. Sprague–Dawley rats were fed a high-fat diet with 10% of *kochujang* or 0.0045% capsaicin (same amount contained in *kochujang*). *Kochujang* was prepared by the traditional method and ripened for 6 months (Figure 14.1). After 6 weeks of feeding an HFD, there was a significant increase in body weight compared with the normal diet (ND). The *kochujang* and capsaicin diet decreased body weight by 72% and 50%, respectively, compared with HFD. (We calculated reduction rate on the basis of the difference between the value of ND and HFD, i.e., HFD increased body weight by 54 g compared with ND, and *kochujang* diet decreased by 39 g compared with HFD.) Therefore, reduction rate (72%) was calculated by 39 g/54 g × 100. The decrease in weight gain by *Kochujang* was more significant than with capsaicin. Triglyceride and cholesterol contents in the serum were significantly reduced in the *kochujang* diet group, and triglyceride levels were lower than that of the ND group. Capsaicin tended to decrease triglycerides and cholesterol, but not as significantly as did *kochujang* (Table 14.6). In this study, both capsaicin and *kochujang* supplementation significantly reduced the weights of organs such as liver, epididymal, and perirenal fat pad, but *kochujang* supplementation reduced the organ weights more than did the capsaicin supplementation. Addition of *kochujang* to HFD resulted in a decrease in total lipids and triglycerides of organs to the same or an even lower level than the ND group. However, capsaicin supplementation in HFD did not show any significant reductions of triglyceride and cholesterol level. Taken together, *kochujang* has significant antiobesity effects, which was higher than that of red pepper or capsaicin alone. Therefore, both studies (Choo, 2000; Kong, 2001) suggested that *kochujang* is able to reduce obesity effectively, and other compounds (fermentation metabolites) along with red pepper may exert the major antiobesity effects.

14.4.3.2 Antiobesity Effect of Traditional and Commercial *Kochujang*

In Table 14.7, the antiobesity effect of traditional *kochujang* (prepared as Figure 14.1) is compared with commercial *kochujang* (prepared as Figure 14.2) and fermented wheat grain which has been used as commercial *kochujang*'s main ingredient. Male Sprague–Dawley rats were fed an HFD with *kochujang* added at 10% of each diet for 6 weeks. Traditional *kochujang* and commercial *kochujang* decreased body weight 100% and 88%, respectively, compared with HFD, which was reduced to the level of the ND group. FiFWG, which is used to prepare commercial *kochujang*, also

TABLE 14.6
Effect of *Kochujang* on Body Weight, Food Intake, Food Efficiency Ratio, and Serum Lipid Contents of Rats Fed Experimental Diets

	Normal Diet	High Fat Diet	Capsaicin Diet[1]	Traditional *Kochujang* Diet[2]
Body Weight				
Initial weight (g)	166.7 ± 11.0	168.2 ± 9.4	166.2 ± 9.1	166.2 ± 7.2
Final weight (g)	309.5 ± 6.3^{d}	363.0 ± 1.3^{a}	336.6 ± 3.4^{b}	324.9 ± 6.9^{c}
Weight gain (g/day)	5.1 ± 0.3^{d}	6.9 ± 0.1^{a}	6.2 ± 0.1^{b}	5.8 ± 0.1^{c}
Food intake (g/day)	19.9 ± 0.6	19.9 ± 0.9	19.1 ± 0.7	19.0 ± 0.7
Food efficiency ratio	0.26 ± 0.01^{b}	0.33 ± 0.0^{a}	0.30 ± 0.0^{a}	0.29 ± 0.0^{ab}
Serum (mg/dL)				
Triglyceride	86.8 ± 5.6^{bc}	101.8 ± 10.1^{a}	95.6 ± 9.8^{ab}	79.0 ± 6.5^{c}
Cholesterol	76.4 ± 1.1^{d}	122.5 ± 3.4^{a}	107.1 ± 3.7^{b}	86.5 ± 2.9^{c}

Source: From Kong, G.R. 2001. Standardization of *kochujang* preparation and its effects of cancer prevention and lipid metabolism in rat. MS thesis, Pusan National University, Busan, Korea. With permission.

[1] Capsaicin 0.0045% diet.

[2] Traditional *kochujang* diet (10%): 6-month fermented *kochujang* (Moonokrye Co.).

[a–d] Means with different letters on the same row are significantly different ($p < 0.05$) by Duncan's multiple range test.

significantly reduced body weight (<0.05) and the reduction was the same for commercial *kochujang*. However, FFWG showed only an intermediate weight-lowering effect. All groups had decreased FER compared with the HFD group. The addition of traditional *kochujang*, commercial *kochujang*, or FiFWG to HFD reduced the effect of HFD on liver weight by almost 100% and the weights were the same as that of the ND group, while FFWG had an intermediate effect. The *kochujang* or fermented wheat grains diet significantly reduced epididymal and perirenal fat pad weights. Traditional *kochujang* appeared to show the greatest effect of all diet groups in reducing fat pad weight: the average weight of epididymal and perirenal fat pad was decreased to the level of the ND group. Commercial *kochujang* also induced a significant reduction of epididymal and perirenal fat pad by 81% and 78%, respectively, compared with HFD. Therefore, the intake of both kinds of *kochujang* can inhibit fat accumulation in rats. A fermented wheat grain diet also decreased the weights of the fat pads, but the reduction effect was not as much as with *kochujang* diets. FiFWG reduced fat pad weights more than did FFWG. From these results, commercial *kochujang* appeared to have an antiobesity effect almost as high as traditional *kochujang*. FiFWG decreased body weight markedly, even though its antiobesity effect was not significant compared with the final product, commercial *kochujang*. However FFWG showed only intermediate antiobesity effect (Kim, 2004). Therefore, the fermentation process may be important for the reduction of body and adipose tissue weight.

TABLE 14.7

Changes in Body Weight, Food Intake, Food Efficiency Ratio, and Organ Weights of Rats Fed Experimental Diets

	ND	HFD	HFD + TK[1]	HFD + CK[2]	HFD + FFW[3]	HFD + FiFW[4]
Body Weight						
Initial weight (g)	167 ± 7^{ns}	168 ± 8	166 ± 6	166 ± 3	165 ± 4	165 ± 4
Final weight	318 ± 10^{b}	354 ± 7^{a}	317 ± 4^{b}	322 ± 2^{b}	336 ± 14^{ab}	328 ± 21^{b}
Weight gain (g/day)	4.7 ± 0.4^{c}	6.0 ± 0.1^{a}	4.7 ± 0.1^{c}	4.9 ± 0.1^{c}	5.4 ± 0.4^{b}	5.2 ± 0.5^{bc}
Food intake (g/day)	19.9 ± 0.6^{ab}	18.9 ± 0.2^{c}	19.1 ± 0.4^{bc}	19.6 ± 0.6^{abc}	20.2 ± 0.2^{a}	19.8 ± 0.3^{ab}
Food efficiency ratio	0.24 ± 0.01^{c}	0.32 ± 0.01^{a}	0.24 ± 0.01^{bc}	0.25 ± 0.00^{bc}	0.27 ± 0.01^{b}	0.26 ± 0.03^{bc}
Organ Weight (g/100 gBW)						
Liver	3.61 ± 0.08^{b}	4.08 ± 0.26^{a}	3.62 ± 0.09^{b}	3.63 ± 0.32^{b}	3.99 ± 0.26^{ab}	3.69 ± 0.12^{ab}
Epididymal fat pad	0.47 ± 0.07^{c}	1.05 ± 0.25^{a}	0.43 ± 0.08^{c}	0.58 ± 0.07^{bc}	0.81 ± 0.22^{ab}	0.63 ± 0.10^{bc}
Perinenal fat pad	0.87 ± 0.07^{d}	1.34 ± 0.14^{a}	0.87 ± 0.04^{d}	0.97 ± 0.03^{cd}	1.15 ± 0.01^{b}	1.04 ± 0.09^{bc}

Source: From Kim, J.Y. 2004. Studies on the promotion of anti-obesity and anti-cancer effects of *kochujang*. MS thesis, Pusan National University, Busan, Korea. With permission.

[1] Traditional *kochujang* (Moonokrye Co.).

[2] Commercial *kochujang* (Haechandle Co.).

[3] First-stage fermented wheat grains (Aspergillus oryzae).

[4] Final-stage fermented wheat grains (fermented wheat grains of FFW for 30–40 days).

[a–c] Means with different letters on the same row are significantly different ($p < 0.05$) by Duncan's multiple range test.

[ns] Not significant.

14.4.3.3 Antiadipogenic Effect of *Kochujang* in 3T3-L1 Adipocytes

The antiadipogenic effect of *kochujang* was investigated *in vitro*. The mouse 3T3-L1 fibroblastic cell is capable of being differentiated into adipocytes under proper conditions, specifically, with the addition of hormonal factors (Todaro and Green, 1963; Grimaldi et al., 1979). 3T3-L1 adipocytes show the same phenotypic characteristics as mouse adipocytes, and have been used for *in vitro* obesity research. Therefore, the antiadipogenic effect of *kochujang* (prepared as in Figure 14.2 at H. Corp) was studied in 3T3-L1 adipocytes (Ahn et al., 2006). By an MTT assay, the methanol extract of *kochujang* (KE) showed no cytotoxicity up to 1 mg/mL concentration, and the relative viabilities were over 97%. To investigate the antiadipogenic effect of *kochujang* on 3T3-L1 adipocytes, the leptin levels and sizes of the adipocytes were measured. Leptin is a protein which is exclusively secreted by adipocytes and its expression increases proportionally with fat accumulation (Campfield et al., 1995; Palou et al., 2000; Norman et al., 2003). In the obese state, the circulating leptin level is higher than in the normal state, varying with the extent of obesity (Maffei et al., 1995; Havel, 2000). Therefore, leptin is used as a good indicator of obesity *in vivo* and *in vitro*. KE decreased leptin secretion by 5%, 15%, and 25% as compared with the control when the adipocytes were treated with KE at 10 μg/mL, 100 μg/mL, or 1 mg/mL (Figure 14.5a). The sizes of the adipocytes were reduced by 24% compared with the control when KE was treated at 1 mg/mL concentration for 3 days (Figure 14.5b and c).

It was examined whether the decrease in lipid accumulation with *kochujang* is associated with TNF-α and its causative effect on the development of apoptosis. TNF-α perturbs the normal regulation of energy metabolism and induces apoptosis, and it is shown to induce metabolic diseases (Doerrler et al., 1994; Sztalryd and Kraemer, 1994, 1995; Hauner et al., 1995; Sztalryd et al., 1995; Sethi and Hotamisligil, 1999; David and Michael, 2000). Expression of TNF-α mRNA was reduced by 70% after KE treatment (Figure 14.6a). Moreover, KE-treated adipocytes exhibited no apoptotic body as a result of nuclear fragmentation (Figure 14.6b). These results indicate that *kochujang* has antiadipogenic effects in 3T3-L1 adipocytes. TNF-α and apoptosis appeared not to be involved in the reduction of lipid accumulation by KE.

14.4.3.4 Antiadipogenic Mechanisms of *Kochujang*

In order to examine whether the reduction of fat accumulation was due to HSL-mediated lipolysis, glycerol secretion and HSL mRNA expression were estimated. Hormone-sensitive lipase (HSL) is the key enzyme that breaks down TG to nonesterified fatty acids and glycerol (David and Michael, 2000). KE treatment increased glycerol secretion from 3T3-L1 adipocytes by 0%, 47%, and 60% at the concentrations of 10 μg/mL, 100 μg/mL, and 1 mg/mL, respectively, which means that KE induces lipolysis in a dose-dependent manner (Figure 14.7a). KE also increased HSL mRNA expression of adipocytes by eightfold in 4 h compared with the control (Figure 14.7b). Overexpression of HSL mRNA increased lipolysis and prevented TG accumulation in adipocytes (Sztalryd et al., 1995).

To determine whether KE modulates adipogenesis in adipocytes, mRNA expressions of related transcriptional factors were examined by real-time RT-PCR. Adipogenesis is induced through the action of several enzymes, including fatty acid

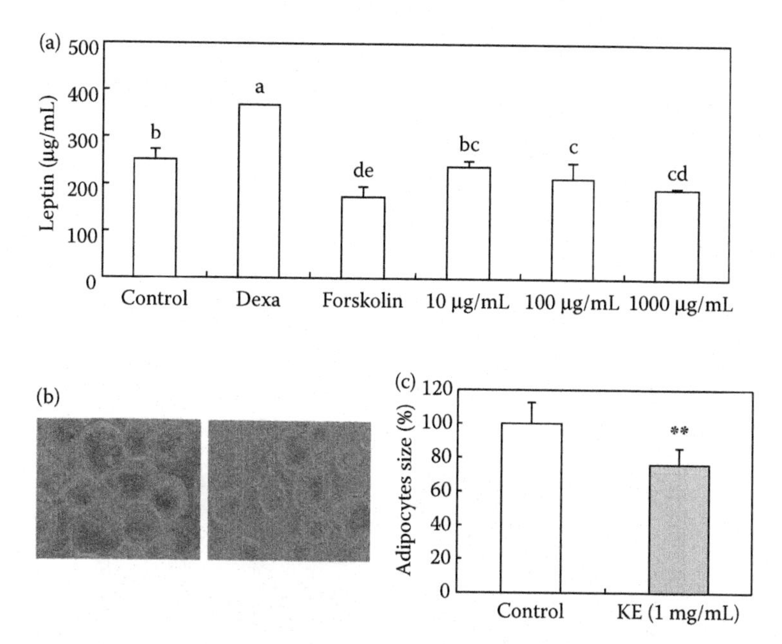

FIGURE 14.5 Effect of *Kochujang* extract (KE) on antiadipogenic effect. (a) Leptin secretion. Adipocytes were treated for 24 h at day 8 after inducing differentiation with vehicle alone (Control), Dexa (0.25 mM, dexamethasone, positive control), Forskolin (3 mM, negative control), KE (10, 100, 1000 μg/mL). Data are expressed as mean ± SD. Means with different letters are significantly different ($p < 0.05$) by Duncan's multiple range test. (b) Adipocyte morphology. Adipocytes were treated for 3 days, at day 8 after inducing differentiation with 1 mg/mL KE. Magnification 200×. (c) Adipocytes sizs. Six fields containing 200 cells were analyzed. Data are expressed as mean ± SD. Means with ** indicates $p < 0.01$. Adipocytes morphology. (From Ahn, I.S. et al., 2006b. *J. Med. Food* 9:15–21. With permission.)

synthase (FAS), ACC, acyl-CoA synthetase, and glycerol-3-phosphate acyltransferase. The expressions of these genes are regulated by transcription factors such as SREBP-1c and PPAR-r (Schoonjans et al., 1995; Ericsson et al., 1997; Lopez et al., 1996; Latasa et al., 2000; Luong et al., 2000). The expressions of SREBP-1c and PPAR-r mRNA were significantly downregulated by 70% and 75%, respectively, after 1 mg/mL KE treatment (Figure 14.8). Taken together, KE has an antiadipogenic effect in 3T3-L1 adipocytes by inhibiting adipogenesis through downregulation of SREBP-1c and PPAR-γ and by stimulation of lipolysis due to increased HSL activity.

Even though we set up a hypothesis that the antiobesity effect of *kochujang* is due to the red pepper and capsaicin, our studies showed that fermentation is a key factor along with red pepper. There are several lines of evidence suggesting that fermentation increases the antiobesity effects in *kochujang*. *Kochujang* ripened for 6 months exhibited higher suppressive effects on body fat gain and serum lipids than RPP or unfermented *kochujang* (Rhee et al., 2003). The fermentation of *kochujang* produces isomalto oligosaccharides (0.8–6.5%) and small peptides, which are reported to have beneficial effects on lipid metabolism in rats (Ahn et al., 1997; Ly et al., 1999; Lee and Chang, 2001).

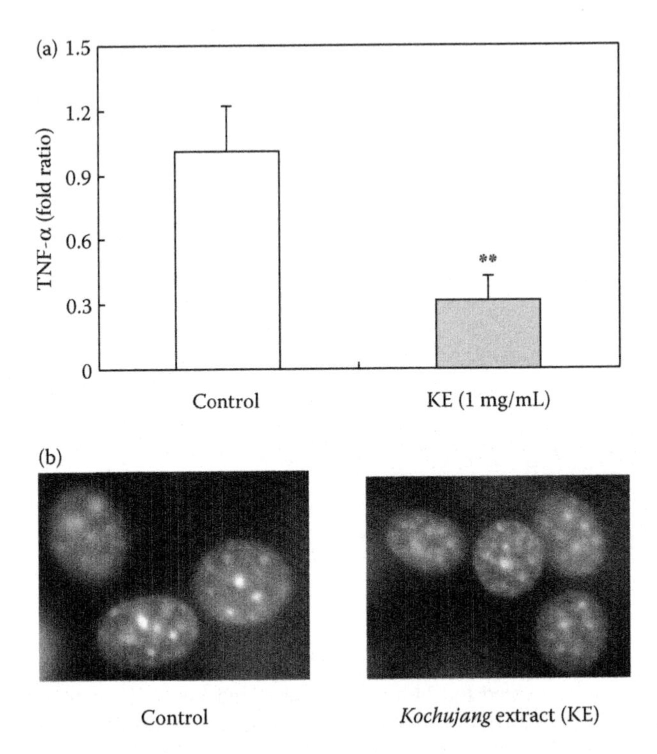

FIGURE 14.6 Effect of KE on mRNA expression of TNF-α (a) and apoptotic assay using Hoechst 33258 staining (b) of adipocytes. Adipocytes were cultured with or without 1 mg/mL KE for 1 or 3 days, respectively. For observation of apoptosis, five fields containing 300 cells were analyzed from three independent experiments. Magnification 400×. (From Ahn, I.S. et al., 2006b. *J. Med. Food* 9:15–21. With permission.)

14.4.3.5 Antiadipogenic Compounds of *Kochujang*

To isolate the active compounds from *kochujang* (Figure 14.2), commercial *kochujang* was selected since its raw material and processing method are tightly controlled. Our experiments (Kim, 2004; Ahn, 2007) revealed that FiFWG has as much antiadipogenic activity as *kochujang*, and FiFWG was used for isolating active compounds. From silica-gel column chromatography, prep TLC, and TLC, 14 compounds were isolated from the ethyl acetate extraction fraction of FiFWG. Six out of them showed strong antiadipogenic effects, which significantly decreased both leptin secretion and OD value of Oil red O stained in 3T3-L1 adipocytes. Compound 2 (K1) and compound 8 (K2) were selected for further examination of their antiadipogenic activities; K1 decreased leptin secretion by 30%, 45%, 50%, respectively, when 3T3-L1 adipocytes were treated with concentrations of 10 μg/mL, 50 μg/mL, and 100 μg/mL. K2 also decreased leptin secretion by 10%, 30%, 40%, at the concentrations of 10 μg/mL, 50 μg/mL, and 100 μg/mL, respectively. K1 and K2 both decreased mRNA expressions of adipogenesis-related genes including PPAR-r, SREBP-1c, FAS, and acyl-CoA (ACS) by over 50% according to real-time RT-PCR. K1 and K2 were further purified and identified to be 1,2,3-propanetriol-1-acetate

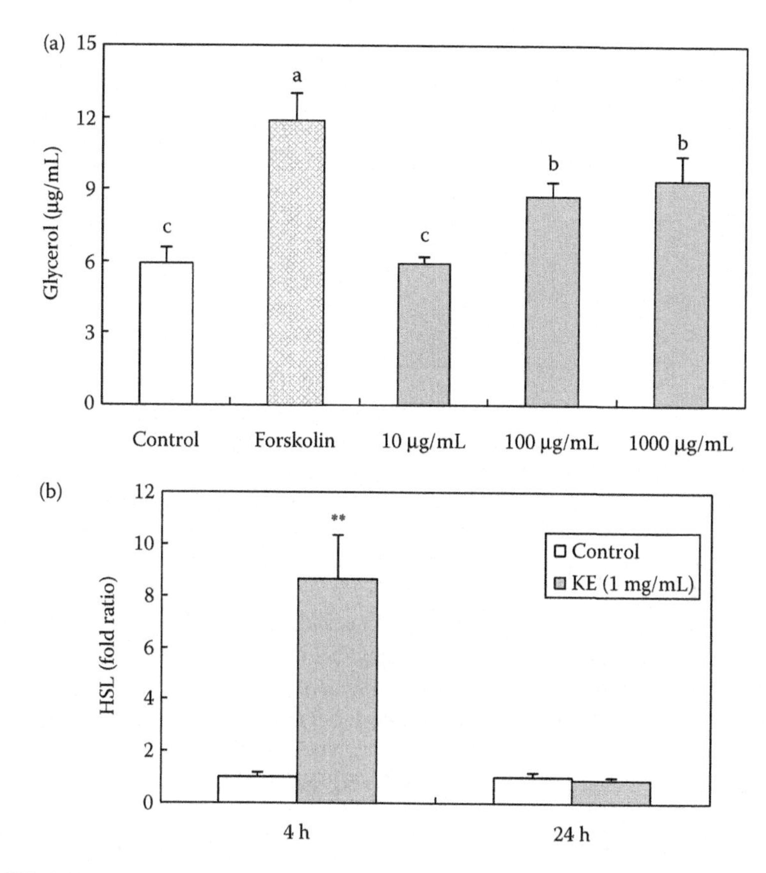

FIGURE 14.7 Effect of KE on glycerol secretion (a) and HSL mRNA expression (b). Starting at day 8 after inducing differentiation, adipocytes were cultured with or without 1 mg/mL KE for 24 h. Media were sampled for glycerol measurement. Data are mean ± standard error values from three independent experiments. *$P < 0.05$ versus control. (From Ahn, I.S. et al., 2006b. *J. Med. Food* 9:15–21. With permission.)

(glycerol-1-acetate) and 1,2,3-propanetriol (glycerol), respectively, on the basis of ^{1}H-NMR and ^{13}C-NMR analysis (Ahn, 2007). Therefore, fermented products including glycerol and glycerol derivatives seem to decrease fat accumulation.

14.4.4 Other Functions of *Kochujang*

The functional activities of *kochujang* come from the main ingredients including meju and RPP as well as some metabolites produced during fermentation by the action of the microorganisms *Mucor, Rhizopus, Aspergillus, Bacillus subtilis*, yeast, and so on. In contrast to the numerous studies on anticancer and antiobesity effects, there has been little research conducted on other effects of *kochujang*.

Chung et al. (1999) reported that *Monascus* koji (*Monascus anka*) prepared with glutinous rice and barley for *kochujang* production inhibited angiotensin converting enzyme (ACE). ACE converts angiotensinogen into angiotensin which contracts

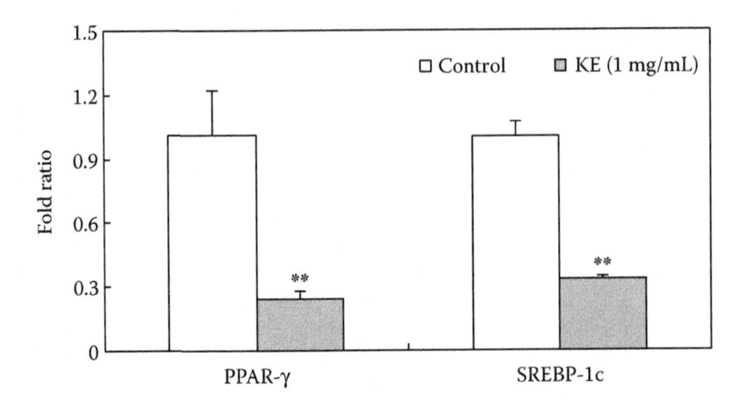

FIGURE 14.8 Effect of KE on mRNA expression of SREBP-1c and PPAR-r. Starting at day 8 after inducing differentiation, adipocytes were cultured with or without 1 mg/mL KE for 24 h. Data are expressed relative to untreated control cells and represent mean ± standard error values (n = 3). $^{**}P < 0.01$ comparing KE-treated with untreated control cells. (From Ahn, I.S. et al., 2006b. *J. Med. Food* 9:15–21. With permission.)

blood vessels and increases blood pressure. Yen et al. (2003) found out that *Aspergillus candidus* acts as an antioxidant, and the ethyl acetate fraction of *Aspergillus candidus* of cultured koji has stronger antioxidant activity than α-tocopherol, a natural antioxidant. Therefore, *Monascus* koji and *Aspergillus* koji prepared by *Monascus anka* and *Aspergillus candidus*, respectively can be used for production of *kochujang*, which helps to reduce high blood pressure and has a strong antioxidant effect. Seo et al. (2007) also prepared *kochujang* by application of *Bacillus stearothermophilus* and *Bacillus amyloliquefacience* which have fibrinolytic activities, which resulted in successfully obtaining the same activity from *kochujang*.

Recently, some food materials have been added to *kochujang* to enhance its functional activities. *Kochujang* prepared with red rice and barley has HMG-CoA reductase inhibitor activity (Hyun et al., 2007). Also, aloe *kochujang* has been reported to reduce allergies. Eleven patients who were allergic to soybeans had a positive reaction with soybeans, but only three patients showed a positive reaction to meju *kochujang*, and only two with aloe *kochujang*. These two kinds of *kochujang* showed less allergenicity. In the case of aloe *kochujang*, the quantities of macromolecular proteins that induce allergy were reduced (Vazquez et al., 1996; Son et al., 2006). Many other functional foods such as green tea, fucoidan, onion, garlic, and sea tangle were added to *kochujang* to increase its functional activity, which showed antiadipogenic, antimicrobial, immunoactivity, anticancer, and antimutagenic effects (Na et al., 1997; Kim J.D. et al., 2005; Kim J.Y. et al., 2005: Kong et al., 2005a, 2005b; Ahn et al., 2006; Ham et al., 2008).

14.5 PRODUCTION, FACTORY SUPPLY, EXPORTING AND IMPORTING OF *KOCHUJANG*

Kanjang (soy sauce), doenjang (soy paste), and *kochujang* (hot-pepper soy paste) are popular factory produced soybean products in Korea. Table 14.8 shows the estimated

TABLE 14.8
Estimated Consumption and Factory Production of Kanjang, Doenjang, and *Kochujang* in Korea

	Kanjang (Soy Sauce)			Doenjang (Soy Paste)			Kochujang		
Year	Est. Consumption	Factory Supply	Ratio (%)	Est. Consumption	Factory Supply	Ratio (%)	Est. Consumption	Factory Supply	Ratio (%)
1980	396,952	108,765	27	274,375	53,995	20	131,317	35,750	27
1990	416,700	167,040	40	278,450	59,300	21	148,600	43,890	30
1995	414,300	178,818	43	280,200	94,444	34	160,100	77,058	48
2000	352,400	179,239	51	266,300	133,476	50	165,700	113,976	69
2005	352,300	201,942	57	288,500	162,694	56	201,100	151,882	76
2008	346,700	225,924	65	279,400	164,469	59	206,000	163,004	79

Source: From Korea Soy Sauce Industrial Cooperative, 2008. With permission.
Note: Unit: k*l*, ton.

consumption and commercial production of kanjang, doenjang, and *kochujang* from 1980 to 2008 in Korea.

The estimated consumption of kanjang was the highest (346,700–416,700 tons) followed by doenjang (266,300–288,500 tons) and *kochujang* (131,317–206,000 tons). The consumption of kanjang has been gradually decreasing and doenjang consumption has been almost constant over the past 18 years. However, the estimated consumption of *kochujang* gradually increased from 131,317 tons in 1980 to 206,000 tons in 2008, and also commercial production has been greatly increased from 27% to 79%. This fact is probably because two big factories, Haechandle Co. and Chungjungwon Co. have been built and have supplied their *kochujang* products.

Table 14.9 shows the export and import of *kochujang* to and from different countries in 2009. The major importing countries are the United States (2891 tons) and Japan (1572 tons). Korean-American and Korean-Japanese may want to purchase the products. Japanese are recently interested in eating *kochujang* similar to kimchi in Japan. People in China, Australia, Canada, and Taiwan are also interested in importing kochujung. The imported kochujung is mainly from China (587 tons). *Kochujang* is a good soybean product that can be globalized similar to kimchi, which is a well-known Korean fermented functional vegetable preparation.

14.6 PACKAGING MATERIAL OF *KOCHUJANG*

The traditionally prepared *kochujang* is fermented in clay pots as shown in Figure 14.9b, but in larger ones for more than 6 months. This can be repacked into smaller containers to sell in the market (Figure 14.9b). The pottery is characterized by small pores, so that small amounts of air can pass through but water does not leak out. It was developed by Korean ancestors who used it for fermenting foods and storing

TABLE 14.9
***Kochujang* Importing and Exporting Countries in 2009**

Country	Importing		Exporting	
	Amount	US $	Amount	US $
USA	2891	5170	–	–
Japan	1572	3586	–	–
China	552	1518	587	260
Australia	408	876		
Canada	327	643	–	–
Taiwan	249	465	–	–
Others	1087	2333	–	–
Total	7086	14,591	587	260

Source: From Korea Soy Sauce Industrial Cooperative, 2008. With permission.
Note: Unit: ton, 1000 US $.

the fermented foods. Regular PE, PP, or PET is employed as containers for commercially manufactured *kochujang* products. The plastic containers are cheaper than the pottery, easy to carry, not easily broken (Figure 14.9a).

Table 14.10 shows the results of sensory evaluation of *kochujang*s fermented for 4 months in different containers. Pottery showed the highest scores in the evaluations ($P < 0.05$) for characteristics such as color, flavor, taste, and overall acceptability compared with other containers—glass bottles, plastic (pp), tupper (pet), or stainless steel (Chung et al., 2005).

Small pots can be used as *kochujang* containers for the market. They are nice looking for traditional products, but can be broken during distribution if handled carelessly. The price of traditional *kochujang* in clay pots compared with that of factory products in plastic is higher by a factor of two or three times.

FIGURE 14.9 Packaging containers of factory (a) and traditional (b) *kochujang*.

TABLE 14.10
Sensory Evaluation of 4-month Fermented *Kochujang* in Different Containers

Contaniner	Color	Flavor	Taste	Overall Acceptability
Glass bottle	3.18[b]	3.18[b]	2.64[b]	2.75[b]
Plastic (pp)	3.36[b]	3.09[b]	2.73[b]	3.25[b]
Tupper (pet)	3.36[b]	3.18[b]	2.73[b]	2.67[b]
Stainless steel	3.00[b]	2.91[b]	2.73[b]	2.75[b]
Clay pot[1]	4.45[a]	4.18[a]	4.00[a]	4.33[a]

Source: From Chung, S.K., Kim, Y.S., and Lee, D.S. 2005. *Korean J. Food Preserv.* 12:292–298. With permission.

pp: polyethylene, pet: polyethylene terephthlate.

[1] Onggi in Korean.

[a–c] Means with the same letter in column are not significantly different at $p < 0.05$ level by Duncan's multiple range test.

14.7 CONCLUSIONS

Kochujang is a food fermented from a starch source, soybeans, and RPP. Starting ingredients can convert into other fermented functional materials and form other fermented products during the fermentation. Fermented *kochujang* shows chemopreventive activities. Antimutagenic effects have been revealed by the SOS chromo test and Ames test. It also exhibited antitumor activities in Sarcoma 180 cell transplanted mice and the natural killer cell activity was significantly increased. Fermented *kochujang* greatly increased the above activities compared with the nonfermented raw materials. *Kochujang* also inhibited metastasis in colon 26-M3.1 cell injected mice. *Kochujang* showed *in vitro* and *in vivo* antiobesity effects. Again, fermentation played a very important role. The fermented wheat grain sample (before adding RPP during manufacturing of factory *kochujang*) increased antiobesity activity. Some of the active compounds identified were glycerol and glycerol-1-acetate. These compounds may be degradation compounds or metabolites or synthesized compounds made by the microorganisms during the fermentation phase. *Kochujang* can be a great seasoning spice or sauce with healthy functionality, and people will like this food. We need to develop different flavors or tastes that might be introduced to foreign dietary cultures around the world.

REFERENCES

Abdel-Aal, E.S., Hucl, P., Sosulski, F.W., Graf, R., Gillott, C., and Pietrzak, L. 2001. Screening spring wheat for midge resistance in relation to ferulic acid content. *J. Agric. Food Chem.* 49:3559–3566.

Adam, A., Lopez, H.W., Tressol, J.C., Leuillet, M., Demigné, C., and Rémésy, C. 2002. Impact of whole wheat flour and its milling fractions on the cecal fermentations and the plasma and liver lipids in rats. *J. Agric. Food Chem.* 50:6557–6562.

Adom, K.K., Sorrells, M.E., and Liu, R.H. 2003. Phytochemical profiles and antioxidant activity of wheat varieties. *J. Agric. Food Chem.* 51:7825–7834.

Adom, K.K., Sorrells, M.E., and Liu, R.H. 2005. Phytochemicals and antioxidant activity of milled fractions of different wheat varieties. *J. Agric. Food Chem.* 53:2297–2306.

Ahn, C.W., and Sung, N.K. 1987. Major components and microorganisms of Korean traditional kochujang fermentation. *J. Korean Soc. Food Nutr.* 16:35–39.

Ahn, I.S. 2007. *In vitro* antiobesity effect of kochujang and isolation and identification of active compounds. PhD dissertation, Pusan National University, Busan, Korea.

Ahn, I.S., Do, M.S., Choi, B.H., Kong, C.S., Kim, S.O., Han, M.S., and Park, K.Y. 2006a. Reduced leptin secretion by fucoidan-added kochujang and anti-adipogenic effect of fucoidan in mouse 3T3-L1 adipocytes. *J. Food Sci. Nutr.* 11:31–35.

Ahn, I.S., Do, M.S., Kim, S.O., Jung, H.S., Kim, Y.I., Kim, H.J., and Park, K.Y. 2006b Antiobesity effect of kochujang (Korean fermented red pepper paste) extract in 3T3-L1 adipocytes. *J. Med. Food* 9:15–21.

Ahn, Y.G., Kim, S.K., and Shin, C.S. 1997. Sugars in kochujang. *Korean J. Food Nutr.* 10:446–452.

Bu, S.Y., Jeon, W.B., Kim, D.S., Heo, H.Y., and Seo, Y.W. 2002. Antioxidant activity and total phenolic compounds in grain extracts of wheat, barly and oat. *Korean J. Crop Sci.* 47:102–107.

Campfield, L.A., Smith, F.J., Guisez, Y., Devos, R., and Burn, P. 1995. Recombinant mouse OB protein: Evidence for a peripheral signal linking adiposity and central neural networks. *Science* 269:546–549.

Choe, M., and Kim, H.S. 2002. Effects of Korean wheat on LDL oxidation and atherosclerosis in cholesterol-fed rabbits. *J. Korean Soc. Food Sci. Nutr.* 31:104–108.

Choi, M.W., Kim, K.H., Kim, S.H., and Park, K.Y. 1997. Inhibitory effects of kimchi extracts on carcinogen-induced cytotoxicity and transformation in C3H/10T1/2 cells. *J. Food Sci. Nutr.* 2:241–245.

Choo, J.J., 2000. Anti-obesity effect of kochujang in rats fed on a high-fat diet. *Korean J. Nutr.* 33:787–793.

Choo, J.J., and Shin, H.J. 1999. Effects of capsaicin through β-adrenergic stimulation in rats fed a high-fat diet. *Korean J. Nutr.* 32:533–539.

Chung, S.H., Suh, H.J., Choi, Y.M., Noh, D.O., and Bae, S.H. 1999. Enzyme activities and inhibitory effect on angiotensin converting enzyme of monascus–koji for the kochujang production. *Food Sci. Biotechnol.* 8:179–183.

Chung, S.K., Kim, Y.S., and Lee, D.S. 2005. Effect of vessel on the quality changes during fermentation of kochujang. *Korean J. Food Preserv.* 12:292–298.

David, L.N., and Michael, M.C. 2000. *Lehninger Principles of Biochemistry.* 3rd ed. Worth Publishers, New York, NY, pp. 449–452.

Doerrler, W., Feingold, K.R., and Grunfeld, C. 1994. Cytokines induce catabolic effects in cultured adipocytes by multiple mechanisms. *Cytokine* 6:478–484.

Ericsson, J., Jackson, S.M., Kim, J.B., Spiegelman, B.M., and Edwards, P.A. 1997. Identification of glycerol-3-phosphate acyltransferase as an adipocyte determination and differentiation factor 1- and sterol regulatory element-binding protein-responsive gene. *J. Biol. Chem.* 272:7298–7305.

Fidler, I.J. 1991. Cancer metastasis. *Br. Med. Bull.* 47:157–177.

Fordyce-Baum, M.K., Langer, L.M., Mantero-Atienza, E., Crass, R., and Beach, R.S. 1989. Use of an expanded-whole-wheat product in the reduction of body weight and serum lipids in obese females. *Am. J. Clin. Nutr.* 50:30–36.

Grimaldi, P., Negrel, R., Vincent, J.P., and Ailhaud, G. 1979. Differentiation of ob 17 preadipocytes to adipocytes. Effect of insulin on the levels of insulin receptors and on the transport of alpha-aminoisobutyrate. *J. Biol. Chem.* 254:6849–6852.

Ham, S.S., Choi, H.J., Kim, S.H., Oh, H.T., and Chung, M.J. 2008. Antimutagenic and cytotoxic effects of kochujang extracts added deep sea water salt and sea tangle. *J. Korean Soc. Food Sci. Nutr.* 37:410–415.

Hauner, H., Petruschke, T., Russ, M., Rohrig, K., and Eckel, J. 1995. Effects of tumour necrosis factor alpha (TNF alpha) on glucose transport and lipid metabolism of newly-differentiated human fat cells in cell culture. *Diabetologia* 38:764–771.

Havel, P.J. 2000. Role of adipose tissue in body-weight regulation: Mechanisms regulating leptin production and energy balance. *Proc. Nutr. Soc.* 59:359–371.

Hyun, K.U., No, J.D., Lim, S.I., Cha, S.K., and Choi, S.Y. 2007. Characteristics and HMG-CoA reductase inhibitory activity of fermented red pepper soybean paste (kochujang) prepared from red-rice and barley. *Korean J. Microbiol. Biotechnol.* 35:173–176

Hwang, K.M. 2004. Studies on the enhancement of chemopreventive and anticancer effects of doenjang. PhD dissertation, Pusan National University, Busan, Korea.

Jung, K.O., Kim, S.J., Yoon, S.K., and Park, K.Y. 2000. Antimutagenic effect of kochujang (Korean red pepper soybean paste) and kochujang ingredients in the Ames test. *J. Korean Assoc. Cancer Prev.* 5:209–216.

Jung, K.O., Park, S.Y., Park, M.S., and Park, K.Y. 2006. Longer aging time increases the anticancer and antimetastatic properties of doenjang. *Nutrition* 22:539–545.

Kawada, T., Hagihara, K.I., and Iwai, K. 1996. Effect of capsaicin on lipid metabolism in rats fed a high fat diet. *J. Nutr.* 116:1272–1278.

Kawada, T., Watanabe, T., Takaishi, T., Tanaka, T., and Iwai, K. 1986. Capsaicin-induced β-adrenergic action of energy metabolism in rats: Influence of capsaicin on oxygen consumption, the respiratory quotient and substrate utilization. *Proc. Soc. Exp. Biol. Med.* 183:250–256.

Kim, C.T., Cho, S.J., Hwang, J.K., and Kim, C.J. 1997. Composition of amino acids, sugars and minerals of domestic wheat varieties. *J. Korean Soc. Food Sci. Nutr.* 26: 229–235.

Kim, D.H., Choi, U., Shin, D.H., Lim, M.S., and Lim, D.K. 1996. Studies on taste components of traditional kochujang. *Korean J. Food Sci. Technol.* 28:152–156.

Kim, J.D., Kim, M.Y., Jung, S.J., Seo, H.J., Kim, E.O., and Lee, S.Y. 2005. Development of kochujang for controlling *V. parahaemolyticus* with green tea and natural products. *J. Life Sci.* 14:783–789.

Kim, J.Y. 2004. Studies on the promotion of anti-obesity and anti-cancer effects of kochujang. MS thesis, Pusan National University, Busan, Korea.

Kim, J.Y., Park, K.W., Yang, H.S., Cho, Y.S., Jeong, C.H., Shim, K.H., Yee, S.T., and Seo, K.I. 2005. Anticancer and Immuno-activity of methanol extract from onion kochujang. *Korean J. Food Preserv.* 12:173–178.

Kim, S.J. 1999. Cancer preventive functions of traditional kochujang (Korean red pepper soybean paste). PhD dissertation, Dongduk Women's University, Seoul, Korea.

Kim, S.O., Kong, C.S., Kil, J.H., Kim, J.Y., Han, M.S., and Park, K.Y. 2005. Fermented wheat grain products and kochujang inhibit the growth of AGS human gastric adenocarcinoma cells. *J. Food Sci. Nutr.* 10:349–352.

Ko, S.H., Ju, H.K., Jung, K.O., and Park, K.Y. 1999. Antimutagenic effect of chungkookjangs prepared with the different kinds of soybeans. *J. Korean Assoc. Cancer Prev.* 4:204–212.

Kong, C.S., Jung, H.K., Kim, S.O., Rhee, S.H., Han, M.S., and Park, K.Y. 2005a. *In vitro* anticancer effect of fucoidan and fucoidan added kochujang (Korean red pepper soybean paste). *Cancer Prev. Res.* 10:264–269.

Kong, C.S., Kim, S.O., Jung, K.O., Kil, J.H., Lee, S.H., Han, M.S., and Park, K.Y. 2005b. Increased anticancer effect of kochujang (Korean red pepper soybean paste) prepared with garlic powder dried with different methods. *Cancer Prev. Res.* 10:128–133.

Kong, G.R. 2001. Standardization of kochujang preparation and its effects of cancer prevention and lipid metabolism in rat. MS thesis, Pusan National University, Busan, Korea.

Kopelman, P.G. 1994. Causes and consequences of obesity. *Med. Int.* 22:385–388.

Latasa, M.J., Moon, Y.S., Kim, K.H., and Sul, H.S. 2000. Nutritional regulation of the fatty acid synthase promoter *in vivo*: sterol regulatory element binding protein functions through an upstream region containing a sterol regulatory element. *Proc. Natl. Acad. Sci.* 97:10619–10624.

Lee, H.M., and Chang, U.J. 2001. Effect of corn peptide on the lipid metabolism in rats. *Korean J. Dietary Culture* 16:416–422.

Lee, K.H., Lee, M.S., and Park, S.O. 1976. Studies on the microflora and enzymes influencing on Korea native kochuzang (red pepper soybean paste) aging. *J. Korean Soc. Agric. Chem. Biotechnol.* 19:82–92.

Lee, T.S., Lee, S.G., Kim, S.S., and Yoshida, T. 1970. Microbiological studies of red pepper paste fermentation (Part I). *Korean J. Microbiol.* 8:151–162.

Lee, T.S., Chun, M.S., Choi, J.Y., and Noh, B.S. 1993. Changes of free sugars and free amino acids in kochujang with different mashing method. *Food Biotechnol.* 2:102–107.

Lopez, J.M., Bennett, M.K., Sanchez, H.B., Rosenfeld, J.M., and Osborne, T.E. 1996. Sterol regulation of acetyl coenzyme A carboxylase: A mechanism for coordinate control of cellular lipid. *Proc. Natl. Acad. Sci.* 93:1049–1053.

Luong, A., Hannah, V.C., Brown, M.S., and Goldstein, J.L. 2000. Molecular characterization of human acetyl-CoA synthetase, an enzyme regulated by sterol regulatory element-binding proteins. *J. Biol. Chem.* 275:26458–26466.

Ly, S.Y., Lee, M.R., and Lee, K.A. 1999. Effects of cakes containing sponge oligosaccharides on blood lipids and intestinal physiology in rats. *J. Korean Soc. Food Sci. Nutr.* 28:619–624.

Maffei, M., Halaas, J., Ravussin, E., Pratley, R.E., Lee, G.H., Zhang, Y., Fei, H., Kim, S., Lallone, R., and Ranganathan, S. 1995. Leptin levels in human and rodent: Measurement of plasma leptin and ob RNA in obese and weight-reduced subjects. *Nat. Med.* 1:1155–1161.

Messina, M., and Messina, V. 1991. Increasing use of soyfoods and their potential role cancer prevention. *J. Am. Diet. Assoc.* 91:836–840.

Moon, T.W., and Kim, Z.U. 1988. Some chemical and physical characteristics and acceptability of kochujang from various starch sources. *J. Kor. Agric. Chem. Soc.* 31:387–393.

Mpofu, A., Sapirstein, H.D., and Beta, T. 2006. Genotype and environmental variation in phenolic content, phenolic acid composition, and antioxidant activity of hard spring wheat. *J. Agric. Food Chem.* 54:1265–1270.

Na, S.E., Seo, K.S., Choi, J.H., Song, G.S., and Choi, D.S. 1997. Preparation of low salt and functional kochujang containing chitosan. *Korean J. Food Nutr.* 10:193–200.

Nagle, B.J., and Burns, E.E. 1979. Color evaluation of selected capsicum. *J. Food Sci.* 44:416–418.

Norman, D., Isidori, A.M., Frajese, V., Caprio, M., Chew, S.L., Grossman, A.B., Clark, A.J., Besser, G.M., and Fabbri, A. 2003. ACTH and alpha-MSH inhibit leptin expression and secretion in 3T3-L1 adipocytes: Model for a central-peripheral melanocortin–leptin pathway. *Mol. Cell Endocrinol.* 200:99–109.

Onyeneho, S.N., and Hettiarachchy, V.S. 1992. Antioxidant activity of durum wheat bran. *J. Agric. Food Chem.* 40:1496–1500.

Palou, A., Serra, F., Bonet, M.L., and Pico, C. 2000. Obesity: Molecular bases of a multifactorial problem. *Eur. J. Nutr.* 39:127–144.

Park, J.M., Lee, S.S., and Oh, H.I. 1995. Changes in chemical characteristics of traditional kochujang meju during fermentation. *Korean J. Food Nutr.* 8:184–191.

Park, K.Y., Baek, K.A., Rhee, S.H., and Cheigh, H.S. 1995. Antimutagenic effect of kimchi. *Foods Biotechnol.* 4:141–145.

Park, K.Y., Kong, K.R., Jung, K.O., and Rhee, S.H. 2001. Inhibitory effects of kochujang extracts on the tumor formation and lung metastasis in mice. *J. Food Sci. Nutr.* 6:187–191.

Parveen, Y., Eric, A.N., and Philip, C.C. 1994. Inhibition of natural killer cell activity by dietary lipids. *Immunol. Lett.* 41:241–247.

Rhee, S.H., Kong, K.R., Jung, K.O., and Park, K.Y. 2003. Decreasing effect of kochujang on body weight and lipid levels of adipose tissues and serum in rats fed a high-fat diet. *J. Food Sci. Nutr.* 32:882–886.

Schoonjans, K., Watanabe, M., Suzuki, H., Mahfoudi, A., Krey, G., Wahli, W., Grimaldi, P., Staels, B., Yamamoto, T., and Auwerx, J. 1995. Induction of the acyl-coenzyme A synthetase gene by fibrates and fatty acids is mediated by a peroxisome proliferator response element in the C promoter. *J. Biol. Chem.* 270:19269–19276.

Seo, M.Y., Kim, S.H., Lee, C.H., and Cha, S.K. 2007. Physiological activity/nutrition: fibrinolytic, immunostimulating, and cytotoxic activities of microbial strains isolated from kochujang. *Korean J. Food Sci. Technol.* 39:315–322.

Sethi, J.K., and Hotamisligil, G.S. 1999. The role of TNF alpha in adipocyte metabolism. *Semin. Cell Dev. Biol.* 10:19–29.

Shin, D.H., Kim, D.H., Choi, U., Lim, D.K., and Lim, M.S. 1996. Studies of the physicochemical characteristics of traditional kochujang. *Korean J. Food Sci. Technol.* 28:157–161.

Smith, M.M., and Harttley, R.D. 1983. Occurrence and nurture of ferulic acid substitution of cell wall polysaccharides in gramineous plants. *Carbohydr. Res.* 118:65–80.

Son, B.K., Huh, Y.E., Kim, J.Y., Noh, G.W., and Lee, S.S. 2006. An evaluation of changes in the allergenicity of kochujang upon preparation using aloe extract. *Nutr. Sci.* 9:317–322.

Sztalryd, C., and Kraemer, F.B. 1994. Regulation of hormone-sensitive lipase during fasting. *Am. J. Physiol.* 266:179–185.

Sztalryd, C., and Kraemer, F.B. 1995. Regulation of hormone-sensitive lipase in streptozotocin-induced diabetic rats. *Metabolism* 44:1391–1396.

Sztalryd, C., Komaromy, M.C., and Kraemer, F.B. 1995. Overexpression of hormone-sensitive lipase prevents triglyceride accumulation in adipocytes. *J. Clin. Invest.* 95:2652–2661.

Terry, L.B., Andrea, R., David, S. and Michael, L.E. 1990. Potentiation of natural killer cell activity and tumor immunity by diacetylputrescine. *Cancer Res.* 50:5460–5463.

Todaro, G.J., and Green, H. 1963. Quantitative studies of the growth of mouse embryo cells in culture and their development into established lines. *J. Cell Biol.* 17:299–313.

Vaaler, S., Hanssen, K.F., Dahl-Jorgensen, K., Frolich, W., Aaseth, J., Odegaard, B., and Aagenaes, O. 1986. Diabetic control is improved by guar gum and wheat bran supplementation. *Diabet. Med.* 3:230–233.

Vazquez, B., Avila, G., Segura, D., and Escalante, B. 1996. Antiinflammatory activity of extracts from *Aloe vera* gel. *J. Ethnopharmacol.* 55:69–75.

Watanabe, T., Kawada, T., and Iwai, K. 1987. Enhancement by capsaicin of energy metabolism in rats through release of catecholamine from adrenal medulla. *Agric. Biol. Chem.* 51:75–79.

Wolk, A., Manson, J.E., Stampfer, M.J., Colditz, G.A., Hu, F.B., Speizer, F.E., Hennekens C.H., and Willett W.C. 1999. Long-term intake of dietary fiber and decreased risk of coronary heart disease among women. *JAMA* 281:1998–2004.

Woo, D.H. and Kim, Z.U. 1990. Characteristics of improved kochujang. *J. Korean Agric. Chem. Soc.* 33:161–168.

Yen, G.C., Chang, Y.C., and Su, S.W.. 2003. Antioxidant activity and active compounds of rice koji fermented with *Aspergillus candidus*. *Food Chem.* 83:49–54.

Zhou, K., Laux, J.J., and Yu, L. 2004. Comparison of Swiss red wheat grain and fractions for their antioxidant properties. *J. Agric. Food Chem.* 52:1118–1123.

15 Antioxidant Functional Factors in Nuts

Yearul Kabir and Jiwan S. Sidhu

CONTENTS

15.1 INTRODUCTION

Recent studies have pointed out that free radicals increase the risk of several human chronic diseases, namely cancer, arteriosclerosis, and neurodisorders, as well as the aging processes (Aruoma, 1998; Sies et al., 2005). Restriction on the use of some of the commercial synthetic antioxidants because of their carcinogenicity (Mahdavi and Salunkhe, 1995) has created a need for identifying alternative natural and safe sources of food antioxidants, especially of plant origin. The growing interest in the substitution of synthetic food antioxidants by natural ones has intensified research on plant sources especially the tree nuts as source of antioxidants. Tree nuts are an important source of beneficial dietary lipids, and a potentially rich source of phenolic compounds that contribute to antioxidant capacity (Shahidi and Miraliakbari, 2006; Alasalvar and Shahidi, 2008).

Nuts come from a diverse group of various plant families. Peanuts (*Arachis hypogaea* L.) belong to the Leguminosae family; pecans and walnuts are distant relatives in the Juglandaceae family. While they are very diverse, the reason they are grouped together as nuts is because their biochemistry and nutritional makeup are similar (Isanga and Zhang, 2007). Nuts are recommended constituents of the daily diet, although their real intake differs remarkably. In particular, they are part of healthy diets such as the Mediterranean diet. Results from several epidemiological studies and traditions suggest that there may be a connection between frequent nut consumption and a reduced incidence of coronary heart disease (CHD) (Kris-Etherton et al., 2001; Simopoulos, 2001). These effects are assumed to be mainly due to the less atherogenic plasma lipid profiles (Abbey et al., 1994; Edwards et al., 1999; Rajaram et al., 2001; Spiller et al., 1998; Zambon et al., 2000). However, emerging proofs indicate that nuts may be a source of many health-promoting compounds that elicit cardioprotective effects (Kris-Etherton et al., 1999b). In particular, nuts are good sources of proteins, unsaturated fatty acids, dietary fiber, plant sterols, and tocopherols (Kris-Etherton et al., 2001).

The main objective of this chapter is to discuss nuts as potential sources of antioxidants, and provide a comprehensive evaluation of the biologically active components present in nuts that could be used in prevention of certain diseases.

15.2 PRODUCTION, PROCESSING, AND UTILIZATION OF NUTS

15.2.1 Cultivation and Production

15.2.1.1 Almond

Almond, scientifically known as *Prunus amygdalus*, belongs to the family Rosaceae. It is the number one tree nut produced on a global basis. It is a native of Central Asia, but soon spread west right through to the Mediterranean. World almond production for the year 2000 has been reported to be about 1.45 Mt of shelled products. The main producing country is the United States, which accounts for 68% of the world production (MirAli and Nabulsi, 2003).

15.2.1.2 Walnut

Walnut (genus *Juglans*, family Juglandaceae) comprises several species and is widely distributed throughout the world. The Persian or common *walnut* (*Juglans regia* L.) is its best-known member, constituting an important species of deciduous trees found primarily in the temperate areas and commercially cultivated in the United States, western South America, Asia, and central and southern Europe. Walnuts' (*Juglans regia*) native range is from southeast Europe and across to China. The Kashmiri/Northern Pakistan area is a particularly important area of biological diversity for walnuts. The oldest archaeological evidence is a find of walnuts in caves in northern

Iraq that date from Mesolithic times. Today, approximately 1.5 Mt of walnuts are harvested every year. The United States is the largest producer of almonds in the world. The United States and China dominate world trade, together producing about 70% of the nuts on the world market.

15.2.1.3 Peanut

Peanut is a pod, or legume, of *Arachis hypogaea* L., of the family Leguminosae, and is cultivated around the world. On a worldwide scale, peanut is grown primarily for its seed oil. Peanut comprises kernels, skins (seed testae, seed coats) and hulls. Total world production of peanut in 2003 was 37.06 Mt (Table 15.1) and China is the world's largest peanut producer accounting for 42% of the total global production in the year 2003.

Peanuts are among the world's major oilseed crops. They occupy a unique position among oilseeds, as they can be utilized in diverse ways and can be consumed directly as well. Approximately two-thirds of the total peanuts produced in the world are utilized in India, China, and the United States of America (USDA, 1992). More than half of the world's peanut production is crushed for oil production (Carley and Fletcher, 1995). The peanut cake, a by-product of the oil industry, is mainly used as livestock feed supplement. However, the pattern of peanut utilization differs among the developed and the developing countries (Freeman et al., 1999). In the United States, the major proportion of produce is processed for direct consumption as peanut butter, salted peanuts, and confectionery, whereas in Asia, particularly in India, peanuts are utilized mainly for oil production (Carley and Fletcher, 1995).

TABLE 15.1
Global Production of Popular Nuts

Types of Nuts	Production in 2003 (Metric Tons)
Almond	1,684,000
Walnut	1,440,000
Peanut	37,060,000
Pistachios	535,000
Pine nut[a]	5,000
Hazelnut	843,000

Source: Adapted from USDA. Foreign Agricultural Service. Production, Supply and Distribution (PSD) database, December 2004. Internet: http:// www.fas.usda.gov/psdonline/psdHome.aspx; Food and Agriculture Organization (FAO). 2004. FAO Stat Book, Web site: http://faostat.fao.org ; International Tree Nut Council (INC). Official Response to WHO/FAO Expert Consultation on Diet, Nutrition and the Prevention of Chronic Diseases, Geneva, Rome, June 15, 2002..

[a] Adapted from International Tree Nut Council (INC). Official Response to WHO/FAO Expert Consultation on Diet, Nutrition and the Prevention of Chronic Diseases, Geneva, Rome, June 15, 2002.

Peanuts may be consumed raw, roasted, pureed, or in a variety of other processed forms, and constitute as a multimillion-dollar crop worldwide (Yu et al., 2005) with numerous potential dietary benefits. Recently, several peanut cultivars were developed with elevated concentrations of the monounsaturated fatty acid (MUFA) oleic acid, in relation to other highly oxidizable polyunsaturated fatty acids (PUFA). The high oleic trait provides peanuts with potentially greater health benefits and serves to prolong shelf-life characteristics (Pattee et al., 2002; Reed et al., 2002; Talcott et al., 2005). Additionally, peanuts were found to lack naturally occurring *trans*-fatty acids that cause various diseases (Awad et al., 2000; Feldman, 1999; Peanut-Institute, 2002).

15.2.1.4 Pistachio

Native to the Middle East, pistachios are one of the oldest flowering nut trees. According to Food and Agriculture Organization (FAO), the pistachio production in the world stood at 535,000 metric tons in 2003 (Table 15.1). Iran is the leading country in pistachio production. The total production of Iran was about 310,000 tons in 2003 and Iran is the largest exporter (about 86%) in the world. Until relatively recently, pistachios were gathered from mountainside village plantations in Turkey, Afghanistan, and Iran. Now, it is cultivated in huge orchards in California, and California has become an important world supplier of this nut. Pistachio production in the United States was 157.3 Mkg in 2004. The United States is the world's second largest producer of pistachios after Iran.

15.2.1.5 Pine Nut

Today, the most common pine nuts in the market place are seeds of the "stone pine," *Pinus pinea*. It is a tree of the northern coastline of the Mediterranean, growing from Portugal and Spain in the west to Lebanon in the east. Despite $100 million in annual sales in the United States, the market for pine nuts and pine nut products is underdeveloped (US Trade Quick-References Tables: December 2003). Data on world production of pine nuts are not available. World pine nut production falls far short of demand. Even in a good crop the year total world production may not be around 20,000 tons of kernel (www.pinenut.com/value.htm). China produces the bulk of the world's supply of pine nuts. A total of 80–90% of pine nuts consumed in the United States are currently imported, primarily from China. Most of the pine nuts imported from China originate in Russia and are brought to China for processing, packaging, and shipping overseas. Global production of popular nuts are given in Table 15.1.

15.2.1.6 Hazelnut

Hazelnut (*Corylus avellana* L.) belongs to the *Betulaceae* family and is a popular tree nut worldwide. Turkey is the world's largest producer of hazelnuts, contributing approximately 70% to the total global production, followed by Italy (12%), the United States (6%), and Spain (2%). Other countries contribute only 10% to the total production. About 700,000 metric tons of hazelnuts are produced worldwide every year (FAOSTAT, 2004). Turkish hazelnut production of 625,000 t accounts for approximately 70% of worldwide production (*World Hazelnut Situation and Outlook*, USDA 2004). Although hazelnut is cultivated in many countries around the world, the number of countries producing hazelnut in international trade scale is limited. The major producer and exporter countries are Turkey, Italy, the United States, and Spain. Other producer countries are not producing at a level that can affect the world hazelnut markets.

The hard hazelnut shell, containing a kernel, is the nut of commerce. After cracking the hazelnut, the hazelnut kernel may be consumed raw (with brown skin) or preferably roasted (without skin). Hazelnut skin, which is a by-product of roasting represents about 2.5% (skin absorbs oil during roasting) of the total hazelnut kernel weight and is discarded upon roasting.

Among tree nuts, hazelnuts have many beneficial health attributes (Alasalvar et al., 2009b) and are among the three most popular and commonly consumed tree nuts in Europe (Jenab et al., 2006) and other Western countries (Fraser et al., 1992).

15.2.2 Nuts as Food

Nuts are widely consumed in both raw and processed forms. They are consumed either as snacks or part of a meal. Nuts are eaten whole (fresh or roasted), in spreads (peanut butter, almond paste) or hidden (e.g., commercial products, mixed dishes, sauces, baked goods, and oils) (Table 15.2). Unlike groundnuts, which are predominantly used for oil and feed, tree nuts are primarily consumed as whole foods, as ingredients in foods or in medicinal preparations. For example, in several Asian cultures, almonds play a significant role in *Ayurvedic* preparations, a philosophy which for thousands of years has promoted the interrelationship of nutrient/diet with healing, prevention, and longevity. As a component in finished foods, they provide crunch, bursts of flavor and other pleasing attributes in baked foods, confections, and now in salads, coatings for meat dishes and other applications. From breakfast to dessert, and all the meals and snacks in between, nut ingredients give foods appealing taste, texture, and appearance.

15.2.2.1 Almond

Almond, with or without the brown skin, is consumed as the whole nut or used worldwide in various confectioneries and chocolates; its discarded components are used as livestock feed (Takeoka et al., 2000). Another popular use of nuts is in ready-to-eat breakfast cereals. The most popular nut is the almond, often cut into thin slices and sometimes glazed along with flakes or clusters. Almonds are the most used nuts in cereal, confectionery, and bakery applications. Almond-based ingredients add flavor and texture to a broad spectrum of products.

Almonds are well known in such applications as rocky road ice cream, chocolate bars, biscotti, granola, and cheese spreads, but also appear in many side dishes, salads, and main entrees particularly in Middle Eastern, Mediterranean, and Asian dishes.

Almonds are used to make almond oil, which is used in many baking recipes to add an intense almond flavor. Ground, sweet almonds are the basis of sweets such as marzipan, nougat, and macaroons. They are especially useful in baking and can be substituted for flour to create a dense, moist texture in cakes and biscuits. Sweet almonds also appear in many savory recipes—from toasted flaked almonds served

TABLE 15.2
Use of Nuts in Traditional Asian Foods

Types of Nuts	Types of Foods
Almond	Almond chicken, almond and rice dessert, almond tofu stir fry, almond tofu, almond tea jelly, almond cream with chowchow, almond tea Asian, almond pudding, almond jelly, almond jello (Chinese gelatin dessert), almond half-moons, almond fudge (*badam halwa*), almond fried ice cream, almond and pistachio brittle, almond cake, almond cookie, almond toffee with chocolate on top, almond crème, almond float, almond float with mandarin oranges
Walnut	Walnut chicken, walnut balls in curry (*akhrot ke kofta*), crispy walnuts, walnut info, pineapple-walnut chutney (*ananas chatni*), walnut fried with sugar, walnut biscuits, walnut cookies (*akhrot ka meetha*)
Peanut	Peanut chicken, peanut and tiny seafish (*teri kacang*), peanut—crusted beef fry, peanut-crusted chicken with Thai hot/sweet sauce, Thai peanut chicken salad, Thai peanut Ramen, Thai peanut salsa, peanut–garlic–chili pasta (*sate*), Thai peanut and veggie lettuce rolls, peanut–ginger–honey dressing or sauce, peanut sauce, peanut sauce (*sot dau phong*), peanut sauce (*sambal kacang*), peanut sauce (*Thai style*), Thai peanut sauce (*Ceideburg*), peanut sauce (Nam Jim Satay), Chinese peanut sauce, peanut butter sauce, peanut butter sauce for satay (*Siamese princess*), peanut line dipping sauce, peanut dipping sauce (*Nuoc Leo*), pineapple– peanut curry (*ananas ki afed kadhi*), Thai peanut dressing, Chinese peanut dressing, Chinese peanut pancakes.
Pistachio	Pistachio ice cream (*pista kulfi*), pistachio kulfi pops, pistachio nuts info
Pine nut	Pine nut-apricot chutney, haricots verts with brown butter–pine nuts, grilled shrimp with pine nut–mint *pesta*, turkey–pine nut dumplins with chili sauce, pine nut–stuffed grape leaves with feta dip, watermelon–pine nut salad with greens, pasta with olive–pine nut salasa
Hazelnut	Roasted hazelnuts, caramel hazelnut bars, hazelnut chicken stir fry, Glico hazelnut dessert pocky, oriental chicken hazelnut salad, hazelnut salmon filets, turkey filbert casserole, hazelnut butter, hazelnut cocoa spread, Espresso hazelnut cupcakes, hazelnut cakes, hazelnut flour/meals, hazelnut cream, chocolate hazelnut Gelato, chocolate hazelnut tart

with trout and brown butter to their role as a thickening agent in romesco sauce. They are often used in Indian cooking to thicken and add texture and flavor to dishes such as *chicken korma*. In India, almonds are ground to a paste for curry bases. They can be fried or blanched as a garnish or used as slivers in sweets. *Biryani's*, milk desserts, and meat curries are dusted with finely chopped almonds. Almond ice cream made with rich creamy milk called *kulfi* is a delicacy. *Badam halwa* made of powdered almonds and sugar is a traditional sweet and almond sherbet is a cool drink made of almonds, milt sugar, and rosewater.

15.2.2.2 Walnuts

Consumers use walnuts most often in baking, as snacks, in cooking, salad dressings, and as ice-cream toppings (Ravai, 1995). When incorporated with other foods, walnuts can add and intensify flavor. They also have rising or puffing properties, and can lighten foods by fluffing or foaming when used in place of egg whites as finely ground walnut meal. Traditional bakery, confectionery, and ice-cream applications represent the largest use of walnuts. Walnuts exhibit an excellent flavor synergy with chocolate-baked products, such as walnut brownies or chocolate cookies with walnuts.

Walnuts are used in desserts, cakes, and confectionery, as well as in many savory dishes such as soups, sauces, stews, and salads including the well-known Waldorf salad. They are also pickled for serving with cold meats and cheeses—they go particularly well with blue cheese—or for adding to meaty stews. Walnut oil has a strong nutty flavor which is excellent when used on fish, steaks, pasta, or salad.

15.2.2.3 Peanut

Peanut is an important source of food worldwide and constitutes an inexpensive source of protein, fat, minerals, and vitamins in the diets of rural populations especially among children (Bankole et al., 2005). Moreover, peanut oil is one of the major oils in the human diet (Ndiaye et al., 1999; Shi et al., 1998). The edible parts of peanuts consist of the kernel and the protective skin. The kernels are used to make peanut butter, roasted snack peanuts, peanut confections, peanut milk, and peanut oil.

Peanuts are one of the most economical and easiest ways to add texture to a food. They are tasty roasted and diced as a salad or dessert topping, or just by the handful. Peanuts and peanut butter appear in ice creams, confections, baked products, snacks, and cereals.

15.2.2.4 Pistachio

Pistachios are known as the "happy nut" in China because of the "smile" appearance of its open shell. The open shell of the pistachio enables it to be roasted and salted while still in its shell, and that is how they are often sold and eaten. They are also found in many Middle Eastern and Mediterranean pastries such as *baklava* and Asian sweets such as *seera*.

The major ingredient use for pistachio is in ice cream, but energy bars, trail mixes, and chocolate bars are some new applications for pistachios. Pistachios make colorful additions to green salads, cooked vegetables, and fruit salads, as well as to cold and hot cereals, muffins, trail mix, rice dishes, pilafs, and pasta. They can also be used ground as a crispy coating for fish filets or poultry breast.

15.2.2.5 Pine Nut

Pine nuts are known throughout the world as a nutritious healthy snack (raw and roasted) and essential ingredient in multiple oriental and Mediterranean dishes such as pine nut milk, pine nut cream, and pine nut butter. They are also added to gourmet chocolates, granolas, and crunch bars. Pine nuts are getting attention from today's cooks and restaurants as a new flavor to add to a variety of dishes. It is most often used in the production of pesto.

Pine nuts have a very delicate taste and texture and are high in protein, which makes them especially useful in a vegetarian diet. They can be eaten raw, when they have a soft texture and a sweet buttery flavor and are especially good in salads.

15.2.2.6 Hazelnut

Hazelnuts have multiple uses and are sold in the market either with shell or as kernel (without shell). Cultivars suited for one use are quite different from that suited for the other. The in-shell market accounts for 10% of the world hazelnut crop. The remaining 90% of the crop is shelled and the kernels are sold to bakers, candy makers and other food processors. Hazelnuts are widely used throughout the world in pastries, breakfast cereals, chocolate, confectionery, and the baking industry.

Hazelnuts are extensively used in confectionery to make praline and also used in combination with chocolate for chocolate truffles and products such as Nutella. In the United States, hazelnut butter is being promoted as a more nutritious spread than its peanut butter counterpart, although it has a higher fat content. In Austria and especially in Vienna, hazelnut paste is an important ingredient in the world famous *torts* (such as Viennese hazelnut tort). Hazelnuts are also the main ingredient of the classic Dacquoise. Vodka-based Hazelnut liqueurs, such as Frangelico, are also increasing in popularity, especially in the United States and Eastern Europe.

Hazelnut is popular as a coffee flavoring, especially in the form of Hazelnut latte. Hazelnut-flavored coffee seems (to many users) a slightly sweetened and less acidic beverage, even though the nut is low in natural saccharides. The reason for such perception is not yet understood. In Australia over 2000 t are imported annually mostly to meet the demand from the Cadbury company for inclusion in its eponymous milk chocolate bar which is the third most popular brand in Australia. Hazelnut oil, pressed from hazelnuts, is strongly flavored and used as cooking oil.

15.2.3 Nuts as Functional Foods

Nuts are rich sources of multiple nutrients and their consumption is associated with health benefits. Many studies as documented in this chapter reported that nuts improve plasma lipid profiles and can have a beneficial effect on cardiovascular disease (CVD) risk (Albert et al., 2002; Almario et al., 2001; Hu et al., 1998; Jenkins et al., 2002; Kris-Etherton et al., 2001; Kushi et al., 1996; Lovejoy et al., 2002; Morgan and Clayshulte, 2000; Zambon et al., 2000). In addition to the cardiac benefits of consuming nuts, the risk of developing type-2 diabetes (Jiang et al., 2002), of advanced macular degeneration (Seddon et al., 2003) and of gallstones (Tsai et al., 2004) have all been found to be lowered by eating nuts. An inverse association between nut intake and clinical gallstone disease (Tsai et al., 2004) could be explained in part by improved insulin sensitivity. Research suggests that daily nut eaters gain an extra 5–6 years of life free of coronary disease (Hu and Stampfer, 1999) and that regular nut eating could increase longevity by about 2 years (Fraser and Shavik, 2001).

Numerous studies have shown that adding nuts to the diet produces favorable health outcomes, including lowering low-density lipoprotein (LDL) cholesterol, improving LDL to HDL ratios, and reducing inflammation associated with increased heart disease risk. The LDL cholesterol-lowering response of nut and peanut studies is greater than expected on the basis of blood cholesterol-lowering equations that are derived from changes in the fatty acid profile of the diet. Thus, in addition to a favorable fatty acid profile, nuts and peanuts contain other bioactive compounds that explain their multiple cardiovascular benefits (Kris-Etherton et al., 2008). Other macronutrients include plant protein and fiber; micronutrients including potassium, calcium, magnesium, and phytochemicals such as tocopherols and phytosterols, other phenolic compounds including flavonoids, stilbenes, and resvertrol, as well as carotenoids and arginine. The total phenolic constituents contribute to the total antioxidant capacity of nuts.

Recently, Kris-Etherton et al. (2008) evaluated the epidemiological and clinical studies of many different tree nuts and peanuts on lipids, lipoproteins, and various CHD risk factors, including oxidation, inflammation, and vascular reactivity. Evidence from these studies demonstrated consistent benefits of nut and peanut consumption on CHD risk and associated risk factors.

For all the healthy effects described, nuts can be classified as natural functional foods, and can be recommended to promote health by their easy incorporation into the usual diet of the people. Thus, eating recommended amounts of nuts may be a simple measure to prevent CVDs, the greatest cause of morbidity and mortality in the western world.

It is clear that nuts contain functional ingredients such as flavonoids, polyphenols, phytosterols, and other phytochemicals that could be extracted and used as food supplements in various formulations (King et al., 2008). They should also be encouraged in the diet as one of the preventive measures against diseases such as cancer, CVD, diabetes, and other degenerative diseases. The combination of functional ingredients and rich nutritional composition of nuts makes them potential functional foods. Although a number of functional properties have been documented in this chapter, a great deal of further research is necessary, including clinical trials, before most of these numerous benefits of nuts are properly measured and substantiated.

15.2.4 Oil of Nuts

15.2.4.1 Almond Oil

The list of almond oil's beauty aiding properties is almost endless. It can be used as a mild and effective bleaching agent when mixed with milk cream and ground with the paste of fresh rosebud. When applied daily, this paste keeps the skin dew fresh and young looking, delays the appearance of wrinkles, blackheads, dryness, and even pimples. The toning effect of the oil works wonders for mature skins. In case of steam burns, application of almond oil immediately cools off the area and protects it from infection of any sort. However, since it has a short shelf life and goes rancid very quickly, it must be consumed fast (Navarro et al., 2002).

Almond oil is widely used in aromatherapy for giving body massage. The oil is easily absorbable and serves as a great emollient, by lending a soft glowing beautiful touch to the skin. It balances the moisture in the body. In case of loss of moisture, it helps to restore it. It makes a great lubricant, thus aids in combating itching and inflammation. It is an excellent natural moisturizer that is suitable for all skin types. Owing to the multitude of benefits that almond oil offers, it is increasingly finding its way in the making of soaps, creams and moisturizers (Knorr, 2004).

15.2.4.2 Walnut Oil

Walnut oil is extracted from walnuts. It is not used as extensively as other oils in food preparation due to its higher cost. Walnut oil is not suitable as a cooking oil, as high heat destroys its delicate flavor and produces a slight bitterness; instead it is used primarily as an ingredient in cold dishes such as salad dressings. Walnut oil was one of the most important and vital oils used by Renaissance painters. Its quickness of drying and lack of yellow tint make it a good oil paint thinner and brush cleaner. Walnut oil is a good source of unsaturated fatty acids, such as linolenic, linoleic, and oleic acids, which are susceptible to oxidation (Tsamouris et al., 2002).

15.2.4.3 Peanut Oil

Peanut oil is considered to be health promoting because of its fatty acid composition and its bioactive compounds such as tocopherols and phytosterols. Peanut oil generally contains 55–65% MUFAs, 26–28% PUFAs, and 17–18% saturated fatty acids (Ory et al., 1992). Long-chain saturated fatty acids (arachidic, behenic, and lignoceric acids) that have been reported to contribute to the development of cholesterol in the human body are present in peanut oil in small quantities (Misra, 2004). Peanut oil is composed of as many as 12 fatty acids, of which about 80% is accounted for by only oleic and linoleic acid (LA) (Savage and Keenan, 1994). Peanut oil mainly contains α-tocopherol (50–373 ppm) and γ-tocopherols (90–390 ppm) (Firestone, 1999).

Plant sterols (phytosterols) are minor components of all vegetable oils constituting major portions of the unsaponifiable fraction of the oil. Peanut oil contains 900–3000 ppm of total phytosterols (Firestone, 1999). Major phytosterols in peanut oil are β-sitosterol (>80% of total phytosterols), campesterol (about 10%, and stigmasterol <5%) (Firestone, 1999; Mishima et al., 2003).

15.2.4.4 Pistachio Oil

Oil extracted from the pistachio nut contained the highest amount of MUFA (60%), which consisted primarily of oleic acid (Ryan et al., 2006). Similarly, Aslan et al. (2002) reported high levels of MUFA in the seed oils of *Pistachia vera* L. of different origins. In a latter study, Ryan et al. (2006) reported the oleic acid to be the dominant MUFA in all *P. vera* L. seeds, and the levels found were also higher (65.6–83%). Perhaps these differences were the result of environmental, processing and genetic factors.

15.2.4.5 Pine Nut Oil

Pine nuts contain up to 68% oil. Pine nut oil is obtained by pressing and is available on the market as one of the expensive gourmet cooking oils or a medicine (in bottles or capsules). While little research was made in the medicinal properties of pine nut oils derived from different species, a number of sources suggest that Siberian pine nuts yield oil with high medicinal value (Megre, 1996), traditionally used to cure a wide array of ailments – ingested (decreasing blood pressure, boosting immune system resistance, etc.) or applied externally (a range of dermatological disorders). Apart from cooking and medicine, pine nut oil is used in cosmetics, beauty products, and as high-end massage oil. It also has a variety of specialty uses such as a wood finish, paint base for paintings, and treatment of fine skins in leather industry.

Ryan et al. (2006) reported that the fatty acids present in the pine nut oil were primarily PUFAs, with LA being the most abundant. Nergiz and Donmez (2004) and Sagrero-Nieves (1992) reported similar findings. Ryan et al. (2006) also reported that the levels of squalene ranged from 39.5 mg/g oil in the pine nut to 1377.8 mg/g oil in the brazil nut. Pine nut oil contains pinolenic, linolenic, and linoleic acids, PUFAs, and is marketed in the United States as a means to stimulate cell proliferation, prevent hypertension, decrease blood lipid and blood sugar, and inhibit allergic reactions.

15.2.4.6 Hazelnut Oil

Hazelnuts provide a ready source of nutritionally valuable oil. The kernels contain approximately 60% oil, which can be recovered effectively by pressing. Hazelnut oil is an increasingly popular nut oil, usually cold or expeller pressed from roasted hazelnuts, called *filberts*. It is not easily available and is fairly expensive too. Most brands of hazelnut oil are exported from Turkey and parts of Asia where the hazel tree is native. Hazelnut oil can be found in the market either as crude (cold-pressed), refined, or a mixture of both. Hazelnut oil is relatively rich in mono- and PUFAs. People use hazelnut oil in salad dressings, in baked goods, and in various cooking applications. It is not the best nut oil for deep-frying as it is high in unsaturated fat. It does have a high amount of polyunsaturated and monounsaturated fat, and these types of fatty acids in combination have desirable health benefits (Miraliakbari and Shahidi, 2008).

Hazelnut oil is considered valuable in giving facial and body massages. It is rich in linoleic acid, which is an essential fatty acid. Hazelnut oil has proved to be very effective in tightening the skin, thereby acting as a superb toner for the skin. Apart from therapeutic properties, this oil is known for its moisturizing qualities as well (Madhaven, 2001). Because of its ability to filter the harmful sun rays, it is used in a number of sun care products. The oil is also used as an additional ingredient in the preparation of creams, soaps, lotions, and hair care products.

Alasalvar et al. (2009a) reported separation of 12 triacylglycerols from five native hazelnut varieties from Turkey. Seven tocol isoforms (four tocopherols and three tocotrienols), seven phytosterols, as well as cholesterol, and one phytostanol were positively identified and quantified; among these, α-tocopherol and β-sitosterol were predominant in all hazelnut oils.

Alasalvar et al. (2006a) identified 16 fatty acids from hazelnut oil among which oleic acid (18:1) contributed 82.78% to the total, followed by LA (18:2) at 8.85%, palmitic acid (16:0) at 4.81%, and stearic acid (18:0) at 2.69%. The remaining 12 fatty acids contributed only 0.87% to the total amount of fatty acids present. The total saturated fatty acids (SFA) made up a small proportion (7.79%) of the fatty acids of hazelnut oil, whereas total MUFA was the highest (83.24%). Unsaturated fatty acids (MUFA + PUFA) accounted for 92.21% of the total fatty acids. As compared with other nut oils, hazelnut oil has been reported to contain the highest proportion of oleic acid (Ebrahem et al., 1994; Kris-Etherton et al., 1999b; Maguire et al., 2004; Venkatachalam and Sathe, 2006). Amaral et al. (2006c) reported that hazelnut oil contained trace amount of *trans*-fatty acids (0.02%).

Miraliakbari and Shahidi (2008) reported that among the nuts samples they tested, hazelnut oil had the highest oleic acid content of 83.4% and 83.3% for hexane and chloroform/methanol extracted oils, respectively. On the other hand, Parcerisa et al. (1997) reported that hazelnut oil contained 98.4% triacylglycerols as nonpolar lipids and less than 0.2% of phospholipids (phosphatidylcholine and phosphatidylinositol) as polar lipids. The 18:1 is a dominant fatty acid in the nonpolar lipid class, whereas 16:0, 18:0, and 18:2 were the most common in the polar lipid class. Alasalvar et al. (2003b) reported that the crude lipid extract from Turkish Tombul hazelnut composed of nonpolar (98.8%) and polar (1.2%) constituents and triacylglycerols was the major nonpolar lipid class and contributed nearly 100% to the total amount. Among polar lipids, phosphatidylcholine, phosphatidylethanolamine, and phosphatidylinositol were present at 56.4%, 30.8%, and 11.7%, respectively. Similar results were reported by Parcerisa et al. (1999).

Erdogan and Aygun (2005) reported that oil content of Turkish tree hazelnuts ranged between 64.48% and 71.92%. Oleic and linoleic acids were the predominant fatty acids, representing 91.7% of the total fatty acids. The amount of palmitic and stearic acid was low while palmitoleic, margaric, margaroleic, linolenic, arachidic, and gadoleic acids were present in trace amounts.

Hazelnut oil contained 1.2–2.2 g/kg of phytosterols primarily in the form of β-sitosterol (Maguire et al., 2004; USDA, 2005). Miraliakbari and Shahidi (2008a) reported that the hazelnut oil contained the highest tocopherol content (462–508 mg/kg oil), followed by pecan oil (454–490 mg/kg) and pine nut oil (399–458 mg/kg).

Alasalvar et al. (2006a) reported the content of total tocols (51.31 mg/100 g) in fresh Tombul hazelnut oil. They detected and quantified seven tocol isoforms (α-, β-, γ-, and δ-tocopherols and α-, β-, and γ-tocotrienols). Among the tocols identified, α-tocopherol was the most abundant (40.40 mg/100 g), accounting for 78.74% of the total, followed by γ-tocopherol (8.33 mg/100 g), β-tocopherol (1.53 mg/100 g), and a small amount of β-tocopherol (0.53 mg/100 g). Tocotrienols contributed only 1.02% to the total tocols present. Earlier, Alasalvar et al. (2003b) have also identified four tocopherols in freshly extracted oil from Tombul hazelnut. Several other researchers have also detected and quantified tocopherols in crude, refined, virgin, mixed, and pressed hazel nuts, among which α-tocopherol was the predominant tocopherol with concentrations ranging from 11.9 to 61.88 mg/100 g (Savage et al., 1997; Parcerisa et al., 2000; Bada et al., 2004; Karabulut et al., 2005; Ebrahem et al., 1994; Kornsteiner et al., 2006).

Small amounts of α-, β-, and γ-tocotrienols together with four tocopherols in hazelnut oil have also been reported (Amaral et al., 2006a; 2006c; 2005; Benitez-Sánchez et al., 2003; Crews et al., 2005).

Among the various sterols (cholesterol, campesterol, stigmasterol, clerosterol, β-sitosterol, Δ^5-avenasterol + β-sitostanol, Δ^7-stigmastenol, and Δ^7-avenasterol) identified and quantified in hazelnut oil, β-sitosterol comprised 81.28% of the total, while Δ^5-avenasterol + β-sitostanol and campesterol were the second and third components of the group with values of 8.45% and 6.02%, respectively (Alasalvar et al., 2006a). The remaining phytosterols contributed only 4.25% to the total phytosterol present. A similar number of sterols in 19 cultivars of hazelnut oil has also been identified by Amaral et al. (2006b).

Alasalvar et al. (2006a) reported that the total content of phytosterol (including cholesterol) in hazelnut oil was 164.92 mg/100 g, which is in good agreement with those reported by others (Amaral et al., 2006b; Benitez-Sánchez et al., 2003; Crews et al., 2005; Karabulut et al., 2005). Earlier, Alasalvar et al. (2003) has identified and quantified only three most common phytosterols (β-sitosterol, campesterol, and stigmasterol) in freshly extracted Tombul hazelnut oil.

Kornsteiner et al. (2006) studied the tocopherol content of oil extracted from different nuts and found that vitamin E content was the highest in hazelnut oil (33.1 mg/100 g), followed by (in the descending order) almond > peanut > pistachio > pine nut > walnut > Brazil nut > pecan > cashew > macademia. Hazelnut oil has been reported to contain the highest α-tocopherol level among nut oils (Ebrahem et al., 1994; Kornsteiner et al., 2006).

15.3 CHEMICAL COMPOSITION OF NUTS

15.3.1 General Composition

Nuts are highly nutritious and of prime importance for people in several regions in Asia and Africa. The caloric content of nuts varies between 5.3 and 6.5 kcal/g of nut. The fat content of all five nut types was high, between 46% and 65% (Table 15.3). On the contrary, in all nuts the levels of saturated fatty acids did not exceed 9.5% (Table 15.4). Although all nuts have high levels of either mono- or polyunsaturated fats, and low levels of saturated fats, the fatty acid composition of each type of nut varies (Anon, 1984).

Analysis of the fatty acid profile of the nuts indicates a favorable high unsaturated to saturated ratio. The main contributing saturated fatty acids for all nuts included stearic acid (C18:0) and palmitic acid (C16:0) with traces of myristic acid (C14:0) and eicosanoic acid (C20:0). The highest levels of saturated fatty acid were found in the pine nuts, followed by peanuts. Nuts are high in fat; however, as greater than 75% of

TABLE 15.3
Key Nutritional Components of Popular Nuts (/100 g)

Nutrients	Almond	Walnut	Peanut	Pistachio	Pine nut	Hazelnut
kcal	575	654	594	557	673	628
Protein (g)	21.2	15.2	17.3	20.6	13.7	15.0
Fat (g)	49.4	65.2	51.5	44.4	68.4	60.8
Carbohydrate (g)	21.7	13.7	25.4	28	13.1	16.7
Dietary fiber (g)	12.2	6.7	9.0	10.3	3.7	9.7
Calcium (mg)	264	98	70	107	16.0	114
Iron (mg)	3.7	2.9	3.7	4.2	5.5	4.7
Zinc (mg)	3.1	3.1	3.8	2.2	6.5	2.5
Potassium (mg)	705	441	597	1025	597	680
Folate (μg)	50	98	50	51	34	113

Source: Adapted from USDA. Agricultural Research Service, 2008. *USDA Nutrient Database for Standard Reference, Release No. 21*. Nutrient Data Laboratory Home Page. Internet: http://www.ars.usda.gov/ba/bhnrc/ndl (accessed 8 November 2008).

the fat present is unsaturated, nuts are thought to have a fatty acid profile that is cardioprotective. The main unsaturated fatty acids in the nuts included oleic acid (C18:1), LA (C18:2), linolenic acid (C18:3) and palmitoleic acid (C16:1). MUFAs are the predominant fatty acids (Table 15.4) and contribute, on the average, approximately 62% of the energy in nuts from fat (Kris-Etherton et al., 1999b). The most abundant MUFA was oleic acid (C18:1), while LA (C18:2) was the most prevalent PUFA. Almonds and pistachios are rich in oleic acid, and for that reason are good sources of MUFAs. Generally, the health benefit of nuts has been attributed to their high level of polyunsaturated fats, a high P/S ratio, and their higher monounsaturated fats content (Edwards et al., 1999). Recently, the favorable fatty acid composition of nuts was discussed in detail by Ros and Mataix (2006) in a review article.

In addition to the fatty acid content, the quantity of components present in nuts makes these products natural foods that provide benefits to health beyond those described up to this time. The quantity of dietary fiber in nuts is appreciable, between 7 and 12 g/100 g of nut, and varies according to different varieties (Table 15.3). As reviewed by Sales-Salvado et al. (2006), nuts provide a significant amount of fiber. Moreover, almonds are a good natural source of vitamin E, which contain around 26 mg/100 g of product (USDA, 2008). In other nuts the vitamin E content is lower, of the order of 3 to 9 mg/100 g. Another vitamin which nuts contribute, in abundance, is folic acid, with values around 50–60 μg/100 g, with the exception of walnut which contains up to 98 μg/100 g. Compared with other common foodstuffs, nuts have an optimal nutritional density with respect to healthy minerals, such as calcium, magnesium, and potassium (Segura et al., 2006). Like that of most vegetables, the sodium content of nuts is very low.

Most nuts have a good protein content (ranging from 10% to 30%), and only a few have a very high starch content (Davidson, 1999) (Table 15.3). In general, many nuts have been identified as a especially rich source of antioxidants (Halvorsen et al., 2002; Wu et al., 2004). Nuts also contain significant amounts of tocopherols, squalene, and phytosterols, and the levels vary greatly among nut species (Maguire

TABLE 15.4
Fatty Acid Composition of Nuts (g/100 g)

Nutrients	Almond	Walnut	Peanut	Pistachio	Pine nut	Hazelnut
Total fat	49.4	65.2	51.5	44.4	61.0	60.8
Saturated fat	3.7	6.1	6.9	5.4	9.4	4.5
Monounsaturated fat	30.9	8.9	31.4	23.3	22.9	45.7
Oleic acid	30.6	8.8	30.8	22.7	21.5	45.4
Polyunsaturated Fat	12.1	47.2	10.8	13.5	25.7	7.9
Linoleic acid	12.1	38.1	10.5	13.2	24.9	7.8
ALA	0.0	9.1	0.19	0.25	0.79	0.09

Source: Adapted from USDA. Agricultural Research Service, 2008. *USDA Nutrient Database for Standard Reference, Release No. 21.* Nutrient Data Laboratory Home Page. Internet: http://www.ars.usda.gov/ba/bhnrc/ndl (accessed 8 November 2008).

et al., 2004) (Table 15.5). Kornsteiner et al. (2006) in a study reported that walnuts, pistachios, and pecans contained the highest total polyphenols and total tocopherol content among all types of nuts, followed by peanuts with skin, hazelnuts, and almonds with skin. Phytosterols, including β-sitosterol, campesterol, and stigmasterol, are integral components of plant cell membranes that have been reported to inhibit the intestinal absorption of cholesterol, thereby lowering total plasma cholesterol and LDL levels (de Jong et al., 2003). Regarding the antioxidant potential (AOP), nuts are an excellent source of tocopherols and polyphenols.

α-Tocopherol was the most prevalent tocopherol in most nuts except walnuts (Table 15.5). The α-tocopherol content ranged between 20.6 μg/g oil in walnuts and 439.5 μg/g oil in almonds. The γ-tocopherol content of the nuts ranged from trace levels to 12.5 μg/g in almonds to 300.5 μg/g in walnuts. Walnuts were the only nuts that contained γ-tocopherol in higher concentrations than α-tocopherol. The total tocopherol content of oil extracted from the five nuts ranged from 148.2 mg/g (peanuts) to 366.8 mg/g (pistachio). γ-Tocopherol was the main tocopherol present in walnut and pistachio nuts, while α-tocopherol was present in a higher amount in almonds (Ryan et al., 2006; Maguire et al., 2004). Therefore, while nuts contain significant amounts of tocopherols, their levels vary greatly among nut species (Table 15.5).

Squalene was identified in all nuts (pine, pistachio, peanut, almond, walnut), with levels ranging from 9.4 mg/g (walnut) to 98.3 mg/g in peanuts (Ryan et al., 2006; Maguire et al., 2004). Squalene levels were high in all nuts with the exception of walnut. It appears that the squalene content varies remarkably between nuts (Table 15.5). According to the USDA (2008) Nutrient Database, tree nuts contain a reasonable amount of phytosterols, in particular β-sitosterol, which mainly competes with the absorption of food cholesterol. Pistachio has the highest amount of β-sitosterol (198 mg/100 g), followed by almond (132 mg/100 g), and walnut (64 mg/100 g).

Maguire et al. (2004) and Ryan et al. (2006) detected considerable amounts of phytosterols in almond, walnut, peanut, pine nut, and pistachio oil (1236–5586 mg/g oil) (Table 15.5). For all five nuts, β-sitosterol was the predominant phytosterol present ranging between 1129.5 and 2071.7 μg/g. Campesterol and stigmasterol were also

TABLE 15.5
Squalene, Tocopherol, and Phytosterol Content (μg/g oil) of Oil Extracted from Nuts

Oil sample	Squalene	α-Tocopherol	γ-Tocopherol	β-Sitosterol	Campesterol	Stigmasterol
Almond	95.0 ± 8.5	439.5 ± 4.8	12.5 ± 2.1	2071.7 ± 25.9	55.0 ± 10.8	51.7 ± 3.6
Walnut	9.4 ± 1.8	20.6 ± 8.2	300.5 ± 31.0	1129.5 ± 124.6	51.0 ± 2.9	55.5 ± 11.0
Peanut	98.3 ± 13.4	87.9 ± 6.7	60.3 ± 6.7	1363.3 ± 103.9	198.3 ± 21.4	163.3 ± 23.8
Pistachio	91.4 ± 18.9	15.6 ± 1.2	275.4 ± 19.8	4685.9 ± 154.1	236.8 ± 24.8	663.3 ± 61.0
Pine nut	39.5 ± 7.7	124.3 ± 9.4	105.2 ± 7.2	1841.7 ± 125.2	214.9 ± 13.7	680.5 ± 45.7
Hazelnut	186.4 ± 11.6	310.1 ± 31.1	61.2 ± 29.8	991.2 ± 73.2	66.7 ± 6.7	38.1 ± 4.0

Source: Adapted from Maguire LS. et al. *Int. J. Food Sci. Nutr*. 55:171–178, 2004; Ryan E. et al. *Int. J. Food Sci*. Nutr. 57:219–228, 2006.

detected. The data indicate that the pistachio nut is particularly a rich source of phytosterols (Table 15.5).

15.3.1.1 Almond

The processing by-products, shells and hulls of almonds, account for more than 50% by dry weight of the fruits (Fadel, 1999; Martinez et al., 1995). Almond hulls contain triterpenoids (Takeoka et al., 2000), lactones (Sang et al., 2002a), phenolics (Sang et al., 2002c), and sterols (Takeoka and Dao, 2003). The isolation and identification of phenolic compounds in almond skins has been reported (Sang et al., 2002b). The water extraction of hulls (Rabinowitz, 2004) and solvent extraction of shells (Pinelo et al., 2004) to produce food ingredients and antioxidants, respectively, has been studied.

The high xylan content of almond shells makes them a suitable substrate for the production of xylose (Pou-Ilinas et al., 1990), furfural (Quesada et al., 2002) or for fractionation into cellulose, pentosans and lignin (Martinez et al., 1995). Almond shell is highly lignified (30–38% of the dry weight) (Martinez et al., 1995) and the guaiacyl to syringyl phenylpropane unit's ratio is similar to that of hardwood (Quesada et al., 2002). Even if most of the lignin is acid-insoluble (Klason lignin), a part of it can be solubilized in acidic media. The AOP of depolymerized lignin fractions produced after mild acid hydrolysis of lignocellulosics has been identified (Cruz et al., 2004, 2005; Garrote et al., 2003; Gonzalez, 2004).

According to the Almond Board of California, almonds are an excellent source of vitamin E and magnesium, a good source of protein and fiber, and offers potassium, calcium, phosphorous, iron, and heart-healthy monounsaturated fat. Sang et al. (2002) reported the isolation of a sphingolipid,1-O-β-D-glucopyranosyl-(2S,3R,4E,8Z)-2-hydroxyhexadecanoyl-amino]-4,8 octadecadiene-1,3-diol and four other constituents, β-sitosterol, daucosterol, uridine, and adenosine, from almonds. In addition to that Sang et al. (2002c) also reported isolation of a new prenylated benzoic acid derivative, 3-prenyl-4-O-β-D-glucopyranosyloxy-4-hydroxylbenzoic acid, and three known constituents, catechin, protocatechuic acid, and urosolic acid from the hulls of almond. All of these compounds, except urosolic acid, have been reported from almond hulls for the first time.

15.3.1.2 Walnut

Walnuts are rich in linoleic acid, which contribute to PUFAs intake in human diet. In addition, walnuts contribute linolenic acid in a proportion of up to 6.8% of fat content. Walnuts are unique compared with other nuts because they are predominantly composed of PUFA (both omega-3 and omega-6) rather than MUFA, which are present in most other nuts. Per serving, walnuts contain a higher amount of omega-3 fatty acids than any of the other tree nuts (almonds, pine nuts, and pistachios) or peanuts (Table 15.4). In fact, the omega-3 fatty acid content of walnuts is 40–500 times greater than most other nuts. Peanuts contain negligible amounts, and almonds have no omega-3 fatty acids at all. Walnuts not only provide the highest amount of α-linolenic acid (ALA), but they also provide the highest concentration of antioxidants relative to other nuts (Simopoulos, 2004).

The walnut seed (kernel) constitutes 40–60% of the nut weight, depending mainly on the variety. The seed has high levels of oil (52–70%) in which PUFA predominate

(Greve et al., 1992; Martinez et al., 2006; Prasad, 2003; Savage et al., 1999). In addition to oil, walnuts provide appreciable amounts of proteins (up to 24%), carbohydrates (12–16%), fiber (1.5–2%), and minerals (1.7–2%) (Prasad, 2003; Savage, 2001; Lavedrine et al., 2000; Sze-Tao and Sathe, 2000).

Fukuda et al. (2003) reported the isolation of three hydrolyzable tannins, glansrins A–C, together with adenosine, adenine, and 13 known tannins from the *n*-butanol extract of walnuts (*Juglans regia* L.). Glansrins A–C was characterized as ellagitannins with a tergalloyl group, or related polyphenolic acyl group. The 14 walnut polyphenols had superoxide dismutase (SOD)-like activity and a remarkable radical scavenging effect against 2,2′-diphenyl-1-picrylhydrazyl (DPPH) (Fukuda, 2008). Reiter et al. (2005) reported the presence of melatonin in walnuts. For more information on tree nuts, the reader is referred to the book edited by Alasalvar and Shahidi (2008).

15.3.1.3 Peanut

The composition of the peanut is 50% oil, which consists of 50–80% oleic acid and 10–25% linoleic acid, 20% protein, and 20% carbohydrates. Peanuts provide 14.3 g of fiber/100 g and are a good source of essential minerals, vitamin E, and B-complex vitamins such as folate. Although classified as a legume, peanuts also have a high lipid content (ca. 46%) that is rich in MUFAs (Higgs, 2003). In addition to their nutrient composition, peanuts contain certain bioactive compounds that may also play a role in the reduction of the risk of the development of chronic diseases such as cancer, diabetes, and CHDs (Higgs, 2003; Hu and Stampfer, 1999).

Peanuts generally contain tocopherols, tocotrienols, phytosterols, and many different flavonoids including isoflavones and quercetin (Haumann, 1998). Isoflavones and *trans*-resveratrol, have previously been identified and quantified in peanuts by several authors (Ibern-Gomez et al., 2000; Liggins et al., 2000; Sanders et al., 2000). Lou et al. (2001) isolated eight flavonoids and two novel indole alkaloids from water-soluble fraction of peanut skins. Two new flavonoid glycosides have been identified as isorhamnetin 3-O-[2-O-β-glucopyranosyl-6-O-α-rhamnopyranosyl]-β-glucopyranoside and 3′,5,7-trihydroxyisoflavone-4′-methoxy-3′-O-β-glucopyranoside. Two alkaloids, namely 2-methoxy-3-(3-indolyl)-propionic acid and 2-hydroxyl-3-[3-(1-N-methy)-indolyl]-propionic acid were also identified.

15.3.1.4 Pistachio

Pistachio serves as a good source of calcium, magnesium, and vitamin A. Pistachio contains a high content of ascorbic acid (30 mg/100 g), but most nuts have none or only tiny amounts. Its protein content is 21%, hence it is a good protein source and has about the same amount of carbohydrate. Ryan et al. (2006) reported that the pistachio and pine nut oils have the highest unsaturated to saturated fatty acids ratio due to their greater content of total unsaturated fatty acids (90.7% and 87.2%, respectively).

15.3.1.5 Pine Nut

Pine nuts are cholesterol-free, contain from 53% to 68% fat (of which 93% is unsaturated fat), multiple micronutrients and vitamins. It contains 13–20% protein depending on the species. Pine nuts in general are a good source of vitamin B1.

15.3.1.6 Hazelnut

Hazelnut is mainly a fruit tree, as its fruit has a pleasant taste and is highly nutritive. Depending on the cultivar, hazelnut contains 55–72% fat (they produce high-quality edible oils but these turn rancid easily); 3–11% digestible carbohydrates; 10–22% protein; 5–7% dietary fiber; 5–6% water, and 2–3% minerals. Energy value of the kernel is 600 cal/100 g. They also contain vitamins B_1, B_2, and C; and are an excellent source of vitamin E (250–550 ppm).

Among the nut species, hazelnut plays a major role in human nutrition and health because of its special fat composition being highly rich in MUFA (primarily oleic acid), protein, carbohydrates, vitamins (vitamin E), minerals, diabetic fibers, tocopherols (α-tocopherol), phytosterols (β-sitosterol), polyphenols, and squalene (Alasalvar et al., 2003b, 2006a; Amaral et al., 2006b, 2006c; Maguire et al., 2004; Miyashita and Alasalvar, 2006; Savage et al., 1997).

Hazelnuts contain a number of proposed cardioprotective compounds, including fiber, vitamin E, arginine, folate, vitamin B_6, calcium, magnesium, and potassium (Griel and Kris-Etherton, 2006). In addition, hazelnuts are a rich source of monounsaturated fats. Hazelnuts contain approximately 91% MUFAs, mostly oleic acid, and less than 4% saturated fatty acids.

Koksal et al. (2006) investigated the chemical compositions of the 17 different hazelnut varieties grown in the Black Sea Region of Turkey and found that the major fatty acids in hazelnut varieties were oleic (79.4%), linoleic (13.0%) and palmitic acid (5.4%). The ratios of polyunsaturated/saturated and unsaturated/saturated fatty acids of hazelnuts varieties were found to be between 1.23 and 2.87, and 11.1 and 16.4, respectively. The average niacin, vitamin B_1, vitamin B_2, vitamin B_6, ascorbic acid, folic acid, retinol, and total tocopherol contents of hazelnut kernels were 1.45 mg/100 g, 0.28 mg/100 g, 0.05 mg/100 g, 0.5 mg/100 g, 2.45 mg/100 g, 0.043 mg/100 g, 3.25 mg/100 g, and 26.9 mg/100 g, respectively. The amount of the essential amino acids, mostly as arginine (2003 mg/100 g) and leucine (1150 mg/100 g), and the nonessential amino acids, mostly as glutamic acid (2714 mg/100 g) and aspartic acid (1493 mg/100 g) were also determined in the hazelnut varieties. Mineral compositions of the hazelnut varieties, for example, K, Mn, Mg, Ca, Fe, Zn, Na, and Cu were (averaged) measured as 863 mg/100 g, 186 mg/100 g, 173 mg/100 g, 5.6 mg/100 g, 4.2 mg/100 g, 2.9 mg/100 g, 2.6 mg/100 g, and 2.3 mg/100 g, respectively.

The quantitative and qualitative determinations of chemical composition (sugars, organic acids, and lipids) of 24 Italian and foreign hazelnut cultivars by Cristofori et al. (2008) revealed a good nutritional and health potential of the hazelnuts, with several differences among cultivars and production years. The total content of oil and sugars ranged from 563.69 to 656.36 g/kg dry weight and from 39.80 to 59.51 g/kg (DW), respectively. Fatty acid profile, sugar, and total phenolic contents varied with the production year. Significantly higher palmitic acid concentration (6.18%) was found in the hot summer of the year 2003; lower saturated fatty acid concentration (8.20%) and higher unsaturated/saturated acid ratio (11.27) were observed in the coolest year 2004. The main fatty acid, oleic acid (18:1), ranged between 78.10% and 84.76%. LA (18:2) showed pronounced differences among cultivars, the lowest content being 6.19% and the highest 14.0%. A negative relationship between oleic and linoleic acids was observed, as previously reported by Parcerisa et al. (1993). Linolenic acid (18:3) ranged between 0.03% and 0.09%. Saturated fatty acid content was less than 10% of the total in all cultivars, ranged between 7.50% and 9.58%. The predominant saturated fatty acid was palmitic acid (16:0), whose content ranged between 5.20% and 6.40%. Starch content ranged between 8.27 and 20.09 g/kg DW, showing significant differences among cultivars (Cristofori et al., 2008).

Oliveira et al. (2008) characterized three hazelnut cultivars (cv. Daviana, Fertille de Coutard and M. Bollwiller) produced in Portugal with respect to their chemical composition, AOP and antimicrobial activity. They reported that the main constituent of fruit is fat that ranges from 56% to 61%, giving a total caloric value around 650 kcal per 100 g of the fruit. Oleic was the major fatty acid varying between 80.67% and 82.63%, followed by linoleic, palmitic, and stearic acids.

Alasalvar et al. (2009a) recently reported that the different varieties of hazelnut served as an excellent source of copper and manganese. Consumption of the recommended daily amount of 42.5 g of hazelnut from different varieties provides 44.4–83.6% of copper and 40.1–44.8% of recommended manganese intake for adults. Alasalvar et al. (2003a) also reported that the total dietary fiber content of Tombul hazelnut was 12.88 g/100 g, of which 2.21 g/100 g was soluble fiber (fresh weight basis). Tombul hazelnut is also an excellent source of selenium (60 μg/100 g).

Alasalvar et al. (2008a) reported that hazelnut is a good source of both essential and nonessential amino acids. Glutamic acid is the most abundant (2.84–3.71 g/100 g) amino acid, followed by arginine (1.87–2.21 g/100 g) and aspartic acid (1.33–1.68 g/100 g). Although hazelnut protein contained all essential amino acids, lysine and tryptophan were the limiting amino acids (Alasalvar et al., 2003a). Amaral et al. (2005b) identified and quantified four phenolic acids, namely 3-caffeoylquinic, 5-caffeoylquinic, caffeoyltartaric, and *p*-coumaroyltartaric acids, in hazelnut leaves from 10 different cultivars grown in Portugal. Emberger et al. (1987) have identified (*E*)-5-methylhept-2-en-4-one (filbertone) as the flavor-impact component of hazelnuts, and its occurrence in raw and roasted fruits. The presence of these compounds in commercially available hazelnut cream has been reported (Guntert et al., 1991; Jauch et al., 1989; Schurig et al., 1990).

15.3.2 Phytochemicals

Phytochemicals are compounds found in plants that are not required for normal functioning of the body but nonetheless have a beneficial effect on health and play an active role in the amelioration of disease. Currently, there is much interest in phytochemicals because of the potential health benefits related to their substantial antioxidant and antiradical activities (Alasalvar et al., 2008). In addition to the macro- and micronutrients, a variety of phytochemicals (i.e., ellagic acid, phenolic compounds, luteolin, and tocotrienols) are present in nuts (Borchers, 2001). Recently, Chen and Blumberg (2008) in a review article reported that phytochemicals from nuts have been associated with numerous bioactivities known to affect the initiation and progression of several pathogenic processes. However, as complete phytochemical profiles are lacking for most nuts, additional studies are needed to characterize their content, bioavailability, metabolism, and elimination in humans.

Phillips et al. (2005) studied the phytosterol composition of nuts consumed in the United States and quantified six phytosterols. Among the 10 economically important nuts (hazelnut, almond, walnut, pistachio, pine nut, Brazil nut, cashew, pecan, macademia, and peanut), the total phytosterol content ranged from 95 to 279 mg/100 g of nut, the lowest being in the Brazil nut and the highest in the pistachio nut.

15.3.3 Flavonoids

15.3.3.1 Almonds

Fraison-Norrie and Sporns (2002) and Sang et al. (2002) have identified and quantified a range of flavonoids in almonds skins. Wijeratne et al. (2006b) revealed the presence of quercetin, isorhamnetin, quercitrin, kaempferol 3-O-rutinoside, isorhamnetin 3-O-glucoside, and morin as the major flavonoids in the extract of defatted almond whole seed, brown skin, and green shell cover. Fraison-Norrie and Sporns (2002) reported identification of four flavonol glycosides, isorhamnetin rutinoside, isorhamnetin glucoside, kaempferol rutinoside, and kaempferol glucoside from almond seed coats. Milbury et al. (2006) reported that the predominant flavonoids in California almond (*Prunus dulcis*) skins and kernels were isorhamnetin-3-O-rutinoside and isorhamnetin-3-O-glucoside (in combination), catechin, kaempferol-3-O-rutinoside, epicatechin, quercetin-3-O-galactoside, and isorhamnetin-3-O-galactoside.

15.3.3.2 Peanut

Procyanidins and catechins are some of the well-studied flavonoids in peanut so far. According to Nepote et al. (2002) limited studies suggest that peanut skin may contain potent procyanidin compounds. Catechins, B-type procyanidin dimmers, procyanidins trimers, tetramers, and oligomers with higher degrees of polymerization were also reported to be present in peanut skin (Lazarus et al., 1999). Recently, Yu et al. (2006) also reported that peanut skin is a rich source of highly active antioxidants including catechins and procyanidins. Peanut skin extract has higher total antioxidant activity (TAA) and free radical scavenging capacity than ascorbic acid solution at equivalent concentration (Yu et al., 2006). Therefore, peanut skin could provide an inexpensive source of natural antioxidants, such as catechins and procyanidin, for use in food and dietary supplement formulations.

Chukwumah et al. (2007) reported that the total flavonoid contents of raw peanuts with skin and commercially boiled peanuts were significantly higher than those of the raw peanuts without skin and those of all roasted peanuts. The higher flavonoid content of the raw (with skin) and boiled peanuts can be attributed to the presence of proanthocyanidins in the peanut skin. Previous studies on the content of peanut skin show that it is rich in proanthocyanidins (Yu et al., 2005, 2006). Lou et al. (1999, 2004) were able to isolate and characterize six proanthocyanidins from mature peanut skins. Karchesy and Hemingway (1986) estimated the procyanidin content of peanut skins to be 17% by weight, 50% of which were low molecular weight oligomers. It was also observed that processing did not affect the total flavonoid content of the peanuts. Thompson et al. (2006) reported that hazelnut contained 198.9 μg/100 g total isoflavones, primarily genistein and 77.1 μg/100 g total lignans, primarily secoisolariciresinol and 107.5 μg/100 g total phytoestrogens.

15.3.4 Polyphenols

Polyphenols are a group of chemical substances found in most plants and occur in nature in free or bound forms. Some of the processing methods such as boiling or heating have been shown to increase the polyphenolic content of foods (Arts and Hollman, 2005). Polyphenols as antioxidants are known to reduce the risk of CVDs and certain types of cancers (Ames, 1983; Gordon, 1996; Middleton, 1998; Pulido et al., 2000; Scalbert et al., 2005; Silva et al., 2004; Steinberg, 1992; Tseng et al., 1997). Great interest has recently been focused on the addition of polyphenols to foods and biological systems, due to their well-known ability to scavenge free radicals (Ningappa et al., 2008; Pinlo et al., 2004; Yadav and Bhatnagar, 2008).

15.3.4.1 Almonds

Almond hulls have been shown to serve as a rich source of triterpenoids, betulinic, urosolic, and oleanolic acids (Takeoka et al., 2000) as well as flavonol glycosides and phenolic acids (Sang et al., 2002c). In addition, Sang et al. (2002c) isolated catechin, protocatechuic acid, vanillic acid, *p*-hydroxybenzoic acid, and naringenin glucoside, as well as galactoside, glucoside, and rhamnoglucoside of 3′-O-methylquercetin and rhamnoglucoside of kaempferol from almonds.

Almond skins, resulting from the hot water blanching process, constitutes about 4% of the almond fruit, and are a readily available source of phenolics (Chen et al., 2005). Four different flavones glycosides: isorhamnetin rutinoside, isorhamnetin glucoside, kaempferol rutinoside, and kaempferol glucoside have been reported in almond seed coats (Fraison-Norrie and Sporns, 2002; Wijeratne et al., 2006a). Other investigators have likewise identified phenolic compounds in almond skins including quercetin glycosylated to glucose, galactose and rhamnose, kaempferol, naringenin, catechin, protocatechuic acid, vanillic acid, and a benzoic acid derivative (Chen et al., 2005; Sang et al., 2002).

High amounts of phenolics, mainly tannins, such as rhamnetin, quercetin, and kaempferol aglycones, have been reported in almond hulls, representing 4.5% of total hull weight (Cruess et al., 1947; Shahidi, 2002). Other phenolic compounds, such as chlorogenic and benzoic acid derivatives were also found in lower quantities (Shahidi, 2002; Takeoka and Dao, 2000). Wijeratne et al. (2006a) reported that extracts of defatted almond whole seed, brown skin, and green shell contained quercetin, isorhamnetin, quercitrin, kaempferol 3-O-rutinoside, isorhamnetin 3-O-glucoside, and morin as the major flavonoids in all extracts.

15.3.4.2 Walnut

The slightly astringent flavor of the walnut fruit has been associated with the presence of phenolic compounds (Colaric et al., 2005; Prasad, 2003). Most phenolic compounds commonly identified in walnuts are phenolic acids and condensed tannins.

Walnut phenolics are found in the highest concentration in the hull (the pellicle surrounds the kernel), and they are reported to have favorable effects on human health owing to their apparent antiatherogenic and antioxidant properties (Anderson et al., 2001; Fukuda et al., 2003; Gunduc and El, 2003; Horton et al., 1999; Lavedrine et al., 1999; Zambon et al., 2000). In spite of these beneficial effects, walnut phenolics may, however, adversely influence the protein solubility (Sze-Tao et al., 2001). Recently, Labuckas et al. (2008) reported that removal of phenolics by solvent extraction improved protein availability, yielding walnut flour with potential applications as a food ingredient.

In walnut leaves, naphthoquinones and flavonoids are considered as major phenolic compounds (Wichtl and Anton, 1999). Juglone (5-hydroxy-1,4-naphthoquinone) is known as being the characteristic compound of *Juglans* spp. and is reported to occur in fresh walnut leaves (Bruneton, 1993; Gîrzu et al., 1998; Solar et al., 2006; Wichtl and Anton, 1999). Several hydroxycinnamic acids (3-caffeoylquinic, 5-caffeoylquinic, *p*-coumaric, 3-*p*-coumaroylquinic and 4-*p*-coumarolquinic acids) and flavonoids (quercetin 3-galactoside, quercetin 3-arabinoside, quercetin-3-xyloside, quercetin 3-rhamnoside, and two other partially identified quercetin 3-pentoside and kaempferol 3-pentoside derivatives) of different walnut cultivars collected at different times have previously been reported (Amaral et al., 2004; Pereira et al., 2007). In addition, the existence of 5-caffeoylquinic acid has also been reported (Wichtl and Anton, 1999). Amaral et al. (2008) also reported that there was no difference in qualitative profiles of phenolic compounds from different cultivars although there was difference in terms of individual compound contents.

Li et al. (2006) reported that the major polyphenolics in walnut were ellagic acid and valoneic acid dilactone. These components were found to contribute to the strong total antioxidant activities measured using ferric reducing antioxidant power and photochemiluminescence methods. The 80% methanol extractable fractions of walnuts contained an average of 0.29 and 1.31 mg of ellagic acid/g nut as the free phenolic acid and acid-hydrolyzable phenolic acid. Fukuda et al. (2003) has reported 16 polyphenols in walnuts, to which, Ito et al. (2007) added 3 new ellagitannins, namely two novel dicarboxylic acid derivatives, glansreginins A (1) and B (2), and a new dimeric hydrolyzable tannin, glansrin D (3).

15.3.4.3 Peanut

So far, numerous polyphenolics and related compounds, such as luteolin (Duh et al., 1992), proanthocyanidins (Lou et al., 1999), resveratrol (Sanders et al., 2000), flavonoids (Lou et al., 2001), ethyl protocatechuate (EP) (Huang et al., 2003), *p*-coumaric acid and its esterified derivatives (Talcott et al., 2005) have been identified in peanuts. Seo and Morr (1985) identified six phenolics from defatted peanut meal amongst which *p*-coumaric acid was the predominant compound, accounting for 40–68% of the total phenolics present. Later on, up to 15 polyphenolics were identified in peanuts by Duke (1992). Fajardo et al. (1995) reported stress-induced synthesis of free and bound polyphenolics in peanuts with *p*-coumaric and ferulic acids present in the highest concentrations.

Yu et al. (2005) reported that three classes of polyphenols were found in peanut skins, including phenolic acids (caffeic acid, chlorogenic acid, ferulic acid, and coumaric acid), flavonoids (catechins and procyanidins), and stilbene (resveratrol). A study by Nepote et al. (2002) found that peanut skin contains about 150 mg of total polyphenols/g of defatted dry skin. Lou et al. (1999) identified 6 A-type procyanidins in peanut skin. These six compounds were found to inhibit the activity of hyaluronidase, an enzyme that is responsible for the release of histamine, which causes inflammation. In addition, resveratrol, a phytochemical found in grapeseed and wine, was also found in peanut skin and peanut kernels (Sanders et al., 2000). Chukwumah et al. (2007) reported that boiled peanuts had significantly higher total polyphenol content than raw and roasted peanuts.

15.3.4.4 Pistachio

Seeram et al. (2006) reported identification of quercetin, luteolin, eriodictyol, rutin, naringenin, apigenin, and the anthocyanins, cyaniding-3-galactoside and cyaniding-3-glucoside phenolics from pistachio skin. Gentile et al. (2007) isolated polyphenolic compounds *trans*-resveratrol, proanthocyanidins, and a remarkable amount of the isoflavones daidzein and genistein from the extract of *Pistachia vera* nuts.

15.3.4.5 Pine Nut

In pine residues, procyanidin oligomers are the predominant phenolics (Pietta et al., 1998; Wood et al., 2002). Nuts therefore constitute one of the most nutritionally concentrated kinds of food available. Most nuts, left in their shell, have a remarkably long shelf-life and can conveniently be stored for winter use.

15.3.4.6 Hazelnut

Hazelnuts have been defined as a good source of total phenolics with high AOP (Kornsteiner et al., 2006). Similar results are reported by Cristoferi et al. (2008) for the kernels of 24 Italian and foreign cultivars. The mean values for phenolics ranged from 1.57 g gallic acid equivalents (GAE)/kg DW to 6.32 g GAE/kg DW. Senter et al. (1983) reported that protocatechnic acid (0.36 mg/kg of hulls) is the predominant phenolic in hazelnut hulls. The levels of other phenolic acids (gallic acid, caffeic acid, vanillic acid, and *p*-hydroxybenzoic acid) do not exceed 10 μg/kg of testa.

Yurttas et al. (2000) isolated and tentatively identified six phenolic aglycones in Turkish and American hazelnuts: gallic acid, *p*-hydroxybenzoic acid, sinapic acid, quercetin and caffeic acid, and epicatechin. Shahidi et al. (2007) studied the total phenolic content of the hazelnut kernel and its by-products. Total phenolic content of hazelnut skin was the highest (577.7 mg catechin equivalents CE/g ethanolic extract) whereas that of hazelnut kernel was the lowest (13.7 mg catechin equivalents CE/g ethanolic extract). Hazelnut skin had ~7.4-fold higher total phenolics (426.7–502.3 mg GAE/g extract depending on the solvent) than that of hazelnut hard shell (56.6–72.2 mg GAE/g extract depending on the solvent) (Contini et al., 2008). They reported a high tannin content in hazelnut by-product extracts (skin and hard shell), and the total tannins alone represented nearly 60–65% of the total phenolic substances of the extracts. Alasalvar et al. (2006b) reported that 80% (*v*/*v*) ethanol extract of hazelnut had a significantly lower total phenolic content compared with those of extracts obtained using 80% (*v*/*v*) acetone.

15.4 EFFECT OF PROCESSING ON THE PHYTOCHEMICAL COMPOSITION OF NUTS

15.4.1 PEANUTS

Yu et al. (2005) in their study on the effect of processing on peanut skin phenolics using 80% ethanol as extraction solvent, reported a 39.5% increase in peanut skin phenolics after roasting. In another study, they found that processing (roasting) of peanut skin had a limited effect on procyanidins (Yu et al., 2006). Boiling had a significant effect on the phytochemical composition of peanuts compared with oil- and dry-roasting techniques (Chukwumak et al., 2007). Boiled peanuts had the highest total flavonoid and polyphenol content. The biochanin A and genistein content of boiled peanut extracts were two- and fourfolds higher than the raw nuts, respectively. *Trans*-resveratrol was detected only in the boiled peanuts (Chukwumak et al., 2007).

The higher flavonoid content of the raw (with skin) and boiled peanuts can be attributed to the presence of proanthocyanidins in the peanut skin. Lou et al. (1999, 2004) were able to isolate and characterize six proanthocyanidins from mature peanut skins. Karchesy and Hemingway (1986) estimated the procyanidin content of peanut skins to be 17% by weight, 50% of which were low-molecular-weight oligomers. It was also observed that processing did not affect the total flavonoid content of the peanuts. However, this was not the case for the total polyphenols. Boiled peanuts had a significantly higher total polyphenol content (36.4–38.6 mg gallic acid equivalents/g) than raw and roasted peanuts (20.1–28.8 mg gallic acid equivalents/g). These values are much higher than those reported earlier by Talcott et al. (2005). The significantly higher total polyphenolic content of boiled peanuts could be explained by the presence of polyphenolic compounds in peanut hulls. Boiled peanuts are processed without dehulling, and several studies have shown that peanut hulls are rich in polyphenolic compounds that increase with peanut maturity (Duh et al., 1992; Yen et al., 1993) giving rise to the high antioxidant capacity of peanut hull extracts.

The presence of vanillin in peanut hulls and kernels of boiled peanuts formed by the hydrolysis of lignin, a major constituent of the peanut hull, was established by Sobolev (2001). This suggests that during the boiling process peanut kernels absorb water that has permeated the hull, water-soluble polyphenols from the hulls are also absorbed by the kernels.

15.4.2 PISTACHIO

Gentile et al. (2007) reported that after roasting, except isoflavones that appeared unmodified, the amounts of other bioactive molecules present in the pistachio nut were remarkably reduced and the TAA was also decreased by about 60%. Seeram et al. (2006) investigated the effects of bleaching on phenolic levels and antioxidative capacities in raw and roasted pistachio nuts. Raw nuts preserved phenolic levels and antioxidant capacity better than roasted nuts, suggesting the contributing effects of other substances and/or matrix effects that are destroyed by the roasting process.

The destruction of bioactive phenolics in pistachio skins may negatively impact the potential health benefits arising from the consumption of roasted pistachio nuts.

15.4.3 Hazelnut

Koyuncu (2004) investigated the change of fat content and fatty acid compositions of three important Turkish hazelnut cultivars during storage at room conditions (21°C and 60–65% RH) and reported that during storage, the total fat, palmitic, and oleic acid content of the oil increased. No significant differences were found for other fatty acids during storage. The effect of storage on shelled and unshelled hazelnuts on their total fat content was significant. While the total saturated fatty acid content increased from 8.14% to 8.30%, the total unsaturated fatty acid content changed from 92.12% to 90.88% during the storage period.

15.5 ANTIOXIDATIVE FUNCTIONS OF NUTS

Nut consumption is associated with a protective effect against CHD, partly due to its high antioxidant content. *In vivo*, it is unclear whether diet-derived polyphenolics can indeed influence the atherogenic process, but it is thought that the AOP of plant-derived foods may be one factor in reducing cardiovascular risk (Fraser et al., 1992).

Blomhoff et al. (2006) identified several nuts among plant foods with the highest total antioxidant content. This suggests that the high antioxidant content of nuts may be the key to their cardioprotective effects. Miraliakbari and Shahidi (2008b) reported that the minor components of tree nut oil extracts possess antioxidant activity as determined by 2,2,azino-bis(3-ethylbenzthiazoline sulfonate) (ABTS) radical scavenging capacity, β-carotene bleaching test, oxygen radical absorbance capacity (ORAC) and photochemiluminescence inhibition assays. Peanuts also contribute significantly to dietary intake of antioxidants.

15.5.1 Almond

It has been reported that the phenolic compounds of almonds act as antioxidants by scavenging free radicals and chelating metal ions in foods (Heim et al., 2002). Antioxidant effect of the phenolic composition of almond skins has also been reported in model systems (Siriwardhana and Shahidi, 2002; Wijeratne et al., 2006a; Wijeratne et al., 2006b). Harrison and Were (2007) reported that γ-irradiation of almond skins extracts increased the yield of total phenolics as well as enhanced their antioxidant activity. The methanol extracts of phenolics obtained from almond hulls showed remarkable radical-scavenging activities (DPPH) and antioxidant capacity (Pinelo et al., 2004).

Extracts of whole almond seed, brown skin, and green shell cover possess potent free radical scavenging capacity (Siriwardhana and Shahidi, 2002). These activities may be related to the presence of flavonoids and other phenolic compounds in nuts. Phenolic compounds, protocatechuic and vanillic acid isolated from almond skins have very strong DPPH radical scavenging activity (Sang et al., 2002). Jia et al. (2006) and Li et al. (2007) reported that almond consumption can enhance antioxidant defenses and has preventive effects on oxidative stress and DNA damage caused by smoking.

Moure et al. (2007) reported that the phenolic fraction of the ethyl acetate extracts of almond shell hydrolysates was composed mainly of vanillic, syringic and *p*-coumaric acids. Vanillin was also found. Vanillic, syringic and *p*-coumaric acids significantly preserved LA and vegetable oils from oxidation (Bratt et al., 2003; Marinova and Yanishlieva, 1996). Oxidation of LDL, rat liver liposomes, rat liver microsomes and emulsions could also be protected by vanillic acid (Natella et al., 1999; Osawa et al., 1987), syringic acid (Anderson et al., 2001; Natella et al., 1999) and *p*-coumaric acid (Natella et al., 1999; Stupans et al., 2002). Vanillin protected efficiently against oxidation of lipids in extruded corn (Camire and Dougherty, 1998) and in cod liver oil (Fujioka and Shibamoto, 2005).

There have been numerous studies indicating that almond consumption is helpful in the prevention of atherosclerosis. Yanagisawa et al. (2006a) reported that almonds have DPPH radical scavenging ability, a prolonged lag time, and a suppression of lyso-PC production. This inhibitory effect against LDL oxidation was mostly derived from the almond skin, which contained a large amount of polyphenols. In addition, cell-mediated LDL oxidation was suppressed after the treatment of whole almonds.

Regular almond consumption may protect against the risk of CHD despite their high caloric and fat content. Yanagisawa et al. (2006b) reported that a modest quantity of almonds (56 g) in the diet each day for 4 weeks did not lead to an increase in the total cholesterol, LDL-cholesterol or apolipoprotein B levels, but led to a decrease in the malondialdehyde-LDL (MDA-LDL) levels. Hyson et al. (2002) compared the effects of whole almond versus almond oil consumption on the LDL oxidation in healthy men and women, and reported that both treatments improved lipid parameters but neither treatment affected *in vitro* LDL oxidation.

15.5.2 Walnut

Walnuts contain more than 20 mmol antioxidants/100 g, mostly in the walnut pellicles. Although the content of α-tocopherol, an antioxidant in walnut is lower than in other nuts, such as almonds, hazelnuts, and peanuts, among others. (Kagawa, 2001), walnut is readily preserved. This implies that the nut contains antioxidants inhibiting lipid autoxidation. Samaranayaka et al. (2008) reported that the phenolics extracted from different fractions of walnut showed marked antioxidant activities in different *in vitro* model systems. They also reported that the trolox equivalent antioxidant capacity (TEAC) and the 2,2′-DPPH radical scavenging ability (at ≥10 ppm GAE concentration) of crude phenolic extracts of whole walnut, skin and kernel fractions correlated well with their phenolic contents.

A walnut extract containing ellagic acid, gallic acid, and flavonoids was reported to inhibit the oxidation of human plasma and LDLs *in vitro* (Anderson et al., 2001). At high intake levels (2–3 servings/day), walnuts have been shown to raise levels of LA and ALA in plasma fatty acid (Abbey et al., 1994; Almario et al., 2001), LDL cholesterol ester fatty acids, LDL phospholipids and triglycerides (TG) (Zambon et al., 2000). In many of these studies plasma levels of oleic acid, palmitic acid and arachidonic acid decrease after the walnut intervention. In the Barcelona Walnut

Trial (Albert et al., 2002), even though LDL particles were enriched with PUFA from walnuts, their resistance to oxidation was preserved. Other intervention studies comparing a walnut-enriched, high PUFA diet with a walnut-free, lower PUFA diet showed no differences in LDL oxidation (Iwamoto et al., 2002; Munoz et al., 2001; Ros et al., 2004) or other measures of oxidative biomarkers (Ros et al., 2004; Tapsell et al., 2004) between diets.

Antioxidant effects of isolated polyphenols obtained from walnuts have previously been reported (Fukuda et al., 2003). They described the SOD-like activity and radical scavenging effect of 14 walnut polyphenols. Later on, they also reported the *in vivo* antioxidative effect of a polyphenol-rich walnut extract on the oxidative stress in mice with type-2 diabetes (Fukuda et al., 2004). Scavenging of hydroxyl radicals (HO) and superoxide radicals is documented for water and methanol extracts of the kernel of walnut (Ohsugi et al., 1999). Walnut liqueur, obtained with green walnuts, also presents antioxidant activity which was correlated with its polyphenolic composition (Alamprese et al., 2005).

Reiter et al. (2005) reported that the amount of melatonin present in walnuts (between 2.5 and 4.6 ng/g) triples the blood levels of melatonin on eating walnuts, which also increases antioxidant activity in the blood stream. The authors theorize that by helping the body resist oxidative stress (free radical damage), walnuts may help reduce the risk of cancer and delay or reduce the severity of CVD and neurodegenerative diseases such as Parkinson's or Alzheimer's disease.

Recently, Pereira et al. (2007) studied the antioxidant activity of different cultivars of walnut leaves by using reducing power assay, the scavenging effect on 2,2′-DPPH radicals and β-carotene linoleate model system, and reported that all studies on walnut leaves showed high antioxidant activity. Their results obtained with walnut leaf extracts indicated a concentration-dependent antioxidant capacity. They also reported that walnut leaves contain a considerable amount of quercetin heterosides. Quercetin, like other flavonoids is able to provide protection against chemically induced DNA damage in human lymphocytes and increase the total antioxidant capacity of plasma (Teippo et al., 2007; Wilms et al., 2005), increase genomic stability in cirrhotic rats, suggesting beneficial effects, probably through its antioxidant properties. The beneficial effects of almond phenolics on the protection of DNA and inhibition of human LDL oxidation have also been reported (Shahidi, 2002).

Almeida et al. (2008) reported that the ethanol–water extract from *Juglans regia* leaves showed a potent scavenging activity against reactive oxygen species (ROS) [such as, HO, superoxide radicals ($O_2^{\cdot-}$), peroxyl radical (ROO·) and hydrogen peroxide (H_2O_2)] and reactive nitrogen species (RNS) [such as, nitric oxide (·NO) and peroxynitrite anion ($ONOO^-$)] and can be used as an easily accessible source of natural antioxidants.

15.5.3 Peanut

In recent years, several investigations were conducted to study the antioxidant properties of peanut, peanut kernels, peanut hulls, and peanut-based products. The extraction and identification of antioxidant components from hulls (Duh et al., 1992), coats (Chang et al., 2002; Muamza et al., 1998), and peels (Larrauri et al., 1998) have been reported. Peanut kernel has been reported to contain antioxidant flavonoids, and dihydroquercetin (Pratt and Miller, 1984). It has been reported that the methanolic extracts from peanut hulls have both strong antioxidant activity (Duh and Yen, 1995) and properties of scavenging free radicals and ROS (Yen and Duh, 1994). The antioxidant component was identified as luteolin (Duh et al., 1992). In addition, peanut seed testa exhibited antioxidant activity and EP was isolated and identified from peanut seed testa (Huang et al., 2003). Hwang et al. (2001) reported that roasted and defatted peanut kernels showed remarkable antioxidative activity in LA emulsions. Nepote et al. (2004) reported that the extracts from peanut skins in honey-roasted peanuts inhibited lipid oxidation.

Yu et al. (2005) reported that the compounds found in peanut skin exhibited potent antioxidant activity, particularly flavonoids and resveratrol. The comparative study of total antioxidant activities of peanut skin extracts and green tea infusions demonstrated that peanut skin extracts had chemically higher AOP than green tea infusions (Yu et al., 2005). O'Keefe and Wang (2006) reported that the phenolic compounds extracted from peanut skins could significantly reduce the oxidation of meat products and extend their storage stability.

Huang et al. (2003) isolated and identified an antioxidant, EP, from peanut skin and showed that EP played an important role in preventing lipid oxidation, and contributed to the antioxidant activity of ethanolic extracts of peanut seed testa (EEPST). The antioxidant activity of EEPST and its antioxidative component, EP, was also examined by Yen et al. (2005) and showed a dose-dependent activity on the inhibition of liposome peroxidation. The inhibitory effect of EEPST in linoleic peroxidation correlated with their polyphenolic contents. EEPST and EP at 100 mg/L showed 92.6% and 84.6% scavenging effects, respectively, on α and α-diphenyl-β-picrylhydrazyl radicals, indicating that they act as primary antioxidants. In addition, at a dose of 200 mg/L, they showed 70.6% and 67.7% scavenging effect, respectively, for HO. These results suggest that the antioxidant mechanism, for both EEPST and EP, could possibly be due to their scavenging effect on free radicals and HO (Yen et al., 2005).

Numerous phytochemical compounds are present in peanuts with potential antioxidant capacity including polyphenolics (Talcott et al., 2005), tocopherols (Hashim et al., 1993), and proteins (Bland and Lax, 2000). Other than contributions from these compounds, mature peanut kernels are likely to possess a few other compounds in significant quantities that would impact antioxidant capacity (Duncan et al., 2006).

The compound *p*-coumaric acid alone has been shown to possess significant radical scavenging activities (Rice-Evans et al., 1996, 1997) but its contribution to the total antioxidant capacity in peanuts has not been reported. Peanuts contain about 25% protein by weight; Tolcott et al. (2005) reported changes in soluble proteins and amino acids following dry roasting that along with moisture loss and formation of roasting by-products may have contributed to increased antioxidant capacity of peanuts. Proteins or amino acids may act to physically trap free radicals or participate in Maillard browning reactions during roasting, resulting in the formation of newer antioxidant compounds (Borrelli et al., 2002; Ehling and Shibamoto, 2005). Yanagimoto et al. (2002) demonstrated that pyrazines formed during peanut roasting had no antioxidant activity, while other classes of Maillard derivatives, namely pyrroles and furans, exhibited minor antioxidant capacity. As demonstrated in a previous study, the antioxidant activity of intact peanuts increased during roasting, possibly from the formation of Maillard reaction derivatives (Talcott et al., 2005).

15.5.4 Pistachio

Goli et al. (2005) reported that pistachio hull extract possess antioxidant properties similar in activity to that of butylated hydroxyanisol (BHA) and butylated hydroxytoluene (BHT) (added at 0.02%), and could be used as alternative natural antioxidants. In addition to the improvement in HDL cholesterol and total cholesterol (TC) to HDL cholesterol ratio, pistachio has beneficial effect on LDL cholesterol oxidation by means of increasing serum antioxidant capacities, suggesting a potential role for pistachio on cardiovascular protection (Aksoy et al., 2007).

15.5.5 Pine Nut

Effects of bioflavonoids, extracted from the pine, on free radical formation have been investigated in murine macrophage cell lines, to have strong scavenging activities against ROS (Cho et al., 2000). Other studies demonstrate that RNS, generated with different kinetics and mechanisms, impair glutathione levels in endothelial cells (Rimbach et al., 1999). Ugartondo et al. (2007) studied the structure–activity–cytotoxicity relationships of polyphenolic fractions obtained from pine bark and reported high antioxidant capacity of the phenolic fractions in a concentration range that is not harmful to normal human cells.

A systematic screening of total antioxidants in dietary plants revealed that walnuts, almonds, and hazelnuts contain antioxidant activity (Halvorsen et al., 2002). The contents of antioxidants in walnuts were found to be much higher than other nuts.

15.5.6 Hazelnut

Shahidi et al. (2007) evaluated the antioxidant efficacies of ethanol extracts of defatted raw hazelnut kernel and hazelnut by-products (skin, hard shell, green leafy cover, and tree leaf) with various methods, such as by monitoring TAA and free-radical scavenging activity tests (hydrogen peroxide, superoxide, and DPPH radical), together with antioxidant activity in a β-carotene-linoleate model system, inhibition of oxidation of human LDL-cholesterol, and inhibition of strand breaking of supercoiled deoxyribonucleic acid (DNA). They reported that the extracts of hazelnut by-products (skin, hard shell, green leafy cover, and tree leaf) exhibited stronger activities than hazelnut kernel at all the concentrations tested. Among the samples tested, extracts of hazelnut skin showed superior antioxidative efficacy and higher phenolic content as compared with other extracts. Five phenolic acids (gallic acid, caffeic acid, *p*-coumaric acid, ferulic acid, and sinapic acid) were tentatively identified and quantified (Shahidi et al., 2007).

The antioxidant and antiradical activities in extracts of Turkish hazelnut kernel, hazelnut green leafy cover (Alasalvar et al., 2006b), and other hazelnut by-products such as hazelnut skin and tree leaf have also been reported by others (Alasalvar et al., 2009b; Oliveira et al., 2007). Oliveira et al. (2008) also reported that the aqueous hazelnut extract exhibited antioxidant activity in a concentration-dependent way for all cultivars they tested. Yurttas et al. (2000) reported that the nonhydrolyzed extracts of hazelnut phenolics exhibited greater antioxidant activities than corresponding hydrolyzed extracts.

15.6 CARDIOVASCULAR HEALTH BENEFITS OF NUT CONSUMPTION

15.6.1 HEART HEALTHY BENEFITS OF NUTS

Recently, in a review article, Kris-Etherton et al. (2008) have reported epidemiological and clinical trial evidence of consistent benefits of nut consumption on CHD and associated risk factors (Feldman, 2002; Fraser et al., 1992; Hu et al., 1998; Kris-Etherton et al., 2001; Kushi et al., 1996; Sabate, 1999). The effects of nuts intake on biomarkers of atherosclerotic CVD or on disease outcome were also evaluated in several controlled intervention studies with normo- or hyperlipidemic human subjects (Almario et al., 2001; Iwamoto et al., 2002; Kendall et al., 2002; Ros et al., 2004; Zambon et al., 2000). Numerous other studies have also shown that including nuts in the diet can reduce the risk of heart disease (Haumann, 1998; Higgs, 2002). Kelly and Sabate (2006) in a review article reported that consuming nuts at least 4 times a week showed a 37% reduced risk of CHD compared with those who never or seldom ate nuts. Each additional serving of nuts per week was associated with an average 8.3% reduced risk of CHD. Substantial reductions in total mortality are also observed on those frequently consuming nuts and peanut butter (Blomhoff et al., 2006).

The beneficial health effects of nuts are assumed to be mainly due to the less atherogenic plasma lipid profiles observed in such studies. However, emerging evidence indicates that nuts may be a source of health-promoting bioactive compounds that elicit cardioprotective effects. In particular, nuts include plant proteins, unsaturated fatty acids, dietary fiber, plant sterols, resveratrol, phytochemicals, and micronutrients like tocopherols (Kris-Etherton et al., 1999b; Kris-Etherton et al., 2001; Pennington, 2002; Sabate et al., 2000). Nuts probably have favorable effects on CVDs through several mechanisms. These effects may be mediated by their fatty acid profiles, phytochemicals, plant protein and fiber, micronutrients or antioxidant contents, or by a combination of these mechanisms. Several studies suggest that nut antioxidants have interesting biological effects that may be related to a favorable effect on CVDs (Blomhoff et al., 2006; Mukuddem-Petersen et al., 2005). Recently, Gebauer et al. (2008) reported that consumption of pistachio in a healthy diet affects CVD risk factors in a dose-dependent manner. The health effects of hazelnuts have been well documented in the literature (Alasalvar et al., 2009b; Mercanligil et al., 2007).

15.6.2 EFFECT OF NUTS ON BLOOD LIPID PROFILE

Many studies have shown that diets enriched in nuts favorably influence serum lipids and lipoproteins (Kris-Etherton et al., 1999b). The cholesterol reduction associated with nut consumption has been attributed to the replacement of saturated fat with MUFA because nuts are high in MUFA content. However, a review of several feeding trials demonstrated that the magnitude of cholesterol reduction found exceeded that predicted on the basis of inputting the changes in dietary fat consumption during nut consumption into equations relating dietary fat composition to plasma lipid levels (Kris-Etherton et al., 1999b).

Replacing half of the daily fat intake with nuts lowered total and LDL cholesterol levels significantly in humans far better than what was predicted according to their dietary fatty acid profiles (Abbey et al., 1994; Kendall et al., 2002; Kris-Etherton et al., 1999a). Epidemiological and intervention studies have shown that the frequent consumption of nuts is associated with reduced incidence of CVD by lowering serum LDL-cholesterol levels and reduces the risk of development of type-2 diabetes (Alper and Mattes, 2002, 2003; Fraser, 1999; Jiang et al., 2002; Kris-Etherton et al., 1999b). Indeed, a number of clinical studies have demonstrated that the addition of nuts to the habitual diet of both normo- and hypercholesterolemic subjects results in a significant reduction in plasma total and LDL cholesterol, whereas HDL either remained unchanged or increased (Morgan and Clayshulte, 2000; Zambon et al., 2000).

Kris-Etherton et al. (1999a) reviewed 18 feeding trials that used diets containing nuts and found that there was a 25% greater cholesterol-lowering response than predicted from equations for blood cholesterol, in response to changes in dietary fatty acids. They concluded that these results suggest that there are nonfatty acid constituents in nuts that may have additional cholesterol-lowering effects. Ros et al. (2004) reported that substituting walnuts for monounsaturated fat in a Mediterranean diet improves endothelium-dependent vasodilation in hypercholesterolemic subjects. This finding might explain the cardioprotective effect of nut intake beyond cholesterol lowering.

A systematic review of well-designed nut intervention studies estimated that consuming a moderate fat diet (approximately 35% of calories) including 1.5–3.5 servings (50–100 g) of nuts/day, especially almonds, peanuts, or walnuts, significantly lowered total cholesterol (216%) and LDL cholesterol (2–19%) levels in normo- and hyperlipidemic individuals compared with control diets without nuts or with a different fatty acid profile (Mukuddem-Petersen et al., 2005).

As reviewed by Griel and Kris-Etherton (2006), numerous controlled feeding trials have convincingly shown that the daily intake of manageable allowances of a variety of nuts for periods of 4–8 weeks has a clear cholesterol-lowering effect. However, the cholesterol-lowering effect observed after nut supplementation has often been higher than that predicted on the basis of the fatty acid profiles of the test diets (Griel and Kris-Etherton, 2006), indicating that nuts may contain other bioactive components capable of reducing blood cholesterol.

15.6.2.1 Almond

Apart from its nutritional value, almond is reported to have beneficial effects on blood cholesterol level and lipoprotein profile in humans (Spiller et al., 1998). According to Hyson et al. (2002) diets containing almond caused a significant reduction in plasma triacylglycerols, and total and LDL cholesterol with increased levels of HDL cholesterol in humans. A cholesterol-lowering effect of almonds compared with typical Western diets in healthy and hypercholesterolemic subjects was reported in two field trials (Abbey et al., 1994; Spiller et al., 1998) and one clinical trial (Jenkins et al., 2002). Total and LDL-cholesterol concentrations declined with progressively higher intakes of almonds, suggesting a dose–response relation (Sabate et al., 2003). The decrease in total and LDL cholesterol observed (Sabate et al., 2003) was greater than those estimated from the fatty acid composition of the diets with the use of predictive equations. Thus nonlipid components of almonds may play a role in lowering serum lipids. Almond, as a part of a dietary approach, was found to be as effective as the starting dose of cholesterol-lowering drugs such as statins in managing cholesterol (Jenkins et al., 2003).

15.6.2.2 Walnut

Consumption of walnuts has a favorable effect on human serum lipid profiles, with a decrease in total and LDL cholesterol as well as triacylglycerols (Abbey et al., 1994; Chisholm et al., 1998; Zambon et al., 2000) and an increase in HDL cholesterol and apolipoprotein A1 (Lavedrine et al., 1999). Experimental studies have shown improvements in the lipoprotein profiles of persons who consume diets high in walnuts (Chisholm et al., 1998; Sabate et al., 1993), almonds (Spiller et al., 1992; Spiller et al., 1998), and peanuts (Kris-Etherton et al., 1999a; O'Byrne et al., 1997).

Four of the studies undertaken on walnuts showed a reduction between 4% and 12% in total cholesterol and 8–16% in LDL cholesterol (Chisholm et al., 1998; Iwamoto et al., 2002; Sabate et al., 1993; Zambon et al., 2000). The effects on the HDL cholesterol were different. Although one study showed a reduction of 5% (Sabate et al., 1993), another detected an increase of 14% (Chisholm et al., 1998) and the other two studies produced no evidence of changes (Iwamoto et al., 2002; Zambon et al., 2000). As far as triacylglycerols are concerned, two studies did not show any variation (Chisholm et al., 1998; Sabate et al., 1993) and in the third study a reduction of 8% in the plasma triacylglycerols was observed (Zambon et al., 2000).

In a crossover trial with hypercholesterolemic patients, Zambon et al. (2000) showed that the mean total cholesterol level and the mean LDL cholesterol level decreased by 9.0% and 11.2%, respectively, when patients were subjected to walnut dietary interventions. Ros et al. (2004) showed that, in moderately hypercholesterolemic patients, walnuts significantly improved oxidative stress-related vascular endothelial function. Iwamoto et al. (2002) reported that in healthy individuals the consumption of walnut diet significantly improved the plasma lipid levels (i.e., total cholesterol and serum apolipoprotein B concentrations and the ratio of LDL cholesterol to HDL cholesterol decreased significantly), However, the LDL oxidizability was not influenced by these diets. These beneficial physiological effects suggest that bioactive compounds of nuts may possess lipid-altering activities due to additive/synergistic effects and/or interactions with each other.

15.6.2.3 Peanut

Alper and Mattes (2003) reported that the addition of peanuts or other MUFA-rich nuts to the diet significantly improved the blood lipid profile. A study conducted by O'Byrne et al. (1997) found that a low-fat diet with a high proportion of monounsaturated fats from high-oleic peanuts, reduced cholesterol levels in women. Another study found that subjects who consumed diets rich in monounsaturated fats, mainly from peanuts, experienced a larger decline in LDL-cholesterol as compared with the low-fat diet. Instead of raising triacylglycerol levels as on the low-fat diet, the peanut-enriched diet eaters lowered their triacylglycerol levels (Colquhoun et al., 1996). Emekli-Alturfan et al. (2007) reported that peanut consumption improved glutathione and HDL-cholesterol levels and decreased thiobarbituric acid reactive substances, without increasing other blood lipids in hyperlipidemia.

15.6.2.4 Pistachio

Edwards et al. (1999) reported that a substitution of 20% of daily fat calories with pistachio nuts as snacks for a consecutive 3-week period led to a significant decrease in total cholesterol, total cholesterol/HDL cholesterol, LDL/HDL cholesterol and increase in HDL cholesterol in subjects with hypercholesterolemia. While the fatty acid profile of the pistachio nut is desirable (low in saturated fatty acids, high in unsaturated fatty acids) and it may, in part, explain its positive effect on blood lipids, the high content of plant sterols may have an added effect.

The consumption of pistachios as a substitute for other sources of fat contributed to reducing total cholesterol by 2% and increasing HDL cholesterol by 12%, while no changes in LDL cholesterol or triacylglycerols were detected (Edwards et al., 1999). Kocyigit et al. (2006) reported that consumption of pistachio nuts in healthy volunteers decreased the levels of plasma total cholesterol, MDA, LDL/HDL ratios, and increased the HDL and AOP levels, and AOP/MDA ratios.

Sheridan et al. (2007) reported that a diet consisting of 15% of calories as pistachio nuts (about 2–3 ounces/day) over a 4-week period can favorably improve some lipid profiles (significantly reduction in TC/HDL-C, LDL-C/HDL-C, B-100/A-1, and increase in HDL-C) in subjects with moderate hypercholesterolemic condition, and thus may reduce the risk of coronary disease. In another study, Aksoy et al. (2007) reported that consumption of pistachio as 20% of daily caloric intake leads to a significant improvement in HDL and TC/HDL ratio and inhibits LDL cholesterol oxidation.

15.6.2.5 Hazelnut

Alphan et al. (1997) evaluated the effect of hazelnuts on blood lipids and lipoproteins in 19 individuals with type-2 diabetes. Individuals consumed a high-carbohydrate diet (60% carbohydrate, 25% total fat, 10% SFA, 10% MUFA, and 5% PUFA) for 30 days, followed by a 15-day washout period before consuming a hazelnut diet (40% carbohydrate, 45% total fat, 9% SFA, 27% MUFA, and 9% PUFA) for 30 days. LDL-cholesterol was significantly reduced following both the hazelnut (26%) and the high-carbohydrate (16%) diets, when compared with baseline values. While both diets reduced the total cholesterol, this reduction was only significant following the hazelnut diet (12%; 5% for high-carbohydrate diet). Compared with baseline, significant differences were observed in apolipoprotein B (+7%, –8%), HDL-cholesterol (+2%, +8%) and triglycerides (–12 %, –16%) following the high-carbohydrate and hazelnut diets, respectively.

Durak et al. (1999) reported that a supplement of 1 g of hazelnuts for each kilo of body weight added to the usual daily diet for a month led to an increase in HDL cholesterol and serum antioxidant activity and a reduction in total cholesterol, LDL cholesterol and triglycerides in healthy humans. Balkan et al. (2003) reported that hazelnut oil supplementation reduced lipid peroxide levels in plasma and apolipoprotein B, and ameliorates aortic atherosclerotic lesions, but did not alter plasma lipid levels in rabbits fed on a high-cholesterol diet. Hazelnut oil supplementation reduced plasma, liver, and aorta lipid peroxide levels and aorta cholesterol levels, without any decreased in the plasma and liver cholesterol levels (Hatipoglu et al., 2004).

Akgul et al. (2007) have reported that the dietary supplementation with hazelnuts significantly improved the levels of triglycerides, HDL cholesterol, and some lipoproteins, although statistically insignificant yet a decreasing trend for LDL and total cholesterol in 15 hypercholesterolemic men was observed. Compared with baseline, the hazelnut-enriched diet decreased the concentrations of VLDL-cholesterol, triacylglycerol, and apolipoprotein B by 29.5%, 31.8%, and 9.2%, respectively, while increasing HDL cholesterol concentrations by 12.6% (Mercanligil et al. 2007). This study demonstrated that a high-fat and high-MUFA-rich hazelnut diet was superior to a low-fat control diet because of favorable changes in plasma lipid profiles of hypercholesterolemic adult men and, thereby positively affecting the CHD risk profile.

15.7 HYPOGLYCEMIC ACTION OF NUTS

Nuts are good sources of unsaturated fatty acids, vegetable proteins, fiber and associated antioxidant flavonoids, which in a limited number of studies, have independently shown to have a number of effects including blunting the postprandial glucose rise, improving carbohydrate tolerance, and reducing risk factors for diabetic complications (Chandalia et al., 2000; Garg, 1998; Kaneto et al., 1999; Paolisso et al., 1993; Sarkkinen et al., 1996). Nuts are also associated with protection from the development of type-2 diabetes in large cohort studies (Salmeron et al., 1997a, b).

Jiang et al. (2002) reported that consuming a half-serving (1 tbsp) of peanut butter or a full serving of peanuts or other nuts (1 oz), five or more times a week, was associated with a 20% or 30% reduced risk of developing type-2 diabetes, respectively. Furthermore, the relationship between consuming peanut butter, peanuts, and other nuts and type-2 diabetes was linear, that is, higher consumption provided a greater protective effect.

Lovejoy et al. (2002) reported that almonds-enriched diets had beneficial effects on serum lipids in healthy adults and produced changes similar to high monounsaturated oils in diabetic patients, although it did not alter insulin sensitivity in healthy adults or glycemia in patients with diabetes. Gillen et al. (2005) reported that when walnuts are eaten as a part of a modified low-fat diet (about 1 ounce/day), the result is a more cardioprotective fat profile in diabetic patients than can be achieved by simply lowering the fat content of the diet.

It has been suggested that nuts may improve insulin sensitivity, partly because of their fiber and other micronutrients content. After reviewing the preliminary evidence from epidemiological studies Rajaram and Sabaté (2006) suggested that frequent nut intake might provide protection from the development of diabetes. More remains to be learnt about thc effects of nuts on postprandial glycemic and insulin response, glycemic control, and improvement of disease risk factors in subjects with prediabetes and diabetes. Recently, in a randomized crossover study, Jenkins et al. (2008a) reported that there was no differences in baseline or treatment values for fasting glucose, insulin, C-peptide, or insulin resistance as measured in hyperlipidemic subjects consuming whole almonds as snacks. However, their 24-h urinary C-peptide output as a marker of 24-h insulin secretion was significantly reduced in almond groups compared with the control, which in the longer term may help to explain the association of nut consumption with reduced CHD risk. Jenkins et al. (2008b) reported that there is justification to consider the inclusion of nuts in the diets of individuals with diabetes in view of their potential to reduce CHD risk, even though their ability to influence overall glycemic control remains to be established.

15.8 EFFECT OF NUTS ON BODY WEIGHT

Hu et al. (1998) concluded from the Nurses' Health Study that persons who consumed more nuts tended to lose weight, indicating that in practice, the energy contained in nuts can readily be balanced by reductions in other sources of energy or by increased physical activity. Finally, it is well known that vegetarians usually consume more nuts than nonvegetarians (3.7 servings/wk versus 2.1 for nonvegetarians) (Rajaram and Wien, 2001).

Fraser et al. (2002) found that incorporating 320 kcal from almonds into the daily diet of 81 free-living subjects for 6 months did not lead, on average, to any statistically or biologically significant changes in their body weight. The available cumulative data demonstrated that nut consumption among free-living people was not associated with higher BMI or increased body weight compared with non-nut consumers despite the fact that nuts are fat- and energy-dense foods (Sabate, 2003).

It has been reported that peanuts promote weight management when consumed as part of a moderate-fat diet as a result of its satiating effect (Jiang et al., 2002). A moderate-fat almond diet resulted in greater weight loss than a low-fat diet containing the same daily calories (Wein et al., 2003). Higgs (2005) also reported that nuts promote weight management when consumed as a part of moderate-fat diet as a result of their satiating effect.

A limited number of studies have specifically looked at the impact of nut consumption on body weight and body composition changes, and all concluded that daily nut consumption posed no risk of significant weight gain. In the most recent study, free-living subjects were instructed to incorporate moderate amounts of walnuts (25 g/day on average) into their regular diet for 6 months, but were given no other dietary advice (Sabate et al., 2005). After adjusting for energy differences between the control and walnut-supplemented diets, no significant changes in body weight or composition were observed. Although the walnut-supplemented diet resulted in a mean increase of 133 kcal/day, which theoretically should have led to a weight gain of 3.1 kg over 6 months, the average weight gain among all participants was only 0.4 kg. These subjects had maintained the same level of physical activity throughout the study. Therefore, it is likely that walnuts partially displaced certain other foods in their diet, perhaps due to the increased satiety levels, and/or affected their rate of energy expenditure.

Although nuts are known to provide a variety of cardioprotective benefits, many avoid them for fear of weight gain. In a prospective study, Bes-Rastrollo et al. (2007)

reported that such fears are groundless. In fact, people who ate nuts, at least twice a week, were much less likely to gain weight than those who almost never consumed nuts. As reviewed by Rajaram and Sabate (2006) and others (Garcia-Lorda et al., 2003; St-Onge, 2005), there is considerable scientific evidence strongly suggesting that frequent nut intake is not associated with weight gain. Recently, Mattes et al. (2008) suggested through a review of the literature pertaining to the association between nut consumption and energy balance that nuts may be included in the diet, in moderation, to enhance palatability and nutrient quality without posing a threat of weight gain.

15.9 OTHER HEALTH BENEFITS OF NUTS

Literature is full of the health benefits associated with the consumption of peanuts, such as weight control (Alper and Mattes, 2002), prevention against CVDs (Feldman, 1999), protection against Alzheimer's disease (Peanut Institute, 2002), and cancer inhibition (Awad et al., 2000). These benefits are mainly attributed to the fact that peanuts contain low levels of saturated fatty acids (Misra, 2004) and do not contain *trans*-fatty acids (Sanders, 2001), while at the same time being rich in mono- and PUFAs (Kris-Etherton et al., 1999b), micronutrients such as vitamin E, folate, minerals (potassium, magnesium, and zinc), fiber, and health-promoting phytochemicals, particularly resveratrol (Sanders et al., 2000; Sobolev and Cole, 1999).

Walnut (*Juglans regia* L.) fruits are highly nutritious foods and are used as a traditional remedy for treating cough, stomachache (Perry, 1980), and cancer in Asia and Europe (Duke, 1989). Walnut leaf has been widely used in folk medicine for the treatment of skin inflammations, hyperhidrosis, and ulcers, and for its antidiarrheic, antihelminthic, antiseptic, and astringent properties (Bruneton, 1999; Proenca da Cunha et al., 2003). Walnut leaves are considered a source of healthcare compounds, and have been intensively used in traditional medicine for treatment of venous insufficiency and hemorrhoidal symptomatology, and for their antidiarrheic, antihelminthic, depurative, and astringent properties (Bruneton, 1993; Van Hellemont, 1986; Wichtl and Anton, 1999). Keratolytic, antifungal, hypoglycemic, hypotensive, antiscrofulous, and sedative activities for walnut leaves have also been described (Gîrzu et al., 1998; Valnet, 1992).

The omega-3 fatty acid (ALA), found in walnuts, promotes bone health by helping to prevent excessive bone turnover, when consumption of foods rich in this omega-3 fatty acid results in a lower ratio of omega-6 to omega-3 fatty acids in the diet (Griel et al., 2007). Green walnuts, shells, kernels, seeds, bark, and leaves have been used in the pharmaceutical and cosmetic industries (Stampar et al., 2006).

Peanut skins were demonstrated to be free of compounds that are toxic to animals, but are rich in phenolics and potentially other health-promoting compounds, that can be extracted for use in food applications (Sobolev and Cole, 2003). Peanut skins have long been used as a traditional Chinese medicine for the treatment of chronic hemorrhage and bronchitis. Recently, it has been shown that water extracts from defatted peanut skins contain antioxidant compounds, and can effectively be used as an ingredient in food applications (Wang et al., 2007). The water-soluble extract of peanut skins containing proanthocyanidins and flavonoids suppressed protein glycation (Lou et al., 2001) and possessed substantial activity against hyaluronidase (Lou et al., 1999).

Inflammation is often a cause or effect of oxidative stress. A recent cross-sectional epidemiological study on the consumption of nuts and seeds found lower levels of the circulating inflammatory markers such as, C-reactive protein, interleukin-6, and fibrinogen with a higher nut consumption (Jiang et al., 2006).

In a specific food-related 59-country study on prostate cancer, researchers concluded that grains, cereals, and nuts were protective against prostate cancer (Hebert et al., 1998). Epidemiological studies have associated the frequency of nut consumption with reduced risk of cancers of the prostate (Mills et al., 1989; Jain et al., 1999) and colorectum (Jenab et al., 2004; Yen et al., 2006). In a review paper, Gonzalez and Salas-Salvado (2006) reported that despite the inconsistent results, a protective effect on cancer of the colon and rectum is possible. However, they suggested that more epidemiological studies based on reliable estimation of nut consumption are required to clarify the possible effects of nuts on cancer.

Various mechanisms have been proposed to explain the cardioprotective effects of nuts. Their unique fatty acid composition and favorable effect on serum lipids and lipoprotein levels is certainly one possibility. Another mechanism of action may involve their high antioxidant capacity due to the presence of several different phytochemicals. Still other constituents present in nuts including fiber, β-sitosterol, ellagic acid, L-arginine, α-tocopherol, vitamin B6, folate, potassium, magnesium, copper, or manganese may also play a role.

15.10 ANTIMICROBIAL ACTIVITY OF NUTS

Antimicrobial activity of walnut products, particularly its bark (Alkhawajah, 1997), and the specific compound juglone (Clark et al., 1990) has been reported. Recently, Pereira et al. (2007) screened the antimicrobial capacity of different cultivars of walnut leaves against Gram-positive (*Bacillus cereus*, *B. subtilis*, and *Staphylococcus aureus*) and Gram-negative (*Pseudomonas aeruginosa*, *Escherichia coli*, and *Klebsiella pneumoniae*) bacteria and fungi (*Candida albicans*, and *Cryptococcus neoformans*). Walnut leaves selectively inhibited the growth of Gram-positive bacteria, *B. cereus* being the most susceptible one (MIC, 0.1 mg/mL). Gram-negative bacteria and fungi were resistant to these extracts at 100 mg/mL.

Darmani et al. (2006) also reported the growth inhibition of various cariogenic bacteria (*Streptococcus mutans*, *Streptococcus salivarius*, *Lactobacillus casei*, and *Actinomyces viscosus*) by walnut aqueous extracts. Walnut may, therefore, be a good candidate for employment as antimicrobial agent against bacteria responsible for human gastrointestinal and respiratory tract infections. Chung et al. (2003) reported that peanut is one of the limited number of plant species that synthesize resveratrol, which is a phytoalexin (an antibiotic produced by a plant that is under attach) with antifungal activity.

Oliveira et al. (2008) reported a high antimicrobial activity against Gram-positive bacteria (MIC 0.1 mg/mL) by hazelnut extracts. Earlier, they had also reported antimicrobial activity of hazel leaves (Oliveira et al., 2007).

15.11 SUMMARY

Nuts not only contain various nutrients, but they are also high in a variety of helpful antioxidants, or phytochemicals that shield against the damaging effects of free radicals. Nuts are also a source of helpful biologically active components or phytochemicals found in plant foods. Some of the phytochemicals in nuts include flavonoids and phenolic compounds. This article discusses nuts as potential sources of antioxidants, and other bioactive compounds that could be useful to protect humans from coronary artery diseases by improving lipid profile and inhibiting lipid oxidation, as well as including some other aspects of the utilization of nuts. Sufficient scientific research data available now state the beneficial effects of nuts as being attributed to the type of fat, especially the low saturated fatty acids and a high contribution of unsaturated fatty acids that is strengthened by a group of bioactive components with antioxidant and cardiovascular protective properties.

This chapter has identified several constituents of nuts as being protective against cardiovascular and perhaps other chronic diseases, but it is not sufficiently specific to ascertain whether such protection is specific only to the antioxidant content of nuts. A great deal of further research is necessary, including clinical trials, to clarify the extent to which the antioxidants as well as other components of nuts may contribute to long-term health.

REFERENCES

Abbey M, Noakes M, Belling GB, Nestle PJ. Partial replacement of saturated fatty acids with almonds or walnuts lowers plasma cholesterol and low density-lipoprotein cholesterol. *Am. J. Clin. Nutr.* 59:995–999, 1994.

Aksoy N, Aksoy M, Bagci C, Gergerlioglu HS, Celik H, Herken E, Yaman A, Tarakcioglu M, Soydinc S, Sari I, Davutoglu V. Pistachio intake increases high density lipoprotein levels and inhibits low-density lipoprotein oxidation in rats. *Tohoku J. Exp. Med.* 212:43–48, 2007.

Alamprese C, Pompei C, Scaramuzzi. Characterization and antioxidant activity of nocino liquer. *Food Chem.* 90:495–502, 2005.

Alasalvar C, Amaral JS, Shahidi F. Functional lipid characteristics of Turkish Tombul Hazelnut (*Corylus avellana* L.) *J. Agric. Food Chem.* 54: 10177–10183, 2006a.

Alasalvar C, Karamac M, Amarowicz R, Shahidi F. Antioxidant and antiradical activities in extracts of hazelnut kernel (*Corylus avellana* L.) and hazelnut green leafy cover. *J. Agric. Food Chem.* 54(13): 4826–4832, 2006b.

Alasalvar C, Amaral JS, Satır G, Shahidi F. Lipid characteristics and essential minerals of native Turkish hazelnut varieties (*Corylus avellana* L.). *Food Chem.* 113(4): 919–25, 2009a.

Alasalvar C, Karamac M, Kosinska A, Rybarczyk A, Shahidi F, Amarowicz R. Antioxidant activity of hazelnut skin phenolics. *J. Agric. Food Chem.* 57(11): 4645–4650, 2009b.

Alasalvar C, Hoffman AM, Shahidi F. Antioxidant activities and phytochemicals in hazelnut (*Corylus avellana* L.) and hazelnut by products. In: *Tree Nuts: Composition, Phytochemicals, and Health Effects.* Alasalvar C, Shahidi F, eds. Nutriceutical Science & Technology Series, CRC Press, Taylor & Francis, Inc., Boca Raton, FL, pp. 215–236, 2008.

Alasalvar C, Shahidi F. Tree nuts: Composition, phytochemicals, and health effects: An overview. In: *Tree Nuts: Composition, Phytochemicals, and Health Effects.* Alasalvar C,

Shahidi F, eds. Nutriceutical Science & Technology Series, CRC Press, Taylor & Francis, Inc., Boca Raton, FL, pp. 1–10, 2008.

Alasalvar C, Shahidi F, Liyanapathirana CM, Ohshima T. Turkish Tombul hazelnut (*Corylus avellana* L.). 1. Compositional characteristics. *J. Agric. Food Chem.* 51: 3790–3796, 2003a.

Alasalvar C, Shahidi F, Ohshima T, Wanasundara U, Yurttas HC, Liyanapathirana CM, Rodrigues FB. Turkish Tombul hazelnut (*Corylus avellana* L.). 2. Lipid characteristics and oxidative stability. *J. Agric. Food Chem.* 51: 3797–3805, 2003b.

Albert CM, Gaziano JM, Willett WC, Manson JE. Nut consumption and decreased risk of sudden cardiac death in the Physicians Health Study. *Arch. Intern. Med.* 162:1382–1387, 2002.

Alkhawajah AM. Studies on the antimicrobial activity of *Juglans regia. Am. J. Chin. Med.* 25:175–180, 1997.

Almario RU, Vonghavaravat V, Wong R, Kasim-karakas SE. Effects of walnut consumption on plasma fatty acids and lipoproteins in combined hyperlipidemia. *Am. J. Clin. Nutr.* 74:72–79, 2001.

Almeida IF, Fernandes E, Lima JLFC, Costa PC, Bahia MF. Walnut (*Juglans regia*) leaf extracts are strong scavengers of pro-oxidant reactive species. *Food Chem.* 106:1014–1020, 2008.

Alper CM, Mattes RD. Effects of chronic peanut consumption on energy balance and hedonics. *Int. J. Obes.* 26:1129–1137, 2002.

Alper CM, Mattes RD. Peanut consumption improves indices of cardiovascular disease risk in healthy adults. *J. Am. Coll. Nutr.* 22(2):133–141, 2003.

Alphan E, Pala M, Ackurt F, Yilmaz T. Nutritional composition of hazelnuts and its effects on glucose and lipid metabolism. *Acta Hort.* (ISHS) 445: 305–310, 1997.

Amaral JS, Casal S, Alves MR, Seabra RM, Oliveira BPP. Tocopherol and tocotrienol content of hazelnut cultivars grown in Portugal. *J. Agric. Food Chem.* 54:1329–1336, 2006a.

Amaral JS, Casal S, Citova I, Santos A, Seabra RM, Oliveira BPP. Characterization of several hazelnut (*Corylus avellana* L.) cultivars based in chemical, fatty acid, and sterol composition. *Eur. Food Res. Technol.* 222: 274–280, 2006b.

Amaral JS, Casal S, Seabra RM, Oliveira BPP. Effects of roasting on hazelnut lipids. *J. Agric. Food Chem.* 54: 1315–1321, 2006c.

Amaral JS, Casal S, Torres D, Seabra RM, Oliveira BPP. Simultaneous determination of tocopherols and tocotrienols in hazelnuts by a normal phase liquid chromatographic method. *Anal. Sci.* 21: 1545–1548, 2005a.

Amaral JS, Cunha SC, Santos A, Alves MR, Seabra RM, Oliveira BPP. Influence of cultivar and environmental conditions on the triacylglycerol profile of hazelnut (*Corylus avellana* L.). *J. Agric. Food Chem.*54: 449–456, 2006d.

Amaral JS, Ferreres F, Andrade PB, Valentao P, Pinheiro C, Santos A, Seabra R. Phenolic profile of hazelnut (*Corylus avellana* L.) leaves cultivars grown in Portugal. *Nat. Prod. Res.* 19: 157–163, 2005b.

Amaral JS, Seabra RM, Andrade PB, Valentao P, Pereira JA, Ferreres F. Phenolic profile in the quality control of walnut (*Juglans regia* L.) leaves. *Food Chem.* 88:373–379, 2004.

Amaral JS, Valentao P, Andrade PB, Martins RC, Seabra RM. Do cultivar, geographical location and crop season influence phenolic profile of walnut leaves? *Molecules* 13:1321–1332, 2008.

Ames BM. Dietary carcinogens and anticarcinogens. Oxygen radicals and degenerative disease. *Science* 221:1256–1264, 1983.

Anderson KJ, Teuber SS, Gobeille A, Cremin P, Waterhouse AL, Steinber FM. Walnut polyphenolics inhibit *in vitro* human plasma and LDL oxidation. *J. Nutr.* 131:2837–2842, 2001.

Anon. *Composition of Foods: Nuts and Seed Products*. United States Department of Agriculture, Agricultural Handbook no. 8–12, Washington, DC, 1984.

Arts IC, Hollman PC. Polyphenols and disease risk in epidemiologic studies. *Am. J. Clin. Nutr.* 81(Suppl. 1):317S–325S, 2005.

Aruoma OI. Free radicals oxidative stress and antioxidants in human health and disease. *J. Am. Oil Chem. Soc.* 75:199–211, 1998.

Aslan M, Orhan I, Sener B. Comparison of the seed oils of *Pistachia vera* L. of different origins with respect to fatty acids. *Int. J. Food Sci. Technol.* 37:333–337, 2002.

Awad AB, Chan KC, Downie AC, Fink CS. Peanuts as a source of β-sitosterol, a sterol with anticancer properties. *Nutr. Cancer* 36:238–241, 2000.

Bada JC, León-Camacho M, Prieto M, Alonso L. Characterization of oils of hazelnuts from Asturias, Spain. *Eur. J. Lipid Sci. Technol.*106:294–300, 2004.

Balkan J, Hatipoglu A, Aykac-Toker G, Uysal M. Influence of hazelnut oil administration on peroxidation status of erythrocytes and apolopoprotein B-100 containing lipoproteins in rabbits fed on a high cholesterol diet. *J. Agric. Food Chem.* 51:3905–3909, 2003.

Bankole SA, Ogunsawa BM, Eseigbe DA. Aflatoxins in Nigerian dry roasted ground-nuts. *Food Chem.* 89:503–506, 2005.

Benitez-Sánchez PL, León-Camacho M, Aparicio R. A comprehensive study of hazelnut oil composition with comparisons to other vegetable oils, particularly olive oil. *Eur. Food Res. Technol.* 218:13–19, 2003.

Bes-Rastrollo M, Sabate J, Gomez-Gracia E, Alonso A, Martinez JA, Martinez-Gonzalez MA. Nut consumption and weight gain in a Mediterranean cohort: The SUN study. *Obesity* 15(1):107–116, 2007.

Bland JM, Lax AR. Isolation and and characterization of a peanut maturity-associated protein. *J. Agric. Food Chem.* 48:3275–3279, 2000.

Blomhoff R, Carlsen MH, Andersen LF, Jacobs Jr DR. Health benefits of nuts: Potential role of antioxidants. *Br. J. Nutr.* 96 (Suppl. 2):S52 – S60, 2006.

Borchers A. Review of the Scientific Literature on Phenolic Compounds and Phytosterols in Tree Nuts. Unpublished 28 page report for *International Tree Nut Council* (INC), 2001.

Borrelli RC, Visconti A, Mennella C, Anese M, Fogliano V. Chemical characterization and antioxidant properties of coffee malanoidins. *J. Agric. Food Chem.* 50:6527–6533, 2002.

Bratt K, Sunnerheim K, Bryngelsson S, Fagerlund A, Engman L, Anderson RE, Dimberg LH. Avenantramides in oats (*Avena sativa* L.) and structure–antioxidant activity relationships principles. *J. Agric. Food Chem.* 50:7022–7028, 2003.

Bruneton J. *Pharmacogosie, phytochimie, plantes medicinales*. Tec. & Doc., Lavoisier, Paris, p. 348, 1993.

Bruneton J. *Pharmacogosie, phytochimie, plantes medicinales*. Tec. & Doc., Lavoisier, Paris, pp. 418–419, 1999.

Camire ME, Dougherty MP. Added phenolic compounds enhance stability in extruded corn. *J. Food Sci.* 63:516–518, 1998.

Carley DH, Fletcher SM. An overview of world peanut markets. In: *Advances in Peanut Science*. Pattee HE. and Stalker TH. Eds., American Peanut Research and Education Society, Inc., Stillwater, OK, pp. 554–577, 1995.

Chandalia M, Garq A, Lutjohann D, von Berqmann K, Grundy SM, Brinkley LJ. Beneficial effects of high dietary fiber intake in patients with type 2 diabetes mellitus. *N. Engl. J. Med.* 342:1392–1398, 2000.

Chang LW, Yen Wj, Huang SC, Duh PD. Antioxidant activity of sesame coat. *Food Chem.* 78:347–354, 2002.

Chisholm A, Mann J, Skeaff M, Frampton C, Sutherland W, Duncan A, Tiszavari S. A diet rich in walnuts favorably influences plasma fatty acid profile in moderately hyperlipidemic subjects. *Eur. J. Clin. Nutr.* 52:12–16, 1998.

Chen CYO, Blumberg JB. Phytochemical composition of nuts. *Asia Pac. J. Clin. Nutr.* 17(S1):329–332, 2008.

Chen CYO, Milbury PE, Lapsley K, Blumberg JB. Flavonoids from almond skins are bioavailable and act synergistically with vitamins C and E to enhance hamster and human LDL resistance to oxidation. *J. Nutr.* 135(6):1366–1373, 2005.

Cho K, Yun C, Ion D, Cho Y, Rimbach G, Packer L, Chung A. Effect of bioflavonoids extracted from the bark of *Pinus maritime* on proinflamatory cytokine interleukin-1 production in lipopolysaccharide-stimulated RAW 264.7. *Toxicol. Appl. Pharmacol.* 168:64–71, 2000.

Chung M, Myoung RP, Chun JC, Yun SJ. Resveratrol accumulation and resveratrol synthase gene expression in response to abiotic stresses and hormones in peanut plants. *Plant Sci.* 164:103–109, 2003.

Chukwumah Y, Walker L, Vogler B, Verghese M. Changes in the phytochemical composition and profile of raw, boiled, and roasted peanuts. *J. Agric. Food Chem.* 55:9266–9973, 2007.

Clark AM, Jurgens TM, Hufford CD. Antimicrobial activity of juglone. *Phytotherapy Res.* 4:11–14, 1990.

Colaric M, Veberic R, Solar A, Hudina M, Stampar F. Phenolic acids, syringaldehyde, and juglone in fruits of different cultivars of *Juglans regia* L. *J. Agric. Food Chem.* 53:6390–6396, 2005.

Colquhoun DM, Hicks BJ, Somerset S, Hamil C. Comparison of the effects of a high-fat diet enriched with peanuts and a low-fat diet on blood lipid profiles. *Abstract: Am. Heart Assoc.* 733, 1996.

Contini M, Baccellon S, Massantini R, Anelli G. Extraction of natural antioxidants from hazelnuts (*Corylus avellana* L.) shell and skin wastes by long maceration at room temperature. *Food Chem.* 110: 659–669, 2008.

Crews C, Hough P, Godward J, Brereton P, Lees M, Guiet S, Winkelmann W. Study of the main constituents of some authentic hazelnut oils. *J. Agric. Food Chem.* 53: 4843–4852, 2005.

Cristofori V, Ferramondo S, Bertazza G, Bignami C. Nut and kernel traits and chemical composition of hazelnut (*Corylus avellana* L.) cultivars. *J. Sci. Food Agric.* 88(6): 1091–1098, 2008.

Cruess WV, Kilbuck JH, Hahl E. Utilization of almond hulls. *Fruit Products J.* 26:363–365, 1947.

Cruz J, Domínguez H, Parajó JC. Assessment on the production of antioxidants from winemaking waste solids. *J. Agric. Food. Chem.* 52:5612–5620, 2004.

Cruz J, Domínguez H, Parajó JC. Anti-oxidant activity of isolates from acid hydrolysates of *Eucalyptus globules* wood. *Food Chem.* 90:503–511, 2005.

Darmani H, Nusayr T, Al-Hiyasat AS. Effects of extracts of miswak and derum on proliferation of Balb/C 3T3 fibroblasts and viability of cariogenic bacteria. *Int. J. Dental Hyg.* 4:62–66, 2006.

Davidson A. *The Oxford Companion to Food.* Oxford University Press, Oxford, UK, 1999.

De Jong N, Plat J, Mensink RP. Metabolic effects of plant sterols and stanols. *J. Nutr. Biochem.* 4:422–427, 2003.

Duh PD, Yen DB, Yen GC. Extraction and identification of an antioxidative component from peanut hulls. *J. Am. Oil Chem. Soc.* 69:814–818, 1992.

Duh PD, Yen GC. Changes in antioxidant activity and components of methanolic extract of peanut hulls irradiated with ultraviolet light. *Food Chem.* 54:127–131, 1995.

Duke JA. *Handbook of Nuts.* CRC Press, London, UK, 1989.

Duke JA. *Handbook of Phytochemical Constituents of GRAS Herbs and other Economic Plants.* CRC Press, Boca Raton, FL, 1992.

Duncan CE, Gorbet DW, Talcott ST. Phytochemical content and antioxidant capacity of water-soluble isolates from peanuts (*Arachis hypogaea* L.). *Food Res. Int.* 39:898–904, 2006.

Durak I, Koksal I, Kacmaz M, Buyukkocak S, Cimen BM, Ozturk HS. Hazelnut supplementation enhances plasma antioxidant potential and lowers plasma cholesterol levels. *Clin. Chim. Acta* 284: 113–115, 1999.

Ebrahem KS, Richardson DG, Tetley RM, Mehlenbacher SA. Oil content, fatty acid composition, and vitamin E concentration of 17 hazelnut varieties, as compared to other types of nuts and oil seeds. *Acta Hortic.* 351: 685–692, 1994.

Edwards K, Kwaw I, Matud J, Kurtz I. Effect of pistachios on serum lipid levels in patients with moderate hypercholesterolemia. *J. Am. Coll. Nutr.* 18:229–232, 1999.

Ehling S, Shibamoto T. Correlation of acrylamide generation in thermally processed model systems of asparagines and glucose with color formation, amounts of pyrazines formed, and antioxidative properties of extracts. *J. Agric. Food Chem.* 53:4813–4819, 2005.

Emberger R, Kopsel M, Bruning J, Hopp R, Sand T. DOS 3 345 784 (27 June 1985), *Chem. Abstr.* 103, 140 657q, 1985; DOS 3 525 604 A1 (22 Jan 1987), *Chem. Abstr.* 106, 155 899 f, 1987.

Emekli-Alturfan E, Kasikci E, Yarat A. Peanuts improve blood glutathione, HDL-cholesterol level and change tissue factor activity in rats fed a high-cholesterol diet. *Eur. J. Nutr.* 46:476–482, 2007.

Erdogan V, Aygun A. Fatty acid composition and physical properties of Turkish tree hazelnuts. *Chem. Nat. Comp.* 41(4):378–381, 2005.

Fadel JG. Quantitative analyses of selected plant by-product feedstuffs, a global perspective. *Animal Feed Sci. Technol.* 79:255–268, 1999.

Fajardo JE, Waniska RD, Cuero RG, Pettit RE. Phenolic compounds in peanut seeds: Enhanced elicitation by chitosan and effects on growth and aflatoxin B1 production by *Aspergillus flavus*. *Food Biotech.* 9:59–78, 1995.

Feldman EB. Assorted monounsaturated fatty acids promote healthy hearts. *Am. J. Clin. Nutr.* 70:953–954, 1999.

Feldman EB. The scientific evidence for a beneficial health relationship between walnuts and coronary heart disease. *J. Nutr.* 132(5):1062S–1101S, 2002.

Firestone D. Physical and chemical characteristics of oils, fats and waxes. In: *Champaign.* AOCS Press, London, 1999.

Food and Agriculture Organization (FAO). 2004. *FAO Stat Book*, web site: http://faostat.fao.org.

Fraison-Norrie S, Sporns P. Identification and quantification of flavonol glycosides in almond seed coats using MALDI-TOF MS. *J. Agric. Food Chem.* 50:2782–2787, 2002.

Fraser GE. Nut consumption, lipids and risk of a coronary event. *Clin. Cardiol.* 22 (Suppl. III): III-11–III-15, 1999.

Fraser GE, Jaceldo K, Bennett H, Sabate J. Changes in body weight with daily supplement of 340 calories from almonds for six months. *J. Am. Coll. Nutr.* 21:275–283, 2002.

Fraser GE, Sabate J, Beeson WL, Strahan TM. A possible protective effect of nut consumption on risk of coronary heart disease. The Adventist Health Study. *Arch. Intern. Med.* 152:1416–1424, 1992.

Fraser GE, Shavik DJ. Ten years of life: Is it a matter of choice? *Arch. Intern. Med.* 161:1645–1652, 2001.

Freeman HA, Nigam SN, Kelley TG, Ntare BR, Subrahmanyam P, Boughton D. *The World Groundnut Economy: Facts, Trends and Outlook.* International Crops Research Institute for Semi-arid Tropics, Andhra Pradesh, India, p. 52, 1999.

Fujioka K, Shibamoto T. Improved malonaldehyde assay using headspace solid-phase microextraction and its application to the measurement of the antioxidant activity of phytochemicals. *J. Agric. Food Chem.* 53:4708–4713, 2005.

Fukuda T. Walnut polyphenols: Structures and functions. In: *Tree Nuts: Composition, Phytochemicals, and Health Effects.* Alasalvar C, Shahidi F, eds. Chapter 19. Nutriceutical Science & Technology Series, CRC Press, Taylor & Francis, Inc., pp. 305–320, 2008.

Fukuda T, Ito H, Yoshida T. Antioxidative polyphenols from walnut (*Juglans regia* L.). *Phytochemistry* 63:795–801, 2003.

Fukuda T, Ito H, Yoshida T. Effect of the walnut polyphenol fraction on oxidative stress in type 2 diabetes mice. *BioFactors*, 21:251–253, 2004.

Garcia-Lorda P, Megias RI, Salas-Salvado J. Nut consumption, body weight and insulin resistance. *Eur. J. Clin. Nutr.* 57:S8–S11, 2003.

Garg A. High monounsaturated fat diets for patients with diabetes mellitus: A meta-analysis. *Am. J. Clin. Nutr.* 67(Suppl.):577S–588S, 1998.

Garrote G, Cruz J, Domínguez H, Parajó JC. Valorisation of waste fraction from autohydrolysis of selected lignocellullosic materials. *J. Chem. Technol. Biotechnol.* 78:392–398, 2003.

Gentile C, Tesoriere L, Butera D, Fazzari M, Monastero M, Allegra M, Livrea MA. Antioxidant activity of Sicilan pistachio (*Pistachia vera* L. var. Bronte) nut extract and its bioactive components. *J. Agric. Food Chem.* 55(3):643–648, 2007.

Gebauer SK, West SG, Kay CD, Alaupovic P, Bagshaw D, Kris-Etherton PM. Effects of pistachios on cardiovascular disease risk factors and potential mechanisms of action: A dose-response study. *Am. J. Clin. Nutr.* 88:651–658, 2008.

Gillen LJ, Tapsell LC, Patch CS, Owen A, Batterham M. Structured dietary advice incorporating walnuts achieves optimal fat and energy balance in patients with type 2 diabetes mellitus. *J. Am. Diet. Assoc.* 105:1087–1096, 2005.

Gîrzu M, Carnat A, Privat AM, Fiaplip J, Carnat AP, Lamaison JL. Sedative effect of walnut leaf extract and juglone, and isolated constituent. *Pharmaceut. Biol.* 36:208–286, 1998.

Goli AH, Barzegar M, Sahari MA. Antioxidant activity and total phenolic compounds of pistachio (*Pistachia vera*) hull extracts. *Food Chem.* 92:521–525, 2005.

Gonzalez CA, Salas-Salvado J. The potential of nuts in the prevention of cancer. *Br. J. Nutr.* 96:S87–S94, 2006.

Gonzalez J, Cruz JM, Dominguez H, Parajo JC. Production of antioxidants from *Eucalyptus globulus* wood by solvent extraction of hemicellulose hydrolysates. *Food Chem.* 84:243–251, 2004.

Gordon MH. Dietary antioxidants in disease prevention. *Natural Prod. Rep.* 13(4):265–273, 1996.

Greve C, McGranahan G, Hasey J, Snyder R, Kelly K, Goldhamerer D, Labavitch J. Variation in polyunsaturated fatty acid composition of Persian walnuts. *J. Soc. Hort. Sci.* 117:518–522, 1992.

Griel AE, Kris-Etherton PM. Tree nuts and the lipid profile: A review of clinical studies. *Br. J. Nutr.* 96:S68–S78, 2006.

Griel AE, Kris-Etherton PM, Hilpert KF, Zhao G, West SG, Corwin RL. An increase in dietary n-3 fatty acids decreases a marker of bone resorption in humans. *Nutr. J.* 6:2, 2007.

Gunduc N, El SN. Assessing antioxidant activities of phenolic compounds of common Turkish food and drinks on *in vitro* low-density lipoprotein oxidation. *J. Food Sci.* 68:2591–2595, 2003.

Guntert M, Emberger R, Hopp R, Kopsel M, Silberzahn W, Werkhoff P. Chirospecific analysis in flavor and essential oil chemistry. Part A. Filbertone—the character impact compound of hazel-nuts. *Z. Lebensm. Unter. Forsch.* 192: 108–110, 1991.

Halvorsen BL, Holte K, Myhrstad MC, Barikmo I, Hyattum E, Remberg SF, Wold AB, et al. A systematic screening of total antioxidants in dietary plants. *J. Nutr.* 132:461–471, 2002.

Harrison K, Were LM. Effect of gamma irradiation on total phenolic content yield and antioxidant capacity of almond skin extracts. *Food Chem.* 102:932–937, 2007.

Hashim I, Koehler PE, Eitenmiller RR. Tocopherols in runner and Virginia peanut cultivars at various maturity stages. *J. Am. Oil Chem. Soc.* 70:633–635, 1993.

Hatipoglu A, Kanbagli O, Balkan J, Kucuk M, Cevikbas U, Aykac-Toker G, Berkkan H, Uysal M. Hazelnut oil administration reduces aortic cholesterol accumulation and lipid peroxides in the plasma, liver, and aorta of rabbits fed on a high cholesterol diet. *Biosci. Biotechnol. Biochem.* 68(10): 2050–2057, 2004.

Haumann BF. Peanuts finds niche in healthy diet. *Information* 9:746–752, 1998.
Hebert JR, Hurley TG, Olendzki BC, Teas J, Ma Y, Hampl JS. Nutritional and socioeconomic factors in relation to prostate cancer mortality: A cross-national study. *J. Natl. Cancer Inst.* 90:1637–1647, 1998.
Heim KE, Taggliaferro AR, Bobilya DJ. Flavonoid antioxidants: Chemistry, metabolism and structure—activity relationships. *J. Biochem.* 13:575–584, 2002.
Higgs J. The beneficial role of peanuts in the diet—an update and rethink! Peanuts and their role in CHD. *Nutr. Food Sci.* 32:214–218, 2002.
Higgs J. The beneficial role of peanuts in the diet—Part 2. *Nutr. Food Sci.* 33:56–64, 2003.
Higgs J. The potential role of peanuts in the prevention of obesity. *Nutr. Food Sci.* 35:353–358, 2005.
Horton KL, Morgan JM, Uhrin LE, Boyle MR, Altomare P, Laskowsky C, Walker KE, Stanton MM, Newman LM, Capuzzi DM. The effect of walnut on serum lipids consumed as part of the national cholesterol educational panel step I diet. *J. Am. Diet. Assoc.* 99(9):A09–A112, 1999.
Hu FB, Stampfer MJ, Manson JE, Rimm JE, Colditz GA, Rosner BA, Speizer FE, Hennekens CH, Willett WC. Frequent nut consumption and risk of coronary heart disease in women: Prospective cohort study. *Br. Med. J.* 317:1341–1345, 1998.
Hu FB, Stampfer MJ. Nut consumption and risk of coronary heart disease: A review of epidemiologic evidence. *Curr. Atheroscler. Rep.* 1:204–209, 1999.
Huang SC, Yen GC, Chang LW, Yen WJ, Duh PD. Identification of an antioxidant, ethyl protocatechuate, in peanut seed testa. *J. Agric. Food Chem.* 51(8):2380–2383, 2003.
Hwang JY, Shue YS, Chang HM. Antioxidative activity of roasted and defatted peanut kernels. *Food Res. Int.* 34:639–647, 2001.
Hyson DA, Schneeman BO, Davis PA. Almonds and almond oil have similar effects on plasma lipids and LDL oxidation in healthy men and women. *J. Nutr.* 132:703–707, 2002.
Ibern-Gomez M, Roig-Perez S, Lamuela-Raventos RM, Carmen de la Torre-Boronat M. Resveratrol and piceid levels in natural and blended peanut butters. *J. Agric. Food Chem.* 48:6352–6354, 2000.
International Tree Nut Council (INC). Official Response to WHO/FAO Expert Consultation on Diet, Nutrition and the Prevention of Chronic Diseases, Geneva, Rome, June 15, 2002.
Isanga J, Zhang G. Biologically active components and nutraceuticals in peanuts and related products: Review. *Food Rev. Intl.* 23:123–140, 2007.
Ito H, Okuda T, Fukuda T, Hatano T, Yoshida T. Two novel dicarboxylic acid derivatives and a new dimeric hydrolysable tannin from walnuts. *J. Agric. Food Chem.* 55(3):672–79, 2007.
Iwamoto M, Imaizumi K, Sato M, Hirooka Y, Sakai K, Takeshita A, Kono M. Serum lipid profiles in Japanese women and men during consumption of walnuts. *Eur. J. Clin. Nutr.* 56:629–637, 2002.
Jain MG, Hislop GT, Howe GR, Ghadirian P. Plant foods, antioxidants, and prostate cancer risk: Findings from case-control studies in Canada. *Nutr. Cancer* 34:173–184, 1999.
Jauch J, Schmalzing D, Schurig V, Emberger R, Hopp R, Kopsel M, Silberzahn W, Werkhoff P. Isolation, synthesis, and absolute configuration of filbertone—the principal flavor component of the hazelnut. *Angew. Chem., Int. Ed. Engl.* 28: 1022–1023, 1989.
Jenab M, Ferrari P, Slimani N, Norat T, Casagrande C, Overad K, et al., Association of nut and seed intake with colorectal cancer risk in the European Prospective Investigation into Cancer and Nutrition. *Cancer Epidemiol. Biomarkers Prev.* 13:1595–1603, 2004.
Jenab M, et al. (40 author's name). Consumption of portion sizes of tree nuts, peanuts and seeds in the European Prospective Investigation into Cancer and Nutrition (EPIC) cohorts from 10 European Countries. *Br. J. Nutr.* 96 (Suppl. 2): S12–S23, 2006.
Jenkins DJA, Kendall CW, Marchie A, Josse AR, Nguyen TH, Faulkner DA, Lepsley KG, Singer W. Effect of almonds on insulin secretion and insulin resistance in nondiabetic

hyperlipidemic subjects: A randomized controlled crossover trial. *Metabolism* 57:882–887, 2008a.

Jenkins DJA, Hu FB, Tapsell LC, Josse AR, Kendall CWC. Possible benefit of nuts in type 2 diabetes. *J. Nutr.* 138:1752S–1756S, 2008b.

Jenkins DJA, Kendall CWC, Marchie A, Parker TL, Connelly PW, Qian W, Haight JS, Faulkner D, Vidgen E, Lapsley KG, Spiller GA. Dose response of almonds on coronary heart disease risk factors: Blood lipids, oxidized low-density lipoproteins, lipoprotein(a), homocysteine, and pulmonary nitric oxide: A randomized, controlled, crossover trial. *Circulation* 106:1327–1332, 2002.

Jenkins DJA, Kendall CWC, Marchie A, Faulkner DA, Wong JMW, de Souza R, Emam A, Parker TL, Vidgen E, Lapsley KG, Yrautwein EA, Josse RG, Leiter LA, Connelly PW. Effects of a dietary portfolio of cholesterol-lowering foods vs. Lovastatin on serum lipids and C-reactive protein. *J. Am. Med. Assoc.* 290:502–510, 2003.

Jia X, Li N, Zhang W, Zhang X, Lapsley K, Huang G, Blumberg J, Ma G, Chen J. A pilot study on the effects of almond consumption on DNA damage and oxidative stress in smokers. *Nutr. Cancer* 54(2):179–183, 2006.

Jiang R, Jacobs DR, Mayer-Davis E, Szklo M. Herrington D, Jenny N, Kronmal R, Barr RG. Nut and seed consumption and inflammatory markers in the Multi-Ethnic study of atherosclerosis. *Am. J. Epidemiol.* 163:222–231, 2006.

Jiang R, Manson JE, Stampfer MJ, Liu S, Willett WC, Hu FB. Nut and peanut butter consumption and risk of Type 2 diabetes in women. *J. Am. Med. Assoc.* 288:2554–2560, 2002.

Kagawa Y. *Standard Tables of Food Composition in Japan*. Kagawa Nutrition University Press, Tokyo, 2001.

Kaneto H, Kajimoto Y, Miyagawa J, Matsuoka T, Fujitani Y, Umayahara Y, Hanafusa T, Matsuzawa Y, Yamasaki Y, Hori M. Beneficial effects of antioxidants in diabetes: Possible protection of pancreatic beta-cells against glucose toxicity. *Diabetes* 48:2398–2406, 1999.

Karabulut I, Topcu A, Yorulmaz A, Tekin A, Ozay SD. Effects of the industrial refining process on some properties of hazelnut oil. *Eur. J. Lipid Sci. Technol.* 107: 476–480, 2005.

Karchesy JJ, Hemingway RW. Condensed tannins 4β → 8; 2 → O → 7-linked procyanidins in *Arachis hypogaea* L. *J. Agric. Food Chem.* 34:966–970, 1986.

Kelly JH Jr., Sabate J. Nuts and coronary heart disease: An epidemiological perspective. *Br. J. Nutr.* 96 (Suppl. 2):S61–S67, 2006.

Kendall CWC, Jenkins DJA, Marchie A, Parker T, Connelly PW. Dose response to almonds in hyperlipidemia: A randomized controlled cross-over trial. *Am. J. Clin. Nutr.* 75(2S):384–388, 2002.

King JC, Blumberg J, Ingwersen L, Jenab M, Tucker KL. Tree nuts and peanuts as components of healthy diet. *J. Nutr.* 138:1736S–1740S, 2008.

Knorr T. Candle and soap making—Almond oil. Web site: http://en.allexperts.com 2004.

Kocyigit A, Koylu AA, Keles H. Effects of Pistachio nuts consumption on plasma lipid profile and oxidative status in healthy volunteers. *Nutr. Metab. Cardiovasc. Dis.* 16(3):202–209, 2006.

Koksal AI, Artik N, Simsek A, Gunes N. Nutrient composition of hazelnut (*Corylus avellana* L.) varieties cultivated in Turkey. *Food Chem.* 99(3):509–515, 2006.

Kornsteiner M, Wagner KH, Elmadfa I. Tocopherols and total phenolics in 10 different nut types. *Food Chem.* 98:381–387, 2006.

Koyuncu MA. Change of fat content and fatty acid composition of Turkish hazelnuts (*Corylus avellana* L.) during storage. *J. Food Quality* 27(4): 304–309, 2004.

Kris-Etherton PM, Hu FB, Ros E, Sabate J. The role of tree nuts and peanuts in the prevention of coronary heart disease: Multiple potential mechanisms. *J. Nutr.* 138:1746S–1751S, 2008.

Kris-Etherton PM, Pearson TA, Wan Y, Hargrove RL, Moriarty K, Fishell V. High monounsaturated fatty acid diets lower both plasma cholesterol and triacylglycerol concentration. *Am. J. Clin. Nutr.* 70:1009–1015, 1999a.

Kris-Etherton PM, Yu-Poth S, Sabate J, Ratcliffe HE, Zhao G, Etherton TD. Nuts and their bioactive constituents: Effects on serum lipids and other factors that affect disease risk. *Am. J. Clin. Nutr.* 70 (Suppl. 3):504S–511S, 1999b.

Kris-Etherton PM, Zhao G, Binkoski AE, Coval SM, Etherton TD. The effects of nuts on coronary heart disease risk. *Nutr. Rev.* 59:103–111, 2001.

Kushi LH, Folsom AR, Prineas RJ, Mink PJ, Wu Y, Bostick RM. Dietary antioxidant vitamins and death from coronary heart disease in women: Prospective cohort study. *Br. Med. J.* 317:1341–1345, 1996.

Labuckas DO, Maestri DM, Perello M, Martinez ML, Lamarque AL. Phenolics from walnut (*Juglans regia* L.) kernels: Antioxidant activity and interactions with proteins. *Food Chem.* 107:607–612, 2008.

Larrauri JA, Ruperez P, Saura-Calixto F. Effect of drying temperature on the stability of polyphenols and antioxidant activity of red grape pomace peels. *J. Agric. Food Chem.* 46:4842–4845, 1998.

Lavedrine F, Ravel A, Villet A, Ducros V, Alary J. Mineral composition of two walnut cultivars originating in France and California. *Food Chem.* 68:347–351, 2000.

Lavedrine F, Zmirou D, Ravel A, Balducci F, Alary J. Blood cholesterol and walnut consumption: A cross-sectional survey in France. *Preventive Med.* 28:333–339, 1999.

Lazarus SA, Adamson GE, Hammerstone JF, Schmitz HH. High-performance chromatography/mass spectrometry analysis of proanthocyanidins in foods and beverages. *J. Agric. Food Chem.* 47:3693–3701, 1999.

Li L, Tsao R, Yang R, Liu C, Zhu H, Young JC. Polyphenolic profiles and antioxidant activities of heartnut (*Juglans ailanthifolia* Var. cordiformis) and Persian walnut (*Juglans regia* L.). *J. Agric. Food Chem.* 54(21):8033–8040, 2006.

Li N, Jia X, Chen CY, Blumberg JB, Song Y, Zhang W, Zhang X, Ma G, Chen J. Almond consumption reduces oxidative DNA damage and lipid peroxidative in male smokers. *J. Nutr.* 137:2717–2722, 2007.

Liggins J, Bluck LJC, Runswick S, Atkinson C, Coward WA, Bingham SA. Daidzein and genistein content of fruits and nuts. *J. Nutr. Biochem.* 11:326–331, 2000.

Lou H, Yuan H, Ma B, Ren D, Ji M, Oka S. Polyphenols from peanut skins and their free radical-scavenging effects. *Phytochemistry* 65:2391–2399, 2004.

Lou H, Yuan H, Yamazaki Y, Sasaki T, Oka S. Alkaloids and flavonoids from peanut skins. *Planta Med.* 67:345–349, 2001.

Lou HX, Yamazaki Y, Sasaki T, Uchida M, Tanaka H. A-type procyanidins from peanuts skin. *Phytochemistry* 51:297–308, 1999.

Lovejoy JC, Most MM, Lefevre M, Greenway FL, Rood JC. Effect of diet enriched in almonds on insulin action and serum lipids in adults with normal glucose tolerance or type 2 diabetes. *Am. J. Clin. Nutr.* 76:1000–1006, 2002.

Madhaven, N. Final report on the safety assessment of *Corylus avellana* (Hazel) seed oil, *Corylus americana* (Hazel) seed oil, *Corylus avellana* (Hazel) seed extract, *Corylus americana* (Hazel) seed extract, *Corylus avellana* (Hazel) leaf extract, *Corylus americana* (Hazel) leaf extract, and *Corylus rostrata* (Hazel) leaf extract. *Int. J. Toxicol.* 20:15–20, 2001.

Maguire LS, O'Sullivan SM, Galvin K, O'Connor TP, O'Brien NM. Fatty acid profile, tocopherol, squalene, and phytosterol content of walnuts, almonds, peanuts, hazelnuts and macadamia nut. *Int. J. Food Sci. Nutr.* 55:171–178, 2004.

Mahdavi DL, Salunkhe DK. Toxicological aspects of food antioxidant. In: *Food Antioxidants.* Mahdavi D.L. Deshpande S.S. and Salunkhe D.K. Eds., Marcel Dekker, New York, 1995.

Marinova EM, Yanishlieva NV. Antioxidative activity of phenolic acids on triacylglycerols and fatty acid methyl esters from olive oil. *Food Chem.* 56:139–145, 1996.

Martinez JM, Granado JM, Montane D, Salvado J, Farriol X. Fractionation of residual lignocellulosics by dilute-acid prehydrolysis and alkaline extraction: Application to almond shells. *Bioresource Technol.* 52:59–67, 1995.

Martinez ML, Mattea M, Maestri DM. Varietal and crop year effects on lipid composition of walnut (*Juglans regia*) genotypes. *J. Am. Oil Chem. Soc.* 83:791–796, 2006.

Mattes RD, Kris-Etherton PM, Foster GD. Impact of peanuts and tree nuts on body weight and healthy weight loss in adults. *J. Nutr.* 138:1741S–1745S, 2008.

Mercanligil SM, Arslan P, Alasalvar C, Okut E, Akgul E, Pinar A, Geyik PO, Tokgozoglu L, Shahidi F. Effects of hazelnut-enriched diet on plasma cholesterol and lipoprotein profile in hypercholesterolemic adult men. *Eur. J. Clin. Nutr.* 61(2): 212–220, 2007.

Middleton Jr. E. Effect of plant flavonoids on immune and inflammatory cell function. *Adv. Exp. Med. Biol.* 439:175–182, 1998.

Milbury PE, Chen CY, Dolnikowski GG, Blumberg JB. Determination of flavonoids and phenolics and their distribution in almonds. *J. Agric. Food Chem.* 54(14):5027–5033, 2006.

Mills PK, Beeson WL, Phillips RL, Fraser GE. Cohort study of diet, lifestyle, and prostate cancer in Adventist men. *Cancer* 64:598–604¸ 1989.

MirAli N, Nabulsi I. Genetic diversity of almonds (*Prunus dulcis*) using RAPD technique. *Sci. Hort.* 98(4): 461–471, 2003.

Miraliakbari H, Shahidi F. Lipid class compositions, tocopherols and sterols of tree nut oils extracted with different solvents. *J. Food Lipids* 15(1): 81–96, 2008.

Miraliakbari H, Shahidi F. Oxidative stability of tree nut oils. *J. Agric. Food Chem.* 56(12): 4751–4759, 2008a.

Miraliakbari H, Shahidi F. Antioxidant activity of minor components of tree nut oils. *Food Chem.* 111(2): 421–427, 2008b.

Mishima K, Tanaka T, Pu F, Egashira N, Iwasaki K, Hidaka R. Vitamin E isoforms α-tocotrienol and γ-tocopherol prevent cerebral infarction in mice. *Neurosci. Lett.* 337:56–60, 2003.

Misra JB. A mathematical approach to comprehensive evaluation of quality in groundnuts. *J. Food Comp. Anal.* 17:69–79, 2004.

Miyashita K, Alasalvar C. Antioxidant and anti-obesity effects of hazelnut and olive oil polyphenols. Presented at the *Second International Congress on Functional Foods and Nutraceuticals*, Istanbul, Turkey, May 4–6, 2006; paper PL06004.

Morgan WA, Clayshulte BJ. Pecans lower low-density lipoprotein cholesterol in people with normal lipid levels. *J. Am. Diet. Assoc.* 100:312–318, 2000.

Moure A, Pazos M, Medina I, Dominguez H, Parajo JC. Antioxidant activity of extracts produced by solvent extraction of almond shells acid hydrolysates. *Food Chem.* 101:193–201, 2007.

Muamza D, Robert R, Sparks W. Antioxidant derived from lentil and its preparation and uses. U.S. Patent US5762936, 1998.

Mukuddem-Petersen J, Oosthuizen W, Jerling JC. A systematic review of the effects of nuts on blood lipid profiles in humans. *J. Nutr.* 135:2082–2089, 2005.

Munoz R, Arias Y, Ferreras JM, Jimenez P, Rojo MA, Girbs T. Sensitivity of cancer cell lines to the novel non-toxic type 2 ribosome inactivating protein Nigrin b. *Cancer Lett.* 167(2):163–169¸ 2001.

Natella F, Nardini M, Di Felice M, Scaccini C. Benzoic and cinnamic acid derivatives as antioxidants: Structure–activity relation. *J. Agric. Food Chem.* 47:1453–1459, 1999.

Ndiaye B, Diop YM, Diouf A, Fall M, Thiaw C, Thiam A, Barry O, Ciss M, Ba D. Measurement and levels of aflatoxins in small-scale pressed peanut oil prepared in the Diourbel and Kaolack regions of Senegal. *Dakar Med.* 44(2):202–205, 1999.

Nepote V, Grosso NR, Guzman CA. Extraction of antioxidant components from peanut skins. Grasas Y. *Aceites* 54(4):391–395, 2002.

Nepote V, Mestrallet MG, Grosso NR. Natural antioxidant effect from peanut skins in honey-roasted peanuts. *J. Food Sci.* 69:S295–S300, 2004.

Nergiz C, Dönmez I. Chemical composition and nutritive valve of *Pinus pinea* L. seeds. *Food Chem.* 86:365–368, 2004.

Navarro VB, Teruel NG, Carratala MLM. Oxidation stability of almond oil. *Acta Hort.(ISHS).* 591:125–131, 2002.

Ningappa MB, Dinesha R, Srinivas L. Antioxidant and free radical scavenging activities of polyphenol-enriched curry leaf (*Murraya koenigii* L.) extracts. *Food Chem.* 106(2):720–728, 2008.

O'Byrne DJ, Knauft DA, Shireman RB. Low fat-monounsaturated rich diets containing high-oleic peanuts improve serum lipoprotein profiles. *Lipid* 32:687–695, 1997.

O'Keefe SF, Wang H. Effects of peanut skin extract on quality and storage stability of beef products. *Meat Sci.* 73:278–286, 2006.

Ohsugi M, Fan W, Hase K, Xiong Q, Tezuca Y, Komatsu K, Namba T, Saitoh T, Tazawa K, Kadota S. Active-oxygen scavenging activity of traditional nourishing-tonic herbal medicines and active constituents of *Rhodiola sacra. J. Ethnopharmacol.* 67:111–119, 1999.

Oliveira I, Sousa A, Morais JS, Ferreira IC, Bento A, Estevinho L, Pereira JA. Chemical composition, and antioxidant and antimicrobial activities of three hazelnut (*Corylus avellana* L.) cultivars. *Food Chem. Toxicol.* 46(5):1801–1807, 2008.

Oliveira I, Sousa A, Valentão P, Andrade P, Ferreira ICFR, Ferreres F, Bento A, Seabra R, Estevinho L, Pereira JA. Hazel (*Corylus avellana* L.) leaves as source of antimicrobial and antioxidative compounds. *Food Chem.* 105:1018–1025, 2007.

Ory RL, Crippen KL, Lovegren NV. Off-flavors in peanuts and peanut products. In: *Developments in Food Science, Vol. 29: Off-flavors in Foods and Beverages.* Elsevier Science Publishers, Amsterdam, The Netherlands, pp. 57–75, 1992.

Osawa T, Ide A, Su JD, Namiki M. Inhibition of lipid peroxidation by ellagic acid. *J. Agric. Food Chem.* 35:808–812, 1987.

Paolisso G, D'Amore A, Giugliano D, Ceriello A, Varricchio M, D'Onofrio F. Pharmacological doses of vitamin E improve insulin action in healthy subjects and non-insulin-dependent diabetic patients. *Am. J. Clin. Nutr.* 57(5):650–656, 1993.

Parcerisa J, Boatella J, Codony R, Farran A, Garcia J, Lopez A, et al., Influence of variety and geographical origin on the lipid fraction of hazelnuts (*Corylus avellana* L.) from Spain: I. Fatty acid composition. *Food Chem.* 43: 411–414, 1993.

Parcerisa J, Casals I, Boatella J, Codony R, Rafecas M. Analysis of olive and hazelnut oil mixtures by high-performance liquid chromatography-atmospheric pressure chemical ionization mass spectrometry of triacylglycerols and gas–liquid chromatography of non-saponifiable compounds (tocopherols and sterols). *J. Chromatogr. A* 881:149–158, 2000.

Parcerisa J, Richardson DG, Rafecus M, Codony R, Boatella J. Fatty acid distribution in polar and nonpolar lipid classes of hazelnut oil (*Corylus avellana* L.). *J. Agric. Food Chem.* 45:3887–3890, 1997.

Pattee HE, Isleib TG, Moore KM, Gorbet DW, Giesbrecht FG. Effect of high-oleic trait and paste storage variables on sensory attributes stability of roasted peanuts. *J. Agric. Food Chem.* 50:7366–7370, 2002.

Peanut-Institute. Antioxidant from food sources, like peanuts and peanut butter, may protect against Alzheimer disease, Press Release, June 26, 2002.

Pennington JAT. Food composition databases for bioactive food components. *J. Food Comp. Anal.* 15:419–434, 2002.

Pereira JA, Oliveira I, Sousa A, Valentao P, Andradc PB, Ferreira IC, Ferreres F, Bento A, Seabra R, Estevinho L. Walnut (*Juglans regia* L.) leaves: phenolic compounds, antibacterial activity and antioxidant potential of different cultivars. *Food Chem. Toxicol.* 45(11):2287–2295, 2007.

Perry LM. *Medicinal Plants of East and Southeast Asia*. MIT Press, Cambridge, 1980.

Phillips KM, Ruggio DM, Ashraf-Khorassani M. Phytosterol composition of nuts and seeds commonly consumed in the United States. *J. Agric. Food Chem.* 53: 9436–9445, 2005.

Pietta P, Simonetti P, Mauri P. Antioxidant activity of selected medicinal plants. *J. Agric. Food Chem.* 46:4487–4890, 1998.

Pinelo M, Rubilar M, Sineiro J, Nunez MJ. Extraction of antioxidant phenolics from almond hulls (*Prunus amygdalus*) and pine sawdust (*Pinus pinaster*). *Food Chem.* 85:267–273, 2004.

Pou-Ilinas J, Canellas J, Driguez H, Excoffier G, Vignon MR. Steam pre-treatment of almond shells for xylose production. *Carbohydr. Res.* 207:1226–1230, 1990.

Prasad RBN. Walnuts and pecans. In: *Encyclopedia of Food Science and Nutrition*. Caballero B., Turgo L.C. and Finglas P.M. Eds., Academic Press, London, UK, pp. 6071–6079, 2003.

Pratt DE, Miller EE. A flavonoid antioxidant in Spanish peanuts (*Arachis hypogaea*). *J. Am. Oil Chem. Soc.* 61:1064–1067, 1984.

Proenca da Cunha A, Silva AO, Roque OR. Plantas e produtos vegetais em fitoterapia. Lisboa, Fundacao Calouste Gulbenkian, pp. 792–793, 2003.

Pulido R, Bravo L, Saura-Calixto F. Antioxidant activity of dietary polyphenols as determined by a modified ferric reducing/antioxidant power assay. *J. Agric. Food Chem.* 48:3396–3402, 2000.

Quesada J, Teffo-Bertaud F, Croue JP, Rubio M. Ozone oxidation and structural features of an almond shell lignin remaining after furfural manufacture. *Holzforschung* 56:32–38, 2002.

Rabinowitz IN. Dietary fiber, process for preparing it, and augmented dietary fiber from almond hulls. US Patent, US 0018255, 2004.

Rajaram S, Burke K, Connell B, Myint T, Sabate J. A monounsaturated fatty acid-rich pecan-enriched diet favorably alters the serum lipid profile of healthy men and women. *J. Nutr.* 131:2275–2279, 2001.

Rajaram S, Sabate J. Nuts, body weight and insulin resistance. *Br. J. Nutr.* 96:S79–S86, 2006.

Rajaram S, Wien M. Vegetarian diets in the prevention of osteoporosis, diabetes, and neurological disorders. In: *Vegetarian Nutrition*. Sabate J., Ed., CRC Press, Boca Raton, FL, pp. 109–134, 2001.

Ravai M. California walnuts: The natural way to a healthier heart. *Nutr. Today*, 30:173–176, 1995.

Reed KA, Sims CA, Gorbet DW, O'Keefe SF. Storage water activity affects flavor fade in high and normal oleic peanuts. *Food Res. Int.* 35:769–774, 2002.

Reiter RJ, Manchester LC, Tan DX. Melatonin in walnuts: Influence on leaves of melanin and total antioxidant capacity of blood. *Nutr.* 21(9):920–924, 2005.

Rice-Evans CA, Miller NJ, Paganga G. Structure—antioxidant activity relationships of flavonoids and phenolic acids. *Free Radical Biol. Med.* 20:933–956, 1996.

Rice-Evans CA, Miller NJ, Paganga G. Antioxidant properties of phenolic compounds. *Trends Plant Sci.* 2(4):152–159, 1997.

Rimbach G, Virgili F, Park YC, Packer L. Effect of procyanidins from *Pinus maritime* on glutathione levels in endothelial cells challenged by 3-morpholinosydnonimine or activated macrophages. *Redox Rep.* 4:171–177, 1999.

Ros E, Ninez I, Perez-Heras A, Serra M, Gilabert R, Casals E, Deulofeu R. A walnut diet improves endothelial function in hypercholesterolemic subjects: A randomized cross-over trial. *Circulation* 109:1609–1614, 2004.

Ros E, Mataix J. Fatty acid composition of nuts—implications for cardiovascular health. *Br. J. Nutr.* 96:S29–S35, 2006.

Ryan E, Galvin K, O'Connor TP, Maguire AR, O'Brien NM. Fatty acid profile, tocopherol, squalene and phytosterol content of brazil, pecan, pine, pistachio and cashew nuts. *Int. J. Food Sci. Nutr.* 57:219–228, 2006.

Sabate J. Nut consumption, vegetarian diets, ischemic heart disease risk, and all-cause mortality: Evidence from epidemiologic studies. *Am. J. Clin. Nutr.* 70 (Suppl.):500S–503S, 1999.

Sabate J. Nut consumption and body weight. *Am. J. Clin. Nutr.* 78 (Suppl.):647S–650S, 2003.

Sabate J, Cordero-Macintyre Z, Siapco G, Torabian S, Haddad E. Does regular walnut consumption lead to weight gain? *Br. J. Nutr.* 94(5):859–864, 2005.

Sabate J, Fraser GE, Burke K, Knutsen SF, Bennett H, Lindsted KD. Effects of walnuts on serum lipid levels and blood pressure in normal men. *N. Engl. J. Med.* 328:603–607, 1993.

Sabate J, Haddad E, Tanzman JS, Jambazian P, Rajaram S. Serum lipid response to the graduated enrichment of a Step I diet with almonds: A randomized feeding trial. *Am. J. Clin. Nutr.* 77:1379–1384, 2003.

Sabate J, Radak T, Brown J. The role of nuts in cardiovascular disease prevention. In: *Handbook of Nutraceuticals and Functional Foods*. Wildman R.E.C., Ed., CRC Press, Boca Raton, FL, pp. 477–495, 2000.

Sagrero-Nieves L. Fatty acid composition of Mexican nine nut oil from three seed coat phenotypes. *J. Sci. Food Agric.* 59:413–414, 1992.

Sales-Salvado J, Bullo M, Perez-Heras A, Ros E. Dietary fibre, nuts and cardiovascular diseases. *Br. J. Nutr.* 96:S45–S51, 2006.

Salmeron J, Ascherio A, Rimm EB, Colditz GA, Spiegelman D, Jenkins DJ, Stampfer MJ, Wing Al, Willett WC. Dietary fiber, glycemic load, and risk of NIDDM in men. *Diabetes Care* 20:545–550, 1997a.

Salmeron J, Manson JE, Stampfer MJ, Colditz GA, Wing Al, Willett WC. Dietary fiber, glycemic load, and risk of non-insulin-dependent diabetes mellitus in women. *JAMA* 277:472–477, 1997b.

Samaranayaka AGP, John JA, Shahidi F. Antioxidant activity of English walnut (*Juglans regia* L.). *J. Food Lipids* 15(3):384–397, 2008.

Sanders TH. Non-detectable levels of *trans*-fatty acids in peanut butter. *J. Agric. Food Chem.* 49:2349–2351, 2001.

Sanders TH, McMichael RW, Hendrix KW. Occurrence of resveratrol in edible peanuts. *J. Agric. Food Chem.* 48:1243–1246, 2000.

Sang S, Cheng X, Fu HY, Shieh DE, Bai N, Lapsley K, Rosen RT, Stark RE, Ho CT. New type sesquiterpene lactone from almond hulls (*Prunus amygdalus* Batsch). *Tetrahedron Lett.* 43:2547–2549, 2002a.

Sang S, Kikuzaki H, Lapsley K, Rosen RT, Nakatani N, Ho CT. Sphingolipid and other constituents from almond nuts (*Prunus amygdalus* Batsch). *J. Agric. Food Chem.* 50(16):4709–4712, 2002.

Sang S, Lapsley K, Jeong WS, Lachance PA, Ho CH, Rosen RT. Antioxidant phenolic compounds isolated from almond skins (*Prunus amygdalus* Batsch). *J. Agric. Food Chem.* 50:2459–2463, 2002b.

Sang S, Lapsley K, Rosen RT, Ho CH. New prenylated benzoic acid and other constituents from almond hulls (*Prunus amygdalus* Batsch). *J. Agric. Food Chem.* 50:607–609, 2002c.

Sarkkinen E, Schwab U, Niskanen L, Hannuksela M, Sarolainen M, Kervinen K, Kesaniemi A, Uusitupa MI. The effects of monounsaturated-fat enriched diet and polyunsaturated-fat enriched diet on lipid and glucose metabolism in subjects with impaired glucose tolerance. *Eur. J. Clin. Nutr.* 50(9):592–598, 1996.

Savage GP. Chemical composition of walnuts (*Juglans regia* L.) grown in New Zealand. *Plant Food Hum. Nutr.* 56:75–82, 2001.

Savage GP, Dutta PC, Mc Neill DL. Fatty acid and tocopherol contents and oxidative stability of walnut oils. *J. Am. Oil Chem. Soc.* 76:1059–1063, 1999.

Savage GP, Keenan JI. *The Composition and Nutritive Value of Groundnut Kernels*. Smartt J, Ed., Chapman & Hall, London, pp. 173–213, 1994.

Savage GP, McNeil DL, Dutta PC. Lipid composition and oxidative stability of oils in hazelnuts (*Corylus avellana* L.) grown in New Zealand. *J. Am. Oil Chem. Soc.* 74: 755–759, 1997.
Scalbert A, Namach C, Morand C, Rémésy C. Dietary polyphenols and the prevention of diseases. *Crit. Rev. Food Sci. Nutr.* 45:287–306, 2005.
Schurig V, Jauch J, Schmalzing D, Jung M, Bretschneider W, Hopp R, Werkhoff P. Analysis of the chiral aroma compound filbertone by inclusion gas chromatography. *Z. Lebensm. Unters. Forsch.* 191: 28–31, 1990.
Seddon JM, Cote J, Rosner B. Progression of age-related macular degeneration: Association with dietary fat, transunsaturated fat, nuts and fish intake. *Arch. Ophthalmol.* 121:1728–1737, 2003.
Seeram NP, Zhang Y, Henning SM, Lee R, Niu Y, Lin G, Heber D. Pistachio skin phenolics are destroyed by bleaching resulting in reduced antioxidative capacities. *J. Agric. Food Chem.* 54(19):7036–7040, 2006.
Segura R, Javierre C, Lizarraga MA, Rose E. Other relevant components of nuts: phytosterols, folate and minerals. *Br. J. Nutr.* 96, S36–S44, 2006.
Senter SD, Horvat RJ, Forbus WR. Comparative GLC–MS analysis of phenolic acids of selected tree nuts. *J. Food Sci.* 48: 788–789, 1983.
Seo A, Morr CV. Activated carbon and ion exchange treatments for removing phenolics and phytate from peanut protein products. *J. Food Sci.* 50:262–263, 1985.
Shahidi F. Phytochemicals of almond and their role in inhibition of DNA niching and human LDL cholesterol oxidation. In: *Abstracts of papers, 224th* ACS National Meeting, Boston, MA, USA, 2002.
Shahidi F, Alasalvar C, Liyana-Pathirana CM. Antioxidant phytochemicals in hazelnut kernel (*Corylus avellana* L.) and hazelnut byproducts. *J. Agric. Food Chem.* 55(4):1212–1220, 2007.
Shahidi F, Miraliakbari H. Tree nut oils and by products: Compositional characteristics and nutraceutical applications. In: *Nutraceutical and Specialty Lipids and Their Co-products.* Shahidi F. Ed., Nutraceutical Science and Technology Series, CRC Taylor & Francis, Chapter 9, pp. 159–168, 2006.
Sheridan MJ, Cooper JN, Erario M, Cheifetz CE. Pistachio nut consumption and serum lipid levels. *J. Am. Coll. Nutr.* 26(2):141–148, 2007.
Shi L, Lu JY, Jones G, Loretan PA, Hill WA. Characteristics and composition of peanut oil prepared by an aqueous extraction method. *Life Support. Biosph. Sci.* 5(2):225–229, 1998.
Sies H, Stahl W, Sevanian A. Nutritional, dietary and postprandial oxidative stress. *J. Nutr.* 135:969–972, 2005.
Silva BM, Andrade PB, Valentaqo P, Ferreres F, Seabra RM, Ferreira MA. Quince (*Cydonia oblonga* Miller) fruit (pulp, peel, and seed) and jam: Antioxidant activity. *J. Agric. Food Chem.* 52:4705–4712, 2004.
Simopoulos AP. The Mediterranean diets: What is so special about the diet of Greece? The scientific evidence. *J. Nutr.* 131 (Suppl. 11):3065S–3073S, 2001.
Simopoulos AP. Health effects of eating walnuts. *Food Rev. Int.* 20(1):91–98, 2004.
Siriwardhana SSKW, Shahidi F. Antiradical activity of extracts of almond and it's by-products. *J. Am. Oil Chem. Soc.* 79(9):903–908, 2002.
Sobolev VS. Vanillin content in boiled peanuts. *J. Agric. Food Chem.* 49:3725–3727, 2001.
Sobolev VC, Cole RJ. *Trans*-resveratrol content in commercial peanuts and peanut products. *J. Agric. Food Chem.* 47:1435–1439, 1999.
Sobolev VC, Cole RJ. Note on utilization of peanut seed testa. *J. Sci. Food Agric.* 84:105–111, 2003.
Solar A, Colaric M, Usenik V, Stampar F. Seasonal variation of selected flavonoids, phenolic acids and quinines in annual shorts of common walnut (*Juglans regia* L.). *Plant Sci.* 170:461–543, 2006.

Spiller GA, Jenkins AJ, Bosello O, Gates JE, Cragen LN, Bruce B. Nuts and plasma lipids: An almond-based diet lowers LDL-C while preserving HDL-C. *J. Am. Coll. Nutr.* 17:285–290, 1998.

Spiller GA, Jenkins D, Cragen LN, Gates JE, Bosello O, Berra K, Rudd C, Stevenson M, Superko R. Effect of a diet high in monounsaturated fat from almonds on plasma cholesterol and lipoprotein. *J. Am. Coll. Nutr.* 11:126–130, 1992.

Stampar F, Solar A, Hudina M, Veberic R, Colaric M. Traditional walnut liqueur-cocktail of phenolics. *Food Chem.* 95:627–631, 2006.

Steinberg D. Metabolism of lipoprotein and their role in the pathogenesis of atherosclerosis. *Atherosclerosis Rev.* 18:1–6, 1992.

Stupans I, Kirlich A, Tuck KL, Hayball PJ. Comparison of radical scavenging effect, inhibition of microsomal oxygen free radical generation, and serum lipoprotein oxidation of several natural antioxidants. *J. Agric. Food Chem.* 50:2464–2469, 2002.

St-Onge MP. Dietary fats, teas, dairy, and nuts: Potential functional foods for weight control? *Am. J. Clin. Nutr.* 81:7–15, 2005.

Sze-Tao KWC, Sathe SK. Walnuts (*Jugulans regia* L.): Proximate composition, protein solubility, protein amino acid composition and protein *in vitro* digestibility. *J. Sci. Food Agric.* 80:1393–1401, 2000.

Sze-Tao KWC, Schrimpt JE, Teuber SS, Roux KH, Sathe SK. Effects of processing and storage on walnut (*Jugulans regia* L.) tannins. *J. Sci. Food Agric.* 81:1215–1222, 2001.

Takeoka G, Dao LT. Antioxidant constituents of almond hulls. In: Abstracts Papers—*American Chemical Society* 2000, 220th AGDF-034, 2000.

Takeoka G, Dao LT. Antioxidant constituents of almond [*Prunus dulcis* (Mill.) D.A. Webb] hulls. *J. Agric. Food Chem.* 51:496–501, 2003.

Takeoka G, Dao L, Teranishi R, Wong R, Flessa S, Harden L, Edwards. Identification of three triterpenoids in almond hulls. *J. Agric. Food Chem.* 48:3437–3439, 2000.

Talcott ST, Duncan CE, Del Pozo-Insfran D, Gorbet DW. Polyphenolic and antioxidant changes during storage of normal, mild, and high oleic acid peanuts. *Food Chem.* 89:77–84, 2005.

Tapsell LC, Gillen LJ, Patch CS, Batterham M, Owen A, Bare M, Kennedy M. Including walnuts in a low-fat/modified-fat diet improve HDL cholesterol-to-total cholesterol ratios in patients with type 2 diabetes. *Diabetes Care* 27:2777–2783, 2004.

Teippo J, Vercelino R, Dias AS, Siva Vas MF, Silveira TR, Marroni CA, Marroni NP, Henriques JAP, Picada JN. Evaluation of the protective effects of quercetin in the hepatopulmonary syndrome. *Food Chem. Toxicol.* 45:1140–1146, 2007.

Thompson LU, Boucher BA, Liu Z, Cotterchio M, Kreiger N. Phytoestrogen content of foods consumption in Canada, including isoflavones, lignans, and coumestan. *Nutr. Cancer* 54:184–201, 2006.

Tsai CJ, Leitzmann MF, Hu FB, Willett WC, Giovannucci EL. Frequent nut consumption and decreased risk of cholecystectomy in women. *Am. J. Clin. Nutr.* 80:76–81, 2004.

Tsamouris G, Hatziantoniou, Demetzos C. Lipid analysis of Greek walnout oil (*Juglans regia* L.). *Z. Zeitforsch.* 57c:51–56, 2002.

Tseng TH, Kao ES, Chu CY, Chou FB, Lin Wu HW, Wang CJ. Protective effects of dried flower extracts of *Hibiscus sabdariffa* L. against oxidative stress in rat primary hepatocytes. *Food Chem. Toxicol.* 35:1159–1164, 1997.

Ugartondo V, Mitjans M, Tourino S, Torres JL, Vinardell MP. Comparative antioxidant and cytotoxic effect of procyanidin fractions from grape and pine. *Chem. Res. Toxicol.* 2(10):1543–1548, 2007.

US. Trade Quick-References Tables: December 2003 Imports. URL: http://www.ita.doc.gov/td/industry/otea/Trade-Detail/Latest-December/Imports/08/080290.html.

USDA. Foreign production, supply and distribution of agriculture commodities, FAS. In: *World Oilseed Situation and Outlook*. USDA, Washington, DC, 1992.

USDA. Foreign Agricultural Service. Production, Supply and Distribution (PSD) database, December 2004. Internet: http://www.fas.usda.gov/psdonline/psdHome.aspx.

USDA. Agricultural Research Service, 2008. *USDA Nutrient Database for Standard Reference, Release No. 21*. Nutrient Data Laboratory Home Page. Internet: http://www.ars.usda.gov/ba/bhnrc/ndl (accessed 8 November 2008).

Valnet J. *Phytotherapie traitement des maladies par les plantes*. Maloine, Paris, pp. 476–478, 1992.

Van Hellemont J. *Compendium de phytotherapie*. Association Pharmaceutique Belge, Bruxelles, pp. 214–216, 1986.

Venkatachalam M, Sathe SK. Chemical composition of selected edible nut seeds. *J. Agric. Food Chem.* 54: 4705–4714, 2006.

Wang J, Yuan X, Jin Z, Tian Y, Song H. Free radical and reactive oxygen species scavenging activities of peanut skins extract. *Food Chem.* 104:242–250, 2007.

Wichtl M, Anton R. *Plantes therapeutiques*. Tec.& Doc., Paris, pp. 291–293, 1999.

Wijeratne SSK, Abou-Zaid MM, Shahidi F. Antioxidant polyphenols in almond and its co-products. *J. Agric. Food Chem.* 54(2):312–318, 2006a.

Wijeratne SSK, Amarowicz R, Shahidi F. Antioxidant activity of almonds and their by-products in food model systems. *J. Am. Oil Chem. Soc.* 83(3):223–230, 2006b.

Wilms LC, Hollman PC, Boots AW, Kleinjans JC. Protection by quercetin and quercetin-rich fruit juice against induction of oxidative DNA damage and formation of BPDE-DNA adducts in human lymphocytes. *Mutation Res.* 582:155–162, 2005.

Wood JE, Senthilmohan ST, Peskin AV. Antioxidant activity of procyanidin-containing plant extracts at different pHs. *Food Chem.* 77:155–161, 2002.

Wu X, Beecher GR, Holden JM, Haytowitz DB, Gebhardt SE, Prior RL. Lipophilic and hydrophilic antioxidant capacities of common foods in the United States. *J. Agric. Food Chem.* 52:4026–4037, 2004.

Yadav AS, Bhatnagar D. Free radical scavenging activity, metal chelation and antioxidant power of some of the Indian spices. *BioFactors* 31:219–227, 2008.

Yanagimoto K, Lee K, Ochi H, Shibamoto T. Antioxidative activity of heterocyclic compounds found in coffee volatiles produced by the Maillard reaction. *J. Agric. Food Chem.* 50:5480–5484, 2002.

Yanagisawa C, Uto H, Tani M, Kishimoto Y, Machida N, Hasegawa M, Yoshioka E, Kido T, Kondo K. The antioxidant activities of almonds against LDL oxidation. *XIV International Symposium on Atherosclerosis*, Rome, Italy, June 18–22, Abstract # We P14:395, p-434, 2006a.

Yanagisawa C, Uto H, Tani M, Kishimoto Y, Machida N, Hasegawa M, Yoshioka E, Kido T, Lapsley KG, Kondo K. The effect of almonds on the serum lipid, lipoprotein and apolipoprotein levels in Japanese male subjects. *XIV International Symposium on Atherosclerosis*, Rome, Italy June 18–22, Abstract # We-P14:396, p-434, 2006b.

Yen C, You S, Chen C, Sung F. Peanut consumption and reduced risk of colorectal cancer in women: A prospective study in Taiwan. *World J. Gastroenterol.* 12:222–227, 2006.

Yen GC, Duh PD. Scavenging effects of methanolic extracts of peanut hulls on free-radical and active-oxygen species. *J. Agric. Food Chem.* 42:62–632, 1994.

Yen GC, Duh PD, Tsai CL. Relationship between antioxidant activity and maturity of peanut hulls. *J. Agric. Food Chem.* 41:67–70, 1993.

Yen WJ, Chang LW, Duh PN. Antioxidant activity of peanut seed testa and its antioxidative components, ethyl protocatechuate. *Food Sci. Tech.* 38(3):193–200, 2005.

Yu J, Ahmedna M, Goktepe I. Effects of processing methods and extraction solvents on concentration and antioxidant activity of peanut skin phenolics. *Food Chem.* 90:199–206, 2005.

Yu J, Ahmedna M, Goktepe I, Dia J. Peanut skin procyanidins: Composition and antioxidant activities as affected by processing. *J. Food Compos. Anal.* 19(4):364–371, 2006.

Yurttas HC, Schafer HW, Warthesen JJ. Antioxidant activity of nontocopherol hazelnut (*Corylus* spp.) phenolics. *J. Food Sci.* 65: 276–280, 2000.

Zambon D, Sabate J, Munoz S, Campero B, Casals E, Merlos M, Leguna JC, Ros E. Substituting walnuts for monounsaturated fat improves the serum lipid profile of hypercholesterolemic men and women: A randomized crossover trial. *Ann. Intern. Med.* 132:538–546, 2000.

16 Functional Foods Based on Sea Buckthorn (*Hippophae rhamnoides* ssp. *turkestanica*) and Autumn Olive (*Elaeagnus umbellata*) Berries

Nutritive Value and Health Benefits

Syed Mubasher Sabir and Syed Dilnawaz Ahmad

CONTENTS

16.1 INTRODUCTION

Hilly areas of Pakistan including Baluchistan, North Western Frontier Province, Kashmir, and the northern areas have a very rich and diverse flora due to their diverse climatic, soil conditions, and multiple ecological regions (Sabir et al., 2003). Medicinal plant resources are not only abundant but are also rich in genetic diversity and biochemical composition. The medicinal plants and herbs are extensively used locally for treating different diseases; however, their commercial exploitation is limited owing to the lack of a scientific basis for their use (Hussain and Khaliq, 1996). The farmers in this area are poor and the cultivated land plots are either very small or unmanageable due to soil degradation and specific topography. Quite recently an initiative was undertaken to characterize the local plants for genetic and biochemical variation in order to improve these plants for commercial purposes. Two plant species, sea buckthorn (*Hippophae rhamnoides* ssp. *turkestanica* L.) from the northern areas of Pakistan and autumn olive (*Elaeagnus umbellata* Thunb.) from Kashmir were analyzed for their chemical and nutritional constituents.

16.1.1 Sea Buckthorns

Sea buckthorn is a shrub or small tree of the genus *Hippophae*. The genus belongs to the family Elaeagnaceae which consists of 6 species and 10 subspecies, among which *Hippophae rhamnoides* L., commonly known as sea buckthorn, is economically most important (Rongsen, 1992). The distribution of sea buckthorn ranges from Himalayan regions including India, Nepal, Bhutan, Pakistan (Skardu, Swat, Gilgit), and Afghanistan, to China, Mongolia, Russia, Kazakstan, Hungary, Romania, Switzerland, Germany, France, and Britain, and northward to Finland, Sweden (Jeppsson, 1999) and Norway (Yao, 1994). The only subspecies found in the northern areas of Pakistan is *H. rhamnoides* ssp. *turkestanica*, widely found in Central and Western Asia, including Afghanistan, Tajikistan, Turkmenistan, Uzbekistan, Kirghisistan, the Xinjiang province of China, and northern India.

Sea buckthorn is locally known as "Booro" in the Shina language particularly in Gilgit while in the Balti language it is called "Chok Foolo" in the Skardu region of Pakistan. It is the only subspecies which can withstand the harsh biophysical conditions characterized by hot arid summers and cold winters (Rongsen, 1992). The *H. rhamnoides* fruit contains 60–80% juice that is rich in sugars, organic acids, amino acids, and vitamins. The vitamin C content is 200–1500 mg/100 g in the fruit which is 5–100 times higher than any other fruit or vegetable (Rongsen, 1992). The oil content range is about 1.5–3.5% in fresh fruit and about 9.9–19.5% in the seeds (Rongsen, 1992). Oil from the juice and pulp is rich in palmitic (16:0) and palmitoleic acids (16:1), while the oil from the seed contains the essential fatty acids linoleic (18:2) and linolenic (18:3) acids. Mark and Peter (2004) reported that the pomace of sea buckthorn which is generally considered waste, contains about 15% palmitoleic acid in its oil and may be utilized as a raw material for the production of a palmitoleic acid methyl ester concentrate. The oil from the seed and juice also contains α-tocopherol and β-carotenoids (Bernath and Foldesi, 1992; Ma and Cui, 1989).

There are 24 chemical elements in sea buckthorn juice including calcium, magnesium, phosphorous, iron, manganese, sodium, potassium, aluminum, and others (Zhang et al., 1989; Tong et al., 1989). In addition, sea buckthorn berries, leaves, and bark contain β-sitosterol, α-tocopherol, lycopene, and malic and quinic acids (Mironov, 1989). Several studies have shown that sea buckthorn is a good and low-cost natural source of phenolic acids. Nine phenolic acids namely protocatechuic, *p*-hydroxybenzoic, vanillic, salicylic, *p*-coumaric, cinnamic, caffiec, and ferulic acids were found in sea buckthorn (SB) (*Hippophae rhamnoides*) berries and leaves. Gallic acid was the predominant phenolic acid both in free and bound forms in sea buckthorn berry parts and leaves (Arimboor et al., 2008).

The health benefits of sea buckthorn berries may be partly attributed to their high content of phenolic compounds, as phenolics possess a wide spectrum of biochemical activities such as antioxidant, antimutagenic, anticarcinogenic, as well as their abilities to modify gene expression (Nakamura et al., 2003).

Negi et al. (2005) screened the crude extracts of sea buckthorn seeds for antioxidant and antibacterial activity and reported the highest activity in the methanolic extract.

16.1.2 History of Sea Buckthorn

Sea buckthorn (*Hippophae rhamnoides* L.), a thorny deciduous shrub/tree, belongs to the family Elaeagnaceae which grows abundantly in the higher Himalayas–Karakoram–Hindu Kush region including Pakistan. It is usually 2–4 m in height with orange or red color berries weighing 0.20–0.35 g. Sea buckthorn has many economical, nutritional, and medicinal benefits. China, Mongolia, and Russia are pioneers among the *Hippophae* growing countries in having harnessed the potential of this plant for various purposes such as food, medicine, and cosmetics. In China alone, the total value of sea buckthorn products was more than US$20 million in 1990. In Pakistan this plant is still not fully exploited for its benefits despite an estimated natural cover of 7000 ha in the northern areas. The sea buckthorn industry has been thriving in Russia since 1940 when scientists there began investigating the biologically active substances found in the fruit, leaves, and bark. The first Russian factory for sea buckthorn product development was established in Bisk. These products were utilized in the diet of Russian cosmonauts and as a cream for protection from cosmic radiation.

The Chinese experience with sea buckthorn fruit production is more recent, although traditional uses date back many centuries. Research and plantation establishment were initiated in 1980. Since 1982 over 300,000 ha of sea buckthorn have been planted in China. In addition, 150 processing factories have been established, producing over 200 products. The sea buckthorn- based sports drinks "Shawikang" and "Jianibao" were designated the official drink for Chinese athletes attending the Seoul Olympic Games.

16.1.3 Traditional Use of Sea Buckthorn

Medicinal uses of sea buckthorn are well documented in Asia and Europe. Investigations on modern medicinal uses were initiated in Russia during the 1950s.

Preparations of sea buckthorn oils are recommended for external use in the case of burns, bedsores, and other skin complications induced by confinement to a bed or treatment with x-ray or radiation. Internally, sea buckthorn was used for the treatment of stomach and duodenal ulcers. In the United Kingdom and Europe, sea buckthorn products are used in aromatherapy.

Research in the late 1950s and early 1960s reported that 5-hydroxytryptamine (hippophan) isolated from sea buckthorn bark inhibited tumor growth. More recently, clinical studies on the antitumor functions of sea buckthorn oils conducted in China have been positive. Sea buckthorn oil, juice or extracts from oil, juice, leaves, and bark have been used successfully to treat high blood lipid symptoms, eye diseases, gingivitis, and cardiovascular diseases such as high blood pressure and coronary heart disease. Sea buckthorn was formally listed in the *Pharmacopoeia of China* in 1977. Health problems (cancer, peptic ulcers, and skin and cardiovascular diseases) are growing concerns worldwide, especially in the developing countries and their therapy by synthetic medicine is expensive and has undesirable side effects. The use of phytochemicals from *Hippophae* is a safe way for combating these diseases.

Cancer therapy: The literature data describing the role of *Hippophae* in the prevention and control of cancer is limited. However, certain analyses of the known experimental research information on anticancer activity by *Hippophae* are now available (Zhang, 1989; Zang et al., 2005). The inhibition of *Hippophae* oil of cancer cells was not as effective as positive medicine, for example, the cancer inhibition rate of phosphamide was found twice as effective compared with *Hippophae* (Nersesyan and Muradyan, 2004). Reports on the potential of a *Hippophae* extract (an alcohol extract, which would mainly contain the flavonoids) to protect the bone marrow from damage due to radiation. This study also showed that the extract might help faster recovery of bone marrow cells (Agrawala and Goel, 2002).

In China, a study was carried out to demonstrate the faster recovery of the hemopoietic system after high dose chemotherapy in mice fed with sea buckthorn oil (Chen, 2003). The direct effects of sea buckthorn on tumorigenesis, in addition to its indirect ones caused by general immunity or other mechanisms, include inhibiting action on the cancer cells and blocking the carcinogenic factors. Experiments on mice transplanted tumors, including sarcoma (S180), lymphatic leukemia (P388), and B_{16} were carried out. It was found that both intraperitoneal injection of sea buckthorn oil and oral administration inhibited the development of tumors. Sea buckthorn juice can both kill the cancer cells of S180 and P388 and inhibit growth of the cell strains of the human gastric carcinoma (SGC7901) and lymphatic leukemia (L1200) (Mingyu, 1994).

Cardiovascular therapy: *Hippophae* is used as anticardio vascular medicine. In a double blind clinical trial, 128 patients with ischemic heart disease were administered total flavonoids of sea buckthorn at 10 mg each time, three times daily, for 6 weeks (Yang and Kallio, 2002). The cholesterol level of the patients was decreased and cardiac function was improved; also they had fewer anginas than those receiving the control drug. No toxic effects of sea buckthorn flavonoids were reported on renal or hepatic functions. The mechanism of action may include reduced stress of cardiac muscle tissue by regulation of inflammatory mediators. In another laboratory animal study, the flavonoids of sea buckthorn were shown to reduce the production of pathogenic thromboses in mice (Zhang, 1987). Some simple formulas (Chen et al., 2003) based on sea buckthorn have been developed recently and are intended for use in the treatment of coronary heart disease and stroke, through improving blood circulation and restoring cardiac function.

There is increasing evidence to support the hypothesis that free radical-mediated oxidative processes contribute to atherogenesis. Sea buckthorn (*Hippophae rhamnoides* L.) is a rich source of antioxidants both aqueous and lipophilic, as well as polyunsaturated fatty acids. It was found that antioxidant rich sea buckthorn juice affects the risk factors (plasma lipids, LDL oxidation, platelet aggregation, and plasma soluble cell adhesion protein concentration) for coronary heart disease in humans (Eccleston et al., 2002).

Gastrointestinal ulcers: Gastric ulcers are fast becoming a major problem in humans, especially in the developing countries like Pakistan, due to unfavorable and nonassessed diet, ignorance, and carelessness. *Hippophae* is traditionally used in the treatment of gastric ulcers and laboratory studies confirm the efficacy of the seed oil for antiulcer activity. Its functions may be to normalize output of gastric acid and reduce inflammation by controlling pro-inflammatory mediators (Xing et al., 2002). The antiulcerogenic effect of a hexane extract from *Hippophae rhamnoides* was tested on indomethacin- and stress-induced ulcer models and was found to be active in preventing gastric injury (Zhou, 1998).

Liver disease: Hepatitis is a major health problem, not only in Pakistan but all over the world. A clinical trial demonstrated that sea buckthorn extracts had the efficacy to normalize liver enzymes (alanine amino transferase (ALT) and aspartate amino transferase (AST)), serum bile acids, and immune system markers involved in liver inflammation and degeneration (Zao et al., 1987). A recent study of *Hippophae* from India demonstrated that leaf extract had significant hepatoprotective activity against carbon tetrachloride-induced liver injury in mice (Geetha et al., 2008).

16.1.4 Harvesting of Sea Buckthorn Berries

In Pakistan, harvesting is carried out in the end of September, when the berries are at the optimum maturity level. The methods used ensured that various techniques were tested in a simple and understandable manner and also guaranteed authenticity of the results (Khan and Kamran, 2004). For picking sea buckthorn berries three simple techniques, that is, handpicking using gloves, cutting of branches followed by clipping of bunches using scissors and beating with wooden sticks, were tested in both managed and unmanaged plots with the same sea buckthorn variety. The products involved are jams, jelly, syrup, and squash prepared at the domestic level. The stick-beating method proved to be the best, with an average of 1949.86 and 1601.53 g of berries collected in one person hour in managed and unmanaged plots, respectively (Khan and Kamran, 2004). It was followed by scissor-picking and hand-picking techniques. The quantity of berries collected in the managed plot was significantly higher than in the unmanaged plot using the same picking techniques, indicating that proper management (spacing and pruning) of even the local varieties of sea buckthorn can make the picking process more cost effective.

16.2 NUTRITIONAL ANALYSIS OF SEA BUCKTHORN

The biochemical constituents and mineral composition of samples collected from eight populations of *Hippophae rhamnoides* from different areas of northern Pakistan are shown in Tables 16.1 and 16.2.

16.2.1 Vitamin C Content in Sea Buckthorn Berries

The vitamin C content of sea buckthorn berries ranged from 240 to 339 mg/100 g (Table 16.1). The highest vitamin C content was found in SBT-1, while the lowest was observed in SBT-8. In previous studies, the vitamin C contents of sea buckthorn berries collected from the Districts Skardu and Khaplu (northern areas) were found to be in the range of 150–250 mg/100 g (Sabir et al., 2003). The vitamin C concentration ranged from 28 to 310 mg/100 g of berries in berries of the European subspecies *rhamnoides* (Yao, 1994; Rousi, 1977), from 460 to 1330 mg/100 g in berries of *fluviatilis* ssp. (Darmer, 1952) and from 200 to 2500 mg/100 g in berries of Chinese *sinensis* ssp. (Zheng and Song, 1992). The vitamin C content varied from 19 to 121 mg/100 mL in sea buckthorn berries grown in Turkey (Sezai et al., 2007). The concentration of vitamin C was highly variable among the populations as reported earlier (Kallio et al., 2002) and was lower than that reported in some other studies. Yao et al. (1992) reported the vitamin C content of Chinese *H. rhamnoides* in the range of 460–1330 mg/100 g. The lower vitamin C concentration in the present investigation could be due to the specific geographical nature of the area where a short growing season prevails (Yao and Tigerstedt, 1995).

TABLE 16.1
Nutritional Analysis of Berries in Eight Populations of Sea Buckthorn from Northern Areas of Pakistan

Populations	Color of Berries	Ascorbic Acid in Berries (mg/100 g)	Fatty Oil in Seed (g/100 g)	Oil in Fruit Pulp (g/100 g)	Phytosterol Content of Seed Oil (g/100 g)	Anthocyanin Content of Fruit Juice (mg/L)
SBT-1	Yellow	339 ± 0.42	12.7 ± 0.2	28.1 ± 0.2	5.2 ± 0.01	5.0 ± 0.12
SBT-2	Yellow	320 ± 0.13	10.4 ± 0.1	25.5 ± 0.3	5.3 ± 0.17	10.0 ± 0.56
SBT-3	Orange-red	250 ± 0.54	7.5 ± 0.5	21.3 ± 0.4	3.1 ± 0.13	1.4 ± 0.03
SBT-4	Light-yellow	273 ± 0.3	11.7 ± 0.2	26.4 ± 1.1	4.13 ± 0.05	1 ± 0.01
SBT-5	Red	280 ± 0.4	5.69 ± 0.3	17.8 ± 0.9	3.9 ± 0.12	22.0 ± 0.12
SBT-6	Red	313 ± 0.28	8.3 ± 0.8	19.2 ± 0.1	4.5 ± 0.20	11.5 ± 0.1
SBT-7	Orange	312 ± 0.37	11.2 ± 0.6	22.3 ± 0.4	5.1 ± 0.15	8.39 ± 0.14
SBT-8	Yellow	240 ± 0.14	12.1 ± 0.1	27.0 ± 1.5	4.2 ± 0.13	15.1 ± 0.3

SBT = sea buckthorn, ± standard deviation (SD).

TABLE 16.2
Mineral Analysis among Different Populations of Sea Buckthorn from the Northern Areas of Pakistan

Populations	Color of Berries	K (g/kg)	Na (g/kg)	Ca (g/kg)	Mg (mg/kg)	Fe (mg/kg)	P (mg/kg)
SBT-1	Yellow	7.9 ± 0.15	0.7 ± 0.01	0.8 ± 0.2	270 ± 0.15	229 ± 0.21	131 ± 0.11
SBT-2	Yellow	6.7 ± 1.1	0.45 ± 0.01	1.21 ± 0.5	225 ± 0.2	140 ± 0.14	125 ± 0.04
SBT-3	Orange-red	3.3 ± 0.17	0.62 ± 0.02	0.98 ± 0.15	149 ± 0.9	120 ± 0.21	138 ± 0.12
SBT-4	Light-yellow	3.4 ± 0.23	0.4 ± 0.01	0.65 ± 0.03	219.5 ± 0.2	115 ± 0.13	115 ± 0.05
SBT-5	Red	2.5 ± 0.1	0.73 ± 0.03	1.11 ± 0.16	150 ± 0.31	45 ± 0.05	117 ± 0.12
SBT-6	Red	5.8 ± 0.32	0.08 ± 0.2	1.17 ± 1.1	192 ± 1.32	110 ± 1.2	121 ± 0.2
SBT-7	Orange	6.2 ± 1.9	0.065 ± 0.1	0.98 ± 0.2	202 ± 0.8	150 ± 0.02	117 ± 0.1
SBT-8	Yellow	6.44 ± 2.1	0.071 ± 1.2	1.22 ± 0.2	210 ± 1.3	170 ± 1.5	126 ± 0.2

SBT = sea buckthorn, ± standard deviation (SD).

16.2.2 Oil Content in Sea Buckthorn Berries

The oil content in sea buckthorn seeds also varied among the different populations examined (Table 16.1). Previous studies reported oil contents of 5.3–15.7% in the seed of *turkestanica* (Yang, 2001). The present investigation reports seed oils in the range of 5.7–12.7% in *turkestanica*, with the maximum amount in the yellow berries of SBT-1, while the minimum amount was found in the red berries of SBT-5. Yellow and yellow-orange fruits have been reported to have higher levels of oil than orange and orange-red fruits (Daigativ et al., 1985), which corresponds to our observations. Dried pulp of *turkestanica* was reported to be the richest source of oil, containing 17.8–34% oil (Yang, 2001). In this study the oil content of dried pulp was in the range of 17.8–28.1% (Table 16.1). The maximum amount of oil was found in SBT-1, while the minimum was in SBT-5. The higher oil content of ssp. *turkestanica* may be very important for its local use in skin diseases.

16.2.3 Sea Buckthorn Oil, a Rich Source of Phytosterol

Phytosterols are plant sterols with structures related to cholesterol and when consumed they can lower plasma cholesterol. Since elevated blood cholesterol is one of the well-established risk factors for coronary heart disease, the lowering of blood cholesterol could reduce the risk of heart disease (Thurnham, 1999). Phytosterols are the major constituents of the unsaponifiable fraction of *H. rhamnoides* oils. The major phytosterol in *H. rhamnoides* oil is β-sitosterol, with 5-avenasterol being the second. Other phytosterols are present in relatively minor quantities.

Our earlier studies reported the phytosterol contents of pulp oils in the range of 1.3–2% (Sabir et al., 2003). In this study the phytosterol contents of the seed oils was higher in the range of 3.9–5.3% (Table 16.1). The maximum amount of sterol was found in SBT-2, while the minimum was found in SBT-5. Thus, oil from *turkestanica* seed contains more phytosterol than the oil extracted from the pulp.

16.2.4 Anthocyanin Content of Sea Buckthorn Berries

Anthocyanin pigments are responsible for the attractive red, purple, and blue colors of most fruits and vegetables. They may play a role in reducing coronary heart disease (Bridle and Timberlake, 1996) and increasing visual acuity (Timberlake and Henry, 1988). They also act as antioxidants (Wang et al., 1997) and anticancer agents (Kamei et al., 1995). Anthocyanins are also used in the food industry as safe and effective food colorants (Strack and Wray, 1994).

Table 16.1 shows that the anthocyanin contents of sea buckthorn berries ranged from 1 to 22 mg/L in berry juice. The maximum amount was estimated in SBT-5, while the minimum was in SBT-4. Red berries (22 mg/L) were found to contain more anthocyanins than yellow (5.3 mg/L), orange (8.39 mg/L), or light yellow (1 mg/L) berries.

16.2.5 Mineral Content of Sea Buckthorn Berries

When the dried sea buckthorn berries from different populations were compared on the basis of mineral contents, a wide range of variation was observed (Table 16.2). These differences could be due to the natural contents of elements in the soil, as well as due to contamination in both the soil and the air.

Potassium was the most abundant element found and this is in line with other studies (Tong et al., 1989; Zhang et al., 1989). The potassium (2.5–7.9 g/kg), sodium (0.065–0.7 g/kg), calcium (0.65–1.22 g/kg), magnesium (149–270 mg/kg), iron (45–229 mg/kg), and phosphorus (115–138 mg/kg) contents were found to be high in dried berries. Kallio et al. (1999) reported that the Chinese sea buckthorn berries contained potassium (6.44–12.2 g/kg), calcium (0.8–1.48 g/kg), magnesium (0.47–0.73 g/kg), iron (64–282 mg/kg), zinc (8.8–27 mg/kg), and copper (3.8–12 mg/kg). The biochemical studies revealed that the fruit berries of sea buckthorn are an important source of valuable nutrients and minerals for local populations.

16.3 SEA BUCKTHORN PRODUCTS

Since the discovery of the nutritional value of sea buckthorn, hundreds of sea buckthorn products have been prepared from the berries, oil, leaves, bark, and their extracts have been developed. In Europe, sea buckthorn juice, jellies, liquors, candy, vitamin C tablets, and ice cream are readily available. Examples of commercial products available are "Biodoat'" sold in Austria; "Exsativa," a vitamin supplement sold in Switzerland; sea buckthorn syrup in France; liqueurs in Finland; and "Homoktovis Nektar" an apple-based fruit juice sold in Hungary. Sea buckthorn jams and jellies are produced on a small scale in Saskatchewan and Canada. At present, the largest producers and consumers of sea buckthorn products are China, Russia, and Mongolia. They all have large-scale processing facilities. Processed products include: oil, juice, alcoholic beverages, candies, ice cream, tea, jam, biscuits, vitamin C tablets, food colors, medicines, cosmetics and shampoos. In Russia, sea buckthorn berries are often used in home-made cosmetics. Recipes for moisturizing lotions, dandruff control and hair loss prevention are widely known and used in Russia. It is generally accepted that sea buckthorn oils have unique antiaging properties and as a result are becoming an important component of many facial creams manufactured in Asia and Europe. In addition, the UV-spectrum of the oil shows a moderate absorption in the UV-B range, which makes sea buckthorn derived products useful in sun care cosmetics. The potential for sea buckthorn oils in face masks, body lotions, and shampoos is excellent.

In India, sea buckthorn berries grow in various areas of Uttarakhand Himalya (Deepak et al., 2007). The moisture contents of these berries varied from 84.9 to 97.6%. Total soluble solid (TSS) content varied between 9.72 to 8.86%. The quantity of starch was the highest at 29.42–85.17% and the titratable acidity varied from 2.64 to 4.54%. Fat content was reported to be high (10.33%) in the fruit pulp. While the protein content was in the range of 7.13 and 28.33% in fruit pulp and seeds, respectively. Similarly, the carbohydrate content in fruits was 0.30–0.40%. Reducing sugars were the highest at 5.0–6.0%. Among minerals, fruit pulp contained 0.67% phosphorus. In the seeds it varied between 0.61 and 0.69% for the populations studied. The concentration of potassium varied between 10.12 and 14.84% in fruit pulp and between 9.33 and 13.42% in seeds. Other macro- and micronutrients, namely sodium, magnesium, iron, copper, and zinc among others are found to be present in low-to-moderate quantities in the fruit pulp and seeds of *H. rhamnoides*. The estimated nutritive value of *H. rhamnoides* fruit pulp varied between 110 and 120 Cal/100 g (Deepak et al., 2007). Deepak et al. (2007) reported that the use of sea buckthorn in central Himalaya is very popular and the rural people traditionally use this fruit as a food for preparing chutney (local jelly) and medicine. The Bhotiya tribes of Niti and Mana Valley mix the fruit juice of *H. rhamnoides* with sugar cubes/gur and boil it for 2–3 h in an iron pan. The thick and dark brown to black-colored cake produced is used as medicine for relief from colds, coughs, and throat infections. Inhabitants of high altitudes in general and *Hippophae* growing areas in particular, also use the fruit berries for veterinary medicine. The juice extracted from the fruits is known to reduce the poisonous effects of some plants grazed by livestock, mainly cattle, sheep, and goats. Besides, the Bhotiya tribe of this region uses the juice and pulp of the fruit berries as a substitute for tomato or curd for vegetable and curry preparation during the winters.

16.3.1 Sea Buckthorn in Northern Areas of Pakistan

In Pakistan, sea buckthorn is found in Kurram Agency, Chitral, Upper Swat, Utror-Gabral, Gilgit, Astore, Skardu, Ganche, Baltistan, Ladak, and all over the northern areas from 1225.9 to 4290.6 m elevation (Rasool, 1998). The northern areas of Pakistan especially Gilgit and Skardu have a tremendous potential for production of the wild sea buckthorn (*Hippophae rhamnoides* ssp. *turkestanica*). The plant is spread over all five districts of the region. According to estimates 3000 ha of land in the northern areas is under natural sea buckthorn cover (Nasir, 1997). If managed and utilized properly the plant can bring positive changes in the socio-economic conditions of the local communities. Though this plant is used there as firewood, hedges (both living and dead), fodder, and compost, yet its uses as medicine, food, and an income-generating source are limited and need to be introduced. Recently, four sea buckthorn products (jams, jelly, syrup, and squash) were first prepared according to the formulae recommended by the Pakistan Agriculture Research Center (PARC). Realizing the importance of sea buckthorn, the government of Pakistan launched a project "Sea buckthorn Exploitation and Development in Pakistan" in 1977 through the National Arid-Land Development Research Institute (NADRI). Under this project the Pakistan Council for Scientific and Industrial Research Institute (PCSIR) was assigned the tasks of developing, introducing, and promoting various products made from sea buckthorn. The Pakistan Agriculture Research Centre (PARC) unit at Gilgit developed a number of products (jams, jellies, chocolates, juices, and squashes, and the most important one sea buckthorn oil) and trained the local communities in their preparations on the domestic level, mainly in Skardu district. The berries are processed into different products by simple techniques. The fruit juice is extracted using a simple hand-press method and sieved with the help of a muslin cloth. The remaining residue (pulp and seeds) is dried in the sun followed by oven drying. The dried pulp and seed samples are made into a fine powder using a plant grinder. Rural people of the northern area use this fruit for preparing jellies and medicine by boiling the berries in water with sugar and use this as a medicine for relief from eye burning, stomach diseases, colds, coughs, and throat infections. The people of Astore use the fruit products of sea buckthorn in whooping cough and its decoction is given for cutaneous eruptions (Shinwari and Gilani, 2003). Sea buckthorn leaves are used in the preparation of herbal teas to relieve indigestion and dyspepsia and as a protective agent in liver diseases. Munawar Industries in Lahore, Pakistan, export sea buckthorn seed oil. It is mainly used in lowering blood cholesterol and hypertension. PCSIR laboratories in Pakistan has developed sea buckthorn jam blended with six prominent fruits from the northern areas namely apple, apricot, plum, cherry, and quince. Russian olive and mulberry were examined with sea-buckthorn like 100% pure sea-buckthorn and blended with other fruits at the ratio of 50:50, 70:30, and 90:10%, respectively. Fruits were thoroughly washed, peeled, pieced, and boiled with the addition of 10–25% water to obtain a uniform pulp. Cane sugar was added at an average of 60 %, citric acid at 0.5–1 % and pectin at 1%, respectively. The mixture was allowed to cook until the temperature reached 104.4–107.2°C. The end product (jam) was then filled in sterilized air tight glass jars, capped, labeled, and stored at ambient temperature. The results revealed that the combination of apple/apricot with sea-buckthorn in the ratio of 90:10 was highly appreciated above all other combinations by the taste panelist and retailers. Pure sea-buckthorn jam was very bitter, and only acceptable as remedies (Shahnawaz et al., 2004).

16.4 AUTUMN OLIVE

A member of the Elaeagnaceae family, also called cardinal olive or autumn elaeagnus (Dirr, 1998), it is a valuable shrub with the inherent ability to grow under natural conditions in Kashmir. It is locally named "Giani or Cancoli." It is a common medicinal shrub found in the wild at elevations of 1371.6–1828.8 m above sea level in Kashmir (Sabir et al., 2003). *E. umbellata* is a large-spreading, spiny-branched shrub often reaching 3.5–5.5 m height and 3.5–5.5 m width. The foliage is light green on the top and silvery green in the bottom (Dirr, 1998). Leaves are alternate and petiolated in small lateral clusters on twigs (Eckardt and Sather, 1987). Leaves are elliptic to ovate-oblong, 4–8 cm long, 1–2.5 cm wide, with upper surface sparsely white lepidote, lower surface densely white lepidote, apex acute to sometime obtuse; petioles are 0.5–1.0 cm long, densely white lepidote. The drupes (fruits) are silvery with brown scales when immature, ripening to a speckled red in September to October (Sternberg, 1982). The fruit is fleshy, subglobose to broadly ellipsoid, 6–8 mm long. Fruits are 1.25–1.5 cm in size and start as a spotted light green in mid-summer turning red in the autumn (Dirr, 1998).

A mature plant can produce 0.9–3.4 kg of fruit per year, with the number of seeds ranging from 20,000 to 54,000 (Eckardt and Sather, 1987). *E. umbellata* berry is an excellent source of vitamins, and minerals, especially vitamins A, C, and E, flavonoids and other bioactive compounds. It is also a good source of essential fatty acids (Chopra et al., 1986). The fruit contains 69.4 g of moisture, 14.5 g of total soluble solids, 1.15 g of organic acids, 8.34 g of total sugar, 8.13 g of reducing sugars, 0.23 g of nonreducing sugars, and 12.04 mg of vitamin C per 100 g of fruit. The total mineral content of the fruit as represented by its ash is 1.05% (Parmar and Kaushal, 1982). The astringent flavor of the berries may be due to the high total phenolic content (1700 mg/kg chlorogenic acid equivalents). The berries were found to be high in flavonols and hydroxybenzoic acids (33 rutin and 31 gallic acid equivalents), while the seeds are high in hydroxycinnamic acids and extremely high in hydroxybenzoic acids (35 chlorogenic acid and 184 gallic acid mg/kg equivalents) (Perkins et al., 2005).

Different parts of the plant possess a broad spectrum of antibacterial activities against Gram-positive bacteria including *S. aureus, B. subtilis* and Gram-negative bacteria including *E. coli* and *P. aeruginosa* (Sabir et al., 2007). It also contains lycopene, β-carotene, lutein, phytofuluene, and phytoene. Its lycopene content ranged from 10.09 to 53.96 mg per 100 g in fresh fruit from the naturalized plants and from 17.87 to 47.33 mg in the cultivars with red-pigmented fruit. Cultivars with yellow fruit had only 0.82 mg/100 g fresh weight of fruit. In contrast, fresh tomato fruit which is the major dietary source of lycopene, has a lycopene content of 0.88–4.20 mg per 100 g. This newly identified source of lycopene may provide an alternative to tomato as a dietary source of lycopene and related carotenoides (Kohlmeier et al., 1997; Fordham et al., 2001). Lycopene is widely believed to protect against myocardial infection (Kohlmeier et al., 1997) and various forms of cancer (Clinton, 1998), including prostrate cancer (Giovannucci et al., 1995). Thus, *E. umbellata* shows potential as a deterrent to heart disease, and cervix and gastrointestinal tract cancers (Matthews, 1994). Fruits (Figure 16.1) can be used in either the raw or cooked form (Hedrick, 1972). The fruit is juicy, pleasantly acidic, and can also be used to prepare jams or other preserves (Reich, 1991).

16.4.1 Traditional Uses of Autumn Olive

The seeds of the plant contain stimulant and its decoction is locally used as a remedy against coughs. The extracted seed oil is used to clear the chest and is used in

FIGURE 16.1 Harvested fruits of autumn olive from Rawalakot, Pakistan.

pulmonary infections. The high content of lycopene suggests the use of *E. umbellata* berries in the treatment of various diseases including prostate cancer, and as a deterrent to heart disease, cervix and gastrointestinal tract cancers (Matthews, 1994). The berries are astringent and the residents of Kashmir utilize the berries to reduce blood pressure. The berry is a rich source of vitamins and minerals and its use is popular among the poor communities suffering from malnutrition. Our recent studies demonstrated the antibacterial activity of *E. umbellata*. The organic and aqueous extracts obtained from leaves, fruits, and flowers displayed broad-spectrum activity against Gram-positive bacteria including *S. aureus*, and *B. subtilis* and Gram-negative bacteria including *E. coli* and *P. aeruginosa* (Sabir et al., 2007). These findings may provide the basis for traditional use of this plant in the treatment of infectious diseases. One hundred grams of autumn olive fruit contained 69.4 g of moisture, 14.5 g of total soluble solids, 1.51 g of acids, 8.34 g of total sugar, 8.13 g of reducing sugars, 0.23 g of nonreducing sugars, and 12.04 mg of vitamin C. The total mineral content of fruit as represented by its ash was 1.045%. The percentage of some of the minerals namely phosphorous, potassium, calcium, magnesium, and iron, was 0.054, 0.346, 0.049, 0.033, and 0.007 ppm, respectively.

16.4.2 Nutritional Analysis of Autumn Olive (AO) Berries

The biochemical constituents of samples collected from five populations of *E. umbellate* are shown in Table 16.3 while the mineral compositions are given in Table 16.4. The vitamin C content ranged from 13.8 to 16.9 mg/100 g among different samples of *E. umbellata* (Table 16.5). Earlier studies have shown 12.04 mg of vitamin C per 100 g of fruit (Parmar and Kaushal, 1982), which was a little lower compared with this study. The oil content in the seeds of *E. umbellata* was in the range of 5.7–6.1% (Table 16.3). Maximum oil was found in AO-1 (6.1%) while the minimum oil was in AO-5 (5.7%). The oil in the pulp was also extracted and was found to be in the range

TABLE 16.3
Concentration of Phytochemicals in Five Populations of Autumn Olive from Rawalakot, Pakistan

Biochemicals	AO-1	AO-2	AO-3	AO-4	AO-5
Vitamin C (mg/100 g)	16.9 ± 0.10	15.6 ± 0.10	13.8 ± 0.12	16.2 ± 0.05	14.4 ± 0.3
Oil in seed (g/100 g)	6.1 ± 0.21	5.91 ± 0.31	6.06 ± 0.01	5.84 ± 0.15	5.7 ± 0.5
Oil in pulp (g/100 g)	8.06 ± 0.15	7.60 ± 0.08	7.63 ± 0.17	8.11 ± 0.01	7.62 ± 0.9
Protein content (g/100 g)	5.3 ± 0.30	3.4 ± 0.6	4.13 ± 1.05	3.2 ± 1.1	2.3 ± 0.6
Reducing sugar (g/100 g)	7.4 ± 1.23	8.4 ± 0.12	8.1 ± 0.21	7.9 ± 0.14	6.8 ± 0.52
Nonreducing sugar (g/100 g)	2.2 ± 0.14	1.7 ± 0.04	1.4 ± 0.15	1.8 ± 0.17	2 ± 0.53

AO = autumn olive, ± standard deviation (SD).

of 7.60–8.1% (Table 16.3). Maximum oil was reported in AO-1 (8.06%) while the minimum level was in AO-2 (7.60%). Consequently, the pulp of *E. umbellata* showed higher quantity of fatty oils compared with the seed. The therapeutic potential of *E. umbellata* against heart and other diseases may be due to the presence of high amounts of oil in fruits. The plant oil and the phytosterols are known to have anti-coagulant properties that are highly suitable for lowering the blood cholesterol and angina (Fordham et al., 2001). *E. umbellata* berries were found to be sweet with reducing sugar content to be in the range of 6.8–8.4 g/100 g. The maximum amount of reducing sugar was reported in AO-2 while minimum was in AO-5. The quantities of nonreducing sugars were in the range of 1.4–2.2 g/100 g. Parmar and Kaushal (1982) reported the reducing and nonreducing sugar content of berries as 8.13 and 0.23%, respectively. This study reported almost equal amounts of reducing sugar (8.4%). However, the amount of nonreducing sugar was quite high, and at 2.2%. The high amounts of sugar make this berry equally good for eating as well as for its use in

TABLE 16.4
Concentration of Minerals in Different Populations of Autumn Olive from Rawalakot, Pakistan

Elements (ppm or mg/L)	AO-1	AO-2	AO-3	AO-4	AO-5
K	175 ± 1.6	375 ± 1.65	240 ± 0.14	340 ± 0.23	185 ± 0.1
Na	30 ± 0.01	25 ± 0.01	30 ± 0.02	20 ± 0.001	40 ± 0.03
Ca	80 ± 0.2	100 ± 0.5	98 ± 0.15	70 ± 0.03	110 ± 0.16
Mg	240 ± 0.15	225 ± 0.2	139 ± 0.9	229.5 ± 0.24	150 ± 0.31
Fe	225 ± 0.21	140 ± 0.14	120 ± 0.21	115 ± 0.13	40 ± 0.05
P	131 ± 0.11	128 ± 0.04	133 ± 0.12	115 ± 0.05	110 ± 0.12

AO = autumn olive, ± standard deviation (SD).

TABLE 16.5
Product/Ingredient Formula of Sea Buckthorn (SBT) Products at Domestic Levels in Pakistan

	Ingredients					
Products	Pulp/Juice (g/mL)	Water (mL)	Sugar (g)	Pectin (g)	Sodium Benzoate (g)	Citric Acid (g)
SBT Jam	1000	—	750	8	1	8
SBT Jelly	1000	4000	1000	16	2	28
SBT Syrup	1000	1000	4000	—	6	—
SBT Squash	2000	1000	3000	—	6	—

Source: Adapted From Khan, M.I. and Kamran, M. 2004. Cost analysis of sea buckthorn berries picking techniques and food products preparation at domestic level: A study conducted in Ghulkin, Upper Hunza, and Northern Areas of Pakistan. WWF-Pakistan, Jutial, Gilgit.

other food products like jams, jellies, and chocolates. The *E. umbellata* berry was also found to be rich in protein (5.3%) and its use should therefore, be encouraged. The protein content of *E. umbellata* berries was found in the range of 2.3–5.3%. AO-5 had the lowest (2.3%) while AO-1 had the highest (5.3%) amount of protein in fruit pulp.

Most of the fruits contain small amount of minerals, but the *E. umbellate* berry was found to be excellent source of minerals (Table 16.4). Potassium was the most abundant of all the elements investigated in the berries or juice. Mineral element composition revealed high contents of potassium (175–375 ppm), sodium (20–40 ppm), calcium (70–110 ppm), magnesium (70–86.6 ppm), iron (78.5–90 ppm), and phosphorus (110–133 ppm). Variations in content of all the elements studied were wide. Differences may originate from the natural contents of elements in the soil as well as from contamination in both the soil and the air.

16.4.3 Autumn Olive Products

This plant is mainly used in Kashmir as fire wood, hedges (both living and dead), fodder, and compost, yet its uses for medicine, and food, as well as an income-generating source, have been limited and need to be introduced. The berries are fully ripened in August and are harvested by handpicking, stick beating, and cutting of branches. Care must be taken in picking fruits as the shrub is thorny. High yields of autumn olive berries suggest their potential use in food and medicine for the poor people of Kashmir. The fruit is attractive and is also a favorite food of birds. The ripe berries are eaten as a food by the local community, and they traditionally use this fruit as food for preparing chutney (local jelly). Some people use the juice and pulp of fruit berries as a substitute for tomato paste. The life shelf of autumn olive berries is about 15 days and it can be easily processed into jams, squashes, and jellies.

In conclusion, the results of these studies provide basic information about the nutritional and medicinal importance of little studied species of sea buckthorn and autumn olive from Pakistan. The high concentration of oil found in *H. rhamminodes* and

E. umbellata can have commercial importance and can help the local communities in marketing their farm produce. Due to their high nutritive value these berries can be processed into fruit juices, squashes, jellies, jams, marmalade, and beverages. Autumn olive berries, being rich source of lycopene, can be used in the preparation of tomato ketchups and pastes. The berries can be used as a flavor ingredient in alcoholic and nonalcoholic beverages, frozen diary desserts, candies, baked foods, gelatins, and puddings. The oil from the flowers of the autumn olive can be used in perfumes and in cosmetic preparations. The foods prepared from these plants will not only provide the basic needs of the body but also help in combating different diseases and may be associated with their use as functional foods. Within this approach, local value-addition in the potential for wild edibles has begun to attract attention as being one of the income-generating components of the nonfarm part of the rural economy.

16.5 SUMMARY

Sea buckthorn and autumn olive are two important multipurpose plants of the Elaeagnaceae family from Pakistan. These plants are locally used in the treatment of different diseases and have a wide potential to be used as a food for economic activity. The chemical composition and nutritive value of sea buckthorn berries from eight populations from different areas of northern Pakistan were compared. The quantity of vitamin C in the berries was 240–339 mg/100 g, the seed oil was 5.69–12.7%, oil in dried pulp was 17.8–28.1% and the pytosterol content of the seed oil was 3.9–5.2%; there were 1–22 mg/L anthocyanins in the fruit juice. Elemental analysis revealed that the potassium (2.5–7.9 g/kg), sodium (0.065–0.7 g/kg), calcium (0.65–1.22 g/kg), magnesium (149–270 mg/kg), iron (45–229 mg/kg), and phosphorus (117–138 mg/kg) contents were high in dried berries. Similarly, five populations of *E. umbellata* from different areas of district Rawalakot Kashmir were compared using fruit characters. Chemical analysis of berries showed vitamin C (13.8–16.9 mg/100 g), seed oil (5.7–6.1%), oil in pulp (7.6–8.1%), reducing sugar (6.8–8.4%), nonreducing sugar (1.4–2.2%, and protein (2.3–5.3%), while the mineral element composition revealed high contents of potassium (175–375 ppm), sodium (20–40 ppm), calcium (70–110 ppm), magnesium (70–86.6 ppm), iron (40–225 ppm), and phosphorus (110–133 ppm). The study established Pakistani sea buckthorn and autumn olive berries as a good source of phytonutrients and mineral elements which may be associated with their potential use as a functional food.

REFERENCES

Agrawala, P.K. and Goel, H.C. 2002. Protective effect of RH-3 with special reference to radiation induced micronuclei in mouse bone marrow. *Indian Journal of Experimental Biology* 40:525–530.

Arimboor, R. Sarin, K.K., and Arumughan, C. 2008. Simultaneous estimation of phenolic acids in sea buckthorn (*Hippophaë rhamnoides*) using RP-HPLC with DAD. *Journal of Pharmaceutical and Biomedical Analysis* 47:31–38.

Bernath, J. and Foldesi, D. 1992. Sea buckthorn (*Hippophae rhamnoides* L.): A promising new medicinal and food crop. *Journal of Herbs, Spices and Medicinal Plants* 1:27–35.

Bridle P. and Timberlake, C.F. 1996. Anthocyanins as natural food colors-selected aspects. *Food Chemistry* 58:103–106.

Chen, Y. 2003. Study on the effects of the oil from *Hippophae rhamnoides* in hematopoiesis. *Chinese Herbal Drugs* 26:572–575.

Cheng, J., Kondoa, K., Suzuki, Y., Ikeda, Y., Meng, X., and Umemura, K. 2003. Inhibitory effects of total flavones of *Hippophae rhamnoides* on thrombosis in mouse femoral artery and *in vitro* platelet aggregation. *Life Sciences* 72:2263–2271.

Chopra, R.N., Nayar, S.L., and Chopra, L.C. 1986. *Glossary of Indian Medicinal Plants.* Council of Scientific and Industrial Research, New Delhi.

Clinton, S.K. 1998. Lycopene: Chemistry, biology, and implications for human health and disease. *Nutrition Reviews* 56:35–51.

Daigativ, D.D., Muratchaeva, P.M., and Magomedmirzaev, M.M. 1985. Correlation of some fruit characteristics with lipid and tocopherol content in *Hippohae rhamnoides* L. [in Russian]. *Rastit Resur* 21:283–288.

Darmer, G. 1952. Der sanddorn als Wild-und Kulturepflaze. Leipzig, pp. 89.

Deepak, D., Maikhuri, R.K., Rao, K.S., Kumar L., Purohit, V.K., Manju, S., and Saxena, K.G. 2007. Basic nutritional attributes of *Hippophae rhamnoides* (Seabuckthorn) populations from Uttarakhand Himalaya, India. *Current Science* 92:1148–1152.

Dirr, M.A. 1998. Manual of Woody Landscape Plants. Their Identification, Ornamental Characteristics, Culture, Propagation and Uses. Stipes, Champaign, IL.

Eccleston, C., Baoru, Y., Tahvonen, R., Kallio, H., Rimbach G.H., and Minihane, A.M. 2002. Effects of an antioxidant-rich juice (sea buckthorn) on risk factors for coronary heart disease in humans. *Journal of Nutritional Biochemistry* 13:346–354.

Eckardt, E. and Sather, A. 1987. The nature conservancy element stewardship abstract for *E. umbellata* practice. Prelim. Report. 111. Department of Conservation, USA, pp. 1–4.

Fordham, I.M., Clevidence, B.A., Wiley, E.R., and Zimmerman, R.H. 2001. Fruit of Autumn olive; A rich source of lycopene. *Hort-Science Alexandria*, 36:1136–1137.

Geetha, S., Jayamurthy, P., Pal, K., Pandey, S., Kumar, R., and Sawhney, R.C. 2008. Hepatoprotective effects of sea buckthorn (*Hippophae rhamnoides* L.) against carbon tetrachloride induced liver injury in rats. *Journal of the Science of Food and Agriculture* 88:1592–1597.

Giovannucci, E.A., Ascherio, E.B., Rimm, M.J., Stampfer, G.A., Colditz, G.A., and Willett, W.C. 1995. Intake of carotenoids and retinol in relation to risk of prostate cancer. *Journal of the Natural Cancer Institute* 87:1767–1776.

Hedrick, U.P. 1972. *Sturtevant's Edible Plants of the World.* Dover Publications, New York.

Hussain, F. and Khaliq, A. 1996. Ethnobotanical studies on some plants of Dabargai Hills Swat. *Proceedings of the First Training Workshop on Ethnobotany and Its Application to Conservation.* NARC, Islamabad, pp. 207–215.

Jeppsson, N. 1999. Progress in research and development of sea buckthorn in Sweden. *Proceedings of International Symposium on Sea Buckthorn (Hippophae rhamnoids* L.), Beijing, China.

Kallio, H., Yang, B.R., and Peippo, P. 2002. Effects of different origins and harvesting time on vitamin C, tocopherols, and tocotrienols in sea buckthorn (*Hippophaë rhamnoides*) berries. *Journal of Agricultural and Food Chemistry* 50:6136–6142.

Kallio K., Yang B.R., Tahvonen R., and Hakala M. 1999. Composition of sea buckthorn berries of various origins. *Proceedings of International Symposium on Sea Buckthorn (Hippophae rhamnoids* L.) Beijing, China.

Kamei, H., Kojima, T., Hasegawa, M., et al. 1995. Suppression of tumor cell growth by anthocyanins *in vitro. Cancer Investigation* 13:590–594.

Khan, M.I. and Kamran M. 2004. Cost analysis of sea buckthorn berries picking techniques and food products preparation at domestic level: A study conducted in Ghulkin, Upper Hunza, and Northern Areas of Pakistan. WWF-Pakistan, Jutial, Gilgit.

Kohlmeier, L., Kark J.D., Gomez, G.E., Martin, B.C., and Steck S.E., 1997.Lycopene and myocardial infarction risk in the EURAMIC study. *American Journal of Epidemiology* 146:618–626.

Ma, Z. and Cui, Y. 1989. Studies on the fruit character and bio-chemical composition of some forms within the Chinese Sea buckthorn (*H. rhamnoides* ssp. *sinesis*) in Shanxi, China. *Proceedings of the International Symposium on Sea Buckthorn (H. rhamnoides L.)*, Xian, China, pp. 106–112.

Mark, R.G. and Peter, U.M. 2004. A palmitoleic acid ester concentrate from sea buckthorn pomace. *European Journal of Lipid Science Technology* 106:412–416.

Matthews, V. 1994. *The New Plantsman*. Royal Horticultural Society, London.

Mironov, V.A. 1989. Chemical composition of *Hippophae rhamnoides* of different populations of USR. *Int. Symp. Sea Buckthorn (H. rhamnoides* L.). Xian, China, p. 67.

Mingyu, X. 1994. Anticancer effects of and direction of research on Hippophae. *Hippophae* 7: 41–43.

Nakamura, Y., Watanabe, S., Miyake, N., Kohno, H., and Osawa, T. 2003. Dihydrochalcones: Evaluation as novel radical scavenging antioxidants. *Journal of Agricultural and Food Chemistry* 51:3309–3312.

Nasir, 1997. Sea buckthorn a valuable medicinal plant. Available at: http://www.wwfpak.org/nap/dnap_medicinalplants_seabuckthorn.php (accessed May 17, 2010).

Negi, P.S., Chauhan, A.S., Sadia G.A., Rohinishree, Y.S., and Ramteke R.S. 2005. Antioxidant and antibacterial activities of various sea buckthorn (*Hippophae rhamnoides* L.) seed extracts. *Food Chemistry* 92:119–124.

Nersesyan, A., and Muradyan, R. 2004. Sea-buckthorn juice protects mice against genotoxic action of cisplatin. *Experimental Oncology* 26:153–155.

Parmar, C. and Kaushal, M.K. 1982. *Elaeagnus umbellata*. In: *Wild Fruits*. Kalyani Publishers, New Delhi, India, pp. 23–25.

Perkins Veazie, P.M., Black, B.L., Fordham, I.M., and Howard, L.R. 2005. Lycopene and total phenol content of autumn olive (*Elaeagnus umbellata*) selections [abstract]. *HortScience* 40 (30):883.

Rasool, G. 1998. Medicinal Plants of the Northern Areas of Pakistan: Saving the Plants that Save Us. Gilgit, Pakistan.

Reich, L. 1991. *Elaeagnus: Gumi, Elaeagnus Umbellata, and Russian Olive, and Uncommon Fruits Worthy of Attention*. Addison-Wesley Publ. Co., Reading, Massachussets, pp. 113–120.

Rongsen, A. 1992. Sea buckthorn a multi-purpose plant species for fragile mountains. ICIMOD occasional paper No. 20, pp. 6–18. Kathmandu, Nepal.

Rousi, 1977. Variation of vitamin C concentration and character correlation between and within natural sea buckthorn (*Hippophae rhamnoides* L.) populations. *Acta Agriculturae Scandinavica* 42 (1):12–17.

Sabir, S.M., Ahmad, S.D., Imtiaz, H., and Tahir, M.K. 2007. Antibacterial activity of *E. umbellata* (Thunb.), a medicinal plant from Pakistan. *Saudi Medical Journal* 28:477–481.

Sabir, S.M., Ahmad, S.D., and Lodhi, N. 2003. Morphological and biochemical variation in Sea buckthorn *Hippophae rhamnoides* L. ssp. *turkestanica*, a multipurpose plant for fragile mountains of Pakistan. *South African Journal of Botany* 69:587–592.

Sezai, E., Emine, O., Ozlem O., and Memnune S. 2007. The genotypic effects on the chemical composition and antioxidant activity of sea buckthorn (*Hippophae rhamnoides* L.) berries grown in Turkey. *Scientia Horticulturae* 115:27–33.

Shahnawaz, M., Khan, T.U., and Tariq, M. 2004. Development of value added sea-buckthorn jam blended with prominent local fruits of northern Pakistan. *Sarhad Journal of Agriculture* 20:643–660.

Shinwari, Z.K. and Gilani, S.S. 2003. Sustainable harvest of medicinal plants at Bulashbar Nullah, Astore (Northern Pakistan). *Journal of Ethnopharmacology* 84:289–298.

Sternberg, G. 1982. *Elaeagnus umbellata* in Illinois conservation practice. Report. 111. Dept. of Conservation, Virginia, pp. 251–278.

Strack, D. and Wray, V. 1994. The anthocyanins. In: *The Flavonoids: Advances in Research Since 1986* (J.B. Harborne, ed.). Chapman and Hall.

Thurnham D.I. 1999. Functional foods: Cholesterol-lowering benefits of plant sterols. *British Journal of Nutrition* 82:255–256.

Timberlake, C.F. and Henry, B.S. 1988. Anthocyanins as natural food colorants. *Prog. Clin. Biol. Res.* 280:107–121.

Tong, J., Guo, C., Zhao, Z., Yang, Y., and Tian, K. 1989. The determination of physical—chemical constants and sixteen mineral elements in sea buckthorn raw juice. *Proceedings of International Symposium on Sea Buckthorn* (*H. rhamnoids.* L), Xian, China.

Wang, H., Cao, G., and Prior, R.L. 1997. Oxygen radical absorbing capacity of anthocyanins. *Journal of Agricultural and Food Chemistry* 45:304–309.

Xing, J.,Yang, B., Dong, Y.,Wang, B., Wang, J., and Kallio, H. 2002. Effects of sea buckthorn (*Hippophaë rhamnoides* L.) seed and pulp oils on experimental models of gastric ulcer in rats. *Fitoterapia* 73:644–650.

Yao, Y. 1994. Genetic diversity, evolution and domestication in sea buckthorn (*Hippophae rhamnoides* L.). PhD dissertation, Helsinki University, Finland.

Yao, Y. and Tigerstedt, P.M.A. 1995. Geographic variation of growth rhythm, height and hardiness and their relations in *Hippophae rhamnoides. Journal of American Society of Horticulture Science* 120:691–698.

Yao, Y., Tigerstedt, P.M.A., and Joy, P. 1992. Variation of vitamin C concentration and character correlation between and within natural sea buckthorn (*H. rhamnoides.* L.) populations. *Acta Agriculturae Scandinavica* 42:12–17.

Yang, B.R. 2001. Lipophilic components of Sea buckthorn (*Hippophae rhamnoides*) seeds and berries and physiological effects of sea buckthorn oils. PhD dissertation, Turku University, Finland.

Yang, B. and Kallio, H. 2002. Supercritical CO_2 extracted sea buckthorn (*Hippophaë rhamnoides*) oils as new food ingredients for cardiovascular health. *Proc. Health Ingred. Europe 2002*. Paris, September 17–19, p. 7.

Zao, T.D., Cheng, Z.X., Liu, X.Y., Shao, J.Y., Ren, L.J., Zhang, L., and Chen, W.C. 1987. Protective effect of the sea buckthorn oil for liver injury induced by CCl_4. *Zhongcaoyao* 18:22–24.

Zhang, M. 1987. Treatment of ischemic heart diseases with flavonoids of *Hippophae rhamnoides. Chinese Journal of Cardiology* 15:97–99.

Zhang, P. 1989. The anti-cancer activities of Hippophae seed oil and its effect on the weight of the immunological organs. *Hippophae* 3:31–41.

Zhang, P., Mao, Y.C., Sun, B., Qian, M., and Qu, W.J. 2005. Changes in apoptosis-related genes expression profile in human breast carcinoma cell line Bcap-37 induced by flavonoids from seed residues of Hippophae Rhamnoides L [in Chinese]. *Ai Zheng* 24:454–460.

Zhang, W., Yan, J., Duo, B., Ren, A., and Guo, J. 1989. Preliminary study of biochemical constitutes of berry of sea buckthorn growing in Shanxi Province and their changing trend. *Proceedings of International Symposium on Sea Buckthorn (H. rhamnoides.* L.*)*, Xian, China.

Zheng, X.W. and Song, X.J. 1992. Analysis of the fruit nutrient composition of nine types of sea buckthorn in Liaoning, China. *Northern Fruits of China* 3:22–24.

Zhou, Y. 1998. Study on the effect of *Hippophae* seed oil against gastric ulcer. Institute of Medical Plants Resource Development, The Chinese Academy of Medical Sciences, Beijing China.

17 Traditional Medicinal Wines

John Shi, Xingqian Ye, Bo Jiang, Ying Ma, Donghong Liu, and Sophia Jun Xue

CONTENTS

17.1 INTRODUCTION

The health benefits of medicinal wines such as herbal wines have a long history of being recognized in Asian countries, and are now getting attention from all around the world. Medicinal wines (liquors) refer to a transparent medicinal liquid obtained by using wine as a solvent to soak out the effective components from herbs, animal or insect parts, or from other medicinal materials. The purpose of medicinal wine is to fortify the medicinal herbal function by extracting the functional components with wine and then condensing the extract, concentrating the effective agent. Because wine or liquor itself has an effect of stimulating blood circulation and relaxing muscles and joints, it can be used to treat general asthenia (loss of strength), rheumatic pain, and traumatic injury. Wines (or liquors) are used not only as beverages, but also as vehicles to preserve medicinal herbal activity. In addition, alcohol in wine is a good solvent, which may extract a higher proportion of ingredients from the medical material. Asian people like to use precious medicinal materials to make medicinal wines (or liquors), which can reinforce body fluids and blood. Most medicinal wines (or liquors) are taken orally, while some are for external use. To improve the taste, crystal sugar or honey can be added to medicinal wines. However, though medicinal wines (liquors) are good for the human body, it cannot be drunk superfluously. A frequency of 2 or 3 times a day with each dosage measuring 10–50 mL is considered being appropriate. For example, "Spirit of Ginseng" and "Gecko and Cordyceps" are famous medicinal wines which are used to treat bronchial asthma at the remission stage. As wines (or liquors) are warm and dispersing in nature, they are contraindicated in the case of flaring of fire due to *Yin* deficiency, according to traditional Chinese medicine tradition.

17.2 HEALTH BENEFITS OF MEDICINAL WINES (OR LIQUORS)

Traditional Chinese herbal medicine has its own theory and philosophy relating to illness, which says that a healthy body has a delicate balance of *Yin-Yang*. If the balance is disturbed, illness will happen, and it will get worse. Insomnia, and loss of appetite among others, are generally caused by the imbalance of *Yin-Yang* in the human body. Stagnation of energy and blockage of blood circulation are caused by injuries. These pains can be easily overcome by individuals during their youth, but these blockages build up as one gets older. By the time people have a deficiency of energy, they can no longer fight these resistances to good health, resulting in pains, becoming more severe as time progresses (Beijing Traditional Chinese Medical College and Hospital, 1981; You, 1996; Zhang, 1997). Medicinal wines (or liquors) can provide for general health and effective treatments for arthritis, backache, improve vitality, insomnia, invigorating sexual competence, joint pain, lost appetite, muscular pain, numbness, menstrual period pain, rheumatism, athletic injury, sprains, and strain (You, 1996; Williams, 1999; Ma, 2002).

When pain occurs, there are two explanations for it. The first is its physical damage to the area, that is, in growing bone, torn muscle, or if on the back, slipped disks. The second is the blockages of circulation in very severe conditions. This cause can be treated with medicinal wines (or liquors) containing special herbs. Medicinal wines such as herbal wines act as cleansing agents to relieve the entire blockage in the veins and arteries, also to balance the energy deficiency or *Yin-Yang* in the patient's organs, and to improve blood circulation. This flow will correct many complaints such as arthritis, backache, rheumatic pain, and many other types of joint pains. Once the circulation of both energy and blood are established, other problems like insomnia, appetite loss, and numbness, among others, will be autocorrected by these changes. The key to a healthy body is to keep the circulation flowing constantly. Generally, the prescribed herbs are tonics for deficiencies in organs, clearing blockage of blood circulation, and improving energy flows in the body systems.

When taking medicinal wines, it is necessary to follow recommended dosage instructions and to be aware of cautionary procedures (Li, 1578; You, 1996; Zhang, 1997; Williams, 1999). Generally, herbal wines are considered as being safe. According to traditional Chinese medicine, when a person suffers arthritis, rheumatism, or joint pain, this is because the liver and kidneys are under long-term weakening from exterior or extrogenous factors, that is, being constantly in cold and damp areas. When some people suffer backache, if it is because of physical damage, and Western medicine is able to replace the damaged area. However, if the cause has no physical or rational explanation, then traditional Chinese medicine explains that there are deficiencies in the kidneys.

When some people suffer insomnia or loss of appetite, it is caused by a history of worries, excessive thinking, and stress. When some people feel muscular pain, menstrual period pain, numbness, aches, and pains, it is caused by restrictions in the blood circulation (Yeung, 1985; Zhou, 1986; Lu, 1991). When some people suffer poor vitality, it is all caused by kidney deficiency. When people have constant mild pains, it is caused by the deficiency of the body's organs (Beijing Traditional Chinese Medical College and Hospital, 1981; Zhou, 1986; Zhang, 1997). Medicinal wines (or liquors) containing special herbal materials can treat all of the above health problems.

The effect can vary according to the individual, as some will be affected by a small dose of alcohol, whereas others will need more doses to enable the circulation to take place properly. The majority will feel a flush in the face followed by sleepiness as the alcohol affects the whole system. Muscles, joints, or back are the major areas in which patients will feel more aches or pains during the course of treatment. There is often a steady improvement from the first few dosages; sometimes there is an initial worsening, followed by rapid improvement (Sun, 1955; Beijing Traditional Chinese Medical College and Hospital, 1981; Lu, 1991).

17.3 HISTORY OF MEDICINAL WINES (OR LIQUORS)

Throughout history, the human civilization has developed numerous ways to prolong health and enhance beauty. The traditional Chinese medicine approach has close to a 5000 year history of prolonging the age and improving of the health conditions for the emperors, queens, and nobleman. Throughout the numerous dynasties of the Chinese civilization, as early as the Yellow Emperor, the traditional Chinese medicine approach has always been first to strengthen the body and second to treat diseases. The approach to strengthening the body enabled the emperors to extend their reign through taking various kinds of medicinal herbs, elixirs, and dietary supplements. Medicinal wines (or liquors) are one of the important dietary supplements for health purposes.

Our common ancestors began to make wine several thousand years ago, and wine making culture is a heritage until now. Ancient Chinese doctors recognized the health care function of wine a long time ago. In the book of *Shi Jing*, doctors recommented that drinking wine was good for people's long life. In the book of *Han Shu*, people regarded wine as something that God gave them. Wine was usually the Monarchs' favorite drink, and people knew it was good for health. Grapes were described in the book of *Shen Nong Ben Cao Jing* which said that sweet grapes helped people become stronger, and that moderate consumption of grape wine was good for people's long life, and grapes were used to make wine one thousand years ago (Sun, 1955; Ling et al., 1984; Zhang, 1997; Williams, 1999).

Wines were also mentioned in the book of *Bencao Gangmu* written by *Li Shizhen* (Li, 1578, new version, 1991). It was said that wines were good for the kidneys and for staying young. When people proposed a wine toast, they used to wish each other a long life. When *Hu xihui*, the writer of the book *Yinshan Zhengyao*, talked about wines in his book, he indicated that wines were good for people's blood circulation. In the book of *Gujin Tushu Jicheng*, wines were mentioned as being good to recover from fatigue. Wines with some medicinal herbs were helpful for the appetite and digestion. They were also good for the skin (Sun, 1955; Beijing Traditional Chinese Medical College and Hospital, 1981; Ling et al., 1984; Lu, 1991).

In China, wine could also be called the "Water of History," because stories about wine can be found in almost every period of China's long story. The origin of alcoholic beverages from fermented grains in China is believed to have a 4000-year history. A legend said that *Yidi*, the wife of the first dynasty's King *Yu* (about 2100 BC) invented the method. At that time, millet was the main grain for wine making, the so-called yellow wine, then rice wine became more popular. It was not until the nineteenth century that distilled liquors become more popular. Traditionally, Chinese distilled liquors are consumed together with food, rather than simply drunk by themselves.

Jiu is the Chinese word that refers to all alcoholic beverages (wines and liquors), from beer (*Pi Jiu*), to liquors (*Jiu*), to grape wine (*Putao Jiu*). This word has often been translated into the English language as "wine," although the meaning is not the same. Many Chinese "wines" are made from grains and herbs and distilled to high concentrations. The same character is used for Japanese and Korean wines. This lumping together of all intoxicating beverages gives us great insight into the traditional uses for wines (or liquors) for health purposes. Traditional Chinese wines are rarely made of fruits. Chinese wines from southern China are mostly made of rice, but those from northern China are mostly made of wheat and sorghum. Most are colorless clear liquids unless medicinal herbs are added to give a different color. Additionally, grape wine is increasingly produced and consumed in China, Korea, and Japan due to the influence of Western culture. "Chinese-style wine" and "Chinese wine" are often used interchangeably and inaccurately to refer to Chinese-made "wines" made of fermented fruit juices, especially mulberries or grapes, as well as grape wine (*Putao Jiu*), rice wines (*Huang Jiu*), or to distilled sorghum-based hard liquors (such as erguotou or more generally *Bai Jiu*). Alcoholic drinks are identifiable in Chinese by the suffix *Jiu* (as in *Pi Jiu* or beer), which can refer to many things other than drinks that contain alcohol (such as rubbing alcohol).

In *Jewang ungi*, a history book written in 1287 during the *Goryeo* Dynasty, Korea, a myth regarding the origin of alcoholic drinks appears. It was about a king who enjoyed using alcohol to tempt a woman to want to have many children. *Su* means water and B*ul* means fire, that is, "firewater" originated from the boiling liquid.

Rice wines had been the most popular alcoholic drinks for the Chinese in ancient times, and are still one of the popular alcoholic beverages, especially in Southern China, Korea, and Japan. "Yellow Wine," a kind of rice wine, is popular among all classes of the native population, and the most popular wine in China. It has a clear orange-yellow color with a fragrant smell. It contains 17 amino acids required by the human body and it is a low-density nourishing wine. According to historical records, rice wine has been made in China since 2500 BC "Shaoxing Rice Wine" (or Yellow Wine) is made from a brown rice from the Shaoxing area, and is considered the best rice wine. The health benefits of this wine are legendary in Chinese culture history and medicine. A little wine like yellow rice wine every day will do wonders for your body and spirit. Most Korean traditional alcoholic beverages have been made from rice, of both the glutinous and nonglutinous varieties, which is fermented with the aid of yeast and *nuruk*, a wheat-based source of the enzyme amylase. Additionally, Koreans often use fruits, flowers, herbs, and other ingredients to flavor these wines.

The wine container is popularly used in the traditional Chinese doctor's office, since the medicinal wines use lots of medicinal herbs mixed with wines. It is necessary to keep the mixture of wine and herbs for a while to be effective. The ancient wine containers have a spout for pouring out the wine to use, during the storage time.

17.4 WINE (OR LIQUOR) SELECTIONS

Wines (or liquors) represent the major portion of medicinal wine products. Chinese wines can be generally classified into two types, namely yellow liquors or clear (white) liquors. Chinese yellow liquors are fermented wines that are brewed directly from grains such as rice or wheat. Such liquors contain less than 20% alcohol, due to the inhibition of fermentation by ethanol at this concentration. These wines are traditionally pasteurized, aged, and filtered before their final bottling for sale to consumers. White liquors are also commonly called *Shao Jiu*, which means "hot liquor" or "burned liquor," because of the burning sensation in the mouth during consumption. Liquors of this type typically contain more than 30%, some even up to 60% alcohol by volume since they have undergone distillation concentration. There are many varieties of wines (or liquors) originating from China, Korea, and Japan that are used for the preparation of medicinal wine (liquor) as listed below.

Fen Jiu: It is the original Chinese white wine made from sorghum. Its alcohol content by volume is 63–65%.

Zhu Ye Qing Jiu: This wine is *Fen Jiu*, brewed with a dozen or more selected Chinese medicinal herbs. One of the ingredients is bamboo leaves, which gives the wine a greenish color and its name. Its alcohol content by volume is 46%.

Mao Tai Jiu: It is named after its origin at *Mao Tai* town in Guizhou Province, China. It is made from wheat and sorghum with a unique distillation process that involves seven iterations of the brewing cycle. This wine was made famous in the Western world when the Chinese government served it at state banquets. Its alcohol content by volume is 54–55%.

Gao Liang Jiu: Gao Liang is the Chinese name for sorghum. Besides sorghum, the brewing process also uses barley and wheat, and so on. Its alcohol content by volume is 61–63%.

Mei Gui Lu Jiu (rose essence wine): A variety of *Gao Liang Jiu* with distillate from a special species of rose, and crystal sugar. Alcohol content by volume is 54–55%.

Wu Jia Pi Jiu: A variety of *Gao Liang Jiu* with a unique selection of Chinese herbal medicines added to the brew. Its alcohol content by volume is 54–55%.

Da Gu Jiu: This wine is made of sorghum and wheat by fermenting in the cellar by a unique process for a long period of time. Its alcohol content by volume is 52%.

Yuk Bing Shiu Jiu: It is a Cantonese rice wine that is made of steamed rice. It is stored for a long period with submerged pork fat after distillation. The pork fat is removed before bottling. Its name is probably derived from the brewing process. *Yuk* is a homophone of meat in Cantonese and *Bing* means ice, which describes the appearance of the pork fat floating in the wine. Cantonese rice wine breweries have prospered since the Northern *Sung* dynasty, when the *Foshan* area was exempted from alcohol tax. Its alcohol content by volume is 30%.

Sheung Jing Jiu (through double distillations) and *San Jing Jiu* (through triple distillations)*:* Two varieties of rice wine by distilling twice and three times, respectively. Its alcohol content by volume is 32 and 38–39%, respectively.

San Hua Jiu (three flowers)*:* A rice wine made in Guilin, China, with allegedly over a thousand years of history. It is famous for the fragrant herbal addition and the use of spring water from Mount Elephant in the region. Its Alcohol content by volume is 55–57%.

Fujian Glutinous Rice Wine: It is made by adding a long list of expensive Chinese medicinal herbs to glutinous rice, then distilled to get a low alcohol rice wine. The unique brewing technique uses another wine as a raw material, not starting with water. The wine has an orange red color. Its alcohol content by volume is 18%.

Hua Diao Jiu: It is a variety of yellow wine originating from Shaoxing, Zhejiang. It is made of glutinous rice and wheat. Its alcohol content by volume is 16%. *Hua Diao Jiu* literally means flowery carving. The name describes the appearance of the pottery that stored the wine. This wine evolved from the Shaoxing tradition of burying the wine underground when a daughter was born. The wine would be dug up for the wedding banquet when the daughter got married. The containers would then be decorated with bright colors as a wedding gift. To make the gift more appealing, people started to use pottery with carvings and patterns, and hence the name *Huadiao*. Depending on the timing of the girl's marriage, the wine was usually aged for years. *Huadiao Jiu* or *Shaoxing Jiu* are basically made of the same wine except that they are named differently depending on the age, the container, and how they are used.

Yakju (medicinal alcohol)*: Yakju* is a refined rice wine in Korea, and is made from steamed rice that has gone through several fermentation stages. It is also called *Myeongyakju* or *Beopju*, and is distinguished from *Takju* by its relative clarity. Varieties include *Baekhaju*, which is made from glutinous rice and Korean *nuruk* and *Heukmeeju* (black rice wine), which is made from black rice.

Cheongju: Cheongju (clear wine or clear liquor) is a clear Korean rice wine similar to Japanese sake. One popular brand of *Cheongju* is *Chung Ha*, There are various local variations, including *Beopju*, which is brewed in the ancient city of Gyeongju.

Some Distilled liquors in Korea: Korean distilled liquors such as *Goryangju* are made from sorghum and are similar to Chinese "Gaoliang Jiu." *Okroju* is made from rice and Job's tears. *Munbaeju* is a traditional distilled liquor made of malted millet, sorghum, wheat, rice, and *nuruk* (fermentation starter) with strength of 40% alcohol by volume. It originates in the Pyongyang region of North Korea and is noted for its fragrance, which is said to resemble the flower of the munbae tree (similar to a pear). *Munbaeju* is popular in South Korea.

Soju: Soju is a clear, slightly sweet, distilled spirit that is by far the most popular Korean liquor. It is made from grain or sweet potatoes. It typically has an alcohol content of 20% by volume. The *Soju* from Andong, Kyongsangbuk-do, is a distilled liquor produced from fermented *nuruk*, steamed rice and water. The *Soju* was a very valuable commodity in olden times, and records show that it was used for medicinal purposes as well. Even today in the Andong, Korea, it is used to treat injuries and various digestive problems as well as to improve one's appetite. The *Soju* also has a high alcohol content of 45% as it is aged in a storage tank for more than 100 days after fermenting for 20 days. Despite its potency, it is known for its smooth taste and rich flavor. The traditional maturation method was to store the distilled liquor in a jar placed underground in a cave with a temperature of under 15°C for 100 days.

Sogokchu: The *Sogokchu* has an alcohol content of 15–16%.

Takju: *Takju*, better known as *Makgeolli*, is a milky, off-white, sweet alcoholic beverage made from rice. It is also called *Nongju* (farmers' alcohol). A regional variant, originally from Gyeonggi-do, is called *Dongdongju*. Another variety, called *Ihwaju* (pear blossom wine) was so named because it was brewed from rice with rice malt that had fermented during the pear blossom season.

Fruit Wines in Korea: Korea has a number of traditional fruit wines, produced by combining fruits or berries with alcohol. *Podoju* is made from rice wine that is mixed with grapes. The most popular fruit wines are made from maesil plums, Chinese quinces, cherries, pine fruits, and pomegranates (such wines are called *Maesilju, Mae Hwa Su, Mae Chui Soon, or Seol Joong Mae*), *Bokbunja* (Korean black raspberries, 15% alcohol). *Bokbunja Ju* (bokbunja wine) is said by many to be especially good for sexual stamina.

Flower Wines in Korea: There are a number of Korean traditional wines produced from flowers. These include wines made from chrysanthemums, peach blossoms, honeysuckle, wild roses, and sweet briar petals and berries. *Dugyeonju* is a wine made from azalea petals, produced in Chungcheong Province. It is sweet, viscous, and a light yellowish brown in color, with a strength of about 21% alcohol. Another variety of flower wine, called *Baekhwaju* is made from 100 varieties of flowers.

Rice Wine in Japan (Nihonshu or Sake): Nihonshu or *Sake* is commonly called *Sake* outside of Japan, and is brewed using rice, water, and white *koji* mold as the main ingredients. Besides major brands, there are several varieties of local rice wines (*Jizake*). The alcohol content of *Nihonshu* is typically about

10–20%. It is drunk either hot or cold, and it is usually filtered, although unfiltered *Nihonshu* (*Nigorizake*) is also popular.

Shochu or Awamori: Shochu is a distilled liquor with an alcohol content usually between 20 and 40%. It is commonly made from rice, sweet potatoes, wheat, and sugar cane. *Awamori* is the Okinawan version of *Shochu.* It differs in that it is made from long-grained thai-style rice instead of short-grained Japanese-style rice, and uses a black *koji* mold indigenous to Okinawa.

Chuhai: Chuhai (*Shochu Highball*) are fruit-flavored liquors with alcohol contents of 5–8%. Common flavors include lemon, ume, peach, grapefruit, lime, and mikan (mandarin orange). In addition, there are many seasonal flavors that come and go. Recent ones include winter pear, pineapple, and *nashi* (Japanese pear). They are usually shochu based, and are available in cans anywhere alcohol is sold.

Plum wine (Umeshu): Umeshu is made of Japanese plums (*ume*), sugar, and *shochu* or *nihonshu*. Its sweet, fruity, juice-like flavor and aroma can appeal to those who normally dislike alcohol.

Happoshu: Happoshu is a relatively recent invention by Japanese brewing companies. It has a similar flavor and low alcohol content, but it is made with less malt, which gives it a different, lighter taste.

17.5 SELECTIONS OF HERBALS AND OTHER MEDICINAL MATERIALS

According to traditional Chinese medicine, a key step is to maintain the *Yin* and *Yang* balance in the human body with medicines and medicinal diet (wine) treatments. Medicinal wines (or liquors) have the function of nourishing *Yin* or *Yang*, based on different medicinal herbs or other medicinal materials. The following natural materials are commonly used for medicinal wine preparation (Wu, 1982, 1996; Wu and Zhong, 1999).

1. *Animals and insects:* snake, tiger bone, tiger bile, ants, deer horn, bear bile, musk moschus, honeycomb, and dog organs.
2. *Fruits:* Hawthorn, Lily bulb and Mulberry fruits.
3. *Medicine herbs:* Angelia, Bamboo leaf, Ginger, American ginseng, Korea ginseng, Cassia bark, Cinnamon bark, Cordyceps, Poria, Pilose antler, Antler glue, Gastrodia tuber, Chrysanthemum, Eucommia bark, Acanthopanax bark, Honeysuckle flower, Indian bread, Soloman seal rhizome, Roxburgh rose, Fleece flower root, Barbary wolfberry fruits, Cherokee rose fruit, Honeysuckle flower, Nutmeg, Magnolia-vine fruit, Pine leaf, and Pine root.

17.6 SOME FAMOUS MEDICINAL WINES (LIQUORS) WITH ANIMAL AND INSECT MATERIALS

Some medicinal wines (liquors) with specific animal and insect material are used for prevention of the aging processes. According to traditional Chinese medicine, the

FIGURE 17.1 Medicinal wine aging in a cold storage room in Aunt Meng Resturant, Xitang Town, Jiashan, Zhejiang, China.

prevention of the aging process is to nourish *Yin*, to facilitate blood circulation, and to eliminate excessive *Yang*. Medicinal wines are usually aged in a cool storage room or in an underground cell for several months, then used for diet purposes (Figure 17.1). During the aging process, health-promoting components from herbs or animal parts are completely soaked out, and some interactions might occur for health benefits.

17.6.1 Snake Wines (or Liquors)

Snake wines (or liquors) generally include a whole snake in the bottle, or one or more submerged snakes in the wine or liquor bottle (Figure 17.2). This kind of wine can be found in many restaurants. Snake wine is considered to have the functionality of alleviating arthritis. The species of snakes are carefully selected for these medicinal properties. Other varieties are snake bile wine, where snake bile is soaked in wine or liquor, and snake skin wine where snake skin is soaked in wine or liquor.

During the *Tang* Dynasty (618–907 CD), remittance of all taxes could be reduced or eliminated if a person sent two or three golden serpents (*Chin She*) to the Emperor. In earlier days and today, snakes, particularly poisonous ones, are considered *Pu* which means they are good for strengthening and restoring, also for supplementing and heating. They are also consumed to improve poor pallor, to ward off chills (particularly in pregnant women), and for other weaknesses in both sexes. These wines are also considered to be good for vision.

Three to five snakes, some poisonous and the rest not, are most often used together when making snake wines. All snakes are considered edible, including the so-called rat snake (*Ptyas mucosus*), rattlesnakes, boa constrictors, the cobra and king cobra, sea snakes, and common garden-type snakes. One traditional way to prepare snake wine is to put a venomous snake or two into either a wine or a liquor and soak for a long period of time. There are some popular medicinal wines (or liquors) called

FIGURE 17.2 Snake wines (or liquors) generally include a whole snake in the bottle, or one or more submerged snakes in the wine or liquor bottle, from Chinese market.

"Dragon and Phoenix Wine" that are made using one venomous snake and a pheasant. After a person consumes snake fat wine, he might find that his penis shrivels, but when he drinks snake liver wine, his own liver might be helped. When a person takes snake bile wine, he might improve his virility and aid his heart.

17.6.2 Tiger Bone Wine

There is a more holistic approach that is centered on a healthy lifestyle, and often recommends the consumption of plants and animal products. Some of the most prized animal parts are from the tiger. Tiger parts are thought to cure a variety of ills including impotence, convulsions, skin disease, and fevers. Tiger bone wine is made from tiger bones soaked in wine (or liquor) as a medium, and have the function of being an elixir of life. Products containing tiger parts have been part of traditional Chinese medicine for centuries have been sold throughout the world. The demand for tiger medicinal wine remains high, and hunters still shoot tigers for their bones and other parts.

The Wildlife Convervation Society (WCS), Asia Conservation Communication Program conducts workshops in China to educate people about the role of the traditional Chinese medicine in tiger conservation. In China, a group of businessmen wants to mass produce tiger bone wine, a tonic produced from the skeletons of captive tigers which died on their tiger farms and in the wildlife fields. Now, Chinese medicine researchers are searching for alternatives to using tiger parts (tiger skins, bones, bile, and other body parts) in medicinal wines.

17.6.3 Other Medicinal Wines (Liquors) with Animal and Insect Materials

a. *Ant Wine:* Ants (20 g, dried) in liquor (500 mL). Ant wines are similar and have a reputation for reducing rheumatism.
b. *Deer Horn Wine:* Deer horn (50 g), liquor (500 mL).
c. *Bear Bile Wine:* Bear bile (20 g), liquor (500 mL).
d. *Yuju or Mayuju: Yuju* or *Mayuju* is made from fermented horse milk, and was introduced to Korea from Mongolia.

17.7 SOME FAMOUS HERBAL WINES

17.7.1 Ginseng Wine

Ginseng roots are soaked in the wine (liquor) as shown in Figures 17.3 and 17.4.

Insamju of *Kumsan: Insamju* of *Kumsan* is a famous ginseng wine in Korea and information about its brewing method and beneficial effects are mentioned frequently in publications since the Choson Dynasty (1392–1910), notably *Imwon shimnyuk-chi* (Sixteen Treatises Written in Retirement, 1827) by 56 *Yu-gu* and *Poncho kangmok* (Encyclopedia of Herbs). Many ancient records indicate that ginseng wine was first

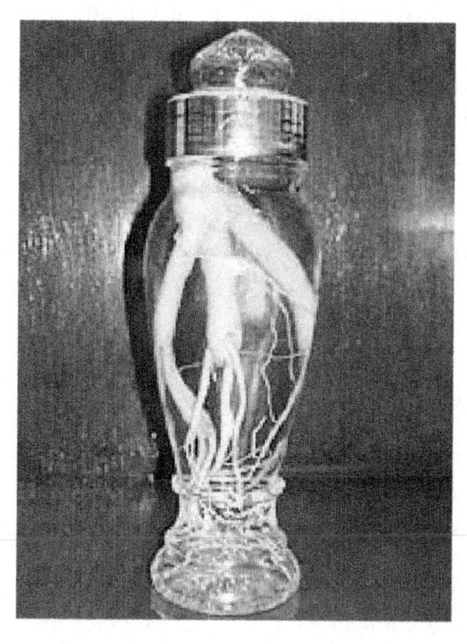

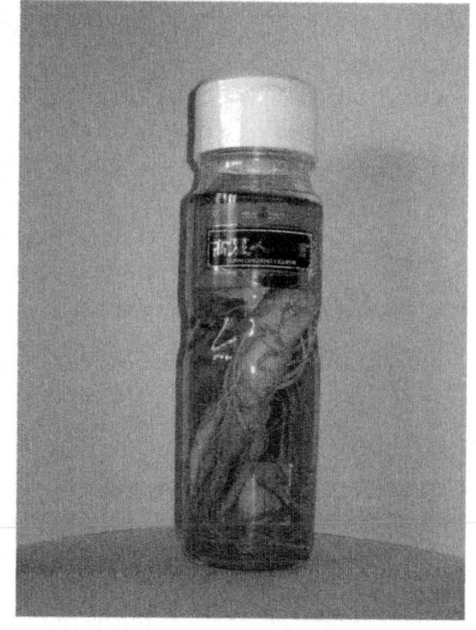

FIGURE 17.3 Korean ginseng wine, from Korean market.

FIGURE 17.4 Chinese ginseng roots are soaked in the wine (liquor), from Chinese market.

developed during the Paekche period (18 BC–AD 660). "*Kumsan* Ginseng Wine" in Korea is made using a unique method. The *nuruk* is first made by mixing wheat and ginseng. Once the *nuruk* is ready, tiny ginseng roots, rice, and water are added to it to make the wine starter. This is then fermented with a mixture of steamed rice, tiny ginseng roots, pine leaves, and mugwort. It takes about 10 days to make the wine starter, 60 days for the fermentation process, and 30 additional days for the aging process. The longer the brew matures, the more flavorful it becomes. The wine has a unique flavor that originates from the blending of pine leaves, mugwort, and ginseng. It has traditionally been believed that drinking a certain amount of ginseng wine strengthens the body. The ginseng wine of *Kumsan* is completely different from the liqueur-type drink made by immersing ginseng in alcohol. The latter is visually appealing, but lacks taste, not having undergone the fermentation process.

Korean ginseng wines (liquors) have been considered as multipurpose remedies for hundreds of years and their medicinal efficacies related to a wide range of health concerns have been scientifically demonstrated. Ginseng wines (liquors) help relieve stress, fatigue, and depression, and are effective in treating heart disease, high blood pressure, hardening of the arteries, anemia, diabetes, and ulcers. They also induce lustrous skin by preventing dryness.

17.7.2 "*Kugijaju* Wine" (Barbary Wolfberry Wine)

The ingredients of this wine are *nuruk*, rice, malt, water and barbary wolfberry including its berries, roots and leaves. The mixture is stored in a cave for five to seven days for fermentation. "*Kugijaju* wine" is clear yellowish brown in color. It is a bit sticky, with a rich aroma and refreshing taste. Its alcohol content is about 16%, which makes its storage for a long time difficult. It can be stored for about a month at 15°C, and in a cooler cave or in a refrigerator, it lasts longer. If it is stored in a clay jar, it maintains its original taste better, and if it is warmed before serving, it tastes much smoother. According to ancient documents, barbary wolfberry wine is good, regardless of one's constitution. The barbary wolfberry wine has been traditionally touted as a miracle longevity drug. Barbary wolfberry has no known toxicity and is good for strengthening bones and muscles as well as relieving fatigue and increasing energy. It is also known for being a good medicine for the stomach, liver, and heart troubles. Its components include rutin that strengthens capillaries and betaine that normalizes liver functions, as well as essential fatty and amino acids, vitamin B, and vitamin C.

17.7.3 Other Herbal Wines

The ingredients of over twenty herbs are especially selected for nourishment and to enhance the energy requirements of the human body, and soaked in wine for extraction of the herbs' natural chemicals (Wang, 1983; Wu, 1996). In some case, the mixture of wine and herbs is filtered to obtain a clear wine.

a. Bee Pollen Wine: Bee pollen (30 g) in liquor (500 mL).
b. Walnut Wine: Walnut (20 g), crystal sugar (50 g), rice wine (500 mL).
c. Nutmeg Wine: Nutmeg (50 g) in liquor (500 mL).
d. Magnoliavine Wine: Magnoliavine fruit(100 g) in liquor (500 mL).
e. *Songsunju*: *Songsunju* is Korean rice wine, made from glutinous rice and soft, immature pine cones or sprouts.
f. *Ogalpiju*: *Ogalpiju* in Korea is made from the bark of *Eleutherococcus sessiliflorus* soaked in wine, blended with some sugar.
g. *Jugyeopcheongju*: *Jugyeopcheongju* is a traditional medicinal liquor in Korea, with bamboo leaves soaked in liquor.
h. *Chuseongju*: *Chuseongju* is a traditional medicinal rice wine in Korea and made from glutinous and nonglutinous rice, herbs including omija (*Schisandra chinensis*) and *Eucommia ulmoides.*
i. *Daeipsul*: *Daeipsul* is a traditional medicinal wine from Damyang County, South Jeolla Province, Korea, and made from glutinous rice, brown rice, and bamboo leaves, along with another 10 medicinal herbs.
j. *Bek Se Ju*: *Bek Se Ju* is a commercial variant of medicinal wine. It is a rice wine infused with ginseng and 11 other herbs including licorice, omija (*Schisandra chinensis*), *gugija* (Chinese wolfberry), astragalus, ginger, and cinnamon, and contains 13% alcohol.
k. *Sansachun*: *Sansachun* is a commercial Korean medicinal wine made from the red fruits of the sansa and Chinese hawthorn fruits, and claiming therapeutic effects.

17.8 FINAL REMARKS

Medicinal wines (liquors) have been used as functional foods to promote health in Asian countries for a long time. They are now distributed in Europe and North America and other parts of the world. Manufacture-processing stages and quality control of medicinal wines still exhibit a lack of standardization, such as the normalization of manufacture-processing, storage, and stability because of the intricacy and diversity of their herbal constituents.

REFERENCES

Beijing Traditional Chinese Medical College and Hospital, Ed. *A Collection of Herbal Prescriptions*. Beijing Traditional Chinese Medical College Publisher, Beijing, 1978–1981.

Li, S. C. *The Chinese Pharmacopoeia*. People's Hygiene Publisher, Beijing, China, 1587. (new version published in 1991).

Ling, I. Q., Zhong, C. Y., and Yian, J. *Chinese Herbal Studies*. Shanghai Science Technology Publisher, Shanghai, 1984.

Lu, H. C. *Chinese Foods for Longevity*. Yuan-Lion Publishing Co., Ltd., Taipei, Taiwan, 1991.

Ma, B. L. Ed. *"Ying-Yang" Balance and Heath Care*. People's Military Medical Publisher, Beijing, China, 2002.

Sun, S. *Prescription Worth a Thousand Gold for Emergencies, the Tang Dynasty*. The People's Medical Publishing House, Beijing, 1955.

Wang, Y. S. Ed. *The Pharmacology of Chinese Herbs and Their Uses*. People's Public Health Publisher, Beijing, 1983.

William, T. *Chinese Medicine, A Comprehensive System for Health and Fitness. Element Books* (Paperback). Rockport, Nutrition Review, Massachusetts, USA, 1999.

Wu, D. X. *Review on Healthy Liquors in China*. Publishing House of Shanghai Science and Technology, Shanghai, 1996.

Wu, J. and Zhong, J. J. Production of ginseng and its bioactive components in plant cell culture: Current technological and applied aspects. *Journal of Biotechnology*, **68**, 89–99, 1999.

Wu, P. J. *The Pharmacology of Chinese Herbs*. People's Public Health Publisher, Beijing, 1982.

Yeung, H. *Handbook of Chinese Herbs and Formulas*, Vol. 1. Institute of Chinese Medicine, Los Angeles, California, USA, 1985.

You, J. *Preliminary Explore of Yin-Yang*. China Overseas Chinese Publishing House, Beijing, China, 1996.

Zhang, E. *Basic Theory of Traditional Chinese Medicine*. Publishing House of Shanghai College of Traditional Chinese Medicine, 1997.

Zhou, J. H. *Chinese Herbs Pharmacology*. Shanghai Science Technology Publisher, Shanghai, China, 1986.

18 Quality Assurance and Safety Protection of Traditional Chinese Herbs as Dietary Supplements*

Frank S. C. Lee, Xiaoru Wang, and Peter P. Fu

CONTENTS

* Contents in this article pertaining to regulatory activities are based on information taken from the public domain. This is a scientific article containing no official guidance and policy statements from, or official support or endorsement by, any governmental agencies including SFDA of China or US FDA.

18.1 INTRODUCTION

Traditional Chinese herbs (TCHs) are gaining increasing popularity worldwide in the development of dietary supplements or pharmaceutical products. Quality control and standardization of TCHs is a challenging task because of (1) the large variations in the sources and properties of raw herbs and (2) the wide diversity in process types and manufacturing conditions leading to the products. The good agriculture practice (GAP) and good manufacturing practice (GMP) guidelines are designed to address, respectively, the quality assurance issues involved in the above two areas. Although GMP guidelines have been well established in manufacturing processes, the scope and operational specifics of GAP guidelines in the agricultural production of herbal plants are still in the developmental stage. In recent years, the development of GAP-based farming in many countries, including China, is accelerating; and the adoption of GAP as an international standard for the marketing and trading of food or herbal products is gaining momentum in international communities. This chapter uses case studies to provide an overview of the type of work and research involved in several GAP programs led by the authors' research team in China (Lee, 2003; Wang, 2006). The technical work in a typical GAP program covers a wide range of different topics including plant science, phytochemistry, biology, and environmental science. Our discussion here thus has to be selective, with emphasis placed primarily on biochemical and analytical related topics. Additional information can be found in several review articles (Lee and Wang, 2002), and in the references listed throughout the text. The major herbs discussed include Danshen (Radix *Salvia miltiorrhiza*), Licorice (Radix Glycyrrhiza), American ginseng (*Panax quinquefolius*), Alisma (*Alisma gramineum* Lej.), and Taizishen (*Pseudostellaria heterophylla* (Miq.) Pax.), and their common names will be used in the following text.

18.2 GAP AND THE "5P" QUALITY ASSURANCE SYSTEM

The "5Ps" (GAP, GMP, GLP, GcLP, and GSP) is the most widely accepted quality assurance system for the development and production of consumer products intended for therapeutic applications. These guidelines cover the entire lifecycle of a product from initial raw material supply, through the manufacturing processes, to the final stage of consumer consumption. Table 18.1 outlines the objectives and major activities of these guidelines. Of the 5Ps, good laboratory practice (GLP) and good clinical laboratory practice (GcLP) safeguard the quality of laboratory and clinical testing. They are the gatekeepers at the front end of the product lifecycle during its research and development stages. At the other end is good supply practice (GSP), which deals with product surveillance activities at the final stage. In between are good agricultural practice (GAP) and GMP, which constitute the heart of the 5P system focusing on quality and safety issues during the manufacturing processes.

For the manufacturing of synthetic chemicals with well-defined properties, GMP is sufficient to safeguard the quality involved in the entire production chain from raw materials to the final products. For the agriculture production of natural products such as herbal plants; however, a different set of problems exists. There, aside from internal factors such as the intrinsic properties of the herbs, the quality of the TCHs

TABLE 18.1
"5Ps" Quality Assurance System for the Production of Medicinal Herbal Products

QC/QA Guidelines	GAP	GLP	GCP—Good Clinical Practice	GMP	GSP—Good Sales (Supply) Practice
Product development stage	Preparation of raw herbs or crude drugs	Laboratory biochemical assay and animal tests	Clinical testing for product registration	In-plant product formulation and manufacturing	Surveillance of finished products
Major activities	Species authentication, plant cultivation, and the processing of crude herbs	Bioassay, chemical analysis, and animal tests	Clinical trials of premarketing products	Quality control of raw materials, in-plant process, and final products	Implementation of inspection, monitoring, and reporting activities for commercial products
Objectives	Ensure the sustainable production of contamination-free herbs with controllable bioactivity and yield	Examination, confirmation, and quantification of active components in herbs	Acquire clinical data to quantify dose/response relationship	Ensure the quality, efficacy, and safety of manufacturing products	Enforcement of regulatory activities for commercial OTC or prescription products

Note: Priority Chinese medicinal herbs selected for 5Ps by Chinese Ministry of Science and Technology (2000) include: Salvia Miltiorrhiza, Radix Astragoli, Radix Ophiopogonies, Bullbus Fritillariae Cirrhosea, Flos Chrysanthemi, Radix Glycyrrhizae, Radix Ginseng, Fructus Lycii, Herba Ephedrae, Radix Coicis, Radix Rehmanniae, Rhizoma Ligustici Chuanxiong, Radix Aconiti Praeparata, Rhizoma Gastrodiae, Cornu Cervi Pantotrichum, Radix Achyranthis Bidentatae, Rhizoma Pinelliae, and Tuber Dioscoreae.

is affected also by external factors including the genetic variations of the plant species, environmental conditions and climate fluctuations. These external variables are hard to control because they vary from grower to grower, from crop to crop, and with the geographic location of the production site. Thus, the stability of the produce is more difficult to monitor and control, and benchmarking standards are more difficult to establish compared with synthetic drugs. It is against this background that GAP has evolved. In recent years, the adoption of GAP as a quality standard for agricultural or food products is gaining increasing international recognition.

Although the overall 5Ps were designed originally for drug manufacturing, the general principle of these quality guidelines should be applicable to dietary supplements or health food products as well. By implementing quality standards from the farm to the factory, TCHs would be better prepared to meet the needs of increasingly discerning domestic consumers and the international demand for botanicals. Table 18.2 outlines the major GAP-related regulatory activities taking place in different countries in recent years. In the 1980s and early 1990s, GLP, GcLP, and GMP guidelines have all been officially promulgated by State Food and Drug Administration in China (SFDA of China) (SFDA, 2006). Meanwhile, the concept of GAP has evolved in Europe, Japan, and North America since the 1990s. From 1998 to 2003, the US Food and Drug Administration (US FDA) published a series of regulations to ensure the microbial safety of fresh produce by defining the GAP and GMP guidelines that producers and handlers should follow (US FDA, 2003). In 1998, EAEM (European Agency for the Evaluation of Medicinal Products) announced official GAP guidelines for botanical drugs and herbal products (EMEA/HMPWG, 1998). Between the years 2000 and 2002, scientists and governmental officials in China carried out extensive studies on the feasibility of implementing GAP in China (Ren and Zhou, 2003; Lee et al., 2008; Chang et al., 2001). In June 2002, GAP for Chinese Crude Drugs (Interim) was passed into effect by SFDA and World Health Organization (WHO) also published guidelines on good agricultural and collection practices (GACP) for medicinal plants in 2003 (WHO, 2003).

The GAP guidelines of China address quality and safety requirements for TCHs in areas including (1) ecological and environment conditions of the production site, (2) germ plasma and propagation material, (3) management for cultivation of medicinal plants, (4) packaging, transportation, and storage, and (5) managerial and technical aspects of quality management. The official GAP program for Chinese medicinal herbs started its trial period in China on June 1, 2002. The GAP certificate, which is usually awarded to a private enterprise, is valid for 5 years. The follow-up monitoring activities are carried out by the expert teams organized by SFDA. In addition, provincial, city, and regional FDA offices of China provide the needed assistance to handle the application and monitoring activities at the provincial or local levels. In the original regulations set out by SFDA in 1998, all manufacturers must have complied with the GMP guidelines by April, 2004; while farms producing raw ingredients have until 2007 to meet the guidelines specified in Good Agricultural Practices.

Work in GAP encompasses a wide range of topics in different disciplines. The type of work and major activities involved are outlined in Table 18.3. More than just the implementation of a quality assurance system on existing practices, the current

TABLE 18.2
Regulatory Milestones Pertaining to GAP Development for Herbal Products

Regulatory Agency Initiated (Year Published)	Regulation/Guidelines Promulgated
Ministry of Health of China (1982)	GMP Guidelines
Ministry of Health of China (1988)	GLP guidelines—Regulations on experimental animals
SFDA of China (1992)	Regulations on registration and approval of new CM drugs (GCP, GLP)
US FDA (1998)	Guidelines for the microbial safety of fresh produce defining the GAP and GMP guidelines that producers and handlers should follow
SFDA of China (2000)	Article (2000)157 published emphasizing the importance of well managed TCM (Traditional Chinese Medicines) herb Farms, and the requirements for fingerprinting of CM Injection fluid products
Secretary of States of China (2000)	Published regulations prohibiting the collection and sales of wild licorice, ephedra sinica stapt
European Agency for the Evaluation of Medicinal Products (2000)	GAP guidelines for botanical drugs and herbal products
SFDA (2002)	Issuance of GAP regulation for medicinal plants and animals
SFDA (2003)	GAP certification system starts operation—certificate valid for 5 years; status followed and monitored by SFDA
WHO (2003)	Published guidelines on GACP for medicinal plants in 2003 (WHO, 2003)

TABLE 18.3
Scope of GAP Work

Area of Work	Major Activities	Objectives
Environmental monitoring and impact assessment	Collection and analysis of environmental quality parameters for air, water, and soil The development and application of environmental friendly practices for pest prevention Environmental impact assessment of the production site	Ensure the meeting of environmental quality standards, the absence of potential sources of contaminants including natural or man-made toxins/pollutants, heavy metals, and pesticides/herbicides residues, and the sustainable development of the production site and the surrounding area
Selection, identification, and authentication of plant species	Seed selection and preservation Plant species identification through DNA fingerprinting and chemical compositional fingerprinting	Establish the authenticity and correct genetic identity of the plant species
Plantation, cultivation, and harvesting technology	Standardized production based on modern science in combination with traditional wisdom; pest prevention; specified fertilizers; and best harvest time determination	Establish standardized farming practice to grow quality and safe produce
Processing, storage, and transportation	Harmonize traditional method with modern science for the field and factory processing of raw plant, and the storage and transport of crude products after primary treatment	Establish modernized and standardized field and factory processing technology for the preparation of contamination-free crude herbal products
Management, training, and documentation	Establish SOPs for the technical operations including site selection, environmental monitoring, cultivation practices, quality control, and primary processing Establish management systems including product registration, personnel training; SOP documentation and updating, and facility maintenance	Establish standardized technical manuals, managerial systems quality control procedures, and qualified personnel for GAP operation

GAP program in China also calls for the application and development of updated technology for the modernization of traditional herb farming practices. Furthermore, the developed techniques have to be user-friendly enough that they can be practiced by farmers on a routine basis. All the methods developed thus have to be standardized and documented in Standard Operational Procedures (SOPs); and personnel training is necessary to facilitate technology transfer.

In the application of the GAP certificate, the species should have completed at least one growth cycle. Documented information should include site selection and selection criteria, historical data, scale of production, and environmental conditions of the surrounding area. In cultivation practice, information should include species authentication/identification, speciation of wild or cultivated varieties, seeding and growth conditions, harvest practice, fertilization, pest prevention, and field and farm management practices. Also to be included are management and operation practices involved in quality control and assessment methods, personnel training and maps showing detailed cultivation area and experimental farms (scale, production yield, and scope). Besides regular GAP studies, the work also emphasizes ways to (1) maintain ecological balance (biodiversity; sustainable development of environment) and (2) facilitate the transition from wild to cultivated farming.

Since the inauguration of the GAP program in 2004, a series of TCH farms have already been awarded the GAP certificates issued by SFDA of China. Based on published information from SFDA (2006), the location of these farms and the herbs which received GAP certifications from 2004 to 2006 are summarized in Table 18.4. These certificates were awarded to the sponsoring party of the program which is usually a business enterprise. The technical work of the program is carried out by a working team of experts and professionals, generally from a research institution or university. For our purpose here, only the locations of the GAP farms are listed in Table 18.4 while the names of the private companies are omitted. Also listed in Table 18.4 are the locations of "genuine" herbs as specified in Chinese Pharmacopoeia (Pharmacopoeia of China, 2005). The concept of "genuine herbs" (Hu, 1997) is rooted deeply in traditional Chinese medicine, meaning that only species grown in specific geographic locations are the authentic species with the best quality. The GAP farms are in general, but not always, located in sites with the reputed "genuine herbs."

With a full GMP/GAP certification scheme in place, and with both industrial and agricultural sectors understanding what is required of them, progress is being achieved at a rapid rate. Quality has improved because raw materials via the GAP system are being controlled; and the supply of raw materials also becomes more stable with less price fluctuation. High-quality and contaminant-free raw materials produced under GAP principles are a prerequisite for the making of quality and safe Chinese Medicines (CM) products based on modern GMP production.

The critical challenge of GAP establishment is the difficulty involved in the quality control and standardization of herb plants. The two main problems are the lack of scientific-based conventions to define and standardize quality, and the lack of comprehensive toxicological data. To date, although GMP guidelines have been well established in manufacturing processes, the GAP for efficacy assurance and safety of Chinese herbal plants used for functional foods and dietary supplements is still in the development stage requiring continuing research.

TABLE 18.4
GAP Herbs Certified in China

Chinese Name[a] of Herb (pinying)	Latin Name[b] of Herb	Functional Part of Herb	Location of Genuine Herb Defined in Chinese Pharmacopoeia[c]	Location of GAP Farm in China[b]	Year SFDA GAP Certificate Received
Baizhi	Radix angelicae dahuricae	Root	Hangzhou, Zhe Jiang province; Sui Ning, Si Chuan province	Si Chuan province	2006
Banlangen	Radix isatidis	Root	An Guo, He Bei province, Nan Tong, Jiang Su province	Bai Yun Shan, Fu Yang	2006
Chuanxiong	Rhizoma chuanxiong	Root and stem	Guan Xian, Si Chuan, Yun Nan, He Bei	Si Chuan	2006
Danggui	Radix angelicae sinensis	Root	Gan Su, Shan Xi	Gan Su	In process
Danshen	Radix salviae miltiorrhizae	Root	Shan Xi, SiChuna	Tian Shi Li Co., Shan Xi	2004
Guanghuoxiang	Herba pogostemonis	Entire grass	Guang Dong, Hai Nan	Guang Dong	2006
Huangqi	Radix astragali	Root	Shan Xi, Inner Mongolia	Inner Mongolia	2006
Maidong	Radix ophiopogonis	Root	Si Chuan, Zhe Jiang	Ya An San Jiu Co in SiChuna	2004
Qinghao	Herba artemisiae annuae	Entire grass	Unspecified	SiChuna	2004
Renshen	Radix ginseng	Root	Ji Lin, Hei Long Jiang	Ji Ling	2004
Saqi	Radix notoginseng	Root	Yun Nan, Guan Xi	Yun Nan	2005
Taizishen	Radix pseudostellariae	Root	Jiang Su	Gui Zhou	2006
Tianma	Rhizoma gastrodiae	Root and stem	Yun Nan, Si Chuan	Shan Xi	2006
Xiyangshen	Radix panacis quinquefolii	Root	Unspecified or unknown	Ji Lin	2004
Yuxingcao	Herba houttuyniae	Whole grass	Unspecified or unknown	Ya An San Jiu Co.	2004
Shanzhuyu	Fructus corni	Fruit	He Nan	He Nan	2006

Jingjie	Fineleaf Schizonepeta herb	Whole grass	Jiang Su, Zhe Jiang, Jiang Xi, Hu Be, He Bei	He Bei	2006
Kudiding	Bungeanae Corydalis herb	Whole grass	Gan Su, Shan Xi, Shan Xi, Shan Dong	He Bei	2006
Yinxingye	Folium Ginkgo	Leaf	Jiang Su	Jiang Su	2006
Heshouwu	Radix Polygoni Multiflori	Root	He Nan, Hu Bei, Guang Xi, Guang Dong, Gui zhou, Si chuan, Jiang Su	Gui Zhou	2005
Jiegeng	Radix platycodi		Shan Dong, Jiang Su, An Hui, Zhe Jiang, Si Chuan	Shan Dong	2005
Dangshen	Radix codonopsis	Root	Shan Xi	Shan Xi	2005
Yiyiren	Semen Coicis	Fruit	Unspecified or unknown	Zhe Jiang (Zhe Jiang Tai Shun)	2005
Jiaogulan	Herba Gynostemmatis pentaphylli	Whole grass	Unspecified or unknown	Shan Xi	2005
Zhizi	Fructus gardeniae	Fruit	Jiang Xi	Jiang Xi	2004
Qinghao	Herba artemisiae Annuae	Whole grass	Unspecified or unknown	Chong Qing, Si Chuna	2004
Huanglian	Rhizoma coptidis	Root	Si Chuan, Hu Bei	Chong Qing, Si Chuna	2004
Chuanxinlian	Herba Andrographis	Whole grass	Guang Dong, Fu Jian	Guang Dong	2004
Dengzhanxixin	Herba erigerontis	Whole grass	Unspecified or unknown	Yun Nan	2004
Yuxingcao	Herba houttuyniae	Whole grass	Unspecified or unknown	Si Chuan Ya An	In process
Xihonghua	Stigma croci	Flower	Originated in Spain and Holand, now cultivated in Shanghai, Zhe Jiang, He Nan, Beijing, Xin Jiang	Shanghai	In process

Source: Based on published data by SFDA from year 2004 to 2006.

[a] Pinging and Latin names are listed in Chinese Pharmacopoeia, 2005 edition.

[b] For locations of "genuine herbs" and "GAP Farms," only the name of the province in China is listed.

18.3 AN OVERVIEW OF GAP PROGRAM: CASE STUDY OF DANSHEN

Our research team has in the past few years worked on 4 different GAP programs including Danshen (*Radix salvia* Miltiorrhiza) in SiChuan (Lee, 2003), Licorice (*Radix glycyrrhiza*) in Inner Mongolia, Taizishen (*Pseudostellaria heterophylla* (Miq.) Pax and Alisma (*Alisma gramineum* Lej.), both in FuJian (Wang, 2006). Part of the technical information in these studies can be found in the references listed in later discussions. In this section, we will use Danshen as a case study to provide an overview of the work involved in a typical GAP study.

18.3.1 Program Implementation and the Development of SOPs

Danshen is the dried root of Radix *Salvia miltiorrhiza*. It is a high-value medicinal plant which has been used in China for many years for the treatment of cardiovascular diseases including angina pectoris and other deleterious effects caused by coronary heart disease (Lin et al., 1988; Chang et al., 1990; Li et al., 1991; Kasimu et al., 1998; Li and Chen, 2001). Many pharmacological studies have also reported that the active components in Danshen demonstrate excellent anticoagulant and antibacterial activities, and have a beneficial effect in patients with chronic renal failure (Lou et al., 1985; Tanaka et al., 1989; Du and Zhang, 1995 ; Lu and Foo, 2002; Liu et al., 2008). In China alone, more than 40 million kg of Danshen is sold annually, with most of it going to manufacturers for the preparation of herbal medicines or dietary supplements.

Funded by the Hong Kong Industrial and Innovation fund, the two-year Danshen GAP program was conducted from 2000 to 2002 (Lee, 2003). The production site was located in ZhongJiang, near the provincial capital of Cheng Du in Sichuan, China. The site was in a hilly area consisting of a total of 3500 acres of Danshen farmland. The field work of the GAP study was carried out on clusters of Danshen experimental plots where routine operations involving seeding, cultivation, and harvesting were carried out by independent local farmers. The team consisted of three separate groups responsible respectively, for cultivation and pest control, field work and farm management, and bioassay and chemical testing. The project team members made regular trips to the production site to perform sampling, inspection, monitoring, and other on-site experiments. The scheduling of these trips and the preparation of supporting activities were synchronized with farming activities in the field, for example, seeding or transplanting activities in Spring, fertilization and pesticide application in Summer and Fall, and harvest operation in late Fall, among others.

Environmental assessment and field monitoring are important parts of the GAP activities, that is, water/air/soil testing, analysis of pesticide and herbicides residues, fertilizer usage. Information regarding the environmental conditions of the Danshen production site and the adjacent area were collected from local agencies, and their impact on Danshen quality assessed from the analysis of historical data. Potential problem areas were identified to be followed up by a systematic monitoring exercise. In the Danshen site, all environmental parameters met the required standards except for a few areas with slightly high metal contents in the soil. A systematic metal analysis on soils and plant samples were therefore carried out (Lee and Wang, 2002; Huang et al., 2003).

Topological, climatic, and environmental factors dictate the growth of herbs of different quality at different sites, and with it the need for different pest prevention, control practices, and corrective measures. To facilitate field applications under highly diverse environmental conditions, and to meet the need to process large sample volumes, on-site analysis by portable analyzers or chemical testing kits were used as much as possible. In the Danshen study, a simple colorimetric test which could be performed in the field using a hand-held spectrophotometric analyzer was developed for the measurement of the total content of tanshinons, a class of major active ingredient in Danshen, in plant samples (Lee and Wang, 2002). These handheld analyzers were used extensively in other GAP programs including the screening of triterpenoid saponins in licorice (Wang et al., 2004a, 2004b), and the assay of active components alisol A and B in alisma (Wang et al., 2003). In all cases, light absorption of the native species or secondary products after derivatization reactions were utilized to quantify the active ingredients of interests. Also developed during the Danshen projects were several high-throughput chemical tests utilizing portable equipment, for example, the analysis of trace Hg contamination in plant and soil samples (Huang et al., 2005, 2006) and the identification and classification of plant species by near-infrared (NIR) techniques (see later discussions).

The primary task of GAP development is to develop and implement a set of managerial and technical SOPs to guide routine operations and testing. A compilation of the SOPs in a GAP operational manual is the major deliverable of the program for GAP certification application. The training of local farmers to implement the optimized cultivation technology and testing methods and equipment is a major task. A field lab was established to perform sample preparation tasks and simple testing. A local office with the participation of local governmental officers was also set up to promote the concept of GAP on a continuous basis.

18.3.2 Species Authentication and Seed Selection

Seed selection including species authentication is the first and the most important step in safety and the production of quality herbs. The establishment of a genetic resources database for Danshen allowed us to select high-quality species for cultivation and transplanting. Among the scattered Danshen farms in the area, different species of Danshen were planted throughout the years, and the genetic history of the species grown in the area was unclear. To identify the true species of "genuine Danshen," the plant samples were authenticated by both morphological examination, compositional analysis and DNA RPD (random amplified polymorphic DNA) fingerprinting.

NIR is an analytical technique which has several attractive features including fast analytical speed, ease of operation and a nondestructive nature (Rodriguez-Otero et al., 1997; Blanco et al., 1998). The technique in its different versions has been used widely in the industry for the routine monitoring of feed/product properties, for example, the determination of water or protein contents in wheat. We have applied NIR techniques for the authentication and classification of a series of herbal products including Danshen. Figure 18.1 shows that a combination of NIR with principal component cluster analysis can distinguish Zhong Jiang Danshen clearly from species originating from other geographic locations (Lee and Wang, 2002). Another technique developed for the authentication of Danshen samples is by high performance liquid chromatography (HPLC) fingerprinting analysis. Figure 18.2 compares the HPLC fingerprints of different plant species of Danshen found commonly in the study area (Li et al., 2003a, 2003b). In the figure, the peaks in the original HPLC chromatogram were converted into segmented, fixed-width "bins." This was for ease of visualization since the relative heights of the "bins" after normalization were then proportional to the observed peak area. The ZhongJiang Danshen, as shown in the figure, shows a distinguishably different pattern from the other plant species in the *Salvia* family.

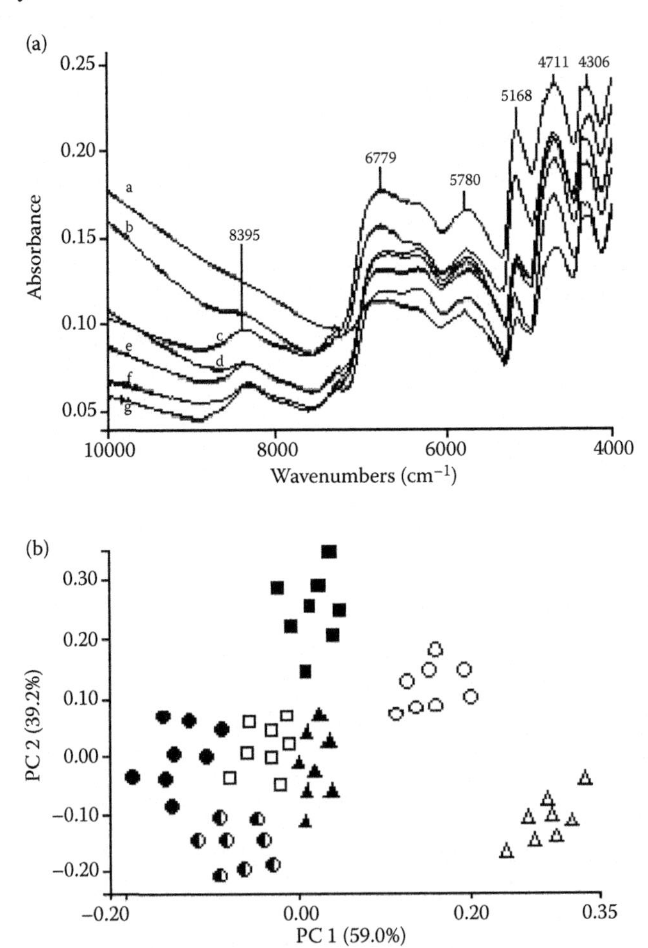

FIGURE 18.1 (a) NIR spectra of Danshen originated from different geographic locations: a. XinJiang, b. GanSu, c. AnHui, d. JiangXi, e. ShangDong, f. ZhongJiang, and g. ShanXi. (b) Principle component cluster analysis of NIR spectra of Danshen samples in Figure 18.1a. (■) aAnHui, (○) GanSu, (▲) JiangXi, (◐) ZhongJiang, (□) ShanDong, (●) ShanXi, (△) XinJiang.

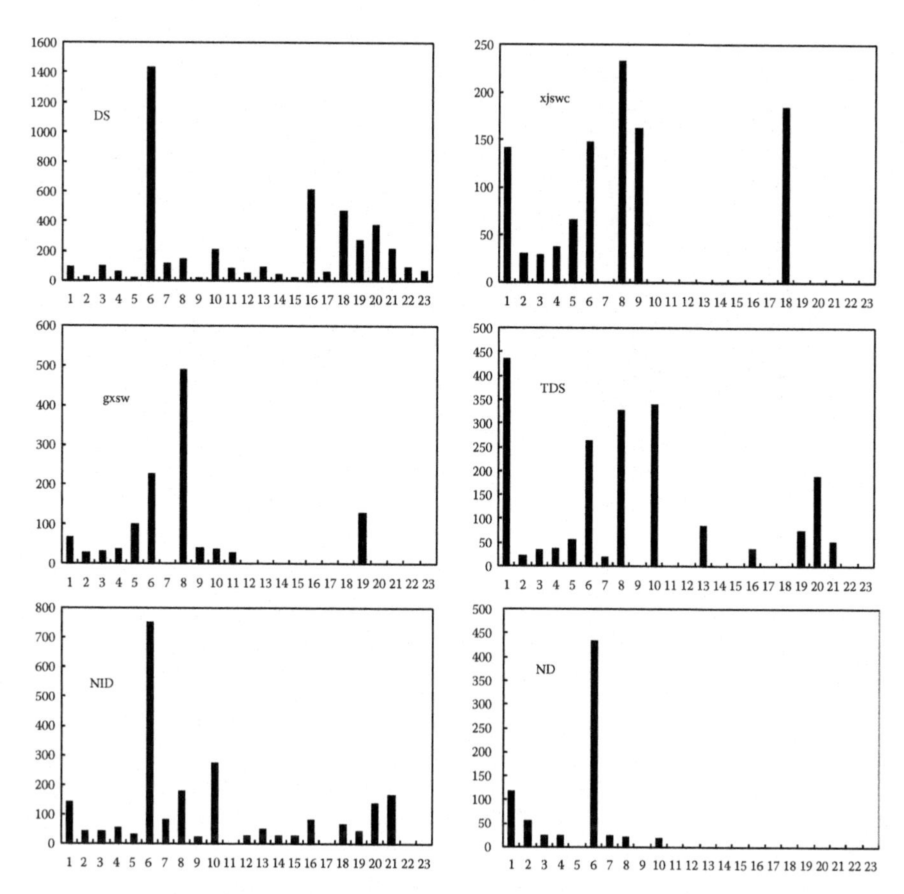

FIGURE 18.2 Bar-coded HPLC fingerprints of Danshen originated from different geographic locations: authentic Danshen from ZhongJiang (*Salvia miltiorrhiza* Bge); GXSW (*Salvia przewalskii*); NID (*Salvia paramiltiorrhiza*); XJSWG (*Salvia deserta from XinJiang*); TDS (*Salvia yunnanensi*) and ND (*Salvia bowleyana*). HPLC separation was on a reversed-phase C18 column; UV 254 nm detection.

After preliminary screening of large numbers of samples grown locally, a small group of samples of which the authenticity remained difficult to confirm was subjected to RAPD analysis. These include the danshen sample of *Salvia miltiorrhiza* Bge from Zhong Jiang (SMB2), a sample of *Salvia miltiorrhiza* Bge obtained from ShanXi (SMB9), a sample of *Salvia bowleyana* Dunn obtained from ZhongJiang (SBD3) and a *Salvia paramiltorrhiza* sample (SP). They all showed highly polymorphic RAPD profiles with the two primers used as given in Figure 18.3. Similarity index (S.I.) was used to reveal the relatedness between the sample pairs. An S.I. of 1 implies that the two samples are genetically identical, and 0 means a complete mismatch. The relatedness among the four Salvia sp. was revealed by the mean SI values listed in Table 18.5. The data indicate that the two samples originating from Zhong Jiang, SMB2 and SBD3, show the closest genetic relatedness of 0.715. On the other hand, the two well-known Danshen samples, SMB2 (from Zhong Jiang) and SMB9

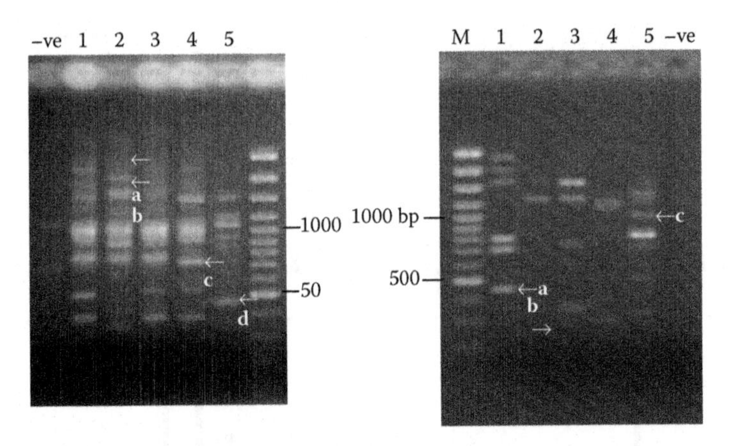

FIGURE 18.3 RAPD profile of Danshen generated by OPC4 (left) and SMB2 (right) primer. Lanes 1–5 are SMB2 (*Salvia miltiorrhiza* Bge from ZhongJiang), SMB9 (*Salvia miltiorrhiza* Bge from ShanXi), SBD3 (*Salvia bowleyana* Dunn from ZhongJiang), SP (*Salvia paramiltiorrhiza*), and SPM (*Salvia przewalskii* Maxim, an adulterant of SMB), respectively. M is 100 bp DNA ladder (MBI). The arrow indicated some of the polymorphic bands. For the left figure, symbols a and b represented the polymorphic bands unique in SMB9, c and d were the polymorphic bands unique in SP and SPM, respectively. For the right figure, symbols a, b, and c indicated the polymorphic band unique in SMB2, SBD3, and SPM, respectively.

(from Shan Xi), show a low similarity index of only 0.439, even though they actually belong to the same species but from different locations. By using RAPD authentication, we were able to establish the true authenticity of Danshen species grown in the area, and to preserve its seed for cultivation by local farmers.

18.3.3 Quality Evaluation

The active ingredients in Danshen and their associated therapeutic properties, as indicated earlier, have been extensively studied in the literature. These ingredients fall

TABLE 18.5
Mean Similarity Index of Five *Salvia* Samples

Sample	SMB2 (ZJ)	SMB9 (SL)	SBD3	SP
SMB2 (ZJ)	—			
SMB9 (SL)	0.4390	—		
SBD3	0.7153	0.4275	—	
SP	0.4565	0.3783	0.4496	—
SPM	0.3520	0.3050	0.3670	0.3686

Note: SMB2 and SMB9 are Salvia miltiorrhiza Bge from ZhongJiang and ShanXiShangLuo, respectively. SBD3 is Salvia bowleyana Dunn from ZhongJiang. SP and SPM are Salvia paramiltorrhiza H.W. Liet X.L. Huary sp now and Salvia przewalskii Maxim, respectively.

into two major classes: the lipid soluble and the water soluble fractions. The lipid soluble, normally obtained by extraction with alcohol solvents, is rich in abietanoids and diterpene quinone pigments. More than 30 diterpenoid tanshinones have been isolated and identified from Danshen, and among them, the three representative bioactive components in the fraction are tanshinone IA, tanshinone IIA and cryptotanshinone. The major active ingredients in the water soluble include many plant phenolic acids which are mostly caffeic acid derivatives. The caffeic acid monomers include caffeic acid itself, danshensu, ferulic acid, and the ester forms of caffeic acids. The dimmers and trimers are the most abundant components and they include rosmarinic acid, protocatechualdehyde, protocatechuic acid, salvianolic acids, lithospermsic acids, rosmaric acid, and so on. In recent years, the water soluble fraction of Danshen has attracted increasing attention because of its effectiveness in improving the renal function of rats with adenine-induced renal failure, as an antioxidant for the removal of free radicals, and their potential in treating Alzheimer's disease (Lu and Foo, 2002).

The biochemical studies of Danshen suggested that the use of "multiple quality indicators" is a better representation of the multifunctional therapeutic effects and bioactivities of Danshen. A series of extractions, chemical functional fractionation, and analytical methods were developed to identify and quantify the above active species in Danshen for quality evaluation. In the GAP study, representativeness of sampling is of primary importance, requiring careful planning. This was because of the wide variations in plant samples produced under different environmental and cultivation conditions. Furthermore, the herb sample could have been prepared from different parts of the plant, which also differed greatly in their composition. A well-designed sampling plan is therefore of utmost importance. For instance, depending on the part taken from the Danshen root, the contents of active components could vary by a large factor. This is illustrated by the distribution of tanshinone IIA yield in different sections along the Danshen root (thin layer chromatography (TLC) analysis) as shown in Figure 18.4

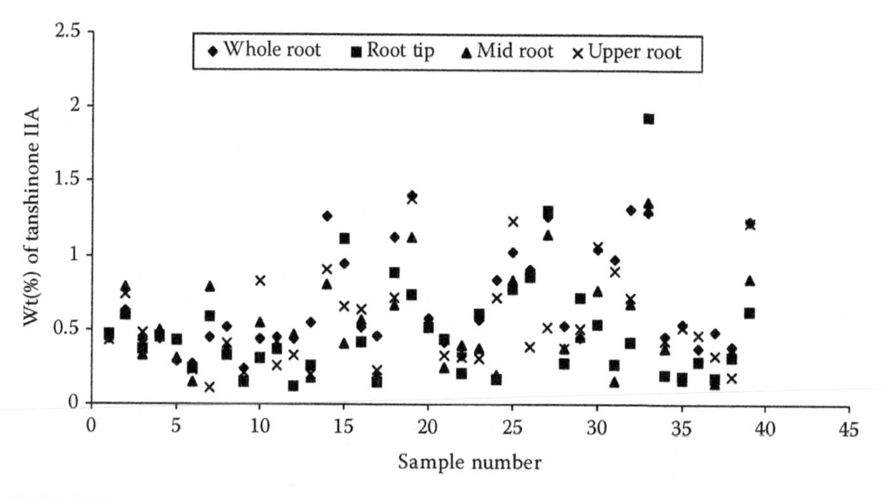

FIGURE 18.4 Distribution of tanshinone IIA along different parts of Danshen root.

(Li et al., 2003a, 2003b). To obtain statistically valid analytical results, a large number of samples thus had to be analyzed, and this dictated the need for fast and high-throughput analytical techniques. One example of such a technique developed for the program is the Time of Flight Mass Spectrometric (TOFMS) technique. Figure 18.5 shows the TOFMS results comparing the contents of the three tanshinones among Danshen samples originated from different geographic locations (Yu et al., 2003). The analysis could be completed in a few minutes through the direct injection of Danshen extract into the mass spectrometric (MS) instrument.

The quality and production yield of Danshen grown in GAP versus non-GAP farms were compared. The yield comparison included the yields of bioactive components, in addition to biomass. The results are shown in Figure 18.6. The yields of both biomass and lipid soluble tanshinone IA in GAP Danshen were significantly

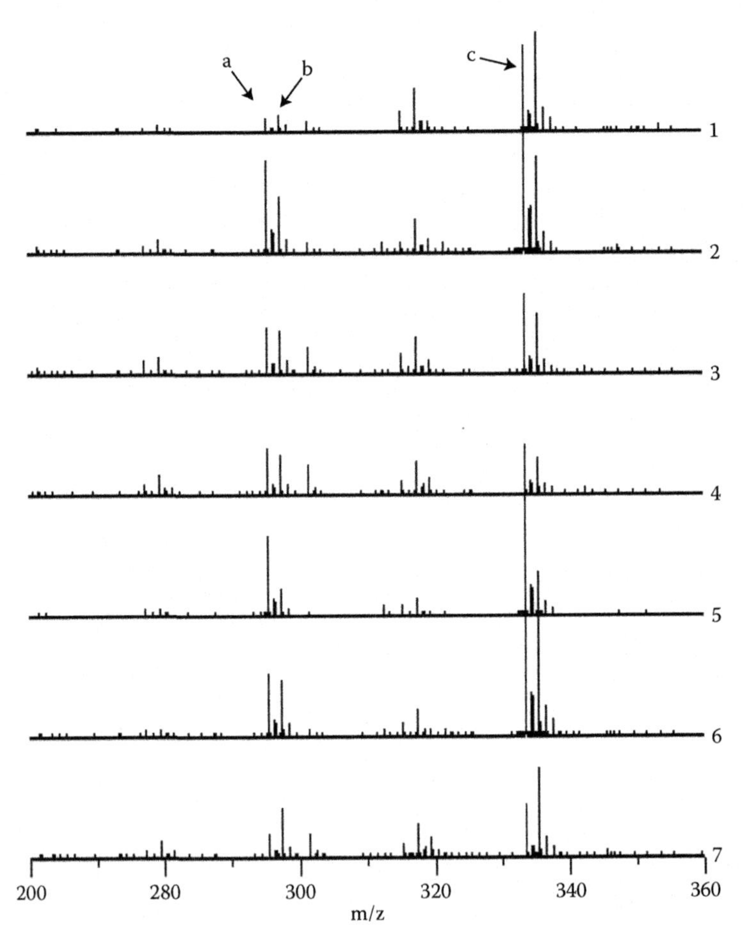

FIGURE 18.5 Time-of-flight mass spectrometric (TOFMS) analysis of three isomeric tanshinones: 1 JiangSi; 2 SiChuan (GAP Danshen); 3 He Nan; 4 AnHui; 5 SiChuna B; 6 SiChuan C; 7 Standard Reference Herb (Beijing CM Institute). a-tanshinone IIA$[M + H]^+$; b-cryptotanshinone $[M + H]^+$; c-tanshinone IIB $[M + Na]^+$.

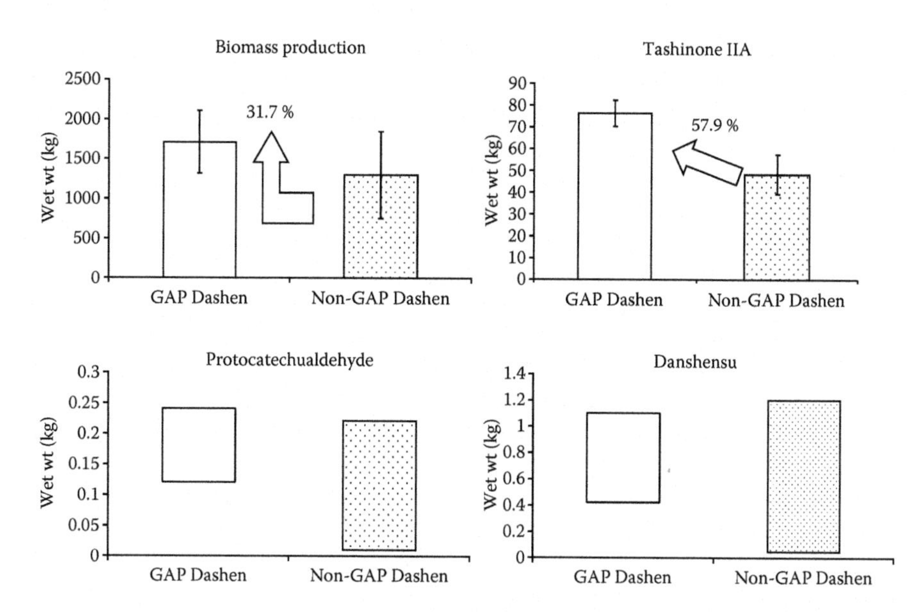

FIGURE 18.6 Comparison of the yield and quality of Danshen produced in GAP versus non-GAP farms.

higher than the corresponding ones grown in non-GAP farms. A similar yield increase was also observed for the two water-soluble components, protocatechualdehyde and danshensu. The yield measurements for the latter two species were subject to large uncertainties because of their small yields. The bar charts for the two species in the figure actually represent the "ranges" of yield measurements. As can be seen, the mean values of the yields of the two species in GAP Danshen are noticeably higher than those of the non-GAP Danshen.

18.3.4 Best Harvest Time

The cultivation of Danshen in the study area has a long history. Prior to our study, the farming operations, including seeding, cultivation, and harvest activities practiced by local farmers, were based primarily on tradition and experience. Our objective in this study was to see whether one can use scientific information and experimental data to guide and optimize these operations. At harvest time selection, it was our intention that the criteria for determination should be the yields of bioactive components, rather than biomass. This could be accomplished through a study of the accumulation of bioactive components in the Danshen plant with time during its growth cycle.

Studying the dynamic accumulation of bioactive components in a plant during its growth cycle actually has broader implications than just harvest time determination. The bioactive components in plants cultivated for TCHs are usually low and unstable. This is because, in contrast to the primary metabolites such as sugars and amino acids, the bioactive components of natural products are mostly secondary metabolites in the plants. They do not play a major role in normal plant functions and are not necessarily beneficial to plant growth. It is conceivable that these secondary metabolites are stress induced, and may actually grow faster under environmentally stressful conditions such as climatic fluctuation or nutritional limitation (Kaufman, 1999). Thus, although the genotypes of a particular species may determine the chemical spectrum of these bioactive components, the yields of these components are largely the result of environmental conditions under which the plant is growing. It is therefore worthwhile to explore the possibility of enhancing the contents of bioactive components in plants through environmental manipulations. Studies along this direction were pursued in several laboratories (Zhang, 2003).

The accumulation of bioactive components in Danshen plants as a function of growth time is shown in Figure 18.7. The accumulation of biomass along with three bioactive components, namely, tanshinone IA, protocatechuic acid, and danshensu were plotted as a function of growing time in months. For biomass, the yield increased steadily after seeding (March) and peaked after about eight months of growth (November). The best harvest time for maximum biomass yield, based on the data, was in mid-November. This was consistent with the traditional practice by local farmers. The three bioactive components behaved somewhat differently. Those of danshensu and tashinone IA showed a similar accumulation behavior. The yields of both peaked in 7.5 months, slightly ahead of those of the biomass. Afterwards, however, their yields dropped much more rapidly than those of the biomass. Thus, a slight delay in harvest time would not much affect the biomass, but could significantly reduce the yields of danshensu and tashinone IA.

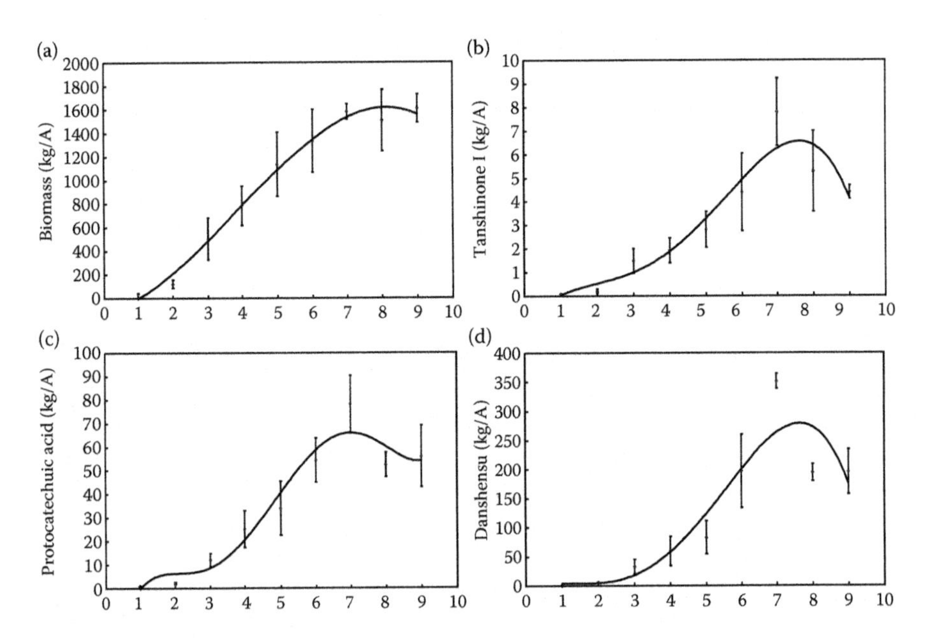

FIGURE 18.7 Accumulation of active ingredients in Danshen (kg/Acre) during plant growth (number of months after seeding). (a) Biomass; (b) tanshinone I; (c) protocatechuic acid; and (d) danshensu.

The dynamic accumulation of individual bioactive components with growing time for TCHs provides valuable information for the selection of the best harvest time in order to maximize yields of individual ingredients. In our GAP studies, similar information was obtained for other TCHs including licorice (Wang et al., 2004a, 2004b), Alisma (*Alisma gramineum* Lej.) (Jiang et al., 2006), and Taizishen (*Pseudostellaria heterophylla* (Miq.) Pax.) (Han et al., 2007).

18.4 THE APPLICATION OF COMPOSITIONAL FINGERPRINTING TECHNIQUES

Although many Chinese medicines are effective in treating diseases, their remedial mechanisms are not well understood. The analysis of active components in Chinese medicine extracts is a key step to unlock the secret of their effectiveness. Because of the complexity of natural products such as TCHs, a comprehensive compositional analysis of the material is very difficult and time consuming. Marker compound analysis simplifies the analytical process but could be misleading sometimes, because the bioactivity of TCHs often arises from the combined actions of a group of multiple components rather than a single compound. Fingerprinting analysis is an effective compromise of the above two approaches. As described earlier in Table 18.2, SFDA of China in 2000 officially published the requirement to use fingerprinting techniques for the authentication and quality appraisal of TCH derived injection fluids (Ren, 2001). Similar suggestions can also be found in the documents published by WHO (2003), US FDA, EMEA (1999), and the British Herbal Medicine Association (BHMA, 2006).

In fingerprinting, the chemical profiles of the active fractions, normally the functional extracts of the TCHs, are assayed. These profiles can have several different applications. The first is to assess the quality of the TCH of concern. In this application, the closeness of the matching between the profiles of the target TCH and the reference herb with known quality is used to rank the quality of the target species. The second application involves the identification and classification of a particular TCH material. Here the profile of the target TCH is compared with those of a set of reference TCHs with known sources or properties. Statistics-based cluster analysis or principal component analysis was then used to quantify the similarities or dissimilarities between the sample and the references, or to discriminate a particular sample against a class of TCHs. The checking for adulterants in TCHs is another area where fingerprinting can provide valuable information. Adulteration is a common problem in TCH products, in which the high-cost ingredients in a formulation are replaced by cheaper substitutions. It is not only an economic fraud, but also a health risk to the consumers.

A number of different fingerprinting techniques were developed during the course of our GAP studies. One example involved the application of NIR for the discrimination of different plant parts of licorice as shown in Figure 18.8a, or the classification of licorice samples collected from different locations (Figure 18.8b) (Wang et al., 2007). Moreover, NIR can also be used to assess the quality of licorice through the determination of its GA (glychrrhetinic acid) content, a known bioactive marker compound in licorice. Different techniques have been applied successfully for

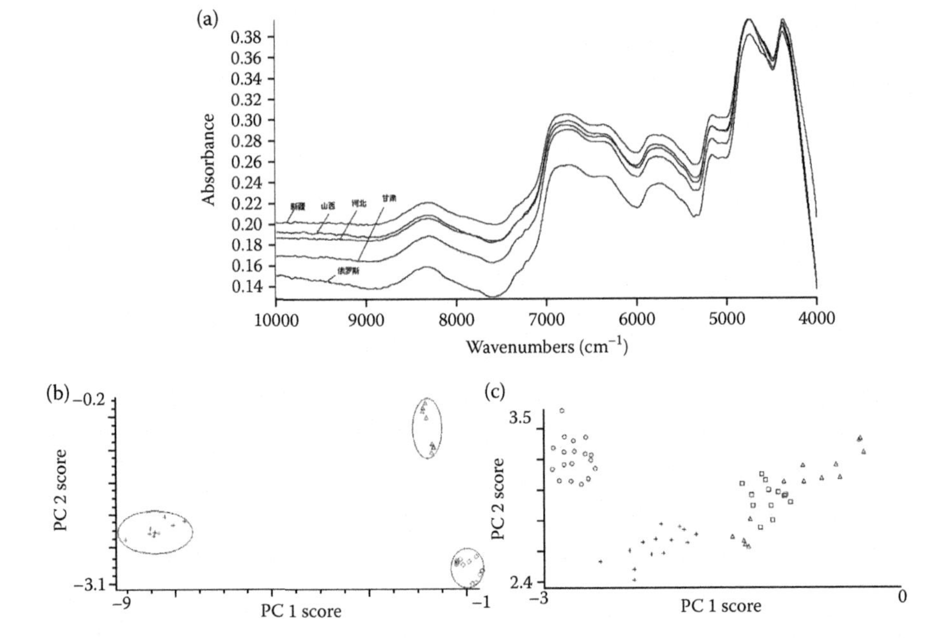

FIGURE 18.8 (a) NIR spectra of licorice collected from different geographic locations. (b) PCA analysis of NIR spectra of different plant parts of licorice (+) root, (○) stem, (△) leaf. (c) PCA analysis of NIR spectra of licorice collected from different geographical locations: (○) XinJiang, (+) GanSu, (□) XanXi, and (△) HeBai.

quantification of GA, and among them the most popular one is by HPLC. However, HPLC is a time-consuming method compared with NIR, and the latter is thus an attractive alternative because of its speed and ease of operation. Figure 18.9 illustrates the validity of the NIR technique in such applications as demonstrated by the good correlations observed between the results of the NIR and HPLC methods.

Figure 18.10 presents another example of fingerprinting application involving the authentication of *Panax quinquefolium* L. (P.Q) by HPLC (Chen et al., 2006). P.Q is the North American variety of ginseng with a reputed bioactivity for reducing stress, lowering high blood sugar, and adjusting immunity (Meng and Li, 2003). In recent years, P.Q and its extracts have been used widely as an ingredient in functional foods, herbal drugs, and as an additive in foods. The dammarane-type saponins including ginsenosides and notoginsenosides are generally considered to be the most important bioactive ingredients of the plant (Huang, 1993; Attele et al., 1999; Ma et al., 1999). In the 2005 edition of the Chinese Pharmacopoeia, ginsenoside Rb1, Re, and Rg1 were selected as the marker species for evaluating the quality and authenticity of P.Q. However, there are actually a total of seven major ginsenosides in P.Q, and thus a profile analysis including all the seven species would be a better representation of the bioactivity of P.Q. This is illustrated in Figure 18.10 (Chen et al., 2006), in which P.Q samples from different sources are compared. Figure 18.10a presents the HPLC profiles of the different P.Q samples, whereas Figure 18.10b shows the Dendrogram diagram from hierarchical cluster analysis of

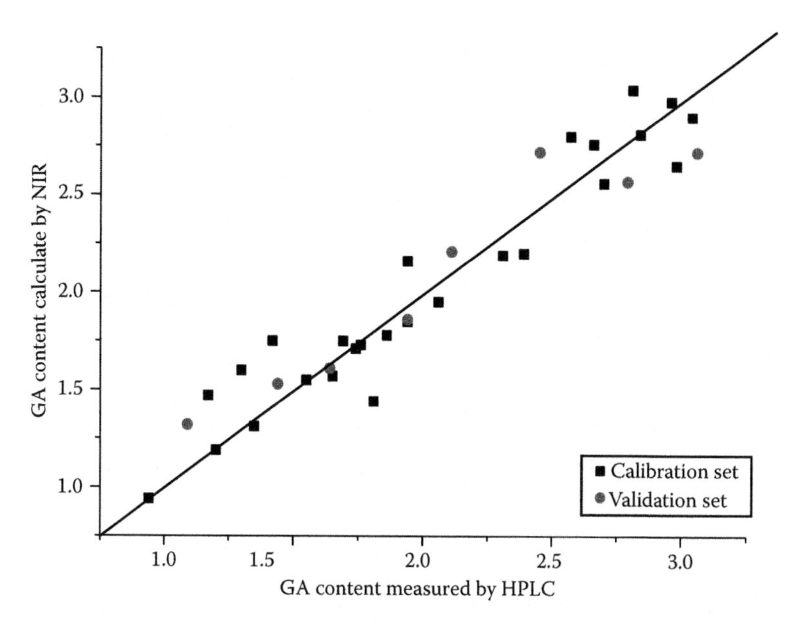

FIGURE 18.9 Glychrrhetinic acid content in the dry licorice: Correlation plot of HPLC measured values and NIR predicted ones ($n = 34$, $R^2 = 0.96$, RMSECV = 0.43).

the sample set. As demonstrated by the diagram, the P.Q samples originating, respectively, from Northern (Jinlin) and Southern (Guang Dong) China, Singapore, and Canada are clearly separated into individual groups of different origin.

Using a similar approach but applying proton nuclear magnetic resonance (NMR) instead of HPLC fingerprinting techniques to discriminate Huangqi (Radix scutellariae p.e) samples is illustrated as another example shown in Figure 18.11 (Chen, 2006), with the NMR spectra given in Figure 18.11a and cluster score plot based on principal component analysis illustrated in Figure 18.11b. The primary objective here was to identify an adulterant sample from genuine Huangqi, and this could indeed be achieved as shown in Figure 18.11.

Besides compositional profiles, fingerprinting based on biochemical activities has also been developed. One example is illustrated in Figure 18.12 (Fu, 2006), in which the antioxidation capabilities of the different chemical constituents in licorice are presented as the fingerprint. The fingerprinting was accomplished on an HPLC instrument equipped with a UV detector and an online 2,2-diphenyl-1-picrylhydrazl (DPPH) mini-reactor downstream. The separated peaks from the HPLC were first quantified by UV absorption. They then flew individually through the downstream mini-reactor along with a constant flow of DPPH doped on line. The DPPH-carrying flow, because of the visible color of DPPH, gave a constant, elevated baseline in the mini reactor equipped with a visible light absorption detector. Negative peaks started to appear, however, when materials with antioxidation capabilities were present in the flow because of their reaction with DPPH. The antioxidation capability of the peak was proportional to the amount of DPPH depleted,

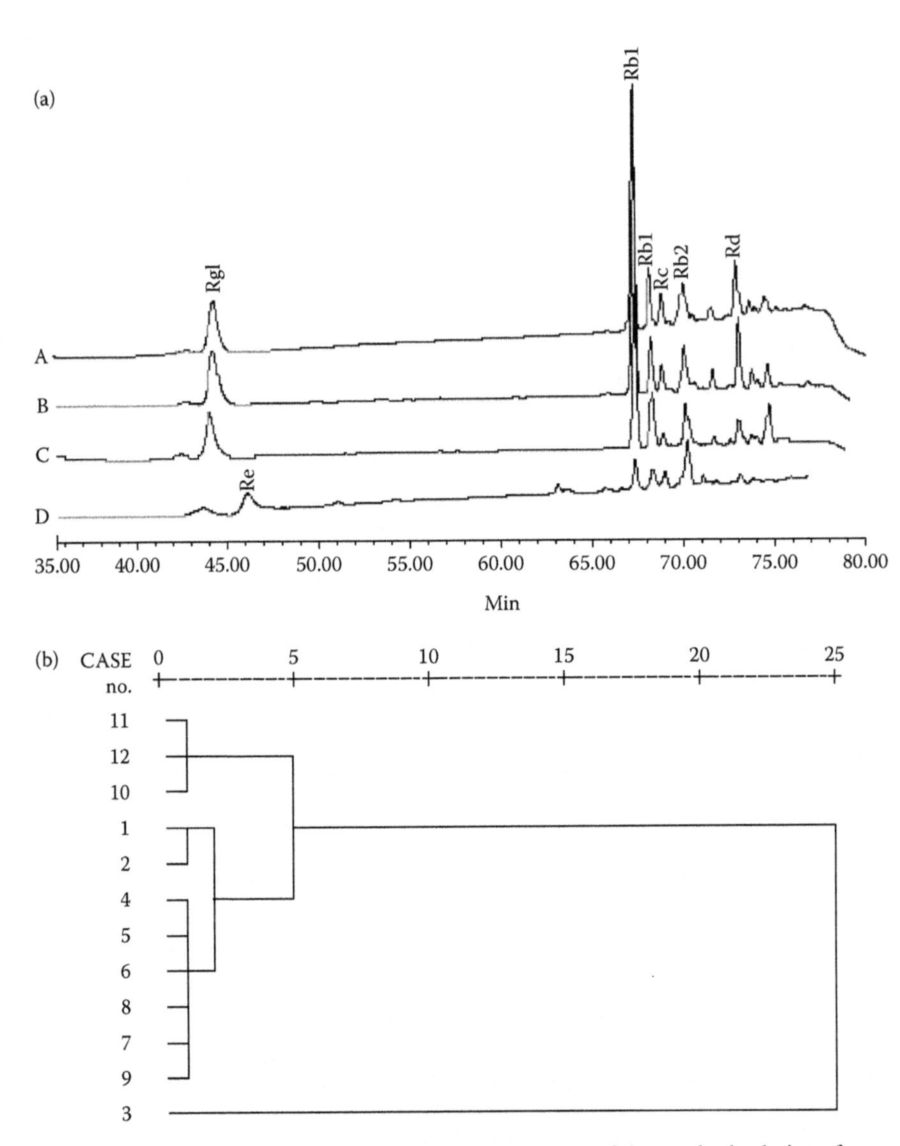

FIGURE 18.10 (a) Representative HPLC chromatograms of A. standard solution of seven ginsenosides; B. P.Q Canada; C. P.Q Jilin, China; D. P.Q Singapore. (b) Dendrogram of clustering analysis of saponin chromatographic fingerprint for 12 P.Q samples. Samples 1,2 were from Singapore; Sample 3 from Canada; Samples 10,11,12 from Guangdong, China, and Samples 4 through 9 from Jilin, China.

or the size of the negative peak, resulting in a bioactivity profile representing the distribution of antioxidation components in the licorice sample.

The use of fingerprinting analysis for the quality control and standardization of TCHs has attracted intense interest in Chinese medicinal research in recent years. A compilation of recently published reports on the subject is given in Table 18.6. The table is not meant to be exhaustive, but only to provide an overview of the recent

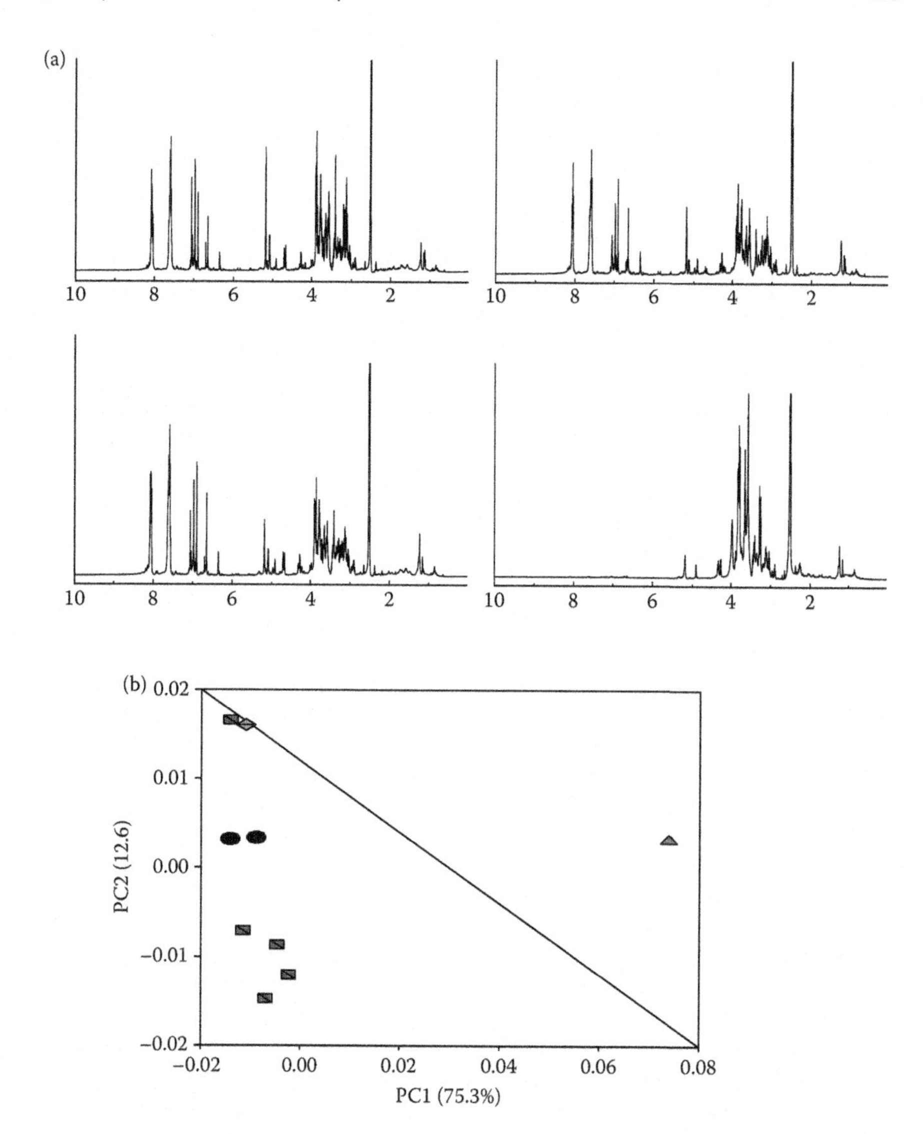

FIGURE 18.11 (a) NMR fingerprint of Radix scutellariae p.e collected from different geographic locations. Top left: HeBei; top right: Inner Mongolia; bottom left: ShangDong; bottom right: adulterant. (b) PCA analysis of NMR spectra of Radix scutellariae p.e collected from different geographic locations (▲ adulterant; ● ShangDong; ◆ Inner Mongolia; ■ He Bei).

activities in this active research area. The table shows that fingerprinting techniques based on chromatographic techniques of HPLC, TLC or GC, or spectroscopic techniques such as Fourier transform infrared (FTIR), NIR, NMR, or different versions of MS techniques have all been successfully applied to the fingerprinting of a wide variety of TCHs or their derived products. Articles exemplify major chemometric techniques for the analysis of fingerprints can be found in the literature (Li et al., 2004a, 2004b; Gong et al., 2005; Xua et al., 2006).

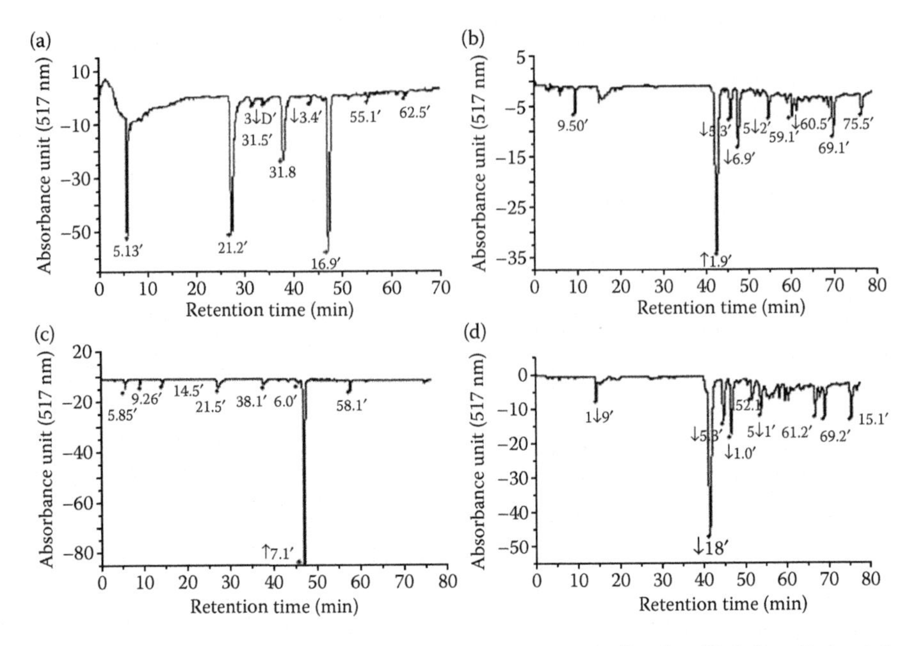

FIGURE 18.12 Fingerprints of anti-oxidation components in licorice (G. Inflata Bat) originated from different geographic locations. (a) Nanjiang, XinJiang; (b) Uzbekistan; (c) Kuerle, Xingjiang; and (d) Turkmenistan. Inversed peak intensity proportional to anti oxidation activity is measured by its reactivity with radical scavenger DPPH in an online HPLC system.

18.5 SAFETY ISSUES

The two most important elements of quality are efficacy and safety. The efficacy of a TCH product can be judged by its authenticity, effectiveness, and batch-to-batch uniformity. The different QC practices and control measures in GAP and GMP programs are clearly mapped out to achieve precisely these quality goals. The situation about product safety is somewhat different. While the general principle of safety protection is emphasized in both GAP and GMP, the strategic approach and action steps designed to meet these objectives are somewhat vague and passive. The concern in these guidelines has been placed primarily on external contaminations such as microbial toxins and environmental pollutants. Other safety concerns such as undesirable side effects, toxicity, and product stability are not adequately addressed, and toxicological data on the identification of possible genotoxic and tumorigenic ingredients in TCHs are also lacking. These shortcomings in the current GAP and GMP systems are areas needing further development, and some related research and regulatory activities along this direction are briefly discussed below.

18.6 REPORTS ON HEAVY METAL AND PESTICIDE RESIDUE CONTAMINATIONS IN TCHs

The contamination of Chinese herbal plants by heavy metals and pesticides remains an issue of continuous concern. The subject has been discussed since the nineties in

TABLE 18.6
Literature Reported Fingerprinting Studies on Chinese Herbs Commonly Used in Traditional Chinese Medicine or Functional Foods

Chinese Name (Pinying)	English Name (Latin Name)	Active Fractions or Marker Compounds	Fingerprinting Technique	Major Objective of Fingerprinting	Reference
Gancao	Radix et Rhizoma Glycyrrhizae	Inflacoumarin A, Licochalcone A, 18β-Glychrrhetinic acid, Liquiritin, licuraside, glycyrrhizin	TLC, NIRS, ME-TLC	Authentication, Quality control	Wang et al. (2005, 2007), Cui et al. (2005)
Taizishen	Radix Pseudostellariae	Pseudostellarin B	HPLC	Authentication	Han et al. (2006, 2007)
Zexie	Rhizoma Alismatis	24-Acety1-alisol A, 23-acety1-alisol B	HPLC	Quality control	Jiang et al. (2006)
Xiyangshen	Radix Panacis Quinquefolii	Ginsenosides	HPLC,	Quality control	Chen et al. (2006), Li et al. (2004a, 2004b)
Renshen	Radix et Rhizoma Ginseng	Ginsenosides, pseudoginsenoside F11	FT-IR, HPLC, HPLC-APCI-MS	Authentication, herb identification	Ma et al. (2004, 2006), Xie et al. (2006)
Danggui	Radix Angelicae Sinensis	Ferulic acid, ligustilide, senkyunolide I, H, A, coniferyl ferulate, butylphthalide, butylidenephthalide, levistolide A	HPLC, FT-IR 2D-IR	Quality control, authentication	Yang et al. (2008), Gong et al. (2003), Fang et al. (2006), Lu et al. (2005)
Danshen	Radix et Rhizoma Salviae Miltiorrhizae	Danshensu, salvianolic acids, tanshinones	HPLC, TLC,TOFMS, NIR, IR,	GAP study, authentication and quality evaluation	Yang et al. (2008), Gong et al. (2003), Fang et al. (2006), Lu et al. (2005)
Zhizi	Fructus Gardeniae	Volatile and semivolatile oil	GC-MS	Quality control	Yan et al. (2006)
Xixin	Radix et Rhizoma Asari	Essential oils	GC	Quality control	Zhang et al. (2004)
Huangqi	Radix Astragali	Flavones	HPLC	Active fraction separation and identification	Xu et al. (2005)

continued

TABLE 18.6 (continued)
Literature Reported Fingerprinting Studies on Chinese Herbs Commonly Used in Traditional Chinese Medicine or Functional Foods

Chinese Name (Pinying)	English Name (Latin Name)	Active Fractions or Marker Compounds	Fingerprinting Technique	Major Objective of Fingerprinting	Reference
Zhiqiao	Fructus Aurantii	Neoeriocitrin, isonaringin, naringin, hesperidin, neohesperidin,	HPLC	Active fraction separation and identification	Zhou et al. (2006), Zhao et al. (2005)
Qingpi	Pericarpium Citri Reticulatae Viride	Neopon-cirin	HPLC, GC	Quality control	Yi et al. (2004)
Yinxingye	Folium Ginkgo	Total flavonoids, luteolin	HPLC	Quality control	Qian and Xie (2004), Gong et al. (2003)
Xiqingguo	Terminalia chebula	Gallic acid, hebulamin	HPLC	Quality control	Yan et al. (2004)
Baishao	RadixPaeoniae Alba	Total glycosides of Paeony	TLC, HPLC	Quality control	Xie and Lin (2004)
Gaoliangjiang	Rhizoma Alpiniae Officinarum	Essential oils, flavonoids	GC, TLC	Authentication, quality control	Qian et al. (2001)
Chuanxiong	Rhizoma chuanxiong	Essential oils, ferulic acid, bultylidene, dihydrophthalide	HPLC, GC/MS	Quality control, authentication	Yang et al. (2008)

China, Taiwan, Singapore, Hong Kong, and other Asian countries (Chi et al., 1992; Wong et al., 1993; Koh and Woo, 2000; Lai et al., 2006a, 2006b), but the problems persist until today. Thus, for instance, a dietary supplement product labeled as ephedra-free and purchased from various retail locations in San Francisco was found to contain significant concentrations of lead, arsenic, cadmium, and mercury by the Food and Drug Laboratory of the California Department of Health Services (Tam et al., 2006). Another survey conducted in 2006 (Lai et al., 2006a, 2006b) analyzed 400 Chinese herbal medicines in Taiwan and found that about 13% of them were contaminated with copper, and 106 out of 400 Chinese herbal medicines were found to have exceeded the allowable limit for cadmium set by WHO. A Chinese herbal medicine, Lan Chou (*Forsythiae fructus*) was found to contain an excessive amount of lead (Lai et al., 2006a, 2006b), and another herbal medicine, Ding Shun (*Caryophylli flos*), was found to contain copper at an alarmingly high concentration of 556 ppm.

18.6.1 Pyrrolizidine Alkaloids: Tumor

Pesticide residue contamination in raw herbs and down stream products has also been a problem as evidenced by many recent reports in China, Taiwan, Hong Kong, and other Asian countries (Wu and Li, 2004; Hao and Xue, 2005; Leung et al., 2005; Lai et al., 2006a, 2006b). One study (Lai et al., 2006a, 2006b) showed that 80% of Red Ginseng (*Ginseng bubra* Radix et Rhizom) samples contained excess amounts of the three organochlorine pesticides, BHCs, PCNB, and hexachlorobenzene at concentrations of 1.90, 1.00, and 0.14 ppm, respectively. Three samples of Loquat leaf (*Eriobotyae folium*) and one sample of Platygala root contained 0.06 ppm of *o,p*-DDT and 0.22 ppm of *p,p*-DDE, respectively. A recent study surveyed the levels of organochlorine pesticide residues in Chinese herbal plants including Radix *Angelicae sinensis*, Radix Notoginseng, Radix *Salviae miltiorrhizae*, and Radix Ginseng cultivated in China or processed in Hong Kong (Leung et al., 2005). It was found that all except Radix *Angelicae sinensis* were contaminated with the pesticides quintozene and hexachlorocyclohexane at varying concentration levels. Hexachlorobenzene and lindane were detected in Radix Ginseng, and DDT and its derivatives were also detected in one of the Radix Notoginseng samples.

18.7 REGULATORY ACTIVITIES BY US FDA ON DIETARY SUPPLEMENTS

For safety and protection, a series of regulations has been promulgated by US FDA on dietary supplement products. A brief introduction of these regulations follows.

18.7.1 Dietary Supplement Health and Education Act (DSHEA) and Safety of Herbal Dietary Supplements

Prior to 1994, FDA regulated herbal medicines as drugs. In 1994, the US Congress passed the DSHEA that amended the US Federal Food, Drug, and Cosmetic ACT

(FFDCA) and created a new regulatory category, safety standard, and other rules for the US FDA to regulate dietary supplements.

According to the DSHEA Act, a dietary supplement is considered unsafe only if it presents a significant or unreasonable risk of illness or injury under conditions of use recommended or suggested in the labeling, or if no conditions of use are suggested or recommended in the labeling, under ordinary conditions of use. Since then, herbal products represent the fastest growing segment of the vitamin, mineral supplements, and herbal products industry. It was reported that in 1999, the United States consumed about US$31 billion worth of dietary supplements and functional foods. Expenditures for dietary supplements and functional foods are projected to go up to US$49 billion by 2010. It is estimated that there are approximately 1500 herbal plants used as herbal dietary supplements or ethnic traditional medicines. Some of the more important herbal dietary supplements prepared by using Chinese herbal plants as source materials and marketed in the US market are listed in Table 18.7.

Since the manufacturers do not have to submit safety reports of the herbal dietary supplement products to the US FDA, the US FDA does not know the concentrations and safety aspects of their botanical ingredients. Accordingly, the US FDA must determine the safety as well as the claimed nutritious functions from the scientific literature, the reports from the media, and other sources. Under this circumstance, the consumers who take dietary supplements are not safely protected. This problem is certainly a burden laid on the FDA.

TABLE 18.7
Chinese Medicinal Herbs Currently Sold in the United States as Dietary Supplements[a]

Alfalfa leaf	Cloves	Ginger root	Nutmeg
Alfalfa seed	Coltsfoot flower	*Ginkgo biloba*	Myrrh Gum
Angelica root	Corn silk	Ginseng Chinese	Nettle leaf
Anise seed	Cumin seed, black	Green Tea	Olive leaf
Aloe vera	Dandelion root	Hawthorne berry	Orange Peel
Astragalus root	Dan Shen	Hops flowers	Peppermint leaves
Bee pollen	Dang Shen	Horehound	Plantain herb
Bee propolis	Dong Quai root	Hydrangea root	Rhubarb root
Bayberry bark	Elecampane root	Hyssop herb	Rosemary leaf
Bistort root	Ephedra	Juniper Berries	Schisandrae Berries
Blue Cohosh root	Eyebright herb	Kelp	Senna pods
Burdock root	Eucalyptus leaf	Licorice root	Spearmint leaves
Calendula flower	False Unicorn root	Lily of Valley root	Thyme leaf
Chamomile flowers	Fennel seed	Lobelia herb	Motherwort herb
Catnip leaf	Gentian root		

Note: This table lists only the most common, but not all, Chinese medicine herbs used as dietary supplements in the United States.

18.7.2 GMP and Food Labeling

To ensure quality control on dietary supplement products, in 1997 the US FDA published initial GMP regulations that manufacturers of herbal products must follow. In 1999, the FDA also enacted the Food Labeling Act. The US FDA issued final GMP regulations in 2003, specifying the conditions for preparation, packing, and storing of dietary supplements, and to ensure that dietary supplements are unadulterated and accurately labeled. This GMP enables dietary supplement products to comply with the safety and sanitation standards, rather than the quality of the products. There is no standardization requirement for dietary supplements in the United States. Thus, although food labeling is required for any food products, it is difficult to determine the quality of dietary supplement products based on the label, even though the word "standardized" may be labeled on the dietary supplement products.

18.7.3 Official Actions by US FDA on Herbal Dietary Supplements

In 2001, US FDA advised dietary supplement manufacturers to remove comfrey products from the market (US FDA, 2001). In 2004, FDA ruled declaring dietary supplements containing ephedrine alkaloid adulterated because they present an unreasonable risk (US FDA, 2004).

18.7.4 The Dietary Supplement and Nonprescription Drug Consumer Protection Act (S. 3546)

In December 2006, the US founded a new Act, the Dietary Supplement and Nonprescription Drug Consumer Protection Act (S. 3546), which was cleared by the Congress on December 9, 2006 and signed by the President on December 22, 2006. This new S. 3546 Act requires the US FDA "to establish systems for collecting data about serious adverse reactions that people experience while using certain nonprescription drugs and dietary supplements." The legislation also requires "manufacturers, packers, or distributors of such products to submit reports to FDA about serious adverse events involving such products based on specific information that they receive from the public."

By definition, serious adverse events include death, a life-threatening experience, inpatient hospitalization, a persistent or significant disability or incapacity, or a congenital anomaly or birth defect. This new act also imposes new requirements for records retention, specifying that "The responsible person must maintain records of all adverse reports it receives, whether serious or not, for 6 years." In addition, S. 3546 also mandates that manufacturers will also have to provide a domestic telephone number or a domestic address on product labels so that consumers can contact them. There are about 30,000 dietary supplements currently sold on the market. This new requirement will apply to all these products. This Act will become effective in December 2007, and should enhance US FDA's ability to fulfill its public health mission to more effectively monitor the medicines and nutritional supplements it regulates.

18.7.5 United States National Toxicology Program

As described previously, under the 1994 DESHE Act, proof of the safety of herbal dietary supplements and herbal medicines is not required prior to market entry. To focus research on the most critical public health issues, the NTP has been conducting a series of long-term studies on the toxicity of herbal medicines or related dietary supplement products nominated by the public and Federal agencies. These nominated herbs and active ingredients are among the most sold and/or most potentially toxic products, and the objective of these studies is to characterize the potential adverse health effects of the products, including reproductive toxicity, neurotoxicity, immunotoxicity, and tumorigenicity. Aiming towards the objective, these studies will also investigate potential herb/herb and herb/drug interactions, and determine responses of the sensitive subpopulations including pregnant women, the young, the developing fetus, and the elderly, among others, towards these herbal products.

Table 18.8 lists the names of the herbs and active or toxic ingredients under study by the NTP. Among the list, echinacea, golden seal, ginseng, kava, ginkgo, and aloe vera are the most sold herbal dietary supplements. The safety of these products is therefore of particular concern.

TABLE 18.8
Herbs and Active or Toxic Ingredient under Study by the US NTP

Echinacea—Most commonly used medicinal herb in the United States
Golden Seal—The second or third most popular herbal dietary supplement in the United States
Ginseng and Gensenosides—The fourth most widely used herbal dietary supplement in the United States; gensenosides the active ingredients
Kava Kava—The fifth most widely used herbal dietary supplement in the United States; used as a calmative and an antidepressant
Ginkgo Biloba Extract—The fifth or sixth most frequently used herbal dietary Supplement in the United States
Aloe Vera—The seventh most widely used herbal dietary supplement in the United States; also used as a component of cosmetics
Comfrey—Herb consumed in teas and as fresh leaves for salads; contains potent hepatotoxic and genotoxic pyrrolizidine alkaloids
Berberine—An active ingredient in golden seal
Milk Thistle Extract—Used to treat depression and several liver conditions and to increase breast milk production
Pulegone—A toxic component of pennyroyal
Thujone—A toxic component of worm wood
Quercetin—A component in blueberries, red onions, apples, and spinach; used as herbal dietary supplement in the United States
Lasiocarpine—A potent hepatotoxic and tumorigenic pyrrolizidine alkaloid
Riddelliine—A potent hepatotoxic and tumorigenic pyrrolizidine alkaloid
Coumadin—A herbal dietary supplement in the United States
Resveratrol—A herbal dietary supplement in the United States
d-Carvone—A herbal dietary supplement in the United States
Furfural—A herbal dietary supplement in the United States

18.7.6 Case Study of Toxic and Tumorigenic Pyrrolizidine Alkaloids

Pyrrolizidine alkaloids, as a class of genotoxic and tumorigenic phytochemicals, are common constituents of hundreds of plant species distributed in many regions of the world. More than 660 pyrrolizidine alkaloids and their N-oxide derivatives have been identified in over 6000 plants, and about half of them exhibit toxicity (Roeder, 2000; Fu et al., 2002, 2004). Pyrrolizidine alkaloids have been found to contaminate human food sources, including herbal medicines, herbal teas, wheat, milk, and honey. More than 15 pyrrolizidine alkaloids, including lasiocarpine and riddelline, have been found to induce liver tumors in experimental animals by the NTP chronic bioassays (Natl Toxicol Program, 2003; NIH Publication, 1991). Ames and Gold (1998) reported that 35 out of the 64 tested phytochemicals were carcinogenic in experimental rodents. Among these 35 rodent carcinogens, six were pyrrolizidine alkaloids, including clivorine, lasiocarpine, monocrotaline, petasitenine, senkirkine, and symphytine. Because of their widespread distribution in the world, the risk to human health posed by the exposure to these compounds has been a major concern.

Several Chinese herbal plants that contain tumorigenic pyrrolizidine alkaloids are used frequently as medicines in China and some Asian countries. The dietary supplement Coltsfoot flower (*Tussilago farfara*; Kuan Dong Hua), as an example, is commonly used in China to moisten lungs, arrest coughing, and reduce phlegm. However, it has been reported that coltsfoot is a suspected liver carcinogen (Fu et al., 2002).

To date, more than 90 pyrrolizidine alkaloids have been identified in herbal plants in China. Among these species, 15 of them have been found to induce tumors in experimental animals (Fu et al., 2002). Tumorigenic pyrrolizidine alkaloids have also been detected in several dietary supplement products and Chinese herbal plant extracts sold in the United States (Chou and Fu, 2006). Although a systematic survey has not been carried out, it is likely that more Chinese herbal plants may contain pyrrolizidine alkaloids (Fu et al., 2002). Since the use of dietary supplements and functional foods is increasing rapidly in the world, the risk of human exposure to them requires careful assessment.

Comfrey and coltsfoot are Chinese herbal medicines produced in many countries including China. Both comfrey and coltsfoot contain tumorigenic pyrrolizidine alkaloids and have been sold commercially as dietary supplements (Fu et al., 2002; Roeder, 2000). It is worth noting that comfrey is a pyrrolizidine alkaloid-containing herbal plant that is used in teas and salads in many countries. Lasiocarpine and riddelliine are pure pyrrolizidine alkaloid chemicals and have been found to be liver carcinogens in experimental rodents by the NTP study (NIH Publication, 1991; Natl Toxicol Program, 2003). Their nomination was based on the finding that pyrrolizidine alkaloids are naturally occurring compounds that are found worldwide. More than 660 pyrrolizidine alkaloids have been identified in over 6000 plants of these three families, and about half of them exhibit potent hepatotoxic and genotoxic activities (Fu et al., 2004). Because pyrrolizidine alkaloids present in staple foods and herbal medicines can result in human poisoning and death, the International Programme on Chemical Safety (IPCS) determined that pyrrolizidine alkaloids present in food are a threat to human health and safety (Fu et al., 2002; IPCS, 1988). It has been reported that Chinese herbal plants cultivated in China (Fu et al., 2002), and several herbal dietary supplements (Chou and Fu, 2006) sold in the United States contain pyrrolizidine alkaloids.

In 1992, the Federal Health Department of Germany restricted "the manufacture and use of pharmaceuticals containing pyrrolizidine alkaloids with an unsaturated necine skeleton." The herbal plants may be sold and used only if daily external exposure is limited to no more than 100 μg pyrrolizidine alkaloids or internal exposure to no more than 1 μg/day for less than 6 weeks a year (Roeder, 2000). Although the US government has not taken any regulatory action on pyrrolizidine alkaloid-containing dietary supplements, in 2001 the US FDA advised dietary supplement manufacturers to remove comfrey products from the market.

18.8 SUMMARY

Being the first among the "5Ps" in the total quality management system, the objective of GAP is to ensure the production of contamination-free crude TCHs with good and uniform quality, controllable yield, and in an environmentally sustainable manner. In recent years, GAP is rapidly developing into an international convention for the trading of agricultural produce such as food and fruits. It is recommended that herbal products derived from similar agriculture processes also adopt the GAP guidelines to ensure the quality and safety of raw herbs.

Although GMP guidelines have been well established for the quality assurance of manufacturing processes, the scope and operational specifics of GAP guidelines for the production of herbal plants is still in the development stage. The major challenges in a GAP program, besides cultivation methodologies involved in agricultural production, are the development of biochemical and analytical technologies for the definition and standardization of the quality of TCHs. Case studies of Danshen, licorice, and several other TCHs have been used to illustrate the different techniques used for species authentication, quality evaluation, and compositional fingerprinting of herbal materials.

In current GAP and GMP programs, safety is a major concern, but the subject has been focused primarily on external contaminations such as microbial toxins and environmental pollutants of metals and pesticide residues. Other safety issues such as undesirable side effects and product stability are not adequately addressed, and toxicological data on the identification of genotoxic and tumorigenic ingredients in many TCHs are also lacking (Fong, 2002). Currently, the United States National Toxicology Program (NTP) is conducting long-term research projects to determine the toxicity of a number of dietary supplements and active ingredients nominated by the FDA and NIH. An organized effort with international participation should be actively pursued to ensure the safety of TCHs and their derived products.

REFERENCES

Ames, B.N. and Gold, L.S., The prevention of cancer. *Drug Metab. Rev.*, 30(2), 201–223, 1998.

Attele, A.S., Wu, J.A., and Yuan, C.S., Ginseng pharmacology: Multiple constituents and multiple actions. *Biochem. Pharmacol.*, 58, 1685–1693, 1999.

BHMA (British Herbal Medicine Association) website: www.medicines.co.uk, 2006.

Blanco, M., Coello, J., Iturriaga, H., Maspoch, S., and Pezuela, C., Near-infrared spectroscopy in the pharmaceutical industry. *Analyst*, 123, 135–150, 1998.

Chang, H.M., Cheng, K.P., Choang, T.F., Chow, H.F., Chui, K.Y., Hon, P.M., Lau, F.W., and Zhong, Z.P., Structure Elucidation and total synthesis of new tanshinones isolated from *Salvia miltiorrhiza* Bunge (Danshen). *J. Org. Chem.*, 55, 3537–3543, 1990.

Chang, N.P., Shiao, X.Y., and Lin, R.C., Current status of GAP production for Chinese medicinal herbs in China. *Res. Inf. Trad. Chin Med.*, 3, 15–17, 2001.

Chen, B., M.S. Thesis, Xiamen University, Xiamen, China, 2006.

Chen, J., Xie, M., Fu, Z., Lee, F.S.C., and Wang, X., Development of a quality evaluation system for *Panax quinquefolium* L. based on HPLC chromatographic fingerprinting of seven major ginsenosides. *J. Microchem.*, 2(85), 2001–2008, 2006.

Chi, Y.W., Chen, S.L., Yang, M.H., Hwang, R.C., and Chu, M.L., Survey of heavy metals in traditional Chinese medicinal preparations. *Zhonghua Yi Xue Za Zhi (Taipei)*, 50(5), 400–405, 1992.

Chou, M.W. and Fu, P.P., Formation of DHP-derived DNA adducts *in vivo* from dietary supplements and Chinese herbal plant extracts containing carcinogenic pyrrolizidine alkaloids. *Toxicol. Ind. Health*, 22, 321–327, 2006.

Cui S., Fu, B., Lee, F.S.C., and Wang, X., Application of microemulsion thin layer chromatography for the fingerprinting of licorice (*Glycyrrhiza* spp.). *J. Chromatogr.* B, 828, 33–40, 2005.

Du, G. and Zhang, J.T., Protective effects of Salvianolic acid A against important memory induced by cerebral ischemia–reperfusion in mice. *Acta Pharm. Sin.*, 30, 184, 1995.

EMEA/HMPWG/18/99/, Draft on the document "Good Agriculture Practice (GAP) from the European Herbs Growers and Producers Association, 5 August 1998; EMEA, Final Proposals for Revision of the Note for Guidance on Quality of Herbal Remedies, 1999.

Fang, K.T., Liang, Y.Z., Yin, X.L., Chan, K., and Lu, G.H., Critical value determination on similarity of fingerprints. *Chemometr. Intell. Lab. Systems*, 82, 236–240, 2006.

Fong, H.H.S., Integration of herbal medicine into modern medical practices: Issues and prospects. *Integr. Cancer Ther.*, 1(3), 287–293, 2002.

Fu, B., Chemical characterization and fingerprinting of licorice (Radix et Rhizoma Glycfyrrhizae). PhD Thesis, Xiamen University, Xiamen, China, 2006.

Fu, P.P., Xia, Q., Lin, G., and Chou, M.W., Genotoxic pyrrolizidine alkaloids—mechanisms leading to DNA adduct formation and tumorigenicity. *Drug Metab. Rev.*, 36, 1–55, 2004.

Fu, P.P., Yang, Y.C., Xia, Q., Chou, M.W., Cui, Y.Y., and Lin, G., Pyrrolizidine alkaloids—tumorigenic components in Chinese herbal medicines and dietary supplements. *J. Food Drug Anal.*, 10, 198–211, 2002.

Gong, F., Liang, Y., Xie, P., and Chau, F., Information theory applied to chromatographic fingerprint of herbal medicine for quality control. *J. Chromatogr. A*, 1002, 25–40, 2003.

Gong, F., Wang, B.T., Chau, F.T., and Liang, Y.Z. Data preprocessing for chromatographic fingerprint of herbal medicine with chemometric approaches. *Anal. Lett.*, 38 (14), 2475–2492, 2005.

Han, C., Chen, J., Chen, B., Lee, F.S.C., and Wang, X. Fingerprint chromatogram analysis of *Pseudostellaria heterophylla* (Miq.) Pax root by high performance liquid chromatography. *J. Sep. Sci.*, 29, 2197–2202, 2006.

Han, C., Chen, Junhui, Liu, Jie, J., Lee, Frank Sen-Chun, and Wang, Xiaoru, Isolation and purification of Pseudostellarin B (cyclic peptide) from *Pseudostellaria heterophylla* (Miq.) Pax by high-speed counter-current chromatography. *Talanta*, 71, 801–805, 2007.

Hao, L.L., and Xue, J., Multiresidue analysis of organochlorine pesticides and the application to Chinese herbal medicine. *Zhongguo Zhong Yao Za Zhi*, 30(6), 405–409, 2005.

Hu, S.L., *A Study of Chinese Genuine Herbs*. Chinese Medicine Publishing Inc., Beijing, China, 1997.

Huang, K.C., *Herbs with Multiple Actions in the Pharmacology of Chinese Herbs* (2nd ed.), CRC Press, Boca Raton, FL, 1993.

Huang, R.J., Zhuang, Z.X., Tai, Y., Huang, R.F., Wang, X.R., and Lee, F.S.C., Direct analysis of mercury in Traditional Chinese Medicines using thermolysis coupled with on-line atomic absorption spectrometry. *Talanta*, 68, 728–734, 2006.

Huang, R.J., Zhuang, Z.X., Wei, J.F., Zhang, S.Q., Shen, J.C., and Wang, X.R., Determination of mercury in Chinese medicinal material and biological samples using pyrolysis atomic absorption spectrometry. *Spectroscopy and Spectral Analysis*, 25(10), 1708–1710, 2005.

Huang, Z.Y., Zhuang, Z.Z., Wang X., and Lee, F.S.C., The preparation and characterization of heavy metal standard reference materials for then quality control of Danshen in GAP farms. *J. Chin. Med.*, 9, 808–811, 2003.

IPCS, *International Programme on Chemical Safety, Pyrrolizidine Alkaloids*. Environmental Health Criteria 80. WHO, Geneva, 1988.

Jiang, Y., Han, C., Qiu, Z., and Wang, X., Dynamic study of zexie by HPLC fingerprinting technique. *Chin. J. Anal. Lab.*, 3(25), 70–74, 2006.

Kasimu, R., Tanaka, K., Tezuka, Y., Gong, Z.N., Li, J.X., Basnet, P., Namba, T., and Kadota, S., Comparative study of seventeen Salvia plants: Aldose reductase inhibitory activity of water and MeOH extracts and liquid chromatography–mass spectrometry (LC–MS) analysis of water extracts. *Chem. Pharm. Bull.*, 46, 500–504, 1998.

Kaufman, P.B., editor, *Natural Products from Plants*, CRC Press, Boca Raton, FL, 1999.

Koh, H.L. and Woo, S.O., Chinese proprietary medicine in Singapore: Regulatory control of toxic heavy metals and undeclared drugs. *Drug Safety*, 23(5), 351–362, 2000.

Lai, L., Chin, L., Chen, Y.H., Lo, C.F., and Lin. J.H., Survey of organochlorine pesticide residues in Chinese herbs and pharmaceutical products (II), Annual Scientific Report of Bureau of Food and Drug Analysis, Department of Health, Executive Yuan, Vol. 24, 265–273, 2006a.

Lai, L., Liu, F.S., Hsu, Y.H., Yu, C.L., Hsiao, S.H., Lo, C.F., and Lin, J.H. Survey of heavy metals in raw material of traditional Chinese medicine (I). Annual Scientific Report of Bureau of Food and Drug Analysis, Department of Health, Executive Yuan, Vol. 24, 228–241, 2006b.

Lai, L., Tseng, J.H., Chen, Y.H., Lo, C.F., and Lin, J.H., Survey of heavy metals in raw material of traditional Chinese medicine (II), Annual Scientific Report of Bureau of Food and Drug Analysis, Department of Health, Executive Yuan, Vol. 24, 242–256, 2006.

Lee, F.S.-C., Development of Good Agriculture Practice and Quality Index for Danshen (Radix Salviae Miliorrhizae) from Zhong Jiang, Sichuan, Program code UIM/6 funded by Hong Kong Innovation and Technology Fund (ITF) and New World Bioscience, Ltd., 2000–2003. For program details, refer to www.itf.gov.hk, 2003.

Lee, F.S.C. and Wang, X., *Key Analytical Technologies for the Quality Control of Chinese Medicinal Herbs Produced according to GAP Guidelines*. Xiamen University Press, Xiamen, China, 2002.

Lee, F.S.C., Wang, X.R., and Li, L., Good agricultural practice (GAP) for the quality assurance of traditional chinese herbs used in dietary supplements, ACS Symposium Series. In: Shibamoto, T., Ho, C.T., Kaezava, K., and Shahidi, F., ed., *Functional Food and Health.* Oxford University Press, USA, 2008.

Leung, K.S., Chan, K., Chan, C.L., and Lu, G.H., Systematic evaluation of organochlorine pesticide residues in Chinese materia medica. *Phytother. Res.*, 19(6), 514–518, 2005.

Li, B., Hu, Y., Liang, Y., Xie, P., and Du, Y., Quality evaluation of fingerprints of herbal medicine with chromatographic data. *Anal. Chim. Acta*, 514, 69–77, 2004a.

Li, H.B. and Chen, F., Preparative isolation and purification of six diterpenoids from the Chinese medicinal plant *Salvia miltiorrhiza* by high speed counter current chromatography. *J. Chromatogr. A*, 925, 109–114, 2001.

Li, L., Lee, F.S.-C., and Wang, X.R., Bar coded HPLC fingerprinting for the quality evaluation of Danshen. *J. Chin, Med. Herbs*, 34, 649–653, 2003a.

Li, W., Zhuang, Z.X., and Wang, X., Extraction and expression of fingerprint characteristics for Danshen by the plate chromatography. *Modern. Trad. Chin. Med. Mater. Med.*, 1(5), 59–63, 2003b.

Li, Y., Sun, S., Zhou, Q., Qin, Z., Tao, J.X., Wang, J., and Fang, X., Identification of American ginseng from different regions using FT-IR and two-dimensional correlation IR spectroscopy. *Vibr. Spectrosc.*, 36, 227–232, 2004b.

Li, Z., Yang, B., and Ma, G., Chemical studies of *Salvia miltiorrhiza* f. *alba. Acta Pharmaceut. Sin.*, 26, 209–213, 1991.

Lin, X. Wang, X., Huang, Y., Huang, and Yang, B. A new diterpenoid dehydroquinone. *Acta Pharmaceut. Sin.*, 23, 273–275, 1988.

Liu, J., Wang, X.R., Cai, Z.W., and Lee, F.S.C., Effect of Tanshinone IIA on the noncovalent interaction between warfarin and human serum albumin studied by electrospray ionization mass spectrometry. *J. Am. Soc. Mass Spectrom.*, 19, 1568–1575, 2008.

Lou, H.W., Wu, B.J., Wu, M.Y., Yong, Z.G., Niwa, M., and Hirata, M.Y., Pigments from *Salvia miltiorrhiza* Bunge. *Phytochemistry* 24, 815–820, 1985.

Lu, G.H., Chan, K., Liang, Y.Z., Leung, K., Chan, C.L., Jiang, Z.H., and Zhao, Z.Z., Development of high-performance liquid chromatographic fingerprints for distinguishing Chinese Angelica from related umbelliferae herbs. *J. Chromatogr. A*, 1073, 383–392, 2005.

Lu, Y. and Foo, L.Y., Polyphenolics of Salvia—a review. *Phytochemistry*, 59, 117–140, 2002.

Ma, W.G., Mizutani, M., Malterud, K.E., Lu, S.L., Ducrey, B., and Tahara, S., *Phytochemistry* 52, 1133–1139, 1999.

Ma, X., Wang, L., Xu, Q., Zhang, F., Xiao, H., Liang, X., Optimization method of multi-segment gradient for separation of ginsenosides in reversed phase-high performance liquid chromatography. *Chin. J. Anal. Chem.*, 2(25), 238–242, 2004.

Ma, X., Xiao, H., and Liang, X., Differentiation of Ginsenoside Rf and pseudoginsenoside F11 by in-source collision-induced dissociation high performance liquid chromatography atmospheric pressure chemical ionization mass spectrometry. *Chin. J. Anal. Chem.*, 9(34), 1273–1277, 2006.

Meng, F.Z. and Li, B., *Panax quinquefolium* L. Science and Technology Publishing House, Beijing, China, 2003.

National Toxicology Program: Toxicology and carcinogenesis studies of riddelliine (CAS No. 23246-96-0) in F344/N rats and B6C3F1 mice (gavage studies). Natl Toxicol Program Tech. Rep. Ser. 508, 1–280, 2003.

NIH Publication No. 91–3140. NTP Technical Report (No. 409) on the Toxicology and Carcinogenesis Studies of Quercetin in F344/N Rats. U.S. Department of Health and Human Services, Public Health Service, National Toxicology Program, Research Triangle Park, NC, 1991.

Pan, X., Niu, G., and Liu, H. Microwave-assisted extraction of tanshinones from *Salvia miltiorrhiza* with the analysis by high performance liquid chromatography. *J. Chromatogr. A*, 922, 371–375, 2001.

Pharmacopoeia Editorial Board, *Pharmacopoeia of the People's Republic of China*, Vol. 1. Chemical Industry Press, Beijing, 2005.

Qian, H., Li, C., and Xie, P., On GC fingerprint of rhizome of *Aplinia officinarum* (Gao Liang Jiang) and its congeners, Traditional Chin. *Drug Res. Clin. Pharma.*, 3(12), 179–183, 2001.

Qian, H. and Xie, P., Methodology study of HPLC fingerprint of herbal medicine—A case analysis of the fingerprint of total flavonoids extracted from *Ginkgo biloba* leaves. *J. Instr. Anal.*, 2004, 5(23), 7–11, 2004.

Ren, D.Q., The significance and application of fingerprinting in Chinese medicine. *J. Chin. Trad. Herbal Drugs*, 24(4), 235–239, 2001.

Ren, D.Q. and Zhou, R.H., *An Operational Manual for the Implementation of GAP*. China Agriculture Publishing, Beijing, China, 2003.

Rodriguez-Otero, J.L., Hermida, M. and Centeno, J., Analysis of dairy products by near-infrared spectroscopy: A review. *J. Agric. Food Chem.*, 45(2), 815–819, 1997.

Roeder, E., Medicinal plants in China containing pyrrolizidine alkaloids. *Pharmazie*. 55, 711–726, 2000.

SFDA:website: www.sfda.gov.cn., 2006.

Tam, J.W., Dennehy, C.E., Ko, R., and Tsourounis, C. Analysis of ephedra-free labeled dietary supplements sold in the San Francisco Bay area in 2003. *J. Herb Pharmacother.*, 6(2), 1–19, 2006.

Tanaka, T., Morimoto, S., Nonaka, G., Nishioka, I., Takako, Y., Chung, H.Y., and Oura, H. Magnesium and ammonium-potassium lithospermates B, the active principle having a uremia-preventive effect from *Salvia miltiorrhiza. Chem. Pharm. Bull.* 37, 340–344, 1989.

US FDA final GMP regulations, 2003.

US FDA Guidance for Industry—Botanical Drug Products (Draft Guidance), US Food and Drug Administration, 2000; USFDA Public Health News, Federal Register, FDA Announced major initiatives for dietary supplements. November 5, 2004.

US FDA, FDA advises dietary supplement manufacturers to remove comfrey products from the market, July 6, 2001. http://www.cfsan.fda.gov/~ dms/supplmnt.html.

Wang, Q.E., Choi, C.Z., Jin, B.H., and Wang, X.R., Rapid testing method for total triterpens in alisma. *Research report of Xiamen University (Natural Science)*, 42(6), 571–575, 2003.

Wang, Q.E., Wang, W.S., Wu, C., Lee, F.S.C., and Wang, X.R., Rapid testing method for the quality control of licorice in GAP research. *J. Research and Practice of Chinese Medicines*, 18(1), 37–41, 2004a.

Wang, Q.E., Wang, W.S., Zhang, J.S., Lee, F.S.C., and Wang, X.R., A study of best harvest time for Leung Wai licorice. *J. Chin. Med. Herbs*, 27(4), 235–237, 2004b.

Wang, X., GAP programs on Licorice (Radix Glycyrrhiza,), Taizishen (*Pseudostellaria heterophylla* (Miq.) Pax.) and Alisma (*Alisma gramineum* Lej.), For program details, refer to www.sfda.gov.cn., 2006.

Wang, L., He, Y., Qiu, Z.C., and Wang, X.R., Determination of glycyrrhizic acid in glycyrrhiza uralensis fisch by fiber optic near infrared spectroscopy. *Spectrosc. Spectral Anal.*, 25(9), 1397–1399, 2005.

Wang, L., Lee, F.S.C., and Wang, X., Near-infrared spectroscopy for classification of licorice (*Glycyrrhizia uralensis* Fisch) and prediction of the glycyrrhizic acid (GA) content. *LWT—Food Sci. Technol.*, 40, 83–88, 2007.

WHO Guidelines on Good Agricultural and Collection Practices for Medicinal Plants, http://www.who.int/medicines/library/trm/medicinalplants/agricultural.shtml, 2003.

Wong, M., Tan, K.P., and Wee, Y.C. Heavy metals in some Chinese herbal plants. *Biol. Trace Elem. Res.*, 36(2), 135–142, 1993.

Wu, J. and Li, L., Determination of the residues of eleven organophosphorus insecticides in Job's tears by gas chromatography with a nitrogen–phosphorus detector. *JAOAC Int.*, 87(5), 1260–1263, 2004.

Xie, P., Chen, S., Liang, Y.Z., Wang, X., Tian, R., and Upton, R., Chromatographic fingerprint analysis—a rational approach for quality assessment of traditional Chinese herbal medicine. *J. Chromatogr. A*, 1112, 171–180, 2006.

Xie, P. and Lin, Q., TLC fingerprinting analysis of total glycosides of paeony (TGP). *Traditional Chin. Drug Res. Clin. Pharma.*, 3(15), 171–173, 2004.

Xu, Q., Wang, L., Zhang, X., Jin, G., Xiao, H., and Liang, X., Quick optimization of high performance liquid chromatographic conditions for separation of flavones of astragalus. *Chin. J. Chromatogr.*, 6(23), 630–632, 2005.

Xua, H., Liang, Y., Chau, F., and Vander Heyden, Y., Pretreatments of chromatographic fingerprints for quality control of herbal medicines. *J. Chromatogr. A*, 1134, 253–259, 2006.

Yan, S.K., Xin, W.F., Luo, G.A., Wang, Y.M., and Cheng, Y.Y., Chemical fingerprinting of *Gardenia jasminoides* fruit using direct sample introduction and gas chromatography with mass spectrometry detection. *J. AOAC Internat.*, 89(1), 40–45, 2006.

Yan, Y., Xie, P., Song, L., Liu, X.X., Lu, P., and Huang, X., Application of chromatographic fingerprint of immature fruits of *Terminalia chebula* and its extracts. *Chin. Trad. Patent Med.*, 8(24), 603–607, 2004.

Yang, B.J., Chen, J.H., Yin, Y.F., Lee, F.S.C., and Wang. X.R. GC-MS fingerprints for discrimination of Ligusticum chuanxiong from Angelica, *J. Sep. Sci.*, 18(8), 3231–3237, 2008.

Yi, L., Xie, P., Liang, Y., and Zhao, Y., Studies on fingerprints of Pericarpium citri Reticulatae Viride. *Trad. Chin. Drug Res. Clin. Pharma.*, 6(15), 403–406, 2004.

Yu, W.J., He, J., and Wang, X., Determination of lipid soluble active components in Danshen by high resolution TOF-MS. *J. Chem. Chin. Univ.*, 24(4), 621–623, 2003.

Zhang, F., Fu, S., Xu, Q., Xiao, H., Cai, S., and Liang, X., Study on GC fingerprint of the constituents in Herba Asari. *China J. Chin. Mater. Med.*, 5(29), 411–414, 2004.

Zhang, J.H., Department of Biology, Hong Kong Baptist University, Personal communications, 2003.

Zhao, Y., Xie, P., Liang, Y., Yang, H., and Yi, L., HPLC fingerprint analysis and chemical pattern recognition of *Fructus aurantii. Chin. Pharma. J.*, 11, 812–815, 2005.

Zhou, D., Qing, X., Xue, X., Zhang, F., and Liang, X., Analysis of flavonoid glycosides in fructus Aurantii by high performance liquid chromatography–electrospray ionization mass spectrometry. *Chin. J. Anal. Chem.*, 34, s31–s35, 2006.

Index

B

E

F

G

H

I

J

K

N

O

P

T

U

Z